Fermented Foods
Part II: Technological Interventions

Books Published in *Food Biology* series

Food Biology Series

Fermented Foods
Part II: Technological Interventions

Editors

Ramesh C. Ray
Principal Scientist (Microbiology)
ICAR - Central Tuber Crops Research Institute
Bhubaneswar, Odisha, India

and

Didier Montet
Food Safety Team Leader
UMR Qualisud, CIRAD
Montpellier, France

CRC Press is an imprint of the
Taylor & Francis Group, an **informa** business
A SCIENCE PUBLISHERS BOOK

CRC Press
Taylor & Francis Group
6000 Broken Sound Parkway NW, Suite 300
Boca Raton, FL 33487-2742

First issued in paperback 2020

© 2017 by Taylor & Francis Group, LLC
CRC Press is an imprint of Taylor & Francis Group, an Informa business

No claim to original U.S. Government works

ISBN-13: 978-1-138-63784-9 (hbk)
ISBN-13: 978-0-367-78225-2 (pbk)

Visit the Taylor & Francis Web site at
http://www.taylorandfrancis.com

and the CRC Press Web site at
http://www.crcpress.com

Preface to the Series

Food is the essential source of nutrients (such as carbohydrates, proteins, fats, vitamins, and minerals) for all living organisms to sustain life. A large part of daily human efforts is concentrated on food production, processing, packaging and marketing, product development, preservation, storage, and ensuring food safety and quality. It is obvious therefore, our food supply chain can contain microorganisms that interact with the food, thereby interfering in the ecology of food substrates. The microbe-food interaction can be mostly beneficial (as in the case of many fermented foods such as cheese, butter, sausage, etc.) or in some cases, it is detrimental (spoilage of food, mycotoxin, etc.). The *Food Biology* series aims at bringing all these aspects of microbe-food interactions in form of topical volumes, covering food microbiology, food mycology, biochemistry, microbial ecology, food biotechnology and bio-processing, new food product developments with microbial interventions, food nutrification with nutraceuticals, food authenticity, food origin traceability, and food science and technology. Special emphasis is laid on new molecular techniques relevant to food biology research or to monitoring and assessing food safety and quality, multiple hurdle food preservation techniques, as well as new interventions in biotechnological applications in food processing and development.

The series is broadly broken up into food fermentation, food safety and hygiene, food authenticity and traceability, microbial interventions in food bio-processing and food additive development, sensory science, molecular diagnostic methods in detecting food borne pathogens and food policy, etc. Leading international authorities with background in academia, research, industry and government have been drawn into the series either as authors or as editors. The series will be a useful reference resource base in food microbiology, biochemistry, biotechnology, food science and technology for researchers, teachers, students and food science and technology practitioners.

Ramesh C. Ray
Series Editor

Preface

Fermentation is applied to a broad range of food substrates (cereals, vegetables, fruits, legumes, milk and meat) and there is an array of food products originated artisanally or at industrial scale. At the heart of most fermented foods are ancient processes that date back to the introduction of agriculture and animal husbandry, approximately 10,000 years ago. Today, fermentation technology has moved from artisanal practices and empirical sciences to the industrialized technology. Currently, fermented foods and beverages are estimated to make approximately one-third of the human diet. The impact of benevolent microorganisms on our ability to harvest energy and nutrients from food is extended to the gastro-intestinal tracts focusing primarily on lactic acid bacteria, *Bacillus* spp., and yeasts. Recent technological advances in sequencing and other omics technologies have provided tools to study the complex gastro-intestinal microbial flora as well as host responses related to the dietary interventions leading to the health benefits of the hosts. Further, the bioactive molecules in various types of fermented foods such as polyphenols, peptides, antioxidants, etc. are of current interest because of their health promoting effects. Most of these aspects have been elaborated in this book in 22 chapters contributed by experts drawn from various parts of the world. We are thankful to these authors for accepting our invitations to contribute and timely submitting their contributions and resolving the queries, if any, very promptly. We are also grateful to the reviewers who have voluntarily agreed and spared their precious time in critically reviewing the individual manuscript.

Ramesh C. Ray
Didier Montet

Acknowledgements

The editors convey their sincere thanks to the following professors/ scientists for critically reviewing the chapters with valuable suggestions in updating the chapters.

1. Prof. B.B. Mishra
Department of Microbiology, Orissa University of Agriculture and Technology, Orissa, India.
E-mail: bb_mishra58@yahoo.com

2. Dr. Amit Kumar Rai
Institute of Bioresources and Sustainable Development, Sikkim Centre Tadong, Gangtok, Sikkim, India.
E-mail: amitraikvs@gmail.com

3. Prof. H.N. Thatoi
Department of Biotechnology, North Orissa University, Takatpur, Baripada, Orissa, India.
E-mail: hn_thatoi@rediffmail.com

4. Prof. Keshab C. Mondal
Department of Microbiology, Vidyasagar University, Midnapore - 721102, India.
E-mail: mondalkc@gmail.com

5. Dr. Nadine Zakhia-Rozis
CIRAD-PERSYST, UMR Qualisud, TA B95/16, 34398 Montpellier Cedex 5, France.
E-mail: nadine.zakhia-rozis@cirad.fr

6. Prof. Pratima Khandelwal
Prof. & Head, Dept. of Biotechnology, New Horizon College of Engineering (Autonomous College Affiliated to VTU and Accredited by NAAC with 'A' Grade), Bengaluru-103.
E-mail: www.newhorizonindia.edu

7. Prof. T.C. Bhalla
Department of Biotechnology, Himachal Pradesh University, Summer hill, Shimla-171 005, India.
E-mail: bhallatc@rediffmail.com

8. Dr. S. Paramithiotis
Laboratory of Food Quality Control and Hygiene, Department of Food Science and Human Nutrition, Agricultural University of Athens, Iera Odos 75, GR-11855, Athens, Greece.
E-mail: sdp@aua.gr

9. Prof. V.K. Joshi
Former Professor and Head of, Department of Food Science, Dr. YSP UHF, Nauni, Solan (HP) India, PIN 173230.
E-mail: vkjoshipht@rediffmail.com
E-mail: fermentvkj@gmail.com

10. Prof. Nevijo Zdolec, Ph.D., D.V.M.
University of Zagreb, Faculty of Veterinary Medicine, Department of Hygiene, Technology and Food Safety, Heinzelova 55, 10000 Zagreb Croatia.
E-mail: nzdolec@vef.hr

11. Prof. Zlatica Kohajdová
Department of Food Science and Technology, Institute of Biotechnology and Food Science, Faculty of Chemical and Food Technology, Slovak University of Technology, Radlinského 9, SK-812 37 Bratislava, Slovakia.
E-mail: zlatica.kohajdova@stuba.sk

Contents

1

Fermented Foods
Microbiology, Biochemistry and Biotechnology

Romain Villéger,[1,a] *Rémy Cachon*[2,]* *and Maria C. Urdaci*[1,b]

1. Introduction

Fermentation is one of the most ancient forms of food preservation technologies in the world that uses microorganisms to convert perishable and sometimes inedible raw materials into safe, shelf-stable and palatable foods or beverages. It can be described as a biochemical change, which is induced by the anaerobic or partially anaerobic metabolism of carbohydrates by microorganisms with the production of acids that result in decrease in pH, or with the production of alcohols. During fermentation, due to the incomplete oxidation of organic molecules, different products such as aromatic compounds are produced. Alkaline fermentation is a special type of fermentation that causes an increase in pH of the food. In that case, microorganisms degrade principally food proteins, leading to the release of ammonia resulting in increase of pH.

[1] Université de Bordeaux, UMR 5248, Bordeaux Sciences Agro, Laboratoire de Microbiologie, 1, cours du Général De Gaulle, 33175 Gradignan cedex, France.
[a] E-mail: romain.villeger@gmail.com
[b] E-mail: maria.urdaci@agro-bordeaux.fr
[2] UMR A 02.102 Procédés Alimentaires et Microbiologiques, équipe Procédés Microbiologiques et Biotechnologiques, AgroSup Dijon, Bâtiment Epicure, 1-esplanade Erasme, F-21000 Dijon, France.
* Corresponding author: remy.cachon@agrosupdijon.fr

There exists a wide variety of fermented foods that have originated from diverse substrates available over the world. Some of the most popular fermented products are derived from milk, cereals, tubers, fruits and vegetables and less from meat and fish (Steinkraus 1997). Foods are usually fermented by microbial interactions including principally bacteria and yeast but occasionally mold as well. Indigenous fermented foods have been prepared and consumed for thousands of years, and are strongly linked to culture and traditions but the microbial and enzymatic processes responsible of many products are poorly understood. Recently, there has been a development in the understanding of these processes and their adaptation for industrialization and commercialization. Traditional fermentation processes are uncontrolled and result in products of low yield and with variable quality. Thus many debates are emerging, opposing repeatability of inoculated fermented products, to local products processed from existing wild ecosystem.

A deep comprehension of fermentation mechanisms is very important to assure the quality and safety of fermented foods. Traditional methods of fermentation monitoring and food safety analysis are generally based on the use of culture, but these techniques are limited by the high time consuming, multi-step procedures and resolution. The application of new biotechnological and molecular methods may improve the knowledge and safety of fermented foods. Indeed, at least two to three days are required for the initial isolation of a microorganism, followed by several days for additional confirmatory assays. Recent methods, usually based on PCR, can provide more accurate results within a relatively short time and for an affordable cost. Moreover, applied science in food processing and food safety should allow in the long term to develop industrialization of fermented products in developing countries.

The very recent emergence of global technologies as 'omics', now applicable to the fermented food products, may help to understand and control complex ecosystems. 'Omics' could play a major role in genome data mining, in deciphering microbial metabolism, for rational metabolic engineering and to better characterize the microbial diversity of products.

Many fermented foods are extremely important in meeting the nutritional requirements of a large proportion of the global population. In addition, health benefits of fermented foods have long been established. Metchnikoff, Nobel Prize winner in 1908, argued the benefits of yogurt in host health. Both, products and microorganisms involved in fermentation can have a beneficial effect in the host. However, the underlying mechanisms and the role of beneficial microorganisms have only been explored during the 20th century, leading to the emergence of the probiotic concept, for example.

2. Microorganisms and Fermented Food Manufacture

In the fermented food process, microorganisms have several major roles such as (1) preservation of food through formation of inhibitory metabolites like organic acids, ethanol, bacteriocins, etc., (2) improvement of food safety *via* pathogens inhibition or by the removal of toxic compounds, (3) enrichment of foods with vitamins, protein, minerals, essential amino acids and essential fatty acids, (4) enrichment of products with a wide diversity of phenolic compounds that improves flavors, aromas and textures, and (5) expansion of the diet for more diversity (Steinkraus 2002, Bourdichon et al. 2012, Ray and Joshi 2014).

Fermented foods can be classified in a number of ways. Figure 1 represents the main fermented foods in eight categories according to Steinkraus (1996).

The great diversity of fermentations is directly linked to the wide variety of microorganisms, raw materials, duration and conditions of fermentation. Microorganisms are highly diverse in their biochemistry, physiology and nutritional modes, which determine the characteristics of the fermented food like acidity, flavor, texture, and the health benefits associated

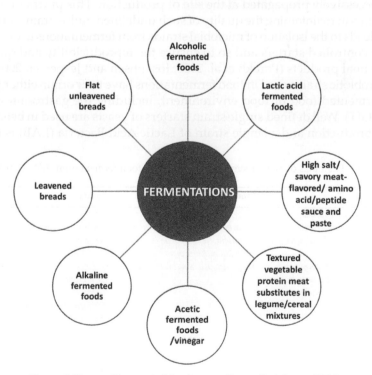

Figure 1. Types of fermented foods according to Steinkraus (1996).

(Vogel et al. 2011). Table 1 summarizes the main genera of microorganisms used in food fermentations, some of them harboring species often used as probiotics, such as *Lactobacillus, Bifidobacterium, Bacillus, Saccharomyces,* etc. (FAO 2009, 2013).

Microorganisms associated with raw food material and processing environment are implicated in spontaneous fermentations, while inoculants containing high concentrations of live microorganisms, referred as starter cultures, are used to initiate and accelerate the conversion rate in non-spontaneous or controlled fermentation processes (Stevens and Nabors 2009). A single species or strain will only rarely be dominant in a fermentation process. For this reason, traditional or artisanal starter cultures are a valuable source of strains and species diversity that are able to ensure the typicality of the numerous traditional fermented foods.

Most of the small-scale fermentations in developing countries, and even some industrial processes that are technically well controlled, are still conducted as spontaneous processes (Holzapfel 2002). Spontaneous fermentation represents a low-cost technology, which is very advantageous in the developing countries to avoid food spoilage (Oyewole 1997). Foods are often inoculated from an earlier fermentation and thus the ferments are successively propagated at the site of production. This practice creates problems in maintaining the quality of such undefined multi-strain cultures, what leads to the isolation of microbial strains from fermentations to develop better controlled starters and so to ensure the reproducibility and quality of the final products (Parekh et al. 2000, Josephsen and Jespersen 2004).

Probiotic strains used in food fermentations have their origin either from the fermented food or food environment, including the gastro-intestinal tract (GIT). Well defined single strain starters of yeasts are used in beer and wine production and a single strain of Lactic Acid Bacteria (LAB) is used

Table 1. Examples of main microorganisms used in food fermentations (FAO 2013).

Bacteria	Moulds	Yeasts
Acetobacter, Arthrobacter, Bacillus, Bifidobacterium, Brachybacterium, Brevibacterium, Carnobacterium, Corynebacterium, Enterobacter, Enterococcus, Gluconacetobacter, Hafnia, Halomonas, Klebsiella, Kocuria, Lactobacillus, Lactococcus, Leuconostoc, Micrococcus, Oenococcus, Pediococcus, Propionibacterium, Staphylococcus, Streptococcus, Streptomyces, Tetragenococcus, Weisella, Zymomonas.	*Actinomucor, Aspergillus, Fusarium, Lecanicillium, Mucor, Monascus, Neurospora, Penicillium, Rhizopus, Scopulariopsis, Sperendonema*	*Candida, Cyberlindnera, Cystofilobasidium, Debaryomyces, Dekkera, Hanseniaspora, Kazachstania, Galactomyces, Geotrichum, Kluyveromyces, Lachancea, Metschnikowia, Pichia, Saccharomyces, Schizosaccharomyces, Schwanniomyces, Starmerella, Torulaspora, Trigonopsis, Wickerhamomyces, Yarrowia, Zygosaccharomyces, Zygotorulaspora*

as a starter in some dairy products and sauerkraut. Though single strain cultures are clearly defined and predictable; but the mixed starters are also advantageous where diverse interactions among the microbes develop the desired product quality and overcome possible infection by bacteriophages (Émond and Moineau 2007). Traditional multi-strains starters exist in different countries as for example the Look-pang, a mix of various fungi (Thailand), the Nuruk, a mix of nine microorganisms (Korea), the Marcha, a mix of 12 fungi, yeast and LAB (India) (Jung 2012, Tamang et al. 2012). Moreover, sophisticated starters with special features exist, as for example bacteriocin-producing bacteria alone or in combination with selected bacteriocin-resistant ferments, which would help to inhibit undesirable bacteria (Holzapfel et al. 2003) or the use of selected probiotic stains.

Over 5000 varieties of fermented foods and beverages exist worldwide but only a few of these products have been subjected to scientific studies (Tamang and Kailasapathy 2010). Microorganisms involved in food fermentations present an enormous diversity and include bacteria, yeasts and molds.

2.1 Bacteria

Several families of bacteria that are present in foods, including dairy, meat, fish, cereals, fruits and vegetables are LAB, the major microorganisms involved in desirable food fermentations. Most LAB are considered as 'generally recognized as safe', GRAS (Silva et al. 2002) and some strains are recognized as probiotics. LAB produce lactic acid and other antimicrobial substances that can inhibit the growth of harmful bacteria (Stiles 1996), thereby prolonging the shelf life of the product. *Lactobacillus* (*Lb. plantarum, Lb. acidophilus, Lb. brevis, Lb. casei*), *Lactococcus lactis*, *Leuconostoc mesenteroides*, *Pediocccus* (*P. acidilactici, P. pentosaceus*), and *Enterococcus* spp. are the most represented species in fermented foods.

Bifidobacterium species are mainly involved in fermentation of milk and possess probiotic properties. *Staphylococcus* spp. are involved in the fermentation of diverse foods (dairy, meat, fish, and soy). In addition, *Bacillus subtilis* and related bacilli may dominate fermentations of legumes and some other raw materials, being responsible for what is called alkaline fermentations. Fermented foods are widely consumed in Southeast Asia and African countries, such as the Japanese Natto and the African Ogiri which are based on fermented beans. They are principally household fermentations but Natto has been commercialized (Steinkraus 1991). In several Asian and African countries, extensive studies have been conducted to the elaboration of *Bacillus* starter cultures for legume fermentations (Kimura and Itoh 2007, Ouoba et al. 2004).

Other important bacteria involved particularly in the fermentation of fruits and vegetables, are the acetic acid producing bacteria (AAB). AAB species are Gram-negative, obligate aerobic bacteria involved in the conversion of the ethanol to acetic acid/vinegar (Bourdichon et al. 2012). Moreover, *Acetobacter* and *Gluconobacter* species are important microorganisms that are present during cocoa and coffee fermentation, and in the manufacture of specialty beers (Kersters et al. 2006).

2.2 Yeasts

Yeasts are unicellular microscopic fungi, widely distributed in nature, and involved in a large variety of fermentations, mainly alcoholic fermentation (Walker 1998). They produce aroma components and alcohols. Although there is a large diversity of yeast and yeast-like fungi (about 500 species), only a small number is implicated in foods fermentation. *Saccharomyces* genera, especially *S. cerevisiae* are the most beneficial yeasts involved in bread making and wine and beer fermentations. *Schizosaccharomyces pombe* is the dominant yeast in the production of traditional fermented beverages, especially those derived from maize and millet (Battcock and Azam-Ali 1998).

Some yeast can tolerate high salt concentrations as *Zygosaccharomyces rouxii*, which is associated with fermentations in which salting is an integral part of the process as the fermentation of soybeans during the manufacture of soy sauce and miso, where it plays an important role in the development of flavors. If *S. cerevisiae* is the king of wine and beer, then *Z. rouxii* is the king of soy sauce. The latter works together with *Aspergillus oryzae* and the lactic bacteria *Tetragenococcus halophilus* (Aidoo et al. 2006). *Z. rouxii* is also present in other fermentation processes such as traditional balsamic vinegar production.

In a large part of fermented foods, a mix of various yeasts and bacteria (principally LAB), are implicated. The *Tepache*, a popular refreshing beverage prepared and consumed throughout Mexico, is produced by the yeasts *Torulopsis insconspicna, S. cerevisiae* and *Candida queretana*, and the bacteria *B. subtilis* and *Bacillus graveolus* (Aidoo 1992). Palm wine is fermented by two yeasts, *S. cerevisiae*, and *S. pombe* and the bacteria *Lb. plantarum* and *L. mesenteroides*. In the Ogi (maize, sorghum or millet in West Africa), the yeasts *Candida mycoderma, S. cerevisiae* and *Rhodotorula*, the bacteria *Lb. plantarum, Corynebacterium* and *Aerobacter* and the molds *Cephalosporium, Fusarium, Aspergillus* and *Penicillium* are present (Akinrele 1970). Kefir, a milk fermentation, involves the yeasts *Kluyveromyces, Saccharomyces* and *Torulaspora* and various LAB species. In cheese manufacturing, *Debaryomyces hansenii, S. cerevisiae, Yarrowia lipolytica, Kluyveromyces lactis* and *Galactomyces geothricum* are the dominant yeasts (Jakobsen et al. 2002). In sourdough, a bread product, a symbiotic combination of the lactobacilli and yeasts (*S. cerevisiae, Candida milleri,* and *Candida humilis*) exists (Zhou and Therdthai 2012).

Generally, yeasts are used as secondary starter cultures because of their ability to enhance the aroma production or to work in synergy with others microorganisms. Yeasts stimulate the growth of LAB by providing essential metabolites such as pyruvate, amino acids and vitamins and moreover influence the nutritional value of products.

2.3 Molds

Molds are also important organisms in the food industry, both as spoilers and preservers of foods. Although molds do not play a significant role in desirable fermentations of fruits and vegetables, their use in others fermentations is practiced on nearly every continent. The diversity of molds used is relatively limited despite their belonging to several orders. Between the Zygomycetes, the genera *Mucor* and *Rhizopus* and for the Ascomycetes the genera, *Monascus* and *Neurospora* are the most represented. The main Deuteromycetes belong to *Aspergillus* and *Penicillium* genera.

Fungi are strict aerobics which limits their occurrence in food. However, they can tolerate high concentrations of salt and sugar, and possess a great array of enzymes, producing both intra- and extracellular proteolytic, amylolytic and lipolytic enzymes that highly influence the flavor and texture of the products (Bourdichon et al. 2012).

In Western cultures, the use of mycelial fungi for fermentation is principally related to mold-ripened cheeses and meat products (Leistner 1986). *Penicillium* is the fungal genus most frequently found on foods, and three *Penicillium* species (*P. camemberti, P. roqueforti* and *P. nalgiovense*) are widely used in the western world as starter cultures/food processors. *P. nalgiovense* is frequently isolated from meat products, mainly dry fermented sausages (Chavez et al. 2011). However, in Asia there exists the greatest variety of fungal fermented foods. Mycelia are used for their enzymatic ability to degrade polymers as well as for texture forming properties and characteristic flavors and colors.

A. oryzae is the most familiar fungi used in various Asian multiple-stage fermentation products as Shoyu (soy sauce) made from soybeans well as the Japanese Miso. *A. oryzae* is also implicated in rice fermented products as Yakju (Korea), Sake (Japan), etc. In Indonesian tempeh, the boiled legume seeds are fermented with *Rhizopus oligosporus* that digests the complex carbohydrates and origin a protein-rich cake that can be used as a meat substitute. *Monascus purpureus* ferments rice giving the specific aroma and purple-red color of the Ang-kak, a popular food of China and Philippines. Fungal fermentations causes the bio-enrichment of food through the vitamins, amino acids, essential fatty acids and are responsible for diverse flavors, aromas and texture, but we must not forget to control them carefully particularly for fungal toxins that may occur, especially in spontaneous fermentations.

3. Food Safety, Quality and Functionality

Over the centuries, fermented foods have been consumed by individuals and population groups, and processing technology has evolved and been refined and diversified. During the last century the industrial production of starters (bacteria, yeasts, and molds) was decisive for the development of different kinds of fermented products all over the world. This change from "local production and local consumption" to the industrial one has reinforced the need of food safety, quality and health benefits of fermented foods. The benefits of microorganisms in food manufacture may result from growth and activity into the product, from production of substances that can be added during the manufacturing process and from growth and activity in the GI tract following ingestion of foods containing them (Bourdichon et al. 2012). So, now-a-days the microorganisms used in the production of fermented foods are characterized on the base of several important functionalities.

3.1 Improve Food Safety

Bio-preservation refers to extended shelf life and enhanced safety of foods using microorganisms or their metabolites (Cizeikiene et al. 2013). In developing countries, food fermentation remains an economic means of preserving food especially for households and small scale food industries (Motarjemi 2002). Microorganisms such as LAB allow rapid fermentation of raw media (i.e., milk, sourdough, cabbage). The result is a decrease in both pH and redox potential (E_h), an increase in organic acid concentration (lactic acid, acetic acid, etc.), and a lowering in available substrates such as carbohydrates, leading to inhibitory physicochemical conditions for spoilage microorganisms by liberating a group of antimicrobial macromolecules like bacteriocins, peptides or proteins. Bacteriocins may help to reduce the use of chemical preservatives and/or the intensity of heat and other physical treatment (Gálvez et al. 2008).

3.2 Improve Nutritional Value

Fermentation may improve nutritional status of foods, changing the nutrient level, nutrient availability and reducing anti-nutritional factors. Phytase activity present in yeasts and LAB increases mineral solubility and bioavailability during sourdough and cereal fermentation (Poutanen et al. 2009). Fermentation processes can increase levels of vitamins principally of the B group such as riboflavin, thiamin and niacin (Capozzi et al. 2012). Fermentation has also the ability to improve anti-oxidative activity of foods (Hur et al. 2014).

3.3 Improve Organoleptic Properties

Microorganisms can modify food texture by the specific proteolytic and lipolytic activities of yeasts, bacteria, and molds, which are advantageously used by technologists in cheese making. They also produce substances that modify the texture such as exo-polysaccharides and organic acids. In yogurt for example, polysaccharides enhance the smoothness and the mouthfeel, while lactic acid accumulation contributes to gel formation. These microbial activities allow to avoid the use of additives (texture modifying agents, thickener), and consequently they increase the "clean label" of food products, and from the producer's point of view they exclude the use of expensive ingredients. The biogenesis of flavor is one of the main modifications in food fermentation, for example, proteins and lipids are involved for texture metabolism of glucosides. Glucose metabolism forms compounds such as organic acids, ethanol, diacetyl (butter flavor), and acetaldehyde. Lipid metabolism leads to the formation of secondary alcohols (mushroom flavor), free fatty acids, lactones (fruity such as coconut, musty, blue cheese note), acids and alcohols. Proteolytic activities form alcohols, sulfur compounds, amines and organic acids. Sulfur compounds are produced in very low quantities but their low detection threshold and their characteristic odors give them an important role in consumer preference (Feron and Wache 2006). In conclusion, the flavor of a fermented product involves a large variety of metabolic products which depends on (i) the raw material, (ii) the microorganism, (iii) the process, and (iv) the storage conditions.

3.4 Improve Technology

Carbon dioxide production and leavening of dough are the main functions of baker's yeasts during bread making. The yeasts produce CO_2 which dissolves in the liquid phase of the dough. The heart of the rising step consists in the mass transfer of CO_2 from the liquid phase into the gaseous phase, leading to the growth of bubbles. Selected yeasts and the physiological state of yeasts may modify fermentation rate, dough extensibility, and therefore affect the gas holding capacity of fermenting dough (Rezaei et al. 2014). CO_2 production by hetero-fermentative LAB such as *Leuconostoc* sp. is essential for "eyes" formation inside the curd in the manufacture of Blue-veined cheeses (Hemme and Foucaud-Scheunemann 2004). Without the formation of cavities, the molds (*Penicicillium roqueforti*) are unable to germinate and to contribute to the formation of texture and flavor through the action of their protease and lipase activities.

3.5 Improve Human Health

Probiotic microorganisms have historically been used to re-balance disturbed intestinal microbiota and to diminish gastrointestinal disorders, such as diarrhea or inflammatory bowel diseases (e.g., Crohn's disease and ulcerative colitis). Recent studies explore the potential for expanded uses of probiotics on medical disorders that increase the risk of developing cardiovascular diseases and diabetes, obesity, hypercholesterolemia, arterial hypertension, and metabolic disturbances such as hyper-homocysteinemia and oxidative stress (Ebel et al. 2014). Molecules produced by microorganisms or extracted from microorganisms may also impact human health. For example, certain forms of β-glucan extracted from yeast cell wall may have medicinal applications (Kwiatkowski and Kwiatkowski 2012). Moreover, several peptides from soy-fermented foods possessed various biological activities such as anti-thrombotic and antioxidant properties (Gibbs et al. 2004).

4. Modern Techniques of Food Investigation and Advanced Research

It is only recently that there is a willingness to understand fermentation mechanisms and their application for commercialization, principally in developing countries. It is important that countries recognize the potential of fermented foods and prioritize actions to assure their safety, quality and availability. The utility of bio/technology-based applications in food processing and food safety is recognized (Fig. 2) (FAO 2013). Moreover, metabolic engineering of microorganisms can be used to improve fermented foods.

Recent developments in molecular methods now enable faster, accurate and more sensitive microorganism characterization than culture-based methods. The study of fermented foods diversity is difficult because of the inability to cultivate most of the viable bacteria or the viable and non-cultivable cells (VNC). Applied to fermented foods, molecular tools are of great help to study the dynamics of microbial communities, their metabolic activity as well as monitoring pathogens. Polymerase Chain Reaction (PCR)-base methods can successfully be applied. Quantitative PCR (qPCR) is an accurate and sensitive detection method for individual microbial species or groups. In qPCR the logarithmic amplification of a DNA target sequence is linked to the fluorescence of a reporter molecule. PCR-based techniques are now routinely being used for the characterization of technological microbiota and the analysis of pathogens in fermented products (Postollec et al. 2011).

PCR was also used to evaluate the general metabolic activity of microorganisms during processes, for understanding which microbiota are active in the transformation of food and their roles during maturing process.

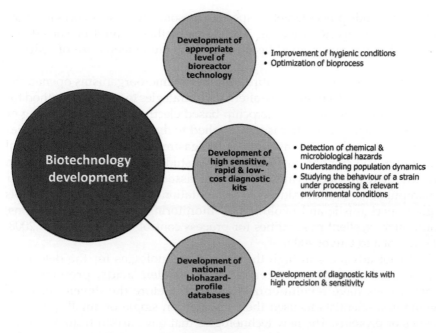

Figure 2. Future implications of biotechnology in development of fermentation process.

One way to distinguish between viable/active and dead microorganism is to use RT-qPCR which targets RNA instead of DNA (Falentin et al. 2010).

It is clear that genetic/phenotypic variations exist among strains of the same species. This variation has a major impact on their performance in food fermentations. Molecular typing of microorganisms is a useful tool to characterize a strain presenting technological properties in order to carry out its monitoring during fermentation and to assure industrial propriety (Kitahara et al. 2005, Bhardwaj et al. 2011). The most used methods are the Pulsed Field Gel Electrophoresis (PFGE), Random Amplification of Polymorphic DNA (RAPD), Ribotyping and the Multilocus Sequence Typing (MLST).

For more than a decade, the genetic fingerprinting technique PCR-DGGE (Polymerase Chain Reaction-Denaturing Gradient Gel Electrophoresis) was used to study the diversity of a microbial community of a specific product (Maiworé et al. 2012, Bigot et al. 2015), to monitor the dynamic of fermentations (Ercolini et al. 2003, Hong et al. 2012, Durand et al. 2013, Hamdouche et al. 2015) or to determine the geographical origin of fruits (El Sheikha et al. 2012), cheeses (Arcuri et al. 2013) and salts (Dufossé et al. 2013).

Some non-PCR based methods are interesting to study food microbial composition as the Fluorescence *In Situ* Hybridization (FISH), based in oligonucleotide probes targeting specific ribosomal RNA sequences in whole micro-organisms (Rohde et al. 2015). This method cannot be considered a high-throughput method because of the simultaneous use of reduced probes.

The sequencing of the entire genomes of microorganisms opened the era of genomics. Genomics use alternative strategies to Sanger's method for DNA sequencing, such as microchip-based electrophoretic sequencing or pyrosequencing. The latter is now applied to dynamic population studies of highly complex systems, such as the gastrointestinal tract. Functional genomics and comparative metagenomic studies have strong application in the field of food fermentations and associated microbes. To better manipulate or control traditional fermentation processes, genomics tools allow mechanistic and evolutionary monitoring in food fermentations, and offer excellent possibilities for process control (Sieuwerts et al. 2008, Humblot and Guyot 2009).

Major advances in high throughput technologies for the detection, quantification and characterization of nucleic acids, proteins, and metabolites have revolutionized research. Before the development of genomics, scientists focused their research on single or small groups of genes or proteins. The new technologies that have arisen from genomics with the objective to understand biological systems; functional genomics enclosing transcriptomics, proteomics, and metabolomics are so called "omics" technologies and are very promising to better understand microbial communities and provide also attractive possibilities for further optimization of food related microorganisms.

Functional genomics is very useful to compare the traits of microorganisms involved in food fermentations, and can also enable the identification of new target genes contributing to specific food characteristics, such as flavor and to functional properties related to health (Kuipers et al. 2000). Comparative analysis of metabolic, regulatory and transport pathways predicted from genome sequences is a valuable tool for predicting strain functionalities. Once template genomes are available, DNA microarrays can be applied for the rapid and cost-effective assessment of gene content in multiple strains.

Quantifying the variability and kinetics of gene expression is very useful in the monitoring of fermentation process. Transcriptomic tools analyze the whole set of transcripts in a cell and their quantity in a specific developmental stage or physiological condition. Several high-throughput RNA measurement tools have been developed for transcriptome analysis, such as DNA microarrays (Han and Wang 2008). They permit to examine whole-genome expression profiles of bacteria grown under various environmental conditions (exposure to alternative carbon sources, evolution of pH, stress conditions, etc.) and during the production of metabolites

(Park et al. 2005). For instance, several studies characterized LAB gene expression during growth in milk or cheese and their adaptation during commercial preparation (Taibi et al. 2011, Azcarate-Peril et al. 2009).

More advanced metatranscriptomics studies aim to analyze the transcriptional response in complex food ecosystems. This requires microarrays of multiple strains. Sieuwerts et al. (2010) used global transcriptome analysis of *Lactobacillus bulgaricus* and *Streptococcus thermophilus*, involved in yogurt fermentation, to study the metabolic interactions between the two strains.

Complex fermentations of plant-derived products have been studied using "omics" methods. A metagenomic analysis of the kimchi, a traditional Korean vegetable product, revealed the presence of 23 LAB related to kimchi fermentation. Metatranscriptome analysis revealed that, with the exception of two microorganisms, all LAB probed in the microarray contributed to kimchi fermentation (Nam et al. 2009). Similar approaches were applied to study the complex and uncontrolled sourdough fermentations. LAB functional gene microarray analysis revealed the activation of different key metabolic pathways as the ability to use starch and maltose, and the conversion of amino acids as a contribution to redox equilibrium and flavor compound generation (Weckx et al. 2011). Actually, new perspectives are opening up in transcriptomic studies by using Next-Generation Sequencing (NGS), applied to determine global transcriptional responses as the sequence depth is increasing (Siezen et al. 2010).

Proteomics studies can help to identifying adaptations that take place over a longer time frame than transcriptomics responses. Usually, 2DE or multidimensional HPLC and tandem MS coupled on-line allow a global and dynamic view of protein expression in bacteria or other organisms. They permit the monitoring of protein expression in microorganisms used for food fermentations, notably for optimization of starter composition or characterization of protein expression of bacteria used for products manufacture or as probiotics (Manso et al. 2005).

In fermented foods, the analysis of relevant metabolites is a standard physiological practice, but global metabolomics studies have not been reported frequently. Metabolomics is an emerging field consisting in high throughput characterization of small molecule metabolites in biological matrices (FAO 2013). Metabolic analyses a precise complex set of biochemical instruments. The techniques used are multidisciplinary and rely on chromatographic separations, often HPLC coupled to mass spectrometry (LC-MS) and HPLC coupled to nuclear magnetic resonance (LC-NMR) (Mozzi et al. 2013). Metabolomics appear as an important tool for development of food product, notably regarding the compliance of regulations, processing, quality, safety, and microbiology in food processes (Cevallos-Cevallos et al. 2009).

Fermented foods are now regarded as part of our staple diet. Recent scientific advances allowed to the establishment of defined microbial starters, leading to the transfer of artisanal fermented food to more controlled and industrialized fermentations. Genomic and proteomic era offers new approaches for the exploitation of micro-organisms for food fermentation and help greatly to enable the safety and quality of products. The rationalized use of micro-organisms in the preparation of fermented foods opens new perspectives, such as the use of functional strains as probiotic strains to improve consumers' health.

5. Technological Interventions and Commercialization of Fermented Foods and Beverages

Traditional fermentations provide an affordable means for alimentation all around the world. In developing countries, fermentations are often considered as the ability to transform undesirable food into one with more nutritional properties, like soy for example, but this definition begins also to take a health dimension. In industrialized countries, usually several versions of functionalized foods like pickles, sauerkraut that are preserved in vinegar instead of the traditional lactic fermentation exist. However, a number of agro-food processing enterprises have invested in the development of a commercial production technology of indigenous fermented foods. For example, Achi reported in 2005 that the production of Tempe has been extended to United-States where about 16 companies were engaged in its process. Indeed, there are interesting prospects for industrial production of fermented foods by harnessing high technology for the tempe process and upgrading the protein content of predominantly starchy foods (Achi 2005).

Increasing of industrialization in many part of the world has lead the preparation of fermented foods to move from a small-scale household level to large-scale operations. The area of biotechnology has allowed to prepare foods with better scientific knowledge and modified procedures (Okafor 1992). Industrial development of fermented food can be divided in four categories according to Achi (2005):

- *Raw material development*

  Selection of more suitable variety, strain, and availability to use as substrate are essential for industrial development. These last decades, the use of biotechnologies for this purpose has led to a considerable increase in the production rates.

- *Starter development*

  Indigenous fermented foods are originally fermented by natural micro-organisms which have been transferred from generations to generations. As described previously in this chapter, it becomes increasingly important to control fermentations: the type of microorganisms and their physiology and metabolism involved in food fermentation have been identified to optimize processes. The use of selected strains for starters preparation is essential, and the use of such starters appears to be one of the most important technological interventions in the manufacturing.

• *Starter culture medium*

The development and commercialization of a starter medium designated to enhance the growth of starter culture for fermentation processes can significantly improve indigenous fermentation technologies. This step allows reducing the number of bacterial or yeast cells used to start the fermentation. Thus, the starter culture is added to the medium to produce an intermediate inoculum later used for the process of fermenting medium.

• *Development of fermentation process*

Technological interventions have first targeted reduction of preparation laboriousness and improvement of hygienic conditions. The increasing demand for safe and high quality food led to the development of quality control processes such as HACCP and the enhancement of fermentation technologies. Moreover, in the industrial scale operations, a quantitative approach is sought. Processing parameters during the fermentation can strongly impact on final products and require optimized fermentative containers, incubation rooms and their heating, cooling and insulator capacities.

Changes in consumer demand also modified commercialization of fermented foods. Indeed, the preference for shopping in supermarkets stimulated the development of improved packaging and reasonable shelf-life (FAO). The consumer is actually more informed regarding the relationship between food and health, leading our society to enter an area in which food products may have a "health" dimension. The search of health benefits in food by residents of developed countries lead to improved fermentation processes with the use of probiotic cultures or the addition of functional ingredients. The first developments of functional foods consisted of the addition of vitamins and/or minerals such as vitamin C, vitamin E, folic acid, zinc, iron and calcium (Sloan 2000). The focus then shifted to foods fortified with micronutrients omega-3 fatty acid, phytosterol, and soluble fiber to promote good health or to prevent diseases such as cancers and to finally develop products that offer multiple health benefits in a single food (Sloan 2004, Siró et al. 2008).

6. Conclusion and Future Perspectives

Fermented foods are now regarded as part of our staple diet. Recent scientific advances have permitted the establishment of defined microbial starters, leading to the transfer of artisanal fermented food to more controlled and industrialized fermentations. The genomic and proteomic era offers new approaches for the exploitation of microorganisms for food fermentation and help greatly to enable the safety and quality of the products. The rationalized use of microorganisms in the preparation of fermented foods opens new perspectives, such as the use of functional strains as probiotic strains to improve host health. In recent years, research follows the concerns of the world for food to be healthy and natural. Consumers desire new products but they should be linked to the tradition or to ethno-food, while the industry looks to meet a better understanding of the mechanisms of action that could justify health claims and technological innovations that may be sometimes inspired by nature or conventional processes. Between consumer trends that are moving towards good health products, natural and "innovative in tradition" and those of the industry to develop new processes, the field of fermented products is a very hot topic and in particular has an appeal for exotic traditional products. This is the origin of Tropical Biotechnology & Bioresources network that combines the activities of several research laboratories and manufacturers to develop new strains, new products or making existing products safer (http://www.umr-pam. fr/relation-internationale/tropical-bioresources-biotechnology.html).

Keywords: Starter cultures, Fluorescence *In Situ* Hybridization (FISH), *Tetragenococcus halophilus, S. cerevisiae, Candida milleri, Candida humilis,* mold, yeast, non-cultivable cells, DNA microarrays, metatranscriptomics, lactic acid bacteria, *Bacillus* spp., Alkaline fermentation, Fermented food, microorganism, beneficial properties, food safety, molecular identification, "omic's", proteomics, metabolomics, Pulsed Field Gel Electrophoresis (PFGE), Random Amplification of Polymorphic DNA (RAPD), Ribotyping and the Multilocus Sequence Typing (MLST), Quantitative PCR

References

Achi, O.K. (2005). The potential for upgrading traditional fermented foods through biotechnology. African Journal of Biotechnology 4(5): 375–380.

Aidoo, K.E., Rob Nout, M.J. and Sarkar, P.K. (2006). Occurrence and function of yeasts in Asian indigenous fermented foods. FEMS Yeast Research 6(1): 30–39.

Aidoo, K.E. (1992). Lesser-known fermented plant foods. *In*: Applications of Biotechnology to Traditional Fermented Foods, Report of an Ad Hoc Panel of the Board on Science and Technology for International Development.

Akinrele, I.A. (1970). Fermentation studies on maize during the preparation of a traditional African starch-cake food. Journal of the Science of Food and Agriculture 21(12): 619–625.

Arcuri, E.F., Rychlik, T., El Sheikha, A.F., Piro-Métayer, I. and Montet, D. (2013). Determination of cheese origin and dominant bacteria by PCR-DGGE: primary application to traditional Minas cheese. Food Control 30(1): 1–6.

Azcarate-Peril, M.A., Tallon, R. and Klaenhammer, T.R. (2009). Temporal gene expression and probiotic attributes of *Lactobacillus acidophilus* during growth in milk. Journal of Dairy Science 92(3): 870–886.

Battcock, M. and Azam-Ali, S. (1998). Fermented fruits and vegetables: a global perspective Food & Agriculture Organization service bulletin n° 134, pp. 107–226.

Bhardwaj, A., Kaur, G., Gupta, H., Vij, S. and Malik, R.K. (2011). Interspecies diversity, safety and probiotic potential of bacteriocinogenic *Enterococcus faecium* isolated from dairy food and human feces. World Journal of Microbiology and Biotechnology 27(3): 591–602.

Bigot, C., Meile, J.C., Kapitan, A. and Montet, D. (2015). Discriminating organic and conventional foods by analysis of their microbial ecology: An application on fruits. Food Control 48: 123–129.

Bourdichon, F., Casaregola, S., Farrokh, C., Frisvad, J.C., Gerds, M.L., Hammes, W.P., Harnett, J., Huys, G., Laulund, S., Ouwehand, A., Powell, J.I.B., Prajapati, J.B., Seto, Y., Schure, E.T., Boven, E.B., Vankerckhoven, V., Zgoda, A., Tuijtelaars, S. and Hansen, E.B. (2012). Food fermentations: microorganisms with technological beneficial use. International Journal of Food Microbiology 154(3): 87–97.

Capozzi, V., Russo, P., Dueñas, M.T., López, P. and Spano, G. (2012). Lactic acid bacteria producing B-group vitamins: a great potential for functional cereals products. Applied Microbiology and Biotechnology 96(6): 1383–1394.

Cevallos-Cevallos, J.-M., Reyes-De-Corcuera, J.I., Etxeberria, E., Danyluk, M.D. and Rodrick, G.E. (2009). Metabolomic analysis in food science: a review. Trends in Food Science and Technology (20): 557–566.

Chavez, R., Fierro, F., Garcia-Rico, R.O. and Laich, F. (2011). Mold-fermented foods: *Penicilium* spp. as ripening agents in the elaboration of cheese and meat products. In: A.L. Monteiro (ed.). Mycofactories, Bentham eBooks, United Arab Emirates 5: 73–98.

Cizeikiene, D., Juodeikiene, D., Paskevicius, A. and Bartkiene, H. (2013). Antimicrobial activity of lactic acid bacteria against pathogenic and spoilage micro-organism isolated from food and their control in wheat bread. Food Control 31(2): 539–545.

Dufossé, L., Donadio, C., Valla, A., Meile, J.C. and Montet, D. (2013). Determination of specialty food salt origin by using 16S rDNA fingerprinting of bacterial communities by PCR-DGGE: An application on marine salts produced in solar salterns from the French Atlantic Ocean. Food Control 32(2): 644–649.

Durand, N., El Sheikha, A.F., Suarez-Quiros, M.L., Gonzales-Rios, O., Nganou, N., Fontana-Tachon, A. and Montet, D. (2013). Application of PCR-DGGE to the study of dynamics and biodiversity of yeasts and potentially OTA producing fungi during coffee processing. Food Control 34(2): 466–471.

Ebel, B., Lemetais, G., Beney, L., Cachon, R., Sokol, H., Langella, P. and Gervais, P. (2014). Impact of probiotics on risk factors for cardiovascular diseases. A review. Critical Reviews in Food Science and Nutrition 54(2): 175–189.

El Sheikha, A.F., Durand, N., Sarter, S., Okullo, J.B. and Montet, D. (2012). Study of the microbial discrimination of fruits by PCR-DGGE: Application to the determination of the geographical origin of *Physalis* fruits from Colombia, Egypt, Uganda and Madagascar. Food Control 24(1): 57–63.

Emond, E. and Moineau, S. (2007). Bacteriophages and food fermentations. pp. 93–124. In: S. Mc Grath and D. van Sinderen (eds.). Bacteriophage: Genetics and Molecular Biology.

Ercolini, D., Hill, P.J. and Dodd, C.E.R. (2003). Bacterial community structure and location in Stilton cheese. Applied and Environmental Microbiology 69(6): 3540–3548.

Falentin, H., Postollec, F., Parayre, S., Henaff, N., Le Bivic, P., Richoux, R., Thierry, A. and Sohier, D. (2010). Specific metabolic activity of ripening bacteria quantified by real-time reverse transcription PCR throughout Emmental cheese manufacture. International Journal of Food Microbiology 144(1): 10–19.

FAO (2013). Status and trends of the conservation and sustainable use of micro-organisms in food processes. Background Study Paper: Alexandraki V., Tsakalidou E., Papadimitriou K. and Holzapfel W., pp. 1–160.

FAO (2009). Scoping study on microorganisms relevant to food and agriculture. Commission on Genetic Resources for Food and Agriculture, Rome, 19–23 October 2009, pp. 1–8.

Feron, G. and Waché, Y. (2006). Microbial biotechnology of food flavor production. pp. 407–441. *In*: K. Shetty, G. Paliyath, A. Pometto and R.E. Levin (eds.). Food Biotechnology. CRC, Taylor and Francis Publishers, New York.

Gálvez, A., López, R., Abriouel, H., Valdivia, E. and Omar, N. (2008). Application of bacteriocins in the control of foodborne pathogenic and spoilage bacteria. Critical Reviews in Biotechnology 28(2): 125–152.

Gibbs, B.F., Zougman, A., Masse, R. and Mulligan, C. (2004). Production and characterization of bioactive peptides from soy hydrolysate and soy-fermented food. Food Research International 37(2): 123–131.

Hamdouche, Y., Guehi, T., Durand, N., Kedjebo, K.B.D., Montet, D. and Meile, J.C. (2015). Dynamics of microbial ecology during cocoa fermentation and drying: Towards the identification of molecular markers. Food Control 48: 117–122.

Hemme, D. and Foucaud-Scheunemann, C. (2004). *Leuconostoc*, characteristics, use in dairy technology and prospects in functional foods. International Dairy Journal 14(6): 467–494.

Han, J.-Z. and Wang, Y.-B. (2008). Proteomics: present and future in food science and technology. Trends in Food Science and Technology 19(1): 26–30.

Holzapfel, W.H. (2002). Appropriate starter culture technologies for small-scale fermentation in developing countries. International Journal of Food Microbiology 75(3): 197–212.

Holzapfel, W.H., Geisen, R., Schillinger, U. and Lücke, F.K. (2003). Food preservatives, 2nd edn. pp. 291–319. *In*: N.J Russell and G.W. Gould (eds.). Starter and Protective Cultures. Kluwer Academic/Plenum Publishers, New York.

Hong, S.W., Choi, J.Y. and Chung, K.S. (2012). Culture-based and denaturing gradient gel electrophoresis analysis of the bacterial community from chungkookjang, a traditional Korean fermented soybean food. Journal of Food Science 77(10): 572–578.

Humblot, C. and Guyot, J.P. (2009). Pyrosequencing of tagged 16S rRNA gene amplicons for rapid deciphering of the microbiomes of fermented foods such as pearl millet slurries. Applied and Environmental Microbiology 75(13): 4354–4361.

Hur, S.J., Lee, S.Y., Kim, Y.-C., Choi, I. and Kim, G.-B. (2014). Effect of fermentation on the antioxidant activity in plant-based foods. Food Chemistry 160: 346–356.

Jakobsen, M., Larsen, M.D. and Jespersen, L. (2002). Production of bread, cheese and meat. pp. 3–22. *In*: H.D. Osiewacz (ed.). The Mycota. A Comprehensive Treatise on Fungi as Experimental Systems and Applied Research. Vol. 10. Industrial Applications. Springer, Berlin Heidelberg New York.

Josephsen, J. and Jespersen, L. (2004). Starter cultures and fermented products. pp. 23–49. *In*: Y.H. Hui, L. Meunier-Goddik, Å.S. Hansen, J. Josephsen, W.K. Nip, P.S. Stanfield and F. Toldrá (eds.). Handbook of Food and Beverage Fermentation Technology. Marcel Dekker, Inc., New York.

Jung, D.H. (2012). Great dictionary of fermented food. Yuhanmunwhasa, Inc., Kayang-dong, Seoul, pp. 146–63.

Kwiatkowski, S. and Kwiatkowski, S.E. (2012). Yeast (*Saccharomyces cerevisiae*) glucan polysaccharides—occurrence, separation and application in food, feed and health industries. pp. 47–70. *In*: D.N. Karunaratne (ed.). Biochemistry, Genetics and Molecular Biology "The Complex World of Polysaccharides". InTech.

Kitahara, M., Sakata, S. and Benno, Y. (2005). Biodiversity of *Lactobacillus sanfranciscensis* strains isolated from five sourdoughs. Letters in Applied Microbiology 40(5): 353–357.

Kersters, K., Lisdiyanti, P., Komagata, K. and Swings, J. (2006). The family *Acetobacteraceae*: the genera *Acetobacter*, *Acidomonas*, *Asaia*, *Gluconacetobacter*, *Gluconobacter*, and *Kozakia*. pp. 163–200. *In*: M. Dworkin, S. Falkow, E. Rosenberg, K.-H. Schleifer and E. Stackebrandt (eds.). The Prokaryotes (3rd edn., Vol. 5). New York: Springer-Verlag.

Kimura, K. and Itoh, Y. (2007). Determination and characterization of IS 4Bsu 1-insertion loci and identification of a new insertion sequence element of the IS 256 family in a natto starter. Bioscience, Biotechnology and Biochemistry 71(10): 2458–2464.

Kuipers, O.P., Buist, G. and Kok, J. (2000). Current strategies for improving food bacteria. Research in Microbiology 151(10): 815–822.

Leistner, L. (1986). Mold-ripened foods. Fleischwirtschaft 66(2): 1385–1388.

Maiworé, J., Tatsadjieu Ngouné, L., Goli, T., Montet, D. and Mbofung, C.M. (2012). Influence of technological treatments on bacterial communities in tilapia (*Oreochromis niloticus*) as determined by 16S rDNA fingerprinting using polymerase chain reaction-denaturing gradient gel electrophoresis (PCR-DGGE). African Journal of Biotechnology 11(34): 8586–8593.

Manso, M.A., Léonil, J., Jan, G. and Gagnaire, V. (2005). Application of proteomics to the characterization of milk and dairy products. International Dairy Journal 15(6): 845–855.

Motarjemi, Y. (2002). Impact of small scale fermentation technology on food safety in developing countries. International Journal of Food Microbiology 75(3): 213–229.

Mozzi, F., Ortiz, M.E., Bleckwedel, J., De Vuyst, L. and Pescuma, M. (2013). Metabolomics as a tool for the comprehensive understanding of fermented and functional foods with lactic acid bacteria. Food Research International 154(1): 1152–1161.

Nam, Y.D., Chang, H.W., Kim, K.H., Roh, S.W. and Bae, J.W. (2009). Metatranscriptome analysis of lactic acid bacteria during kimchi fermentation with genome-probing microarrays. International Journal of Food Microbiology 130(2): 140–6.

Okafor, N. (1992). Commercialization of Fermented Foods in Sub-Saharan Africa. Application of Biotechnology in Traditional Fermented Foods. Chapter 24: 165–169.

Ouoba, L.I.I., Diawara, B., kofiAmoa-Awua, W., Traoré, A.S. and Møller, P.L. (2004). Genotyping of starter cultures of *Bacillus subtilis* and *Bacillus pumilus* for fermentation of African locust bean (*Parkia biglobosa*) to produce Soumbala. International Journal of Food Microbiology 90(2): 197–205.

Oyewole, O.B. (1997). Lactic fermented foods in Africa and their benefits. Food Control 8(5): 289–297.

Parekh, S., Vinci, V.A. and Strobel, R.J. (2000). Improvement of microbial strains and fermentation processes. Applied Microbiology and Biotechnology 54(3): 287–301.

Park, S.J., Lee, S.Y., Cho, J., Kim, T.Y., Lee, J.W., Park, J.H. and Han, M.-J. (2005). Global physiological understanding and metabolic engineering of microorganisms based on omics studies. Applied Microbiology and Biotechnology 68(5): 567–579.

Postollec, F., Falentin, H., Pavan, S., Combrisson, J. and Sohier, D. (2011). Recent advances in quantitative PCR (qPCR) applications in food microbiology. Food Microbiology 28(5): 848–861.

Poutanen, P., Flander, L. and Katina, K. (2009). Sourdough and cereal fermentation in a nutritional perspective. Food Microbiology 26(7): 693–699.

Rohde, A., Hammerl, J.A., Appel, B., Dieckmann, R. and Al Dahouk, S. (2015). FISHing for bacteria in food–A promising tool for the reliable detection of pathogenic bacteria? Food Microbiology 46: 395–407.

Ray, R.C. and Joshi, V.K. (2014). Fermented foods: past, present and future. pp. 1–36. *In*: R.C. Ray and D. Montet (eds.). Microorganisms and Fermentation of Traditional Foods. CRC Press, Boca Raton, Florida.

Rezaei, M.N., Dornez, E., Jacobs, P., Parsi, P., Verstrepen, K.J. and Courtin, C.M. (2014). Harvesting yeast (*Saccharomyces cerevisiae*) at different physiological phases significantly affects its functionality in bread dough fermentation. Food Microbiology 39: 108–115.

Sieuwerts, S., De Bok, F.A., Hugenholtz, J. and van Hylckama Vlieg, J.E. (2008). Unraveling microbial interactions in food fermentations: from classical to genomics approaches. Applied and Environmental Microbiology 74(16): 4997–5007.

Sieuwerts, S., Molenaar, D., van Hijum, S.A., Beerthuyzen, M., Stevens, M.J., Janssen, P.W., Ingham, C.J., de Bok, F.A., de Vos, W.M. and van HylckamaVlieg, J.E. (2010).

Mixed-culture transcriptome analysis reveals the molecular basis of mixed-culture growth in *Streptococcus thermophilus* and *Lactobacillus bulgaricus*. Applied and Environmental Microbiology 76(23): 7775–84.

Siezen, R.J., Wilson, G. and Todt, T. (2010). Prokaryotic whole-transcriptome analysis: deep sequencing and tiling arrays. Microbial Biotechnology 3(2): 125–30.

Silva, J., Carvalho, A.S., Teixeira, P. and Gibbs, P.A. (2002). Bacteriocin production by spray-dried lactic acid bacteria. Letters in Applied Microbiology 34(2): 77–81.

Siró, I., Kápolna, E., Kápolna, B. and Lugasi, A. (2008). Functional food. Product development, marketing and consumer acceptance. A review. Appetite 51: 456–467.

Sloan, A.E. (2004). The top ten functional food trends. Food Technology 58: 28–51.

Sloan, A.E. (2000). The top ten functional food trends. Food Technology 54: 33–62.

Steinkraus, K.H. (1991). African alkaline fermented foods and their relation to similar foods in other parts of the world. pp. 87–92. *In*: A. Westby and P.J.A. Reilly (eds.). Traditional African Foods-Quality and Nutrition. International Foundation for Science, Sweden.

Steinkraus, K.H. (2002). Fermentations in world food processing. Comprehensive Reviews in Food Science and Food Safety 1(1): 23–32.

Steinkraus, K.H. (1997). Classification of fermented foods: Worldwide review of household fermentation techniques. Food Control 8(5): 311–317.

Steinkraus, K.H. (1996). Introduction to Indigenous fermented Foods. pp. 1–5. *In*: K.H. Steinkraus (ed.). Handbook of Indigenous Fermented Foods. 2nd edn. Marcel Dekker, New York.

Stevens, H.C. and Nabors, L.O. (2009). Microbial food cultures: a regulatory update. Food Technology 63(3): 36–41.

Stiles, M.E. (1996). Biopreservation by lactic acid bacteria. Antonie van Leeuwenhoek 70(2-4): 331–345.

Taibi, A., Dabour, N., Lamoureux, M., Roy, D. and Lapointe, G. (2011). Comparative transcriptome analysis of *Lactococcus lactis* subsp. *cremoris* strains under conditions simulating Cheddar cheese manufacture. International Journal of Food Microbiology 146(3): 263–275.

Tamang, J.P. and Kailasapathy, K. (2010). Fermented Foods and Beverages of the World. CRC Press, Florida.

Tamang, J.P., Tamang, N., Thapa, S., Dewan, S., Tamang, B., Yonzan, H., Rai, A.K., Chettri, R., Chakrabarty, J. and Kharel, N. (2012). Microorganisms and nutritional value of ethnic fermented foods and alcoholic beverages of North East India. Indian Journal of Traditional Knowledge 11(26): 7–25.

Vogel, R.F., Hammes, W.P., Habermeyer, M., Engel, K.H., Knorr, D. and Eisenbrand, G. (2011). Microbial food cultures—opinion of the Senate Commission on Food Safety (SKLM) of the German Research Foundation (DFG). Molecular Nutrition & Food Research 55(4): 654–662.

Walker, G.M. (1998). Yeast Physiology and Biotechnology. John Wiley & Sons, England.

Weckx, S., Allemeersch, J., Van der Meulen, R., Vrancken, G., Huys, G., Vandamme, P., Van Hummelen, P. and De Vuyst, L. (2011). Metatranscriptome analysis for insight into whole-ecosystem gene expression during spontaneous wheat and spelt sourdough fermentations. Applied and Environmental Microbiology 77(2): 618–26.

Zhou, W. and Therdthai, N. (2012). Fermented bread. pp. 477–492. *In*: Y.H. Hui and E. Özgül Evranuz (eds.). Handbook of Plant-Based Fermented Food and Beverage Technology (2nd ed.). CRC Press, Florida.

2

Technological Innovations in Processing of Fermented Foods
An Overview

Swati S. Mishra,[1,]* Ramesh C. Ray,[2] Sandeep K. Panda[3] and Didier Montet[4]

1. Introduction

The last 100 decades have witnessed significant changes in the world's food system. In ancient times fermentation of food was meant for food preservation and flavor improvement. Food processing/fermentations use various technologies and operations to convert relatively perishable, typically bulky and inedible raw materials into palatable foods and potable beverages with high stability and added values (Ray and Joshi 2014). Biotechnological aspects in the industrialization of indigenous fermented foods open the possibilities for exploring these technological interventions

---

[1] Department of Biodiversity and Conservation of Natural Resources, Central University of Orissa, Koraput 764020, India.
[2] ICAR - Regional Centre, Central Tuber Crops research Institute, Bhubaneswar 751 019, India.
E-mail: rc_rayctcri@rediffmail.com
[3] Department of Biotechnology and Food Technology, Faculty of Science, University of Johannesburg, PO Box 17011, Doornfontein Campus, Johannesburg, South Africa.
E-mail: sandeeppanda2212@gmail.com
[4] CIRAD-PERSYST, UMR Qualisud, TA B95/16, 34398 Montpellier Cedex 5, France.
E-mail: didier.montet@cirad.fr
* Corresponding author: swatisakambarimishra@gmail.com

for improved production methods and quality of the foods. The assurance of the quality and safety of the final product is the main achievement of the technologies applied. The science of food safety provides assurance about the physical, chemical or microbiological hazards being present at a permissible level with respect to health implications of the consumers (Ray 2013). Biotechnology has played a revolutionary role in production, preservation, nutritional enhancement and value addition of foods. Since time immemorial fermented foods have pleased our palates along with increased nutritive values. Understanding the science of microbiology in food applications with identification of new fermenting species was a boon to enhance the quality of our food in a number of ways. Traditional biotechnology has been helpful for production of functional foods, flavor enhancement, biopreservation, probiotics and enzyme modification of foods for long. With advanced technologies in food biotechnology like genetic engineering we have an upper hand in the field of functional foods.

As a technology, food fermentations date back at least 6000 years. The start of industrialization in 16th century initiated technological intervention in food production sector (Truninger 2013). The industrial revolution and blossoming of microbiology at the same time formed the foundation for preparation of bulk quantities and commercialization of processed food to meet the growing food requirement of the masses (Caplice and Fitzgerald 1999). Some of the conventional food processing technologies include salting, drying and fermentation. However, food processing using microorganisms is the most convenient technology for the development of novel fermented food products of commercial significance. Solid state fermentation is used for processing of vinegar, soy sauce, curing of tea and tobacco leaves, ripening of cheese, etc. (Ghosh 2015). Similarly, wine, beer, distilled beverages and yogurt are developed by submerged fermentation. The current article covers the various technological interventions covering the processing of fermented food, past and present.

2. Fermented Foods

The term fermentation comes from Latin word *"fermentum"* (to ferment). The science of fermentation is called zymology and the first zymologist was Louis Pasteur identifying and applying yeast in fermentation (Dubos 1995). Fermented foods are the ones that undergo microbial or enzymatic alterations for quality and sensorial improvement by biochemical changes (Campbell-Platt 1987). It is one of the oldest and most economical methods of production and preservation of foods and various types of fermentations have been used by different civilizations since prehistoric times. Foods are fermented for many reasons, including the enhancement of nutritive value, removal of anti-nutrients and the improvement of sensory characteristics such as flavor and taste. Fermentation with respect to technology and industrial microbiology can be defined as a process of biotransformation of cooked/uncooked food matrices carried out by microorganisms or their enzymes. Many valuable bio-products around the world are the result of fermentation, either occurring naturally or through addition of starter cultures. Presently, modern technologies and large scale production exploit defined species of starter cultures to ensure consistency and quality in the final product.

2.1 History of Fermented Foods

Food preservation and the art of fermentation is centuries old. Since ancient times fermentation has been used to conserve and alter food without understanding the microbial mechanisms. Greeks even attributed fermentation to Dionysus, the god of fruit fermentation (Stanislawski 1975). Cheese making was developed early when plants and animals were just domesticated, 8000 years back in Tigris and Euphrates (Fox 1993). By 2000–4000 B.C., Egyptians and Sumerians developed wine making and brewing with the start of alcoholic fermentation. The fermented beverages appeared in 5000 B.C. in Babylon, 3150 B.C. in ancient Egypt, 2000 B.C. in Mexico and 1500 B.C. in Sudan (Mirbach and El Ali 2005, Ray and Joshi 2014). Also dough fermentation was started by Egyptians for making bread. Evidence suggests that fermented foods were consumed 7,000 years ago in Babylon (Battcock and Azam-Ali 1998). Though for thousands of years the exploitation of fermentation for food and beverages has been taking place but only in the recent past the microorganisms responsible for the metabolic process have been recognized. The knowledge of microorganisms responsible for fermentation, pasteurization and industrial revolution took place simultaneously as massive migrations of local communities to large cities and industrial sectors increased the demand for bulk food production (Caplice and Fitzgerald 1999). This led to large scale fermentation processes for industrial production of fermented foods and beverages.

2.2 Types of Fermented Foods

Broadly, fermentation can be grouped into those that take place using solid foods, that is solid state fermentation and those, that use liquid raw materials called submerged fermentation. The solid state fermentation requires uses of aerobic microorganisms whereas most submerged fermentations using liquid foods require anaerobic conditions. Different production methods, technologies, equipments and protocols are designed to address the different types of fermentation. For solid state fermentation the simplest technologies are tray, bowl or other containers that are used for incubation in room or in cabinet. For small scale production of fermented pickles, beer and dairy products by liquid or submerged fermentation, it consists of covered containers and drums of plastic, aluminum or stain less steel. Equipment known as "Bioreactor" is used for large scale and industrial production. Different types of batch bioreactors include rotating-drum bioreactors, traditional and zymotis packed-bed bioreactors, packed-bed energy balance, intermittently-mixed forcefully-aerated bioreactor, well-mixed bioreactor, etc. (Banks 1984).

Fermented foods can be classified in a number of ways (Dirar 1993):

- By categories: (Yokotsuka and Sasaki 1985)—(1) alcoholic beverages fermented by yeasts, (2) vinegars/acetic acid fermented by *Acetobacter*, (3) milks fermented by *lactobacilli*, (4) pickles fermented through *lactobacilli*, (5) fish or meat fermented with *lactobacilli* and (6) plant proteins fermented with molds with or without *lactobacilli* and yeasts.
- By classes: (Campbell-Platt 1987)—(1) beverages (alcoholic/lactic acid fermented), (2) cereal bio-products, (3) dairy products, (4) fish products, (5) meat products;
- By commodity: (Kuboye 1985)—(1) cassava and other root crops, (2) cereal, (3) legumes and (4) beverages;
- On functional basis: (Dirar 1993)—(1) Kissar (staples)—porridges and breads such as aceda and kissra, (2) Milhat (sauces and relishes for the staples), (3) marayiss (30 types of opaque beer, clear beer, date wines and meads and other alcoholic drinks) and (4) Akil-munasabat (food for special occasions);
- On basis of types of microorganisms involved, food safety and physical, chemical and nutritive changes occurred: (Steinkraus 1996)—(1) Fermentations producing textured vegetable protein meat substitutes from legume/cereal mixtures. Examples are Indonesian tempe and ontjom (2) High salt/meat-flavored amino acid/peptide sauce and paste fermentations, (3) Lactic acid fermentations, (4) Alcoholic fermentations (5) Acetic acid fermentations, (6) Alkaline fermentations, (7) Leavened breads.

2.3 Microorganisms Associated with Fermented Foods

2.3.1 Yeasts

Yeasts, classified under fungus kingdom are being used since 5000 years ago for the production of bread, wine, beer and other alcoholic beverages. *Saccharomyces* is the most important genus of the yeasts that contribute significantly to the production of fermented foods and beverages of commercial importance (Panda et al. 2014, 2015). *S. cerevisiae*, *S. uvarum* and *S. bayanus* are the popular ones and considered as GRAS (generally regarded as safe). The application of a particular species of the yeast depends on the type of the product to be developed. For example in the production of ale-type beers, top-cropping yeast (ex. *S. cerevisiae*) is applied and in case of lager-type beers bottom-cropping yeast (*S. pastorianus*) is used. Apart from the *Saccharomyces* species other yeasts such as *Hanseniaspora* (*Kloeckera*), *Candida, Pichia, Metschnikowia, Kluyveromyces, Schizosaccharomyces, Issatchenkia Dekkera* (*Brettanomyces*) species and *Schizosaccharomyces pombe* are known to add to the production of various alcoholic beverages. Furthermore, yeasts are used in production of single cell protein, flavor precursors and colors from cheaper substrates, and also play a vital role in soy sauce fermentation (Fleet 2006).

2.3.2 Molds

Molds are certain multicellular filamentous fungi known as a major spoilage agent of food. However, several genera of the molds are used for the production of value added food products (Lasztity 2009). The most prominent application of molds is in the production of cheeses. *Penicillium camemberti* is used to produce white cheese where as *P. roqueforti* is applied for the production of blue cheese. *P. roqueforti* is known to produce roquefortin, a secondary metabolite that acts against Gram positive bacteria containing hemins. Molds are also used in the fermentation of meats. *P. nalgiovense* and *P. chrysogenum* are generally used for the production of mold-fermented meat products (Holzapfel et al. 2003).

2.3.3 Bacteria: Acetic acid bacteria, Lactic acid bacteria (LAB) and Bacillus

Application of bacteria in food processing is practiced especially for the production of vinegars and dairy products. Vinegars are diluted acetic acid, produced in a two stage fermentation process; in the first step the sugar sources are fermented into alcohol using yeasts and further the alcoholic medium (10–15% v/v, ethanol) is fermented to acetic acid using acetic acid bacteria (Stasiak and Blazejak 2009). In the process ethanol is dehydrogenated to acetic acid and the reduced co-substrates are oxidized simultaneously. Although various bacterial species are known to produce acetic acid but species belonging to *Acetobacter* and *Gluconobacter* are adopted in industries for commercial production of vinegars (Raspo and Goranovic 2009). Vinegars are further classified into different types depending upon the substrate used. Wine vinegar is developed by acetous fermentation of wine, cider vinegar from apple wine, honey vinegar from alcoholic fermented medium of diluted honey. Starchy sources are also used for the production of vinegar. Alpha amylase is used for the breakdown of starch to fermentable sugars such as maltose, dextrins and dextrose. The popular vinegars from starchy sources are malt vinegar, rice vinegar and molasses vinegar.

LAB are beneficial microorganisms, popularly used for fermentation of milk. They are gram-positive, catalase-negative, acid tolerant, aerotolerant, non-sporulating, and they are strictly fermentative rods or cocci which produce lactic acid as the major product from the energy-yielding fermentation of sugars (Temmerman et al. 2004, Wessel et al. 2004). The genus *Lactobacillus* is sub-divided into three groups based on sugar fermentation: facultative hetero-fermentative (*Group I*), obligated hetero-fermentative (*Group II*) and obligated homo-fermentative (*Group III*), respectively (Bernardeau et al. 2008). Lactobacilli from *Group I* are known to ferment hexoses to lactic acid and pentoses to lactic acid and acetic acid, and gas is not produced from glucose, but from gluconate. *Group II* bacterial species produce carbon dioxide, lactic acid, acetic acid and/or ethanol from hexoses, and produces gas from glucose. Lactobacilli from *Group III* do not ferment gluconate or pentoses, but ferment glucose to lactic acid. *Lactobacillus* spp. from all three of these groups participate in food fermentation. Keeping in view the health promoting and anti-microbial properties of probiotic LAB, currently they are the prime interest of researchers (Khandelwal et al. 2016). Hence, the modern food market is witnessing the innovative dairy and non-dairy probiotic products. Probiotic juices, yogurts and ice creams are being developed by using probiotic LAB and are getting acceptance in the market.

Similarly, bacteria belonging to *Bacillus* species are involved in preparation of different types of fermented foods. Especially, the unpalatable legumes are fermented using *Bacillus* to improve the organoleptic properties for consumption (Reddy et al. 1983). Several traditional fermented legumes are prepared in different regions such as *natto* of Japan, Nigerian *dawadawa* or *iru*, Nepalese *kinema* and Thai *thua nao*. *Natto* in Northern Japan is prepared by fermentation of cooked soybean seeds by pure cultures of *B. subtilis*. Fermentation of legumes is mostly carried out in solid state fermentation and each product has a unique distinct flavor and aroma. The *Bacillus* fermented foods are generally used as meat substitute and as a flavoring agent in soups or consumed directly. *Dawadawa* is produced and sold in African market by Nestle (Leejeerajumnean 2003).

3. Era of Starter Cultures

A starter culture may be defined as a preparation containing large numbers of desired microorganisms, used for accelerating the fermentation process. The preparations may contain some unavoidable residues from the culture substrates and additives (such as antifreeze or antioxidant compounds), which support the vitality and technological functionality of the microorganisms. A typical starter after being adapted to the substrate facilitates improved control of a fermentation process and predictability of its products (Holzapfel 1997). Basically there are three categories of starter cultures: (1) Single strain culture-contains only one strain of a species, (2) Multi-strain cultures-contain more than one strains of a single species, (3) Multi-strain mixed cultures-contain different strains from different species.

3.1 History and Subsequent Development of Starter Culture

Microorganisms are naturally omnipresent and hence observed in raw food materials. This was the basis of the idea of spontaneous fermentation. Backslopping was the important technological phenomenon used in spontaneous fermentation by inoculating the raw material with a small quantity of a previously performed successful fermentation. Hence, the dominance of the best adapted strains resulted in backslopping. This technology is still used for production in foods where the ecology and concrete knowledge about microbial population and role is not clearly known (Harris 1998). This is also an economical and reliable method of production of fermented foods.

3.2 Application of Functional Starter Cultures in Food Fermentations

The presence of at least one inherent functional property in the starter makes it a functional starter culture. Functional foods refer to the food with health promoting and disease prevention properties. The food fermentation industries have started to explore the application of functional starter cultures in the last two-to-three decades (De Vuyst 2000, Gasper and Crespo 2016). The careful selection of strains as starter cultures or co-cultures and its implementation in fermentation processes can achieve the desired natural and healthy product. Many probiotic strains are also considered for its application in food fermentation as functional starter or co-cultures (Jahreis et al. 2002, Khandelwal et al. 2016). Nowadays most leading starter culture manufacturers produce and market common Gastro Intestinal (GI)-based LAB and bifidobacteria commercially (Picard et al. 2005).

3.3 Recent Technologies and Use of Starter Cultures

The performance of the starter culture is regulated by the strain's growth, type of sugars it ferments and the nature of the final product formed. Biochemical and functional properties of yeasts used in bread, beer and wine manufacturing have been extensively studied (Montet and Ray 2016). LAB are intensively studied for the development of starter cultures

(Ray and Joshi 2014). Some examples of functional starters with technological advances are cited below:

3.3.1 Food grade genetic engineering

Genetically manipulated microorganisms used as starter cultures in food fermentation should be safe and regarded as acceptable ingredients in our food. One example of engineering for food safety was the removal of the D-lactate dehydrogenase (*ldhD*) gene from *Lactobacillus johnsonii* La1, which demonstrated the exclusion of the undesired D-isomer of lactate and leaving only the desired L-lactate (Mollet 1999).

3.3.2 Phage-resistant starters for the dairy industry

In large scale production in dairy industry bacteriophages possess a serious problem. Hence Phage-resistant starters with rotation of use and proper sanitation will overcome the issue. Natural resistance mechanisms (restriction and modification enzymes), phage adsorption and absorptive phage infection that prevent intracellular phage development, intracellular defense strategies are the usual cause of phage resistance (Forde and Fitzgerald 1999). Large scale application of strains with acquired natural mechanisms of phage resistance either by *in vivo* recombination (conjugation) or *in vitro* self-cloning, is highly desirable in dairy industry (Moineau 1999).

3.3.3 Mild yogurt by lactose-negative starters

Lactose-negative mutants of *Lb. delbrueckii* subsp. *bulgaricus* only grow in the presence of actively lactose fermenting *Streptococcus thermophilus* cells and hence enable production of mild yogurts (Mollet 1996). This is a technological achievement against the undesirable bitter taste due to acidification during yogurt production by *Lb. delbrueckii* subsp. *bulgaricus.*

3.3.4 Acceleration of cheese maturation

Aroma, flavor and taste in cheese production are enhanced by lactic acid production by LAB. Hence optimal activity of endogenous and exogenous enzymes that modify the cheese by rational selection of LAB starter and co culture and also *in situ* autolysis gives alternate solutions (Fox et al. 1996).

3.3.5 Metabolic engineering of starter culture

The possibility of using metabolic engineering to alter or optimize various aspects of this metabolic network have been reviewed by several researchers (Swindell et al. 1996, Daly et al. 1998). The currently known metabolic pathways of immediate practical importance are the ones described for the conversion of sugars *via* pyruvate to acids and metabolites with distinct flavors (Ray and Joshi 2014).

3.3.6 Bacterial genomics and high throughput technologies

This is the era of genomics in biological sciences. Acquiring and analyzing biological data and obtaining the whole genome sequence for a large number of microorganisms including LAB (Klaehammer et al. 2002, 2005) will provide a rich source to guide physiological studies, mutant selections and the use of genetic and protein engineering for starter culture preparations. For example, the *Lactococcus lactis* genome sequence revealed a number of unexpected genetic and metabolic potentials (Bolotin et al. 2001).

3.3.7 Safety in microbial cultures

Over the last decade, directly or indirectly microbial food cultures have come under various regulatory frameworks. They emphasise the history of use, traditional foods, or general recognition of safety. Foods and food additives are regulated according to the Food Drug and Cosmetic Act (1958), in the United States. The status of Generally Recognized as Safe (GRAS) was introduced by FDA in 2010. GRAS substance is adequately shown to be safe under the conditions of its intended use generally recognized, among qualified experts. A substance recognized for such use prior to 1958 is by default GRAS (like food used in the EU prior to May 15, 1997, not being Novel Food) (Anon. 1997, ILSI Europe Novel Food Task Force 2003). Microbial food cultures are covered by general food laws. They must be considered safe and suitable for their intended use. An organism or a product with GRAS status is exempt from the statutory premarket approval requirements. If a microorganism is GRAS for one food usage, it does not make it necessarily GRAS for all food usages.

Similarly the Qualified Presumptions of Safety (QPS) started in November 2007 by European Food Safety Authority (EFSA), was applied in evaluating microorganisms that requires a market authorization (like feed cultures, production of enzymes, etc.) (Anon. 2005, Vogel et al. 2011). The QPS assessment is based on the taxonomic level, body of knowledge, history of use and identification of potential safety concerns. After the assessment a list of microorganisms was prepared at the species level that are presumed safe for use, independent of media, fermentation conditions and intended use. The list of microorganisms submitted to EFSA for safety assessment is being updated annually. The factors like undesirable properties of the microorganisms for food fermentation, opportunistic infection, toxic metabolites and virulence factors, antibiotic resistance are taken into consideration when considering a microorganism as safe for human use (Bourdichon et al. 2012).

3.3.8 Recent list of organisms that can be used

In order to document microorganisms, traditionally used as food ingredients, the International Dairy Federation (IDF) in collaboration with European Food and Feed Cultures Association (EFFCA) has compiled a non exhaustive inventory of microorganisms with a documented history of use in foods. The "2002 IDF Inventory" listed 82 bacterial species and 31 species of yeast and molds whereas the present "Inventory of MFC" contains 195 bacterial species and 69 species of yeasts and molds belonging to 62 genera (Bourdichon et al. 2012).

4. LAB as Cell Factories

LAB are used all over the world in a large variety of industrial food fermentations because of their enormous potential for the biosynthesis of a number of compounds as metabolic end products or secondary metabolites (LeBlanc et al. 2012, Ray and Joshi 2014). LAB comprise different groups of microorganisms, such as *Carnobacterium, Enterococcus, Tetragenococcus, Vagococcus, Weissella*. The industrial core species of LAB belong to the genera, like *Lactococcus, Lactobacillus, Streptococcus, Pediococcus* and *Leuconostoc* (Klaenhammer et al. 2002). LAB produce lactic acid as an anaerobic product of glycolysis with high yield and productivity. They play important roles in the production of food and feed and are increasingly used as health-promoting probiotics. LAB also have the ability to enhance flavor, texture and nutrition. Hence this technology is widely used for food processing in dairy industries, fermentation of meat and vegetables.

LAB has a long history of use in food and beverages without any detrimental health effects. Commercially, important LAB strains such as *Lb. plantarum, Lb. fermentum, Lb. rhamnosus* and *Lb. acidophilus*, have been useful due to their high acid tolerance and their ability to be engineered for the production of D- or L-lactic acid (Kyla-Nikkila et al. 2000, Abdel-Rahman et al. 2013). However in specific fermentation, particular species, strain or variant is determined in the food substrate that is used, the temperature of the process, other environmental conditions. *L. lactis* a mesophilic bacterium is used for Gouda cheese that has ability to grow in 35–38°C where as thermoplilic LAB such as *Streptococcus thermophilus* and *Lb. helveticus* are used for production of yogurt and parmesan cheese where heating of 50°C is employed (Delcour et al. 2000).

Lactic acid fermentation, the metabolism of LAB is completely geared towards production of a single metabolite—lactic acid. Such a focused metabolism seems to be a perfect basis for creating cell factories of single metabolites. This potential has been demonstrated by many successful technological interventions and metabolic engineering. The development of the adequate gene expression system called NICE (Nisin controlled gene expression) for the production of the antimicrobial compound nisin is produced by *L. lactis* (Kuipers et al. 1993). Such cell factories can overproduce proteins (enzymes) a thousand folds or more. Production of low calorie sweeteners by conversion of lactose into sugar alcohols like mannitol is accelerated by LAB (Korakli et al. 2000, Ladero et al. 2007). Manufacture of various vitamins and flavors are also enhanced with LAB. Also it plays a novel role of efficient cell factory for the production of functional biomolecules and food ingredients to enhance the quality of cereal-based beverages (Waters et al. 2015). Genomics techniques also have been very productive in utilizing LAB as cell factory. Thus, keeping in view the extensive application of LAB in the food industry coupled with consumer demand for healthier and functional foods, the use of these food grade microorganisms as cell factories would be of great advantage in the near future.

5. Immobilization of Cell and Encapsulation of Bacteria

The application of encapsulated and immobilized cell technologies has been of interest to fermentation industries recently. Encapsulation technique is defined as a process whereby cells are embedded or enrobed within a gel-matrix, wherein the metabolic activity of encapsulated cells is completely responsible to carry out the fermentation process. Several research groups have attempted whole cell immobilization as a viable alternative to the conventional free cells fermentations. Immobilized enzymes play a pivotal role in food processing, for example, lactose hydrolysis, whey processing, skimmed milk production, production of high fructose corn syrups have been greatly facilitated by the use of immobilization technology. Immobilization has been carried out by employing *Bacillus brevis* MTCC 7521 for the production of α-amylase and calcium alginate was used as the immobilization matrix (Ray et al. 2008). There are various advantages in the use of encapsulated and immobilized cell technologies (Hutkins 2008) like re-use of cells for several cycles, continuous extraction of metabolites, low cost of the process as also to produce continuous cell biomass from encapsulated cells. Recent fusion proteins (Ushasree et al. 2012) and nanotechnology are used for encapsulation of bacteria in food processing because of their efficiency in increasing enzyme loading and diffusion properties and reduction in mass transfer limitation. The encapsulation of bacteria is an improved technology in food science that basically concentrates on economic, fast, non destructive and food grade purity which will help the food industries with improved quality, aroma and fine taste to the final product (Doleyres and Lacroix 2005).

The technology of micro-encapsulation has developed from a simple immobilization or entrapment to sophisticated and precise micro capsule formation. Microencapsulation helps to separate a core material from its environment until it is released. There are different methods for micro-encapsulation like spray drying, extrusion, emulsion and phase separation. The advances in the field of nutraceuticals and food ingredients have been tremendous; however, the focus on the micro-encapsulation of live probiotic bacterial cells, in fermented food processing is recent (Mortazavian et al. 2016).

6. Probiotics, Prebiotics and Synbiotics

Human health and efficient nutrition absorption are maintained to a larger extent by the microbes of the gastrointestinal tract. Various end products of nutritional substrates like organic acids, vitamins, short chain fatty acids are metabolized by the gut bacteria through fermentation. Probiotics, prebiotics and synbiotics (a combination of probiotics and prebiotics) are new technologies developed to modulate the target gastrointestinal microflora balance. Probiotic therapy (or microbial intervention) is based on the concept of healthy gut microflora.

6.1 Probiotics

The word "probiotic" was first used in 1954 to indicate substances that were required for a healthy life. Out of several definitions, the most widely used and accepted definition is the one proposed by a joint FAO/WHO panel (FAO/WHO 2001): "Live micro-organisms which, when administered in adequate amounts, confer a health benefit on the host". Probiotics helps in enhancing resistance to colonization by exogenous, potentially pathogenic organisms (Elmer et al. 1996, Helland et al. 2004). They produce compounds such as lactic acid, hydrogen peroxide and acetic acid increasing the acidity of the intestine that inhibit the reproduction of many harmful bacteria; they also out compete the pathogenic organisms preventing latter's survival in the gastro intestinal tract (Reid et al. 1999, Helland et al. 2004). The intestinal microbial balance of the consumer is maintained by these live microbial food supplements (Fuller 1991). Starting about 20 years ago functional foods with probiotics are now well established and presently known to most consumers. Probiotic microorganisms are mostly of human or animal origin. The dairy industry, in particular, has found probiotic cultures. Yogurts and fermented milks are the main vehicles for probiotic cultures (Trabelsi et al. 2013). New fermented products such as milk-based desserts, powdered milk for newborn infants, ice creams, butter, mayonnaise, various types of cheese are also being introduced in the international market (Cruz et al. 2009); however, some studies show that strains recognized as probiotics are also found in non-dairy fermented substrates (Martins et al. 2013, Panda and Ray 2016). With the revolution in sequencing and bioinformatics technologies well under way, it is timely and realistic to launch genome sequencing projects for representative probiotic microorganisms. Increasing knowledge of genomes important for the technological functionality and rapid development of the toolboxes for the genetic manipulation of *Lactobacillus* and *Bifidobacterium* species will help in tailoring the technological properties of probiotic strains in future. In addition to the dietary supplementation approach of directly introducing live bacteria named as probiotics to the colon, another approach to increase the number of beneficial bacteria in the intestinal microbiota is through the use of prebiotics.

6.2 Prebiotics and Synbiotics

The concept of prebiotics emerged during mid nineties of the twentieth century (Gibson and Roberfroid 1995). The prebiotic concept for modulation of gut microbiota was introduced in 1995. After a meeting of the International Scientific Association of Probiotics and Prebiotics (ISAPP) in 2010; the definition summarized was "A dietary prebiotic is a selectively fermented ingredient that results in specific changes, in the composition and/or activity of the gastrointestinal microbiota, thus conferring benefit(s) upon host health" (Chapman et al. 2011). Prebiotics doesn't breakdown by digestive enzymes while passing through digestive system. Hence this non-digestible carbohydrate serves as feast for the probiotic bacteria in the large intestine. It stimulates indigenous beneficial flora of the gut while inhibiting the growth of pathogenic bacteria. Target organisms of species belonging to the *Lactobacillus* and *Bifidobacterium* genera are generally preferred for prebiotics. Multipronged beneficial effects including gut health, higher mineral absorption, lowering of cholesterol, immune system stimulation, pathogen exclusion make prebiotic oligosaccharides the centre of attraction against other functional foods. Also prevention and treatment of hypertension with prebiotics has been proved (Rycroft et al. 2001, Roberfroid et al. 2002, Samanta et al. 2007). Most prebiotics as known today are fermentable, non-digestible carbohydrates. Some examples are lactulose, galacto- and fructo-oligosaccharides and resistant starch. Finding of new source and types of natural prebiotics might explore new areas of research. A combination of these probiotics and prebiotics for human endeavor is often referred to as synbiotics (Gibson and Roberfroid 1995, Collins and Gibson 1999). Basically functional foods with both probiotics and prebiotics are called synbiotics (Roberfroid 1998).

New and improved technological interventions are necessary to study the mechanisms and action of probiotics in the gastro intestinal (GI) tract, its effects in GI-diseases, GI-infections and allergies, to develop diagnostic tools and biomarkers for assessment of the GI-tract, to develop technology for non-diary probiotics (Bansal et al. 2016).

7. Ultrasonic Sounds Applications in Fermented Foods

Ultrasound is an unconventional innovative technology for processing and preservation of food. Ultrasound is a sound energy (really pressure) wave of high frequencies that are too high to be audible by human ears, i.e., above 16 kHz. However, the science of application of the technology in food fermentation is a recent one (Ojha et al. 2016).

Ultrasound can be used in fermentation processes to either monitor the progress of fermentation or to influence its progress. High frequency ultrasound (> 2 MHz) has been extensively reported as a tool for the measurement of the changes in chemical composition during fermentation providing real time information on reaction progress. Low frequency ultrasound (20–50 kHz) can influence the course of fermentation by improving mass transfer and cell permeability leading to improved process efficiency and production rates. It can also be used to eliminate microorganisms which might otherwise hinder the process. Ojha et al. (2016) reviewed the key applications of high and low frequency ultrasound in food fermentation applications.

Milk is often pasteurized prior to its use in various fermented dairy products. Application of low frequency ultrasound alone or in combination with external pressure (manosonication), heat (thermosonication) or both (manothermosonication) is reported to improve the safety profile of milk and can achieve the desired 5 log reduction of pathogenic microorganisms including *Listeria innocua* and *Escherichia coli* (Nguyen et al. 2012). Also, low frequency ultrasound processing of milk is reported to improve homogenization, pasteurization, reduction in yogurt fermentation time (Wu et al. 2000) and improved rheological properties of yoghurt (Vercel et al. 2002). Similarly, low frequency ultrasound applications help in controlling spoilage of wine and malolactic fermentation (Ojha et al. 2016).

8. GMO for Food Processing

Genetically Modified Foods have been around for years, but how many of us actually know that almost 70% of foods in the grocery shelves are genetically modified (Teisl et al. 2003). There are several rising concerns about the upcoming push of genetically modified foods, due mainly to the emergence of new products from GM companies. The use of genetic modification has become relatively common in today's technologically expanding world. By selecting specific long strands of DNA (genes) of our interest and inserting them into other host cells, it is possible for the new cells to carry useful traits. These new cells that emerge with foreign genes are called transgenic organisms, and are also known as genetically modified organisms (GMOs). Genetically modified microbial enzymes were the first application of GMOs in food production and were approved in 1988 by the US Food and Drug Administration (FDA 1990). In the early 1980s, Gist-brocades began investigating the possibility of producing chymosin from a microorganism using the genetic engineering approach (van Dijck 1999). The chymosin preparation, registered under the brand-name Maxiren®, has been commercially produced since 1988.

With the advancement of recombinant deoxyribonucleic acid (rDNA) technology, the metabolic potentials of microorganisms are being explored and harnessed in a variety of new ways. Today, genetically modified microorganisms (GMMs) have found tremendous applications in food industries (Arvanitoyannis and Krystallis 2005). Genetic engineering and different technologies in molecular biology are utilized to manipulate the microorganisms for the expression of desired traits. GMMs will be used to produce enzymes with optimized properties regarding activity, specificity or stability (Roller and Goodenough 1999). Technologies are (1) gene transfer methods to deliver the selected genes into desired hosts; (2) cloning vectors; (3) promoters to control the expression of the desired genes; and (4) selectable marker genes to identify recombinant microorganisms (Arvanitoyannis and Krystallis 2005).

8.1 Pros and Cons

Genetic engineering is still a relatively young technology, about 25 years old, and many of the predictions about it, for better or worse, have yet to be verified in practice. There are various pros and cons that have been debated for years like two sides of the same coin.

8.1.1 Pros

There are many pros or merits of using GMO in food application discussed from day to day. With the discovery of DNA, unreveiling the genetic code it contains and its possibility to be transferred from one organism to another and design organisms at will has been a boon for the industrial aspect of food technology. Genetic engineering allows introducing desirable traits to economically useful organisms by gene targeting and gives desired economic and industrial benefits in bulk production of enzymes or other food applications. It has tremendously alleviated the food and nutritional requirement of the world with other functional properties in pharmaceuticals, agriculture and feed. These are supposed to be beneficial to people in countries that do not have an adequate supply of these nutrients. And it is safe for human consumption too as so far there has been no case to prove them unsafe.

8.1.2 Cons

There has not been any long term testing to detect possible problems of the use of GMO in food application with respect to long term disorder as it is a new concept, a mere 25 years. Major cons may be allergic reactions as the genetic modification mixes or adds proteins that are not endemic which may cause new allergic reactions in consumers. Some GMO's have antibiotic features added to them so as to make them resistant to certain contaminations. Hence human consumption of these products may lead to some antibiotic features persisting in human body that may actually make the antibiotic medication less effective in future. The loss of the diversity of gut microflora endemic in humans is a major concern. Also GMO is not natural and scrambling genomes may lead to total chaos in evolution. Genetic engineering exposes people to the increased dangers of horizontal gene transfer, a process whereby genes are passed not 'vertically' down the generations in the usual way but 'horizontally' from organism to organism and from species to species (Arvanotoyannis and Krystalles 2005). GE is potentially dangerous and therefore involves taking risks. The consequences could be devastating and irreversible. Furthermore, the adverse consequences could take years to show and the company liable for any damages may have long since ceased trading.

8.2 Ethics and Legislation

GMO has been a very controversial topic since many years. Researchers have been accused numerous times of manipulating the genetic make-up of organisms. The food prepared by GMO was named as "Franked-food" and researchers were condemned to be unethical and playing god. They were criticized for hampering nature and for loss of biodiversity and natural immunity. Hence strict regulations were imposed for using GMO, such as safety aspects, thresholds, labeling, detection and coexistence. Many traditional and cultural beliefs were ignored for the sake of commercialization of GMO. Hence with respect to ethics, beliefs, customs and traditions there are a number of organizations who govern and monitor the regulations for use of GMO in food processing.

8.2.1 Structure of pertinent legislation

There is no comprehensive federal legislation specifically addressing GMOs. GMOs are regulated under the general statutory authority of environmental, health, and safety laws. The three main agencies involved in regulating GMOs are the US Department of Agriculture's Animal and Plant Health Inspection Service (APHIS), the Food and Drug Administration (FDA), and the Environmental Protection Agency (EPA). The FDA's primary statutory authority is the Federal Food, Drug, and Cosmetic Act (FFDCA), Under the FFDCA, substances added to food can be classified either as "food additives", which require approval from the FDA that they are safe before they can be marketed (21 U.S.C. §348 2012) and substances added to food classified as "generally recognized as safe" (GRAS), as to which preapproval is not needed (21 U.S.C. §321 2012). The EPA has established regulations specifically for microorganisms that require submission of a Microbial Commercial Activity Notice (MCAN) before they are used for commercial purposes (40 C.F.R. §725.100 2013). The Notice must include information describing the microorganism's characteristics and genetic construction, byproducts of its manufacture, use, and disposal, health and environmental effects data and other information (40 C.F.R. §§725.155, 725.160 2013). All of the various statutory schemes under which GMOs are regulated in the US provide for civil and criminal penalties, and different countries have their own set of rules with respect to the same.

9. Patented Approaches

For improved food safety and enhanced health benefits more attention has been given to standardize the protocols and microorganisms used in fermentation. Fermented foods are patented using novel mechanisms and microorganisms that may enhance nutritional composition and safety of fermented food products. These patented approaches may improve the quality of fermented foods with a promising strategy for the prevention, control and treatment of both infections and chronic diseases. Various cultures are invented for fermented milk products like infant formula, yogurt or ice-cream composition, i.e., LAB culture for *Lactobacillus acidophilus* species that provide multiple uses of the species (Izvekova et al. 2000, 2002). Also from vegetable proteins various inventions have been made and patented (Flambard 2011). Patents involving fermented tea generally focus on the methods involved in the fermentation process that helps to provide a consistent product for mass distribution (Toba et al. 2001, 2005). Also some inventions of fermented tea claim to improve flavor and safety with additional nutritional values (Kwach et al. 2011). Different food crops have been patented and marketed as fermented foods in Soy and Rice (Ghoneum and Maeda 1996). This is comprehensively summarized in Table 1 (Borresen et al. 2012).

Table 1. Patents in foods.

Patent Title	Health Utility	Mechanism/Novelty	Patent ID (Year)
Fermented Milk Products			
Fermented milk Nutraceuticals	Treat diseases or conditions resulting from opportunistic and pathogenic microorganisms	Novel cultures of *Lactobacillus* acidophilus combinations of group Er-2 strain and *L. acidophilus* N.V. Er 317/400	US6,156,320 (2000) and 6, 357,521, B1 (2002)
Fermented milk proteins comprising receptor ligand and uses thereof	Reduce and/or stabilize heart rate in CVD. Treat or relieve benign prostrate hypertrophy	Comprised of LAB strain DSM 14998 and a receptor ligand	EP1796480B1 (2011)
Fermented milk or vegetable protein comprising receptor ligand and uses thereof	Reduce and/or stabilize heart rate in CVD. Treat or relieve benign prostrate hypertrophy	Comprised of LAB strain DSM 14998 and a receptor ligand	US20110195891A1 (2011)
Fermented Tea			
Method of producing fermented milk containing manganese and tea	Prevent diseases caused by active oxygen	Bacteria with catalase activity (*Lactobacillus plantarum*) with manganese-containing natural material	US6228358B1 (2001)
Antioxidant food product, antioxidant preparation, and antioxidation method	Prevent diseases caused by active oxygen	Antioxidation to express superoxides dismutase like activity and catalase activity. Preferred that tea or other natural material is added to the food product in the form of powder	US6884415B2 (2005)
Method of preparing fermented to using *Bacillus* sp. strains	Improve flavor and safety	Fermenting tea leaves by treating with stabilized *Bacillus* sp. Strains from Korean traditional fermented foods	US20110250315A1 (2011)

Fermented Soy			
Fermented Soy nutritional supplements including mushrooms components	Wide ranging-malnutrition to mood related disorders to metabolic support	Comprised of a mushroom grown in fermented soy growth medium and curcuminoids	US20110206721A1 (2011)
Methods for inhibiting cancer growth, infection and promoting general health with a fermented soy extract	Promote general health, prevent and/or treat cancer, prevent infections, reduce incidence of infections, treat infections, asthma, and inflammation. Modulate the immune system and treat immune disorders	Fermented soy extracts	US6855350B2 (2005)
Fermented Rice			
Methods and compositions employing red rice fermentations products	Treat or prevent hyperlipidemia and associated disorders and symptoms (CVD, cerebrovascular diseases, diabetics, hypertension, obesity, etc.)	Fermentation of atleast one *Monascus* strain with red rice products to be used as a dietary supplement	US6046022 (2000)
Method for prevention and treatment of Alzheimer's disease	Treat and prevent Alzheimer's disease	Administration of effective amount of *Monascus* fermented product extracted from red mold rice (powder or beverage)	US8097259B2 (2012)

Source: Borresen et al. (2012) with permission

10. Conclusion and Future Scope

Recent scientific advances have revealed the important role of microorganisms particularly of LAB and yeasts in fermented food industries with novelties and added values. Rationalized use of microorganisms in our diet as evident from either ancient and traditional fermented foods or new advanced patented foods opens up new perspectives. Though this remains undoubtedly promising, one should not forget that man has not yet finished characterizing all traditional fermented foods consumed for centuries, with often numerous isolates belonging to species with undefined roles. The use of novel technological interventions in bioengineering, food processing and fermentation, system biology and bioinformatics shall open vast avenues for new generation microorganisms with enhanced functional features. This will not only provide several health benefits but also the technological advances and marketing shall profusely contribute to the development of small and medium sized enterprises on the one hand, and product diversification of large companies which directly or indirectly contributes to the economy as a whole.

Keywords: Genetically modified foods, food grade GMO, patented approaches, lactic acid bacteria, molds, *Bacillus* sp., fermented foods, bacterial genomics, probiotics, prebiotics, synbiotics

References

21 U.S.C. §321(s) (excluding substances from definition of "food additive" that are "generally recognized. . . to be safe.").

21 U.S.C. §348 (2012) http://uscode.house.gov/view.xhtml?req=granuleid:USCprelimtitle21 section348& num=0&edition=prelim.

40 C.F.R. §§725.155,725.160 (2013) http://www.ecfr.gov/cgibin/textidx?SID=7c33b229782bc8 24885546d 25f6cf057&node=40:32.0.1.1.13&rgn=div5#40:32.0.1.1.13.4.1.5.

40 C.F.R. §725.100 (2013) http://www.ecfr.gov/cgibin/textidx?SID=7c33b229782bc82488554 6d25f6cf057& node=40:32.0.1.1.13&rgn=div5#40:32.0.1.1.13.4.1.1.

Abdel-Rahman, M.A., Tashiro, Y. and Sonomoto, K. (2013). Recent advances in lactic acid production by microbial fermentation processes. Biotechnology Advances 31: 877–902.

Anon (2005). EFSA Scientific Colloquium—Microorganisms in Food and Feed: Qualified Presumption of Safety—13–14 December 2004, Brussels, Belgium. ISSN 1830-4737.

Anon (1997). Regulation (EC) No 258/97 of the European Parliament and of the Council of 27 January 1997 concerning novel foods and novel food ingredients N° 258/97. Official Journal L 043, 1–6.

Arvanitoyannis, I.S. and Krystallis, A. (2005). Consumers' beliefs, attitudes and intentions towards genetically modified foods, based on perceived safety vs. benefits perspective. International Journal of Food Science and Technology 40: 343–360.

Banks, G.T. (1984). Scale up of fermentation processes. pp. 170–266. In: A. Wiseman (ed.). Topics in Enzyme and Fermentation Biotechnology. Chichester: Ellis Horwood Limited.

Bansal, S., Mangal, M., Sharma, S.K. and Gupta, R.K. (2016). Non-dairy Based Probiotics: A Healthy Treat for Intestine, Critical Reviews in Food Science and Nutrition 56(11): 1856–1867.

Battcock, M. and Azam-Ali, S. (1998). Fermented fruits and vegetables: A global perspective (FAO Agricultural services bulletin No. 134). Rome: Food and Agriculture Organisation of the United Nations.

Bernardeau, M., Vernoux, J.P., Henri-Dubernet, S. and Gueguen, M. (2008). Safety assessment of dairy microorganisms: the *Lactobacillus* genus. International Journal of Food Microbiology 126: 278–285.

Bolotin, A., Wincker, P., Mauger, S., Jaillon, O., Malarme, K., Weissenbach, J., Ehrlich, S.D. and Sorokin, A. (2001). The complete genome sequence of the lactic acid bacterium *Lactococcus lactis* ssp. *Lactis* IL1403. Genome Research 11: 731–753.

Borresen, E.C., Henderson, A.J., Kumar, A., Weir, T.L. and Ryan, E.P. (2012). Fermented foods: Patented approaches and formulations for nutritional supplementations and health promotion. Recent Patents on Food, Nutrition and Agriculture 4: 134–140.

Bourdichon, F., Casaregola, S., Farrokh, C., Frisvad, J.C., Gerds, M.L., Hammes, W.P., Harnett, J., Huys, G., Laulund, S., Ouwehand, A., Powell, I.B., Prajapati, J.B., Seto, Y., Schure, E. T., Van Boven, A., Vankerckhoven, V., Zgoda, A., Tuijtelaars, S. and Hansen, E.B. (2012). Food fermentations: Microorganisms with technological beneficial use. International Journal of Food Microbiology 154: 87–97.

Campbell-Platt, G. (1987). Fermented foods of the world—a dictionary and guide, London, Butterworths. ISBN: 0-407-00313-4.

Caplice, E. and Fitzgerald, G.F. (1999). Food fermentations: role of microorganisms in food production and preservation. International Journal of Food Microbiology 50(1-2): 131–149.

Chapman, C.M., Gibson, G.R. and Rowland, I. (2011). Health benefits of probiotics: are mixtures more effective than single strains? European Journal of Nutrition 50: 1–17.

Collins, M.D. and Gibson, G.R. (1999). Probiotics, prebiotics and synbiotics: dietary approaches for the modulation of microbial ecology. American Journal of Clinical Nutrition 69: 1052–1057.

Cruz, A.G., Antunes, A.E.C., Sousa, A.L.O.P., Faria, J.A.F. and Saad, S.M.I. (2009). Ice-cream as a probiotic food carrier. Food Research International 42: 1233–1239.

Daly, C., Fitzgerald, G.F., O'Connor, L. and Davis, R. (1998). Technological and health benefits of dairy starter cultures. International Dairy Journal 8: 195–205.

De Vuyst, L. (2000). Technology aspects related to the application of functional starter cultures. Food Technology and Biotechnology 38: 105–112.

Delcour, J., Ferain, T. and Hols, P. (2000). Advances in the genetics of thermophilic lactic acid bacteria. Current Opinion in Biotechnology 11: 497–504.

Dirar, M. (1993). The Indigenous Fermented Foods of the Sudan. CAB International. University Press, Cambridge.

Doleyres, Y. and Lacroix, C. (2005). Technologies with free and immobilized cells for probiotic bifidobacteria production and protection. International Dairy Journal 15: 973–988.

Dubos, J. (1995). Louis Pasteur: Free Lance of Science, Gollancz. In: K.L. Manchester (ed.). Louis Pasteur (1822–1895)-Chance and the Prepared Mind. Trends in Biotechnology 13(12): 511–515.

Elmer, G.W., Surawicz, C.M. and McFarland, L.V. (1996). Biotherapeutic agents. A neglected modality for the treatment and prevention of selected intestinal and vaginal infections. The Journal of the American Medical Association 275: 870–876.

Europe Novel Food Task Force, I.L.S.I. (2003). The safety assessment of novel foods and concepts to determine their safety in use. International Journal of Food Sciences and Nutrition 54: S1–S32 (Suppl).

FAO/WHO (2001). Joint FAO/WHO Expert Consultation on evaluation of health and nutritional properties of probiotics in food including powder milk with live lactic acid bacteria. Cordoba, Argentina, October 2001. http://www.who.int/foodsafety/publications/fs_management/en/probiotics.pdf.

FDA (2010). Generally Recognized as Safe (GRAS) Notifications. FDAhttp://www.fda. gov/ AnimalVeterinary /Products/AnimalFoodFeeds/GenerallyRecognizedasSafeGRASNotifications/default.htm.

FDA (1990). Approves 1st Genetically Engineered Product for Food. Los Angeles Times. Retrieved 1 May 2014.

Flambard, B. (2011). Fermented milk or vegetable proteins comprising receptor ligand and uses there of US20110195891A.

Forde, A. and Fitzgerald, G.F. (1999). Bacteriophage defence systems in lactic acid bacteria. Antonie van Leeuwenhoek International Journal of General and Molecular Microbiology 76: 89–113.

Fox, P.F., Wallace, J.M., Morgan, S., Lynch, C.M., Niland, E.J. and Tobin, J. (1996). Acceleration of cheese ripening. Antonie Van Leeuwenhoek 70: 271–297.

Fox, P.F. (1993). Cheese: an overview. pp. 1 –36. In: P.F. Fox (ed.). Cheese; Chemistry, Physics and Microbiology. Chapman and Hall, London, England.

Fleet, G.H. (2006). The commercial and community significance of yeasts in food and beverage production. pp. 1–12. In: A. Querol and G.H. Fleet (eds.). Yeasts in Food and Beverages. Springer, Berlin Heidelberg.

Fuller, R. (1991). Probiotics in human medicine. Gut 32: 439–442.

Gaspar, F.B. and Crespo, M.T.B. (2016). Lactic acid bacteria as functional starter in food fermentations. pp. 166–184. In: D. Montet and R.C. Ray (eds.). Fermented Foods. Part 1. Biochemistry and Biotechnology. CRC Press, Boca Raton, Florida.

Ghoneum, M.H. and Maeda, H. (1996). Immunopotentiator and method of manufacturing the same. U.S. Patent #5,560,914.

Ghosh, J.S. (2015). Solid state fermentation and food processing: a short review. Journal of Nutrition and Food Sciences. 6: 453http://dx.doi.org/10.4172/2155-9600.1000453.

Gibson, G.R. and Roberfroid, M.B. (1995). Dietary modulation of the human colonic microbiota: introducing the concept of prebiotics. Journal of Nutrition 125: 1401–1412.

Harris, L.J. (1998). The microbiology of vegetable fermentations. In: B.J.B. Wood (ed.). Microbiology of Fermented Foods. London: Blackie Academic and Professional 1: 45–72.

Helland, M.H., Wicklund, T. and Narvhus, J.A. (2004). Growth and metabolism of selected strains of probiotic bacteria in maize porridge with added malted barley. International Journal of Food Microbiology 91: 305–313.

Holzapfel, W.H., Schillinger, U., Geisen, R. and Lucke, F.K. (2003). Starter and protective cultures. pp. 291–320. In: W.H. Holzapfel, U. Schillinger, R. Geisen and F.K. Lucke (eds.). Food Preservatives, Springer, US.

Holzapfel, W.H. (1997). Use of starter cultures in fermentation on a household scale. Food Control 8: 241–258.

Hutkins, R.W. (2008). Microbiology and Technology of Fermented Foods. John Wiley and Sons Inc., New Jersey.

Izvekova, E., Kornilov, V. and Amirian, E. (2002). Fermented milk nutraceuticals. US Patent#US6358521B1.

Izvekova, E., Kornilov, V. and Amirian, E. (2000). Fermented milk nutraceuticals. US Patent#US6156320.

Jahreis, G., Vogelsang, H., Kiessling, G., Schubert, R., Bunte, C. and Hammes, W.P. (2002). Influence of probiotic sausage (*Lactobacillus paracasei*) on blood lipids and immunological parameters of healthy volunteers. Food Research International 35: 133–138.

Khandelwal, P., Gaspar, F.B., Crespo, M.T.B. and Upendra, R.S. (2016). Lactic acid bacteria: General characteristics, food preservation and health benefits. pp. 1127–147. In: D. Montet and R.C. Ray (eds.). Fermented Foods. Part 1. Biochemistry and Biotechnology. CRC Press, Boca Raton, Florida.

Klaenhammer, T.R., Barrangou, R., Buck, B.L., Azcarate-Peril, M.A. and Altermann, E. (2005). Genomic features of lactic acid bacteria effecting bioprocessing and health. FEMS Microbiology Reviews 29(3): 393–409.

Klaenhammer, T., Altermann, E., Arigoni, F., Bolotin, A., Breidt, F., Broadbent, J., Cano, R., Chaillou, S., Deutscher, J., Gasson, M., van de Guchte, M., Guzzo, J., Hartke, A., Hawkins, T., Hols, P., Hutkins, R., Kleerebezem, M., Kok, J., Kuipers, O., Lubbers, M., Maguin, E., McKay, L., Mills, D., Nauta, A., Overbeek, R., Pe,l H., Pridmore, D., Saier, M., van

Sinderen, D., Sorokin, A., Steele, J., O'Sullivan, D., de Vos, W., Weimer, B., Zagorec, M. and Siezen, R. (2002). Discovering lactic acid bacteria by genomics. Antonie Van Leeuwenhoek 82(1-4): 29–58.

Korakli, M., Schwarz, E., Wolf, G. and Hammes, W.P. (2000). Production of mannitol by Lactobacillus sanfranciscensis. Advances in Food Science 22: 1–4.

Kuboye, A.O. (1985). Traditional fermented foods and beverages of Nigeria. In: Development of Indigenous Fermented Foods and Food Technology in Africa. Proc. IFS/UNU Workshop, Douala, Cameroon, Oct. 1985. International Foundation for Science, Stockholm, Sweden.

Kuipers, O.P., Beerthuijzen, M.M., Siezen, R.S. and de Vos, W.M. (1993). Characterization of the nisin gene cluster nisABTCIPR of *Lactococcus lactis*: Requirement of expression of the nisA and nisI genes for development of immunity. European Journal of Biochemistry 216: 281–291.

Kwach, L., Lee, B., Oh, Y., Chung, J., Lee, T., Suh, K. and Kim, H.K. (2011). Method for preparing fermented tea using *Bacillus* sp. strains (as Amended). US Patent# US20110250315A.

Kyla-Nikkila, K., Hujanen, M., Leisola, M. and Palva, A. (2000). Metabolic engineering of *Lactobacillus helveticus* CNRZ32 for production of pure L(+) lactic acid. Applied Environmental Microbiology 66: 3835–3841.

Ladero, V., Ramos, A., Wiersma, A., Goffin, P., Schranck, A., Kleerebezem, M., Hugenholtz, J., Smid, E.J. and Hols, P. (2007). High-level production of low-calorie sugar sorbitol by *Lactobacillus plantarum* through metabolic engineering. Applied and Environmental Microbiology 73: 1864–1872.

Lasztity, R. (2009). Micro-organisms important in food microbiology. Food Quality and Standards-Volume III. http://www.eolss.net/sample-chapters/c10/e5-08-06-01.pdf.

LeBlanc, J.G., Milani, C., de Giori, G.S., Sesma, F., van Sinderen, D. and Ventura, M. (2012). Bacteria as vitamin suppliers to their host: a gut microbiota perspective. Current Opinion in Biotechnology 24: 160–168.

Leejeerajumnean, A. (2003). Thua nao: Alkali fermented soybean. International Journal of Silpakorn University 3: 277–292.

Martins, E.M.F., Ramos, A.M., Vanzela, E.S.L., Stringheta, P.C., Pinto, C.L.O. and Martins, J.M. (2013). Products of vegetable origin: A new alternative for the consumption of probiotic bacteria. Food Research International 51: 764–770.

Mirbach, M.J. and El Ali, B. (2005). Industrial Fermentation (Chapter 9). In: M.F. Ali, B.M. El Ali and J.G. Speight (eds.). Handbook of Industrial Chemistry. Organic Chemicals. New York: Mc Graw-Hill.

Moineau, S. (1999). Applications of phage resistance in lactic acid bacteria. Antonie van Leeuwenhoek International Journal of General and Molecular Microbiology 76: 377–382.

Mollet, B. (1999). Genetically improved starter strains: opportunities for the dairy industry. International Dairy Journal 9: 11–15.

Mollet, B. (1996). New technologies in fermented milk. Cerevisia 21: 63–65.

Montet, D. and Ray, R.C. (eds.). (2016). Fermented foods. Part 1. Biochemistry and Biotechnology. CRC Press, Boca Raton, Florida, 402 pp.

Mortazavian, A.M., Moslemi, M. and Sohrabvandi, S. (2016). Microencapsulation of probiotics and applications in food fermentation. pp. 185–210. In: D. Montet and R.C. Ray (eds.). Fermented Foods. Part 1. Biochemistry and Biotechnology. CRC Press, Boca Raton, Florida.

Nguyen, T.M.P., Lee, Y.K. and Zhou, W. (2012). Effect of high intensity ultrasound on carbohydrate metabolism of bifidobacteria in milk fermentation. Food Chemistry 130: 855–874.

Ojha, K.S., Mason, T.J., Donnell, C.P., Kerry, J.P. and Tiwari, B.K. (2016). Ultrasonic technology for food fermentation applications. Ultrasonics Sono Chemistry, June 2016, doi: 10.1016/j.ultsonch.2016.06.001.

Panda, S.H. and Ray, R.C. (2016). Amylolytic lactic acid bacteria: Microbiology and technological interventions in food fermentations. pp. 148–165. In: D. Montet and R.C.

Ray (eds.). Fermented Foods. Part 1. Biochemistry and Biotechnology. CRC Press, Boca Raton, Florida.

Panda, S.K., Panda, S.H., Swain, M.R., Ray, R.C. and Kayitesi, E. (2015). Anthocyanin rich sweet potato (*Ipomoea batatas* L.) beer: technology, biochemical and sensory evaluation. Journal of Food Processing and Preservation 39(6): 3040–3049.

Panda, S.K., Sahu, U.C., Behera, S.K. and Ray, R.C. (2014). Fermentation of sapota (*Achras sapota* Linn.) fruits to functional wine. Nutrafoods 13(4): 179–186.

Picard, C., Fioramonti, J., Francois, A., Robinson, T., Neant, F. and Matuchansky, C. (2005). Review article: bifidobacteria as probiotic agents—physiological effects and clinical benefits. Alimentary Pharmacology and Therapeutics 22: 495–512.

Raspo, P. and Goranovic, D. (2009). Biotechnological applications of acetic acid bacteria in food production. http://www.eolss.net/sample-chapters/c17/E6-58-06-08.pdf.

Ray, R.C. (2013). Fermented foods in health related issues. Editorial, International Journal of Food and Fermentation Technology 3: 1.

Ray, R.C. and Joshi, V.K. (2014). Fermented Foods: Past, present and future scenario. pp. 1–36. *In*: R.C. Ray and D. Montet (eds.). Microorganisms and Fermentation of Traditional Foods. CRC Press, Boca Raton, Florida, USA.

Ray, R.C., Kar, S., Nayak, S. and Swain, M.R. (2008). Extracellular α-amylase production by *Bacillus brevis* MTCC 7521. Food Biotechnol 22(3): 234–246.

Reddy, N.R., Pierson, M.D., Sathe, S.K., Salunkhe, D.K. and Beuchat, L.R. (1983). Legume-based fermented foods: Their preparation and nutritional quality. Critical Reviews in Food Science and Nutrition 17(4): 335–370.

Reid, G., Millsap, K. and Bruce, A.W. (1999). Implantation of *Lactobacillus casei* var *rhamnosus* into vagina. Lancet 344: 1229.

Roberfroid, M. (2002). Functional food concept and its application to prebiotics. Digestive and Liver Disease 34(2): 105–110.

Roberfroid, M.B. (1998). Prebiotics and synbiotics: concepts and nutritional properties. British Journal of Nutrition 80: S197–S202.

Roller, S. and Goodenough, P.W. (1999). Food Enzymes. *In*: S. Roller and S. Harlander (eds.). Genetic Modification in the Food Industry: A Strategy for Food Quality Improvement. Aspen Publisher, Frederick, USA.

Rycroft, C.E., Jones, M.R., Gibson, G.R. and Rastall, R.A. (2001). A comparative *in vitro* evaluation of the fermentation properties of prebiotics oligosaccharides. Journal of Applied Microbiology 91: 878–887.

Samanta, A.K., Kolte, A.P., Chandrasekhariah, M., Thulasi, A., Sampath, K.T. and Prasad, C.S. (2007). Prebiotics: The rumen modulator for enhancing the productivity of dairy animals. Indian Dairyman 59: 58–61.

Stanislawski, D. (1975). Dionysus Westward: Early religion and the economic geography of wine. Geographical Reviews 65(4): 427–444.

Stasiak, L. and Blazejak, S. (2009). Acetic acid bacteria-perspectives of application in biotechnology—a review. Polish Journal of Food and Nutrition Sciences 59(1): 17–23.

Steinkraus, K.H. (1996). Handbook of Indigenous Fermented Foods, Second edition. Marcel Dekker, New York.

Swindell, S.R., Benson, K.H., Griffen, H.G., Renault, P., Ehrlich, S.D. and Gasson, M.J. (1996). Genetic manipulation of the pathway for diacetyl metabolism in *Lactococcus lactis*. Applied and Environmental Microbiology 62: 2641–2643.

Temmerman, R., Huys, G. and Swings, J. (2004). Identification of lactic acid bacteria: culture-dependent and culture-independent methods. Trends in Food Science and Technology 15: 348–359.

Teisl, M.F., Garner, L., Roe, B. and Vayda, M.E. (2003). Labeling genetically modified foods: How do US consumers want to see it done? AgriBioForum 6(1 and 2): 48–54.

Toba, M., Uchiyama, S., Ohta, R., Shimizu, S. and Sakamoto, S. (2005). Antioxidation food product, antioxidation preparation and antioxidation method. US Patent# US6884415B2.

Toba, M., Uchiyama, S., Ohta, R., Shimizu, S. and Sakamoto, S. (2001). Method of producing fermented milk containing manganese and tea. US Patent# US6228358B1.

Trabelsi, I., Bejar, W., Ayadi, D., Chouayekh, H., Kammoun, R., Bejar, S. and Salah, R.B. (2013). Encapsulation in alginate and alginate coated-chitosan improved the survival of newly probiotic in oxgall and gastric juice. International Journal of Biological Macromolecules 61: 36–42.

Truninger, M. (2013). The historical development of industrial and domestic food technologies. pp. 82–102. *In*: A. Murcott, W. Belasco and P. Jackson (eds.). The Handbook of Food Research, Bloomsbury, London.

Ushasree, M., Gunasekaran, P. and Pandey, A. (2012). Single-step purification and immobilization of MBP–phytase fusion on starch agar beads: application in dephytination of soy milk. Applied Biochemistry and Biotechnology 167: 981–990.

van Dijck, P.W.M. (1999). Chymosin and phytase. Journal of Biotechnology 67: 77–80.

Vercel, A., Oria, P., Marquina, P., Crelier, S. and Lopez-Buesa, P. (2002). Rheological properties of yoghurt made with milk submitted to manothermosonication. Journal of Agriculture and Food Chemistry 50: 6165–6171.

Vogel, R.F., Hammes, W.P., Habermeyer, M., Engel, K.H., Knorr, D. and Eisenbrand, D. (2011). Microbial food cultures—opinion of the Senate Commission on Food Safety (SKLM) of the German Research Foundation (DFG). Molecular Nutrition and Food Research 55(4): 654–662.

Waters, D.M., Mauch, A., Coffey, A., Arendt, E.K. and Zannini, E. (2015). Lactic acid bacteria as a cell factory for the delivery of functional biomolecules and ingredients in cereal-based beverages: A review. Critical Reviews in Food Science and Nutrition 55: 503–520.

Wessels, S., Axelsson, L., Hansen, E.B., De Vuyst, L., Laulund, S., Lahteenmaki, L., Lindgren, S., Mollet, B., Salminen, S. and von Wright, A. (2004). The lactic acid bacteria, the food chain, and their regulation. Trends in Food Science and Technology 15: 498–505.

Wu, H., Hulbert, G.J. and Mount, J.R. (2000). Effect of ultrasound on milk homogenization and fermentation with yoghurt starter. Innovative Food Science and Emerging Technologies 1: 211–215.

Yokotsuka, T. and Sasaki, M. (1985). Fermented protein foods in the orient, with emphasis on *Shoyu* and *Miso* in Japan. pp. 351–415. *In*: B.J.B. Wood (ed.). Microbiology of Fermented Foods. Elsevier Applied Science Publishers, UK.

3

Functional Properties of Traditional Food Products Made by Mold Fermentation

Jacek Nowak and *Maciej Kuligowski**

1. Introduction

Fermentation is one of the oldest and most economic methods of food production and preservation. It has been used worldwide long before there was any knowledge of microorganisms. Fermentation of food not only preserves but also gives food a variety of tastes, flavors and different textural characteristics. The process often stabilizes the shelf life of food but also offers some health benefits. With increasing knowledge of nutritional, therapeutic and functional properties of fermented food products, indigenous foods have a chance to spread all over the world. The foods having beneficial effect in one or many functions of the human, contributing to the maintenance or improvement of health (or reduction of risk of disease), to the degree exceeding the results obtained from the presence of nutrients, is referred to as the functional food (Kwak and Jukes 2001, Mattila-Sandholm and Saarela 2003, Grajek et al. 2005). As many alternative food preservation methods have been developed over the years, fermentation is no longer the most important method for food preservation

Poznań University of Life Sciences, Faculty of Food Science and Nutrition, Institute of Food Technology of Plant Origin; Department of Fermentation and Biosynthesis, ul. Wojska Polskiego 31, 60-637 Poznań, Poland.
E-mail: jacnow@up.poznan.pl
* Corresponding author: maciek@up.poznan.pl

in most developed countries but it is a way to give food widely appreciated flavor, aroma and textural attributes (Prajapati et al. 2008).

So far, the classification systems of fermented food are incomplete and not uniform, in particular if molds are used along with several other types microorganisms. The most interesting taxonomy was proposed by Steinkraus who separated the fermented foods to seven groups (Steinkraus 1997), but the molds fermented products belong to different groups in this classification:

1. Textured vegetable protein meat substitutes obtained by fermentation from legume/cereal mixtures (*tempeh, sufu (furu), ontjom*),
2. High salt/meat-flavored sauce and paste fermentation (*shoyu*-soy sauce, *miso,* fish sauces and pastes),
3. Lactic acid fermentation (cheeses, kefir, yogurt, fermented dry sausages and ham, pickled cucumbers, *kimchi,* pickled cabbage-sauerkraut, sourdough bread, *idli, enjera, kisra, puto*),
4. Alcoholic fermentation (sake, wines, *tape ketan, lao-chao,* beers, honey wines, *pulque*),
5. Acetic acid fermentations include different types of vinegars from wine, ciders and beers,
6. Alkaline fermentations—products like *dawadawa, ogiri, natto, thua-nao, iru, kenim,*
7. Leavened breads which include Western yeast and sourdough breads.

Another division has been proposed according to the four main types of fermentation processes: alcoholic, lactic acid, acetic acid and alkali fermentation (Blandino et al. 2003). Other possible classifications have been based on the features of technology applied:

- liquid and solid fermentation (form of fermented material),
- continuous and periodic fermentation (time of the process),
- spontaneous (natural) and controlled fermentation (microflora form added),
- bacterial, yeast and mold fermentation (type of microorganisms).

The majority of fermented food products are produced by the activity of lactic acid bacteria (LAB) and fungi (yeast and mold). LAB is the most important group of microorganism in food fermentation both in the West and East cultures. Yeast is the main microorganism used in beverages production containing ethanol, but the process of hydrolyzation of starchy materials for such fermentation in some Asian countries is based on the *koji* system with mold fermentation as the first step. Generally molds (fungi) are much less popular in Western food fermentation than in many food fermentations in the East in which molds play the crucial role.

In traditional methods of food fermentations there is a wide gap between the technologies applied in the Orient and those practiced in the West. In many fermented food products from the East and Southeast Asia the mold activity produces a range of hydrolytic enzymes whose role is decomposition of the macromolecular material present in a raw material into smaller particles. In many fermentation processes from which the most important are those leading to soy sauce, *miso*, or rice wines production, the mold starter is often used for the *koji* stage.

These Asian fermented foods share many similarities. One of them is the development of special hydrolyzation system of starch, lipids and proteins in which fungi (mold) break down the polysaccharides into maltose, glucose and dextrin, lipids to fatty acids and proteins to peptides and amino acids in the substrate such as rice, legumes and other grains. The presence of simple sugars and growth factors allow the activity of yeast and bacteria in later stages of the process. Usually, a thick mycelium is grown on cooked cereals, legumes and other materials to get a saccharification-fermentation agent called *koji* or *qu*. At this step the fungi representing the genera *Aspergillus*, *Rhizopus*, *Monascus*, and others are used, depending on the availability in the close environment. The starters provide molds that produce a wide variety of enzymes capable of hydrolysis of starch, protein and lipid components in the raw material. The commercial *koji* starters can be kept at ambient temperatures for about 6 months with no significant loss of viability (Hsieh et al. 2008).

In the *koji* stage of fermentation, aerobic conditions allow molds to grow on the substrate. *Aspergillus* is the most common mold used in *sake*, *shoyu* and *miso*, while a mixture of strains, mainly *A. oryzae* grow in the bulk of the substrate for 2–3 days at 25–30°C. In the second, mash or *moromi* stage, the conditions are changed into anaerobic so no further mold growth can occur. Similar steps are involved in the production of soy sauce and soya bean pastes known as *miso* in Japan and *chiang* in China. A number of other rice-based mold starters are used in the countries of East and Southeast Asia to play a similar role to *koji* (*look pang* in Thailand, *ragi-tapai* in Malaysia, *nuruk* in Korea or *bakhar* in India). They are used to produce sweetened rice products which can be consumed fresh or added to other products (Jongrungruangchok et al. 2004).

2. Mold Technology in the Western Countries

Molds are the fungi growing in the form of multicellular filaments called hyphae. These fungi are incapable of photosynthesis but derive energy from the matter on which they live. Mold fermented foods are found in food developed by many human cultures. In Asian countries such food products are based on raw materials of plant origin, whereas in Europe substrate of animal origin such as cheese, fermented sausage and ham are used. Molds in the West technologies are used mainly in cheese and dry sausages making.

2.1 Cheeses with Mold Fermentation

Unlike fermented milk, the physical features of cheeses are far from those of milk. As cheese making involves a series of relatively simple unit operations, and each of them bring some specific contribution to the final product, so even the slight variations in the operations and the use of different types of milk combine to generate a vast range of cheeses available, including 78 types of blue cheese and 36 Camembert's alone (Adams and Moss 2008).

Cheeses made with mold fermentation are *Camembert* and *Brie* (with surface growth of *Penicillium camemberti*), and ripened principally by internal mold growth of *Penicillium roqueforti* (*Roquefort*) coming from France, *Stilton* from Britain, *Gorgonzola* from Italy, *Cambazola* from German,

Table 1. Main mold effects in fermented products.

Product	Activity	References
Chesees (Camembert)*	decreasing of the nitrate reductase activity (reduction of pro-carcinogenic factors)	Lay et al. 2004, Firmesse et al. 2008
Meat products	utilization of a safety mold culture in fermented sausages is prevents to avoid growth of molds synthesizing mycotoxins	Galvalisi et al. 2012, Cordero et al. 2015
Botrytized wines	increase total polyphenol content and antioxidative capacity	Pour Nikfardjam et al. 2006
Tempeh	inhibition of diarrhea	Karmini et al. 1997, Karyadi and Lukito 2000, Kiers et al. 2003
	reduction of flatulence	Nowak 1992, Nowak and Szebiotko 1992
	increased content of vitamins	Keuth and Bisping 1993, Wiesel et al. 1997, Denter et al. 1998
	increased content of isoflavoneaglycones	Hutchins et al. 1995, Kuligowski et al. 2016
*Miso**	radioprotective effect	Ohara et al. 2001, Watanabe 2013
	decrease risk of brest cancer	Yamamoto et al. 2003
Soysauce*	hypoallergenicity and antiallergicactivity	Kobayashi 2005
Fermented red rice	improves lipid pattern, high-sensitivity C-reactive protein, and vascular remodeling parameters in moderately hypercholesterolemic	Lin et al. 2011, Cicero et al. 2013

* Effect can be caused by activity of molds and/or other microorganisms

Mycella from Denmark, *Gamalost* from Norwegian and *Hermelin* from Czech Republic. Generally, the moldy cheeses are prepared by using mixed mold-lactic acid fermentation (Robinson and Tamime 1990).

2.1.1 Nutritional properties of mold fermented cheeses

- Soft cheeses are rich sources of calcium and vitamins A, D and E, as well as vitamins from group B and vitamin C,
- Contents of vitamins B1, B2, B6, B12 and PP increase in the course of soft cheese ripening,
- A rich source of readily available proteins, found in the form of peptides and amino acids.

Camembert cheese consumption has a beneficial effect on the intestinal metabolism such as a decrease in azoreductase activity, a decrease in NH_3 concentration and an increase in the proportion of ursodeoxycholic acid (Lay et al. 2004, Firmesse et al. 2008).

2.1.1.1 Production of *Camembert* and *Brie*

The steps of cheese making are as follows.

1. Milk is pasteurized at a temperature of approximately 73°C for 15–20s,
2. Then the milk is cooled to 15–20°C and standardized,
3. At such a temperature LAB and *Pencillium camemberti* cultures are added,
4. Then milk is heated to a temperature of approximately 32°C and rennet is added in order to obtain curd,
5. Formed cheeses are salted in brine of a concentration of approximately 15%, for 40 min,
6. Cheeses are placed in a ripening chamber, where the mold deposit appears. At this time enzymes formed by molds cause oxidation of lactic acid as well as lipolysis of fats and proteolysis of proteins. These enzymes penetrate to the bulk of cheese and cheese ripening starts from the surface (Derengiewicz 1997, Harbutt 1998, www.lactalis.pl).

2.2 Molds in the Manufacture of Fermented Meat Products

In Northern Europe smoking of sausages is a common practice, while in Mediterranean and South-East European countries mold ripened sausages production is an old and preferred tradition. Mold causes characteristic surface appearance and flavor. The latter originates mainly from the proteolytic and lipolytic activity of molds. Further effects of mold growth are associated with prevention of adverse effects of rancidity and discoloration. In addition, the drying of products proceeds more evenly (Geisen et al. 1992).

Because molds only grow aerobically therefore they are chiefly found on the surface of dry sausages or ham. They are usually not very demanding as regards to nutrients and tolerant to low pH and a_w values. The inocula for sausages usually originate from the environment, so a great number of species are found in this product. The "natural" mold flora of meat products includes the species of the genus *Penicilium*, which incorporate many species potentially able to produce mycotoxins. Because of that, starter cultures which are toxicologically harmless were adopted to a dry sausage substrate and revealed technologically attractive properties. As early as in 1972, Minzlaff and Leister identified a non-toxic strain of *Penicillium nalgiovense* which shows good technological properties and which was later used in starter preparation (Mintzlaff and Leistner 1972).

The molds applied in starter preparations for meat fermentations must show:

- no toxigenic and no pathogenic potential,
- competitive ability against microorganisms growing on the surface,
- firm and lasting surface mycelia of white, yellowish or ebony color,
- well balanced proteolytic and lipolytic activity,
- Characteristic moldy aroma.

The gradual shift in sausage production from small producers to large-scale factories and increasing awareness of the risk for consumers' safety has stimulated industrialized production of mold starter cultures. While the technological aspects of mold inoculation have been solved, its influence on production of secondary metabolites is still under study (Sunesen and Stahnke 2003, Galvalisi et al. 2012, Cordero et al. 2015).

In any conditions starter cultures should not show pathogenic or toxigenic properties in chemical and biological tests. Also they should not be able to produce antibiotics. The most popular species selected for starter cultures for ham and sausages are *Penicilium nalgiovense, P. chrysogenum* and *P. camemberti* though a number of other *Penicillium* species were found on the surface of dry sausages among them: *P. olsonii, P. oxalium, P. viridicatum* (Iacunin et al. 2009).

The growth of mold on the surface of sausages leads to intense proteolysis and lipolysis which cause an increase in the concentration of free amino acids and volatile compounds, such as aliphatic aldehydes and alcohols. Its final effect is the improvements in aroma and taste and, as a consequence, in the overall quality of sausages.

2.3 Wines with Molds Action

The most commonly known wine-types in the world which have to be made with the action of grape molds *Botrytis* are: *Tokaji Aszu* (Hungary), *Sauternes* (France) and German quality wines as "Pradikat Auslese, Beerenauslese, Eiswein, and Trockenbeerenauslese". *Botrytis* splits mainly acids and some of the sugar and produce botrytin, antibiotic, which slows down fermentation. This fermentation is good for the formation of aromatic substances (Farkas 1988).

The best known functional effect of wine consumption is known as French paradox. The famous French paradox is defined as a low incidence of coronary heart disease among French consumers, while consuming a diet rich in saturated fat. The cause of this phenomenon is usually the drinking of wine in small quantity, supposedly the consequence of the presence of polyphenols in red wine (Feher et al. 2007). Some authors have suggested that the consumption of resveratrol in red wine protects the French against the health consequences of a high fat diet (Rosenkranz et al. 2002, Cordova et al. 2005, Poussier et al. 2005). Compared to German white wines of normal quality, the wines made from botrytized grapes from Germany had lower concentrations of resveratrol. The Hungarian botrytized wines had slightly higher concentrations of this compound because of their production technology. The total polyphenol content and antioxidative capacity in Hungarian Tokaj and botrytized German wines were higher than in the normal German wines, particularly so for Tokaj. These wines are an excellent source of polyphenols and antioxidative capacity (Pour Nikfardjam et al. 2006).

3. Mold Technology in Asian Countries

In many Asian countries, fermented foods are traditionally prepared in the indigenous way (in the household) using relatively simple techniques that helped to preserve food components or products and endowed them with specific taste. Regional differences in the choice of starting materials, climate, and culture have contributed to the development of unique fermented products in many parts of the world. Understanding the essential role of microorganisms permits devising efficient and controlled processes. Moreover, new possibilities to deeply characterize the microflora of fermentation extend the knowledge of these processes and allow assuming the standardized approach to making traditional fermented food using specific starter cultures and at the same time avoiding undesirable microorganisms.

While functionality of probiotic products is under profound research and functionality traditionally fermented products of Japan are well characterized, there is a number of other fermented products whose pro-health properties are quite obvious, but should be scientifically proved and professionally marketed.

Foods that are fermented by mold alone are very common in Asia, the best known examples include *tempeh* (a soybean fermented food) originated from Indonesia, *douchi* (fermented whole soybean), *furu* or *sufu* (fermented soybean curd or *tofu*) in China, *hongqu* (red mold rice or red fermented rice) or *tian-jiu-niang* (fermented glutinous rice) in China and, *tao-hu-yi* (soybean curd fermented by *Monascus*) in Thailand (Hsieh et al. 2008, Tanasupawat and Visessanguan 2008).

Still a number of fermented foods are produced with involvement of activity of more than one microorganism, either working together or in a sequence. To maximize their growth these microorganisms secret hydrolytic enzymes, assimilate and convert some hydrolyzed substrates into structural components and secondary metabolites. So it is reasonable to suppose that fermented food products may contain a number of functional components originating either from the raw material or formed in the process of fermentation (Gibbs et al. 2004, Kuligowski and Nowak 2007, Kuligowski et al. 2016).

Generally, the health promoting benefits of food fermentation are related to the following five main features (Tanasupawat and Visessanguan 2008):

- enhanced digestibility
- increased bioavailability
- increased nutritional value, e.g., vitamins and cofactors
- probiotic and prebiotic properties
- microbial products such as enzymes, metabolites, and bioactive peptides released after enzymatic digestion of food proteins.

The functional impact of fermented foods on nutrition effects can be direct or indirect. Food fermentation that raise the protein content or improves the balance of amino acids or the bioavailability of vitamins can have profound direct effects on the health of the consumers. This is especially true for people sustaining largely on cereals in which the contents of some vitamins (niacin, nicotinic acid) are limited. Biological enrichment of foods *via.* fermentation can prevent this.

On the other hand, fermentation does not generally increase the calorific value of a given food product unless substrates usually inedible for humans are converted to human quality foods, for example conversion of peanut and coconut press cakes to edible foods such as Indonesian *ontjom* (*oncom*) (Steinkraus 1997).

3.1 Enhanced Digestibility

The digestibility of nutrients in food of plant origin (especially legumes) is often limited by the presence of anti-nutritional factors (ANFs), such as trypsin inhibitors, phytic acid, non-starch polysaccharides (NSPs), oligosaccharides, lectins, and saponins. It has been established that these components cause gastro-intestinal disturbances, intestinal damage, increased disease susceptibility, and reduced growth performance (Liener 1994, Piecyk et al. 2005).

The hydrothermal preparation of raw materials and operation of microorganisms during the fermentation is known to improve the availability of some nutrients thanks to the destruction of ANFs. For example hydrothermal processing as a process preceding fermentation in *tempeh* that results in a reduction of the amount of phytates in pea seeds by 37%, while in broad bean seeds it reduced by 31% (Rozwandowicz 2007), whereas after hydrothermal processing of soybeans, *Macrotylomageocarpa* and common cowpea an increase in the amount of phytic acid by 21, 43 and 34%, respectively was observed (Egounlety and Aworh 2003). Moreover, microorganisms contain certain enzymes that cannot be synthesized in humans. These enzymes hydrolyze complex food components into simple units that can be readily digested or absorbed in the intestines.

It has been reported that a reduction the amounts of phytic acids in soy *tempeh* as a result of fermentation reached from 30% (Egounlety and Aworh 2003) to 60% (Astuti et al. 2000). Owing to the fact that fermentation increases the quantity of soluble proteins in foods, often improves the amino acid profile and reduces the levels of certain anti-nutritional effects, fermented foods tend to be more efficiently utilized than unfermented foods (Paredes-Lopez 1992, Kovac and Raspor 1997). For instance, the protein efficiency ratio (PER), biological value (BV) and true digestibility (TD) in *tempeh* fermented soybean were significantly higher than in the unfermented product (Steinkraus 1996).

3.2 Increased Bioavailability

The term bioavailability is defined as referring to absorption and utilization of the ingested nutrients. Absorption of nutrients is often limited by their chemical character and interaction with other food components. From among isoflavones of legumes recently much attention has been paid to soy isoflavones, used in the pharmaceutical industry as agents alleviating symptoms of hormone deficiency in menopausal women (Messina 2000, Pereira et al. 2006, Bijak et al. 2010). Soybean isoflavones show a number of biological activities, they can prevent chronic diseases, act as phytoestrogen, antioxidants, and have anti-tumoral effect (Anderson and Garner 1997, Constantinou et al. 1998, Messina 1999, Brouns 2002, Lai and Yen 2002, Panat et al. 2016). On the other hand, it is known that the bioavailability of soybean isoflavones in humans is strictly related to their metabolic capacity and to their hydrolysis caused by β-glucosidase enzymes produced by the intestinal microflora.

Daidzein and genistein together and their effect on physiological changes in the human have been most frequently described in literature. In soy seeds other isoflavones have been also detected—biochanin A, formononetin and glycitin (Mazur et al. 1996, Kraszewska et al. 2007). Genistein exhibits a capacity to inhibit growth of prostate cancer cells in men (Adlercreutz 1999). A high concentration of phytoestrogens in systemic fluids of humans results in a reduced risk of coronary arterial disease as well as breast and prostate cancers (Duncan et al. 2003, Jasińska and Michniewicz 2005).

Epidemiological studies on women in Asian countries showed that changes during menopause (hot flushes, sleep disorders, headaches, physical and mental exhaustion as well depressive states) are alleviated by isoflavones taken in the diet (Gruber et al. 2002, Han et al. 2002). Among the individuals eating food containing large amounts of phytoestrogens, e.g., isoflavones, a less frequent incidence of the mammary gland, prostate and the endometrium cancer has been noted (Kustrzewa–Tarnawska 2007).

Many authors have been interested in the use of isoflavonoids contained in fermented and unfermented soy products in relation to human health. Hutchins et al. (Hutchins et al. 1995) have found that the bioavailability of the isoflavones was higher from *tempeh* than from soybean, although the fermented soybean products contained lower amounts of isoflavones. During *tempeh* fermentation isoflavoneaglycones are released (Kuligowski et al. 2016). Studies on the secretion of isoflavones showed that aglycones are more readily absorbed by the human than glycosides (Duncan et al. 2003, Nout and Kiers 2005). Isoflavoneaglycones in the human alimentary tract undergo biotransformation stimulated by intestinal microflora. Through oxidation, reduction, deconjugation and demethylation they are transformed into active forms, capable of acting on estrogen receptors and able to cause a physiological effect, consisting in their binding with hypothalamic neuron receptors and regulation of hormonal metabolism (Chatenoud et al. 1998, Rowland et al. 1999, Bowey et al. 2003).

3.3 *Functional Properties of* Tempeh *from Soybean and Other Legumes*

In Asian countries some civilization-related diseases and cancers occurring in representatives of the Western civilization are not met. Most researchers associate this phenomenon with differences in the diet between people living in both regions. The diet of Asian countries includes a range of indigenous fermented foods that are made at home, in villages, small cottage industries, and even in larger commercial processing establishments. The vastest survey of such products has been presented by Steinkraus in his Handbook of Indigenous Fermented Foods (Steinkraus 1996). Unlike many indigenous products fermented with the use of molds such as *miso,* soy sauces, red mold rice, and *furu* and many others which are mainly used as flavoring agents, coloring agents, side dishes, breakfast items, and condiments, *tempeh* is used as a main dish in Indonesia.

Tempeh, or *tempekedelee,* which comes from Indonesia, is obtained by the fermentation of dehulled, cooked soybeans with *Rhizopus oligosporus.* The mycelia bind the soybean cotyledons together into a firm cake. In the process of fermentation, the product acquires new organoleptic properties and a clean, fresh, and yeasty odor. When sliced and deep-fat fried, *tempeh* has a nutty flavor, pleasant aroma, and a texture much appreciated by most people around the world. Also, *tempeh* does not have the "beany flavor" associated with soybean products found disagreeable by many Westerners. Nout and Rombouts have suggested that the lack of "beany" flavor in *tempeh* was a result of thermal inactivation of the lipo-oxygenases responsible for the formation of such flavors (Nout and Rombouts 1990). Traditional *tempeh* is produced in banana leaves. Nowadays it is produced in various kinds of containers (plastic, wooden, metal or in plastic perforated bags). Preparation of *tempeh* involves several essential steps (Nout and Kiers 2005).

Cleaning of legumes seeds (cleaning in tap water to remove contaminants that can accumulate on storage) and hydration and lactic acid fermentation (the ratio of soaking should be more than 3:1 water; beans, lactic acid prevents growing of undesirable microorganisms) are initial steps for *tempeh* preparation. The other steps are:

1. De-hulling, which is the removal of non-nutritional components, like tannins which are accumulated in hulls,
2. Cooking or autoclaving (pasteurization or sterilization of material, facilitate fungal penetration, raffinose family sugars extraction),
3. Draining and cooling down the cotyledons. Inoculating with *tempeh* starter (the range of spore suspension for inoculum is from 10^4 to 10^6 colony forming units and the optimal concentration of inoculum is 1%),
4. Packing into suitable containers and incubating at 31 to 37°C (the incubation condition should ensure good gas exchange and mycelial growth but should not allow sporulation) to get a good quality product.

According to Shurtleff and Aoyagi (1979) *R. oligosporus* NRRL 2710 is the most widely used strain. Fermentation is considered complete when the soybeans are covered with, and bound together by, the white mycelia of the mold. The well-made raw *tempeh* should look like a firm white cake. The shelf life of *tempeh* can be extended by drying, frying, freezing, dehydration, blanching, steaming, and even canning.

Tempeh is a very versatile product and can be used in combination with many different recipes and dishes. *Tempeh* can be served with grains and eggs for a breakfast item or in salads, sandwiches, burgers, sauces, or soups for a lunch or dinner. When it is to be an ingredient in recipes for salads, soups sauces, or casseroles, it is recommended to subject it to frying at first to ensure a crisp texture. Deep-fat frying and pan frying *tempeh* in vegetable oil yields a crisp, golden brown product. It can be made also from seeds of other plants, e.g., chickpea (Reyes-Moreno et al. 2000), beans (Kuligowski and Nowak 2006), peas (Nowak and Szebiotko 1992), lupine (Fudiyansyah et al. 1995), common cowpea (Egounlety and Aworh 2003), buckwheat (Handoyo et al. 2006), barley (Feng et al. 2005), maize (Cuevas-Rodriguez et al. 2004), oat (Nowak 1992b) and sorghum (Mugula and Lyimo 2000).

The use of raw materials other than soy in *tempeh* type fermentation may lead to the formation of products rich in certain health-promoting ingredients. A twofold increase in the content of L-dihydroxyphenylalanine (L-DOPA) was achieved in the flour made from broad bean seeds subjected to *tempeh*-type fermentation. L-DOPA is a precursor of dopamine, a neurotransmitter responsible for the transmission of impulses between neurons and a precursor in the biosynthesis of noradrenalin and adrenalin. L-DOPA is applied in the treatment of symptoms of Parkinson's disease. Natural sources of this compound are highly valuable in the prevention of bothersome side-effects caused by the administration of synthetic L-dihydroxyphenylalanine (Randhir et al. 2004).

Most functional properties of foodstuffs are connected with the fact that they contain chemical compounds with a specific effect on the human. An increase in the concentration of these compounds or their availability may lead to the formation of food supplying identical amounts of biologically active components as pharmaceuticals, particularly in view of the amount of the consumed medicine and food. However, not all functional properties are related in a simple manner with the presence of a specific ingredient in food. An example of such an instance is *tempeh* and its capacity to control diarrhoea in monogastric mammals.

So far such an effect of soy *tempeh* consumption has been observed in humans (Karyadi and Lukito 2000), rabbits (Karyadi and Lukito 1996, Karmini et al. 1997) and pigs (Kiers et al. 2003). In the 1960s it was reported that extracts from the substrate fermented by fungi *R. oligosporus* show antibacterial properties. Among the microorganisms found as most susceptible to the action of this type of antibacterial substances, are *Streptococcus cremoris* (Wang et al. 1969) or *Bacillus subtilis* (Kobayasi et al. 1992). Later studies on antibacterial activity of soy *tempeh* extracts showed a slight antibacterial activity of only one of the analyzed strains of *Rhizopus* towards *B. subtilis*, while the other used strains exhibited activity towards *Bacillus stearothermpohilus* (Kiers et al. 2002). Peptides found in *tempeh* have been reported to exhibit antibacterial activity towards *Bacillus cereus* (Roubos et al. 2008). However, no capacity was shown for *tempeh* isolates to inhibit the growth of *Escherichia coli*, causing diarrhea in humans and animals (Kiers et al. 2002). In *in vitro* studies it was found that isolates from *tempeh* could have contributed to inhibition of cell adhesion in an enterotoxic strain of *E. coli*. Depending on the strain of *R. oligosporus* applied in fermentation, the effect on cell adhesion varies. However, there are no data that would permit concluding which of the substances contained in *tempeh* affected this process (Kiers et al. 2002). Unfortunately, in most of the studies raw *tempeh* was used, not boiled or fried as it is normally consumed, and the process of digestion was not taken into consideration.

Processing of soybeans and other legumes into *tempeh* brings favorable changes, including reduction in concentration of flatulence-causing oligosaccharides. It is well known that *tempeh* is non-flatulent when eaten by humans and animals. Undoubtedly, an important reason for this important phenomenon is the reduction of stachyose, raffinose, and other flatulence-causing carbohydrates by fermentation. Still fermentation for 24 h is normally not enough to eliminate fully raffinose-family sugars (Nowak and Szebiotko 1992). It was found that pea and soybean *tempeh* extracts inhibit gas-producing *Clostridia* of human intestine origin. This might be another way to study the reasons of anti-flatulent effect of *tempeh* fermentation as the mechanism of decreasing gas production in the gut, which is more complex than just investigation of reduction in the content of raffinose-family sugars and permits additional evaluation of the effect of the product of the first fermentation process (involving *Rhizopus* mold) on the intestines flora (Nowak 1992a).

Tempeh is easily digested and is even tolerated by patients suffering dysentery and nutritional Edema (Ko and Hesseltine 1979). *Tempeh-* and milk-based formulas were evaluated in the rehabilitation of children with chronic diarrhea. It has been reported that recovery from diarrheal disease was faster with *tempeh*-based formula and resulted in better weight gain.

Another subject of interest was the bioavailability of minerals such as iron and zinc. It was evaluated in rat feeding tests. The availability of zinc was found to be 1.22 times better from soybean *tempeh* than from boiled soybeans, moreover it improved by fermenting the soybeans with *R. oligosporus*. It is known that such processing steps as boiling, soaking, and fermenting decrease the level of phytic acid in legumes, which prevents the chelation of minerals by phytic acid. More recently, Kasaoka et al. have studied the effects of *tempeh* on iron bioavailability and lipid peroxidation in rats (Kasaoka et al. 1997). They reported that consumption of fermented soybean *tempeh* increased the level of liver iron, when compared with the effect of intake of unfermented soybeans, without promoting lipid peroxidation in iron-deficient anemic rats.

Another healthy and nutritional benefit of *tempeh* is the increased content of some vitamins of the B-group, related to the activities of fungal and bacterial metabolic activities. The synthesis of vitamin B12 by certain bacteria has met with great interest, especially in the societies in which vegetarianism is popular. The most common mold in *tempeh* fermentation *R. oligosporus* was not found to produce vitamin B12. However, many bacteria grown along with *Rhizopus* were found to produce it. Usually, soybeans contain less than 1 ng of vitamin B12 per gram. The effects of *Klebsiella pneumoniae, Enterobacter* spp. unidentified Gram-positive and Gram-negative rods on vitamin B12 production have been well studied by Keuth and Bisping. They reported the production of vitamin B12 by *Citrobacterf reundii* or *K. pneumoniae* during *tempeh* fermentation (Keuth and Bisping 1993). Denter et al. have observed the ability to produce carotenoids in all 14 strains of *R. oligosporus* used in their studies, of which only in six strains the synthesis of β-carotene, was shown. The highest increase in the content of β-carotene was found between the 36th and 48th hour of fermentation. Moreover, we need to stress here on the synthesis of ergosterol by the strains of *Rhizopus*, whose presence was not observed in soybeans. The highest ergosterol concentration at the 34th hour of fermentation, amounting to 750 µg/g dry matter (dm) has been reported (Denter et al. 1998). They produced *tempeh* using a strain of *R. oligosporus* MS35, which when used by Wiesel et al. in the 36th hour of fermentation yielded 190 µg/g dm ergosterol (Wiesel et al. 1997). In the studies by Denter et al. and Wiesel et al., identical fermentation conditions were applied. In both these experiments performed with the use of pure cultures of filamentous fungi, an increase was observed in the contents of carotenoids and ergosterol, at a slight reduction of contents of vitamin K1. Wiesel et al. have observed also an increase in the contents of riboflavin, pyridoxine, niacin, biotin, folic acid, as well as a slight reduction in the amounts of vitamin E and thiamine. However, after the application of mixed cultures of filamentous fungi and bacteria in the fermentation process, higher amounts of vitamins, i.e., riboflavin and folic acid, contained in *tempeh*, as well as a decrease in the amount of niacin were established.

It has been established that dried *tempeh* is significantly more resistant to lipid oxidation than unfermented soybeans. According to Hoppe et al., the anti-oxidative effect of *tempeh* oil seems to be a result of a synergist effect of tocopherol (present in the soybeans), and amino acids liberated during the fermentation process by *R. oligosporus* (Hoppe et al. 1997).

The effect of *tempeh* fermentation on total nitrogen content is negligible but an increase in the amount of free amino acids has been noted. Practically no changes are observed in the essential amino acid index after 24 h fermentation period, but longer fermentation brings the loss in threonine of 8.9%, lysine of 25%, and arginine of 13.5% (Stilling and Hacler 1965). Nowak and Szebiotko have observed an increase in free amino acid and ammonia content in soy and pea *tempeh* with increasing fermentation time (Nowak and Szebiotko 1992).

During fermentation of soy *tempeh* a strong reduction was also observed in the amount of allergenic proteins, which may prevent the use of functional components contained in soy by individuals suffering from allergies (Kuligowski 2009).

Summarizing the benefits offered by *tempeh* include excellent digestibility, increased content of vitamin B12 and other B vitamins, antioxidative properties, the ability to reduce amount of gases produced in the intestine after consumption of legumes (flatulence), conversion of genistein and daidzein glycosides (naturally occurring in legume foods isoflavones) to their more active aglycone forms, alleviation of intestine and digestion problems (diarrhea) and reduction of allergic activity of soybean proteins. In order to produce a *tempeh*-like functional food containing a high level of isoflavone with a high absorptivity, the soybean germs (hypocotyl) that contained a large amount of isoflavones were added. This procedure permitted Nakajima et al. to prepare a new isoflavone-enriched *tempeh* in the form of a granular fermented soybean-based food product (Nakajima et al. 2005).

Intensive proteolysis may lead to the formation of biologically active amino acids, such as gamma–amino butyric acid, which is a neurotransmitter in the sympathetic nervous system. It was shown in the study that it plays an important role in the lowering of blood pressure and prevention of certain symptoms of menopause, such as insomnia, depression and vegetative neurosis (Aoki et al. 2003a). Through the modification of fermentation conditions, consisting in the application of 5 h storage in a nitrogen-filled incubator following 20 h fermentation, Aoki et al. obtained an increase in the amount of gamma–amino butyric acid from 30 mg/100 g dm in the traditional *tempeh* to 370 mg/100 g dm in the *tempeh* produced under modified conditions (Aoki et al. 2003b).

3.4 Miso

Miso is a traditional Japanese fermented soybean paste, which has become one of the most important seasonings in Japanese cuisine. It is made from steamed soybeans mixed with rice, barley, salt, and *koji*. *Miso* soup is very popular for breakfast in Japan. There are a number of regional differences in flavor and taste, for instance *shiro* (white) *miso* paste has a mild taste and is low in salt, whereas *aka* (red) *miso* is very salty and has a different, stronger smell than *shiro miso*. The ripe varieties are brownish-red. The fermentation involves the microorganisms called *koji* mold (fungi *Aspergillus oryzae*). Yellow *koji* mold, *A. oryzae*, is also used in the process of Japanese *sake* production. Enzymes (amylases) provided by *koji* convert the rice starch to sugar. The *koji* used for soy sauce and *miso* is also high in proteases and peptidases, which convert proteins to amino acids (Sugawara and Sakurai

1999). Rice mold starters prepared from *Aspergillus* species, especially *A. awamori, A. kawachii, A. saitioi* exhibit high anti-oxidative activity.

For many years, the supply of *koji* was controlled by *miso* manufacturers, and this product was not commercially available. Recently, better strains of miso *koji*, have been selected and several standardized *miso* products have been produced. *Koji* for *miso* is prepared from cereal grains (most commonly, rice) alone. When the mold growth is sufficient, salt is added to inhibit further growth. The salted *koji* is mixed with cooked, mashed whole soybeans and water is added to reduce the solids content to about 50%. The mass is inoculated with salt tolerant lactic acid bacteria and yeast cultures and allowed to ferment at 30°C for two weeks to 6 months in closed tanks. During this period, the semi-solid mash is transferred from one fermentation vessel to another for the purpose of mixing the contents and accelerating the fermentation. When the paste is sufficiently ripe (depending on the type of *miso*), it is blended, pasteurized and packaged. Benzoate and/or sorbate are sometimes added as preservatives. *Miso* can be stored and used for several years, and during this time it continues to smell good and retain its yellow-brown color (Minamiyama and Okada 2008).

Long-ripened *miso* had a characteristic mouthfulness and continuity of flavor especially increasing from 11th month of ripening. Protein and polysaccharide hydrolyses were terminated after 3–5 months of ripening and remained constant thereafter (Ogasawara et al. 2006).

Fat is an essential component of *miso*. Therefore, whole soybeans and not defatted meal are used as the starting material. A hot variety, flavored with generous amounts of red hot pepper, is produced in Korea. The best recognized functional properties of *miso* are connected with its anti-oxidative properties, the outstanding medical qualities connected with decreased chance of stomach cancer and heart disease, removing of heavy metals and radioactive materials from the body, neutralization effects of some carcinogens and protection of harmful effects of smoking. Consuming a cup of miso three times a day reduced the occurrence of breast cancer but tofu, natto, soybean and fried bean curd did not have such an effect (Yamamoto et al. 2003).

Miso is characterized by strong antioxidant properties (Esaki et al. 1999). The brown-colored substance in aged *miso* has been identified as melanoidin. It strongly suppresses the production of peroxides derived from fatty acids in the body and prevents aging of the body. Vitamin E, daidzein, saponin, and the brown pigment contained in *miso* act as antioxidants.

It is believed that the Japanese diet and methods of food preparation have contributed to the longevity of the Japanese people. In 1981 Hirayama, from the Japan National Cancer Centre, performed an epidemiologic study whose results permitted him to claim that those who eat *miso* soup daily are significantly less likely to develop stomach cancer and heart disease. It has been known for years that *miso* keeps the body in good condition, and it is said that *miso* is "a detoxicating drug in the morning" or "keeps the doctor away". It has been shown that people who eat *miso* soup regularly (daily) are less susceptible to stomach diseases such as gastritis and gastric and duodenal ulcers, than those who seldom or never eat it (Tsugane et al. 1994). From among the isoflavones contained in *miso*, genistein, which has an inhibitory activity of tyrosine kinase, showed a particularly potent anti-*Helicobacter pylori* activity (Odenbreit et al. 2000, Bae et al. 2001). A large portion of the proteins contained in soybeans are degraded by enzymes and microbes in the fermentation process of *miso*. In addition, *miso* contains highly active enzymes, which help digestion and absorption of other nutrients.

It is now widely known that anticancer effects are associated with *miso* intake on a regular basis (Reddy et al. 1980). Interestingly, a very low incidence of breast cancer has been observed in Japanese women at home and abroad as long as they maintain their native diet, but if they break from this diet this incidence becomes higher. A similar relationship holds for Japanese men and prostate cancer (Yan 1989, Messina et al. 1994). It has been established that the intake of soy and not the intake of fat is responsible for this difference. Isoflavones exhibit anticancer effects and soybeans are the only one commonly consumed food that provides isoflavones in the diet. Certain sugars, in particular oligosaccharides, in soybeans promote the growth of the beneficial bacteria called Bifidobacteria in the colon (Bouhnik et al. 2004). A remarkable discovery was that the consumption of *miso* helps to eliminate radioactive substances from the body. After the Hiroshima and Nagasaki atomic bombing at the end of World War II, it was observed that *miso* factory workers were less affected by radiation than others in the general population. The reason for this protective effect of *miso* is not known (Ohara et al. 2001, Minamiyama and Okada 2008, Watanabe 2013).

Miso contains slinoleic acid, plant sterols, and vitamin E which among others have been shown to be cardioprotective. As a result of clinical and experimental studies it has been established that substitution of soy protein for animal protein or simply adding soy protein to the diet, significantly reduces the concentration of cholesterol irrespective of the type or amount of fat in the diet (Belleville 2002, Messina 2003, Gibbs et al. 2004). A preventive effect of *miso* as to drug-induced liver injury in mice has been observed (Suzuki et al. 2008). Watanabe et al. have reported that a *miso* diet including 2.3% NaCl and a control diet containing 0.3% NaCl did not increase blood pressure in rats, whereas increased blood pressure was noted in the rats consuming the 2.3% NaCl control diet (and no *miso*). According to these results blood pressure was not increased by the *miso* diet (Watanabe et al. 2006).

Many of the beneficial effects of *miso* seem to be connected with bioactive compounds found in the soybeans. However, the fermentation process used in the production of *miso* provides additional benefits.

3.5 Soy Sauce

Soy sauce is a dark brown liquid obtained from a fermented mixture of soybeans and wheat. It has recently become very popular all over the world. Although soy sauce comes originally from China, its production and uses have gradually spread to Japan, Korea and other countries in Asia. The greatest contribution to producing technology and developing high quality soy sauce products has been brought by the Japanese who introduced the use of precisely combined starter cultures under controlled fermentation conditions (Hsieh et al. 2008).

There are two fermentation stages involved in the production of soy sauce. The first stage is aerobic *koji* fermentation, which uses fungi (*Aspergillus oryzae, A. sojae*) to break down the polysaccharides into simple sugars. The second stage is an anaerobic salt mash or *moromi* (in Japanese), in which the mixture undergoes LAB and yeast (*Zygosaccharomyces rouxii*) fermentation (Aidoo et al. 2006).

The chemical composition of soy sauce depends on the raw materials used. In general, good quality soy sauce contains 1.0 to 1.65% total nitrogen (w/v), 2 to 5% reducing sugars, 1 to 2% organic acids, 2.0 to 2.5% ethanol, and 17 to 19% sodium chloride (w/v). About 45% of the total nitrogen consists of simple peptides, whereas 45% is in the form of amino acids (Luh 1995).

Soy sauce contains several bioactive components showing a wide gamut of biological activities, including anti-carcinogenic, anti-microbial, anti-oxidative, anti-platelet, hypoallergenic and anti-allergic activities, moreover the inhibition of an angiotensin I-converting enzyme. Thanks to these collective effects of soy sauce it is a healthy and beneficial food in the human diet (Kataoka 2005).

Proteins of the raw materials are completely degraded into peptides and amino acids by microbial proteolytic enzymes after fermentation, and no allergens of the raw materials are present in soy sauce. In contrast, polysaccharides originating from the cell wall of soybeans are resistant to enzymatic hydrolysis. These polysaccharides are present in soy sauce even after fermentation and are known as *shoyu* polysaccharides (SPS). Soy sauce generally contains about 1% (w/v) SPS and SPS exhibit potent anti-allergic activities *in vitro* and *in vivo*. Furthermore, an oral supplementation of SPS is an effective intervention for patients with allergic rhinitis in two double-blind placebo-controlled clinical studies (Kobayashi 2005). Soy sauce promotes digestion, because the consumption of a cup of clear soup containing soy sauce enhances gastric juice secretion in humans. Moreover, soy sauce shows antimicrobial activity against bacteria such as *Staphylococcus aureus*, *Shigella flexneri*, *Vibrio cholera*, *Salmonella enteritidis*, non-pathogenic *Escherichia coli* and pathogenic *E. coli* O157:H7 (Kataoka 2005).

3.6 Furu

Furu (*sufu*) is made by fungal solid-state fermentation of *tofu* followed by aging in brine containing salt, spices, alcohol, and other ingredients. The starters used for *furu* production include the following fungi: *Actinomucor* spp., *Mucorwutungkino* spp., *Mhiemelis* spp. and *Rhizopus* spp. (Tanasupawat et al. 1998, Tanasupawat et al. 2000). Most of the commercially used starters of *furu* fermentation represent species of *Mucor* and *Actinomucor*, from which, *Actinomucor elegans* and *Actinomucor taiwanesis* seem to be the preferred molds widely used for the commercial production of *furu* in China and Taiwan, respectively. As most of these commercial starter species grow best at temperatures from 25 to 30°C, they are unsuitable for *furu* production at temperatures higher than 30°C. For this reason, *R. oligosporus*, which grows well at temperatures up to 40°C, has been explored as a potential alternative to *Actinomucor elegans* as the starter for *furu* production during the hot seasons (Tanasupawat and Visessanguan 2008).

The production of *furu* involves a few stages. The first one is the preparation of *tofu*, made by soaking, grinding, and filtering the soybeans to obtain soymilk, which is then coagulated by adding calcium salt to form soybean curd. The curd is then pressed into sheets and cut into cubes of *tofu*. At the second stage *tofu* is inoculated with a mold starter and incubated for 2 to 7 days at 12 to 30°C, to give the mycelium covered pieces of *pehtze*. The starter may either be inoculated with a pure culture applied as a sprayed suspension, or by natural house flora inoculated onto the *tofu* by direct contact with utensils (Tanasupawat et al. 2000). The *pehtze* fermentation step is applied to foster the formation of a white cover of mold mycelia around the cubes of *tofu*. At the final stage, the *pehtze* is ripened in a dressing mixture or brine containing rice wine, salt, and other product-specific ingredients determining the color and flavor. It is mainly appreciated for its pleasant creamy taste and intense savory flavor, which are generally used to accent the otherwise bland flavors of rice or bread. In Chinese culture *furu* is recommended as an easily digested and healthy food for children, the elderly and the infirm. It is probably related to a content of calcium, free amino acids, peptides, and enzymes resulting from the fermentation process. The content of water-soluble proteins in *furu* is from 6 to 7 times higher than in *tofu* (Tanasupawat et al. 1991, Namwong et al. 2006). Besides, thanks to fermentation it contains many bioactive components, such as soybean peptides, B vitamins, nucleic glycosides, and aromatic compounds, which do not exist in unfermented soybeans (Tanasupawat et al. 1992). The vitamin B12 content, an essential nutrient for the nervous system, in *chou furu* (strong smelly, stinky *furu*) is much higher (9.8 to 18.8 mg/ 100 g) than that in red *furu* (0.42 to 0.78 mg/100 g) (Tanasupawat et al. 1991), suggesting a high activity of microorganisms during fermentation of the chou *furu*. The anti-oxidative and anti-hypertensive effects of *furu* have also been thoroughly studied (Robert 1982, Takano 2002).

3.7 Fermented Red Rice

Fermented red rice is a common food ingredient used to enhance the color and flavor of various Thai foods. It also has been known as a food preservative and a folk medicine for digestive and vascular functions in China, India and Japan (Journoud and Jones 2004). In addition to the edible pigments with a polyketide structure, various species of *Monascus* have also been shown to produce other bioactive metabolites with different properties. Monoascidin A produced by *Monascus purpureus* exhibited antibiotic action not only against bacteria but also against yeasts and some filamentous fungi (Martinkova et al. 1995). *M. purpureus* MTCC 410—fermented rice lowers blood–lipid levels and monacolins have been proven to be the main active constituents in red mold rice (Kumari et al. 2009). Various metabolites from *Monascus*-fermented rice might have potential implications in clinical artheriolsclerosis disease. Fermented rice, containing naturally occurring statins and various pigments, has lipid-modulating, anti-inflammatory and anti-oxidative effects (Endo et al. 1985, Hossain et al. 1996, Lin et al. 2011, Cicero et al. 2013).

Examples like those provided in this chapter serve to highlight the potential of exploiting fermented foods as a basis for functional food development. Studies on health promotion activity *in vivo* would greatly facilitate the development of a completely new generation of fermented functional foods tailored to contain clinically proven health promoting metabolites and/or microorganisms. Future possibilities to exploit the potential health promoting properties of such food products will continue to expand, as researchers gain a greater understanding of the microorganisms used and how they, or their metabolites, can directly interact in a positive way with the human host.

4. Conclusion

The knowledge of the role of microorganisms and biotechnology methods in scaling up of indigenous fermented production enables the use of the functional properties of this kind of food. Besides offering desirable sensory characteristics, traditional fermented foods and beverages also convey health benefits. Some of the less known, traditional and indigenous techniques of food fermentation, after some research, might be a source of new naturally produced functional food. Recognition of a mechanism of induction of civilization-related diseases, cancers and Parkinson's disease makes it possible to select food rich in specific components, which may prevent them. Fermentation with the participation of molds makes it possible to broaden the assortment of food in order to avoid boredom, at the same time facilitating the elimination of compounds having a negative effect on the human, e.g., strongly reducing allergenic epitopes of proteins. The protective role of molds and their metabolites in relation to certain nutrients is another property of this type of fermentation. There is still a large group of products obtained with the participation of molds, which have not been thoroughly investigated to date in terms of their effect on human health. Additionally, reports have been published on ethnic food from different regions of the world that are fermented among other things by molds, which practically has not been evaluated in terms of their functional properties. The world wide popularity of yogurt (and fermented foodstuff with probiotics) as well as the growing popularity of *tempeh* is the example of expansion of this type of healthy food and prove that the nutrition awareness of people is growing globally. The new possibilities to deeply characterize the microflora of fermentation, transformations taking place on fermentation as well as the undesirable microorganisms, significantly contribute to extend the knowledge of these processes. For instance microbial characterization of dairy ecosystems in a number of traditional European cheeses is currently performed by using conventional culturing and culture-independent molecular techniques. Among the latter techniques, the denaturing gradient gel electrophoresis (DGGE) tracks compositional changes in the microbial communities by sequence-specific separation of PCR-amplified fragments.

Molds would find its important place especially when the mechanisms involved in the interaction between food microorganisms and its products on gastro-intestinal host intestinal bacteria are identified and when the role molds play in prebiotic formation is known. Foodstuff that improve or change the intestinal microflora are of particular interest because of the growing knowledge of the role the intestinal microflora plays in health promotion and disease resistance. Thanks to the application of modifications in traditional technologies it is possible to increase the amount of compounds with a potential positive effect on human health. For example by the selection of the *R. oligosporus* strain it is possible to optimize the amount of isoflavones in soy *tempeh*. In future more fermented food products with health promoting effects will become available on the market. Examples like the ones provided in this chapter serve to highlight the potential of exploiting fermented foods as a basis for functional food development. The planned studies on health promotion activity *in vivo* are expected to facilitate the development of a completely new generation of fermented functional food products, tailored to contain clinically proven health promoting metabolites and/or microorganisms.

Keywords: Mold fermentation, functional properties, *tempeh*, *miso*, soy sauce, *furu* (*sufu*)

References

Adams, M.R. and Moss, M.O. (2008). Food Microbiology, 3rd Edition, RSC Publishing, Cambridge, pp. 330–331.

Adlercreutz, H. (1999). Phytoestrogens. State of the art. Environmental Toxicology and Pharmacology 7: 201–207.

Aidoo, K.E., Nout, R.M.J. and Sarkar, K. (2006). Occurrence and function of yeasts in Asian indigenous fermented foods. Federation of European Microbiological Societies Yeast Research 6: 30–39.

Anderson, J.J.B. and Garner, S.C. (1997). The effect of phytoestrogens on bone. Nutrition Research 17: 1617–1632.

Aoki, H., Uda, I., Tagami, K., Furuya, Y., Endo, Y. and Fujimoto, K. (2003b). The production of a new tempeh-like fermented soybean containing a high level of γ-aminobutyric acid by anaerobic incubation with *Rhizopus*. Bioscience, Biotechnology, and Biochemistry 67(5): 1018–1023.

Aoki, H., Furuya, Y., Endo, Y. and Fujimoto, K. (2003a). Effect of γ-aminobutyric acid-enriched tempeh-like fermented soybean (gaba-tempeh) on the blood pressure of spontaneously hypertensive rats. Bioscience, Biotechnology, and Biochemistry 67(8): 1806–1808.

Astuti, M., Meliala, A., Dalais, F.S. and Wahlqvist, M.L. (2000). Tempe, a nutritious and healthy food from Indonesia. Asia Pacific Journal of Clinical Nutrition 9(4): 322–325.

Bae, E.A., Han, M.J. and Kim, D.H. (2001). *In vitro* anti-*Helicobacter pylori* activity of irisolidone isolated from the flowers and rhizomes of *Puerariathunbergiana*. Planta Medica 67: 161–163.

Belleville, J. (2002). Hypocholesterolemic effect of soy protein. Nutrition 18(7-8): 684–686.

Bijak, M., Połać, I., Borowiecka, M., Nowak, P., Stetkiewicz, T. and Pertyński, T. (2010). Isoflavones as an alternative to menopausal hormone therapy. Przegląd Menopauzalny 6: 402–406.

Blandino, A., Al-Aseeri, M.E., Pandiella, S.S., Pantero, D. and Webb, C. (2003). Cereal-based fermented food and beverages. Food Research International 36: 527–543.

Bouhnik, Y., Raskine, L., Simoneau, G., Vicaut, E., Neut, Ch., Flourié, B., Brouns, F. and Bornet, F.R. (2004). The capacity of non-digestible carbohydrates to stimulate fecal bifidobacteria in healthy humans: a double-blind, randomized, placebo-controlled, parallel-group, dose-response relation study. The American Journal of Clinical Nutrition 80: 1658–1664.

Bowey, E., Adlercreutz, H. and Rowland, I. (2003). Metabolism of isoflavones and lignans by the gut microflora: A study in germ-free and human flora associated rats. Food and Chemical Toxicology 41: 631–636.

Brouns, F. (2002). Soya isoflavones: A new and promising ingredient for the health foods sector. Food Research International 35: 187–193.

Chatenoud, L., Tavani, A., Vecchia, C., Jacobs, D., Negri, E., Levi, F. and Franceschi, S. (1998). Whole grain flood intake and cancer risk. International Journal of Cancer 77: 24–28.

Cicero, A.F., Derosa, G., Parini, A., Maffioli, P., D'Addato, S., Reggi, A., Giovannini, M. and Borghi, C. (2013). Red yeast rice improves lipid pattern, high-sensitivity C-reactive protein, and vascular remodeling parameters in moderately hypercholesterolemic Italian subjects. Nutrition Research 33: 622–628.

Constantinou, A.I., Kamath, N. and Murley, J.S. (1998). Genistein inactivates bcl-2, delays the G2/M phase of the cell cycle and induces apoptosis of human breast adenocarcinoma MCF-7 cells. European Journal of Cancer 34: 1927–1934.

Cordero, M., Córdoba, J.J., Bernáldez, V., Rodríguez, M. and Rodríguez, A. (2015). Quantification of *Penicillium nalgiovense* on Dry-Cured Sausage 'Salchichón' Using a SYBR Green-Based Real-Time PCR. Food Analytical Methods 8(6): 1582–1590.

Cordova, A.C., Jackson, L.S., Berke-Schlessel, D.W. and Sumpio, B.E. (2005). The cardiovascular protective effect of red wine. Journal of American College of Surgeons 200: 428–439.

Cuevas-Rodriguez, E.O., Milan-Carrillo, J., Mora-Escobedo, R., Cardenas-Valenzuela, O.G. and Reyes-Moreno, C. (2004). Quality protein maize (*Zea mays* L.) tempeh flour through solid state fermentation process. Lebensmittel Wissenschaft und Technologie 37: 59–67.

Denter, J., Rehm, H.J. and Bisping, B. (1998). Changes in the contents of fat-soluble vitamins and provitamins during tempe fermentation. International Journal of Food Microbiology 45: 129–134.

Derengiewicz, W. (1997). Technologiaserówmiękkich. Oficyna Wydawnicza HOŻA, Warszawa: 171–189.

Duncan, A.M., Phipps, W.R. and Kurzer, M.S. (2003). Phyto-oestrogens. Best Practice & Research Clinical Endocrinology and Metabolism 17(2): 253–271.

Egounlety, M. and Aworh, O. (2003). Effect of soaking, de-hulling, cooking and fermentation with *Rhizopus oligosporus* on the oligosaccharides, trypsin inhibitor, phytic acid and tannins of soybean (*Glycine max* Merr.), cowpea (*Vigna unguiculata* L. Walp) and ground bean (*Macrotyloma geocarpa* Harms). Journal of Food Engineering 56: 249–254.

Endo, A., Hasumi, K. and Negishi, S. (1985). Monacolins J and L, new inhibitors of cholesterol biosynthesis produced by *Monascus ruberę*. Journal of Antibiotics 38: 420–422.

Esaki, H., Kawakishi, S., Morimitsu, Y. and Osawa, T. (1999). New potent anti-oxidative o-dihydroxyisoflavones in fermented Japanese soybean products. Bioscience, Biotechnology, and Biochemistry 63: 1637–1639.

Farkas, J. (1988). Biochemistry of wine. Gordon and Breach Science Publisher S.A. Montreux, Swizerland, vol. 3: pp. 423.

Feher, J., Lengye, G. and Lugasi, A. (2007). The cultural history of wine—theoretical background to wine therapy. Central European Journal of Medicine 2(4): 379–391.

Feng, X.M., Eriksson, A.R.B. and Schurer, J. (2005). Growth of lactic acid bacteria and *Rhizopus oligosporus* during barley tempeh fermentation. International Journal of Food Microbiology 104: 249–256.

Firmesse, O., Alvaro, E., Mogenet, A., Bresson, J.L., Lemée, R., Le Ruyet, P., Bonhomme, C., Lambert, D., Andrieux, C., Doré, J., Corthier, G., Furet, J.P. and Rigottier-Gois, L. (2008). Fate and effects of Camembert cheese micro-organisms in the human colonic microbiota of healthy volunteers after regular Camembert consumption. International Journal of Food Microbiology 125(2):176–181.

Fudiyansyah, N., Petterson, D., Bell, R. and Fairbrother, A. (1995). A nutritional, chemical and sensory evaluation of lupin (*L. angustifolius*) tempe. International Journal of Food Science & Technology 30(3): 297–305.

Galvalisi, U., Lupo, S., Piccini, J. and Bettucci, L. (2012). *Penicillium* species present in Uruguayan salami. Revista Argentina de Microbiologia 44(1): 36–42.

Geisen, R., Lucke, F.K. and Krockel, L. (1992). Starter and protective cultures for meat and meat products. Fleischwirtsch 72(6): 894–898.

Gibbs, B.F., Zougman, A., Masse, R. and Mulligan, C. (2004). Production and characterization of bioactive peptides from soy hydrolysate and soy-fermented food. Food Research International 37: 123–131.

Grajek, W., Olejnik, A. and Sip, A. (2005). Probiotics, prebiotics and antioxidants as functional foods. Acta Biochimica Polonica 52(3): 665–671.

Gruber, C., Tschugguel, W., Schneeberger, C. and Huber, J. (2002). Production and actions of estrogens. The New England Journal of Medicine 346(5): 340–352.

Han, K., Soares, J., Haidar, M., Lima, G. and Baracat, E. (2002). Benefits of soy isoflavone therapeutic regimen on menopausal symptoms. Obstetrics and Gynecology 99(3): 389–394.
Handoyo, T., Maeda, T., Urisu, A., Adachi, T. and Morita, N. (2006). Hypoallergenic buckwheat flour preparation by *Rhizopus oligosporus* and its application to soba noodle. Food Research International 39: 598–605.
Harbutt, J. (1998). Sery świata – encyclopedia. Wydawnictwo Książkowe Twój Styl, Warszawa, pp. 7–11.
Hoppe, M.B., Jha, H. and Egge, H. (1997). Structure of an antioxidant from fermented soybeans (Tempeh). Journal of the American Oil Chemists' Society 74: 477–479.
Hossain, C.F., Okuyama, E. and Yamazaki, M. (1996). A new series of coumarin derivatives having monoamine oxidase inhibitory activity from *Monascus anka*. Chemical & Pharmaceutical Bulletin 44: 1535–1539.
Hsieh, Y-H., Pao, S. and Li, J. (2008). Traditional chinese fermented foods. pp. 433–463. *In*: R. Farnworth (ed.). Handbook of Functional Fermented Foods. CRC Press, Boca Raton.
Hutchins, A.M., Slavin, J.L. and Lampe, J.W. (1995). Urinary isoflavonoid phytoestrogen and lignan excretion after consumption of fermented and unfermented soy products. Journal of the American Dietetic Association 95: 545–554.
Iacunin, L., Ciesa, L., Boscolo, D., Manzano, M., Cantoni, C., Orlic, S. and Comi, G. (2009). Moulds and oxratoxin A on surfaces of artisinal and industrial dry sausages. Food Microbiology 26: 65–70.
Jasińska, I. and Michniewicz, J. (2005). Fitoestrogeny – znaczenie i metody identyfikacji. Aparatura Badawcza i Dydaktyczna X(2): 120–127.
Jongrungruangchok, S., Kittakoop, P., Yongsmith, B., Bavovada, R., Tanasupawat, S., Lartpornmatulee, N. and Thebtaranonth, Y. (2004). Azaphilone pigments from a yellow mutant of the fungus *Monascus kaoliang*. Phytochemistry 65: 2569–2575.
Journoud, M. and Jones, J.H. (2004). Red yeast rice: A new hypolipidemic drug. Life Sciences 74: 2675–2683.
Karmini, M., Affandi, E., Hermana, H., Karyadi, D. and Winarno, F. (1997). The inhibitory effect of tempe on *Escherichia coli* infection. International Tempe Symposium, Bali, Indonesia.
Karyadi, D. and Lukito, W. (2000). Functional food and contemporary nutrition-health paradigm: tempeh and its potential beneficial effects in disease prevention and treatment. Nutrition 16: 697.
Karyadi, D. and Lukito, W. (1996). Beneficial effects of tempeh in disease prevention and treatment. Nutrition Reviews 54: S94–S98.
Kasaoka, S., Astuti, M., Uehara, M., Suzuki, K. and Goto, S. (1997). Effect of Indonesian fermented soybean tempeh on iron bioavailability and lipid peroxidation in anemic rats. Journal of Agricultural and Food Chemistry 45: 195–198.
Kataoka, S. (2005). Functional effects of Japanese style fermented soy sauce (shoyu) and its components. Journal of Bioscience and Bioengineering 100: 227–234.
Keuth, S. and Bisping, B. (1993). Formation of vitamins by pure cultures of tempe moulds and bacteria during the tempe solid substrate fermentation. Journal of Applied Bacteriology 75: 427–434.
Kiers, J.L., Meijer, J.C., Nout, M.J.R., Rombouts, F.M., Nabuurs, M.J.A. and Meulen, J. (2003). Effect of fermented soya beans on diarrhea and feed efficiency in weaned piglets. Journal of Applied Microbiology 95: 545–552.
Kiers, J.L., Nout, M.J.R., Rombouts, F.M., Nabuurs, M.J.A. and Meulen, J. (2002). Inhibition of adhesion of enterotoxigenic *Escherichia coli* K88 by soya bean tempe. Letters in Applied Microbiology 35: 311–315.
Ko, S.D. and Hesseltine, C.W. (1979). Tempe and related foods. pp. 115–140. *In*: A.H. Rose (ed.). Economic Microbiology, Vol. 4, Microbial Biomass. Academic Press, London.
Kobayashi, M. (2005). Immunological functions of soy sauce: hypoallergenicity and antiallergic activity of soy sauce. Journal of Bioscience and Bioengineering 100(2): 144–151.

Kobayasi, S., Okazaki, N. and Koseki, T. (1992). Purification and characterization of an antibiotic substance produced from *Rhizopus oligosporus* IFO 8631. Bioscience, Biotechnology, and Biochemistry 56(1): 94–98.

Kovac, B. and Raspor, B. (1997). The use of the mould *Rhizopus oligosporus* in food production. Food Technology and Biotechnology 35: 69–73.

Kraszewska, O., Nynca, A., Kamińska, B. and Ciereszko, R. (2007). Phytoestrogens. I. Occurrence, metabolism and biological effects in females. PostępyBiologii Komórki 34(1): 189–205.

Kuligowski, M. (2009). Analysis of functional properties products fermented by *Rhizopus oligosporus*. Ph.D. thesis, Poznań University of Life Sciences, Poznań, Poland, pp. 69–121.

Kuligowski, M. and Nowak, J. (2007). The possibility of modeling functional properties of legume food products through the tempeh type fermentation. Biotechnologia 4(79): 113–124.

Kuligowski, M. and Nowak, J. (2006). Antibacterial activity of isolates from after growth moulds *Rhizopus oligosporus*. Food Science. Technology. Quality 2(47) (Supp.l): 182–189.

Kuligowski, M., Pawłowska, K., Jasińska-Kuligowska, I. and Nowak, J. (2016). Isoflavone composition, polyphenols content and antioxidative activity of soybean seeds during tempeh fermentation. CyTA-Journal of Food (in press).

Kumari, H.P.M., Naidub, K.A., Vishwanathab, S., Narasimhamurthy, B, K. and Vijayalakshmi, G. (2009). Safety evaluation of *Monascus purpureus* red mould rice in albino rats. Food and Chemical Toxicology 47(8): 1739–1746.

Kustrzewa-Tarnawska, A. (2007). Chemoprewencja nowotworów hormonozależnych za pomocą przeciwutleniaczy. pp. 295–310. *In*: W. Grajek (ed.). Przeciwutleniacze w żywności, Aspekty zdrowotne, technologiczne, molekularne i analityczne. Wydawnictwo Naukowo-Techniczne, Warszawa.

Kwak, N. and Jukes, D. (2001). Functional foods. Part 2: the impact on current regulatory terminology. Food Control 12: 109–117.

Lai, H.H. and Yen, G.C. (2002). Inhibitory effect of isoflavones on peroxynitrite-mediated low-density lipoprotein oxidation. Bioscience, Biotechnology, and Biochemistry 66(1): 22–28.

Lay, C., Sutren, M., Lepercq, P., Juste, C., Rigottier-Gois, L., Lhoste, E., Lemée, R., Le Ruyet, P., Doré, J. and Andrieux, C. (2004). Influence of Camembert consumption on the composition and metabolism of intestinal microbiota: a study in human microbiota-associated rats. British Journal of Nutrition 92: 429–438.

Liener, I.E. (1994). Implications of antinutritional components in soybean foods. Critical Reviews in Food Science and Nutrition 34: 31–67.

Lin, C.-P., Lin, Y.-L., Huang, P.-H., Tsai, H.-S. and Chen, Y.-H. (2011). Inhibition of endothelial adhesion molecule expression by *Monascus purpureus*-fermented rice metabolites, monacolin K, ankaflavin, and monascin. Journal of the Science of Food and Agriculture 91: 1751–1758.

Luh, B.S. (1995). Industrial production of soy sauce. Journal of Industrial Microbiology 14: 467–471.

Martinkova, L., Jůzlova, P. and Veselý, D. (1995). Biological activity of polyketide pigments produced by the fungus *Monascus*. Journal of Applied Bacteriology 79: 609–616.

Mattila-Sandholm, T. and Saarela, M. (2003). Functional dairy foods. International Dairy Journal 13: 1003.

Mazur, W., Fotsis, T., Wahala, K., Ojala, S., Salakka, A. and Adlercreutz, H. (1996). Isotope dilution gas chromatographic–mass spectrometric method for the determination of isoflavonoids, coumestrol, and lignans in food samples. Analytical Biochemistry 233: 169–180.

Messina, M. (2003). Potential public health implications of the hypocholesterolemic effects of soy protein. Nutrition 19(3): 280–281.

Messina, M. (2000). Soyfoods and soybean phyto-oestrogens (isoflavones) as possible alternatives to hormone replacement therapy (HRT). European Journal of Cancer 36: S71–S77.

Messina, M. (1999). Soy, soy phytoestrogens (isoflavones), and breast cancer. American Journal of Clinical Nutrition 70: 574–575.

Messina, M.J., Persky, V., Setchell, K.D. and Barnes, S. (1994). Soy intake and cancer risk: A review of the *in vitro* and *in vivo* data. Nutrition and Cancer 21: 113–131.

Minamiyama, Y. and Okada, S. (2008). Miso: Production, properties, and benefits to health. pp. 321–332. *In*: R. Farnworth (ed.). Handbook of Functional Fermented Foods. CRC Press, Boca Raton.

Mintzlaff, H.J. and Leistner, L. (1972). Untersuchungen zur Selektion eines technologisch geeigneten und toxikologisch unbedenklichen Schimmelpilzstammes fur die Rohwurstherstellung. Zentralblatt furveterinarmedizin reihe b-journal of veterinary medicine series b-infectious diseases immunology food hygiene veterinary public health 19: 291–300.

Mugula, J.K. and Lyimo, M. (2000). Evaluation of the nutritional quality and acceptability of sorghum based tempe as potential weaning foods in Tanzania. International Journal of Food Sciences and Nutrition 51: 269–277.

Nakajima, N., Nozaki, N., Ishihara, K., Ishikawa, A. and Tsuj, H. (2005). Analysis of isoflavone content in tempeh, a fermented soybean, and preparation of a new isoflavone-enriched tempeh. Journal of Bioscience and Bioengineering 100(6): 685–687.

Namwong, S., Hiraga, K., Takada, K., Tsunami, M., Tanasupawat, S. and Oda, K. (2006). A halophilic serine proteinase from *Halobacillus* sp. SR5-3 isolated from fish sauce: purification and characterization. Bioscience, Biotechnology, and Biochemistry 70: 1395–1401.

Nout, M.J.R. and Kiers, J.L. (2005). Tempe fermentation, innovation and functionality: update into the third millennium. Journal of Applied Microbiology 98: 789–805.

Nout, M.J.R. and Rombouts, F.M. (1990). Recent development in tempe research. Journal of Applied Bacteriology 69: 609–633.

Nowak, J. and Szebiotko, K. (1992). Some biochemical changes during soybean and pea tempeh fermentation. Food Microbiology 9: 37–43.

Nowak, J. (1992b). Oats tempeh. Acta Biotechnologica 12: 345–348.

Nowak, J. (1992a). Effect of pea and soybean extracts on growth of 5 *Clostridium* strains. Acta Biotechnologica 12(6): 521–525.

Odenbreit, S., Püls, J., Sedlmaier, B., Gerland, E., Fischer, W. and Haas, R. (2000). Translocation of *Helicobacter pylori* CagA into gastric epithelial cells by type IV secretion. Science 287: 1497–1500.

Ohara, M., Lu, H., Shiraki, K., Ishimura, Y., Uesaka, T., Katoh, O. and Watanabe, H. (2001). Radioprotective effects of miso (fermented soy bean paste) against radiation in B6C3F1 mice: increased small intestinal crypt survival, crypt lengths and prolongation of average time to death. Hiroshima Journal of Medical Sciences 50: 83–86.

Ogasawara, M., Yamada, Y. and Egi, M. (2006). Taste enhancer from the long-term ripening of miso (soybean paste). Food Chemistry 99: 736–741.

Panat, N.A., Maurya, D.K., Ghaskadbi, S.S. and Sandur, S.K. (2016). Troxerutin, a plant flavonoid, protects cells against oxidative stress-induced cell death through radical scavenging mechanism. Food Chemistry 194: 32–45.

Paredes-Lopez, O. (1992). Nutrition and safety considerations. pp. 153–156. *In*: Applications of Biotechnology to Traditional Fermented Foods. Report of an Ad Hoc Panel of the Board on Science and Technology for International Development. National Academy Press, Washington.

Pereira, I.R., Faludi, A.A., Aldrighi, J.M., Bertolami, M.C., Saleh, M.H., Silva, R.A., Nakamura, Y., Campos, M.F., Novaes, N. and Abdalla, D.S. (2006). Effects of soy germ isoflavones and hormone therapy on nitric oxide derivatives, low-density lipoprotein oxidation, and vascular reactivity in hypercholesterolemic postmenopausal women. Menopause 13(6): 942–950.

Piecyk, M., Klepacka, M. and Worobiej, E. (2005). The content of trypsin inhibitors, oligosccharides, and phytic acid in the bean seed (*Phaseolus vulgaris*) preparations obtained by crystallization and classical isolation. Food Science Technology Quality 3(44): 92–104.

Pour Nikfardjam, M.S., Laszlo, G. and Dietrich, H. (2006). Resveratrol-derivatives and antioxidative capacity in wines made from *botrytized* grapes. Food Chemistry 96: 74–79.

Poussier, B., Cordova, A.C., Becquemin, J.-P. and Sumpio, B.E. (2005). Resveratrol inhibits vascular smooth muscle cell proliferation and induces apoptosis. Journal of Vascular Surgery 42(6): 1190–1190.e14.

Prajapati, J.B. and Nair, B.M. (2008). The history of fermented foods. pp. 2–24. *In*: R. Farnworth (ed.). Handbook of Functional Fermented Foods. CRC Press, Boca Raton.

Randhir, R., Vattem, D. and Shetty, K. (2004). Solid-state bioconversion of fava bean by *Rhizopus oligosporus* for enrichment of phenolic antioxidants and L-DOPA. Innovative Food Science and Emerging Technologies 5: 235–244.

Reddy, B.S., Sharma, C., Darby, L., Laakso, K. and Wynder, E.L. (1980). Metabolic epidemiology of large bowel cancer. Fecal mutagens in high- and low-risk population for colon cancer: A preliminary report. Mutation Research 72: 511–522.

Reyes-Moreno, C., Romero-Urias, C.A., Milan-Carrillo, J. and Gomes-Garza, R.M. (2000). Chemical composition and nutritional quality of fresh and hardened chickpea (*Cicer arietinum* L.) after the solid state fermentation (SSF). Food Science and Technology International 6: 251–258.

Robert, D. (1982). Bacteria of public health significance. pp. 356–367. *In*: M.H. Brown (ed.). Meat Microbiology. Applied Science, London.

Robinson, R.K. and Tamime, A.Y. (1990). Microbiology of fermented milks. pp. 291–343. *In*: R.K. Robinson (ed.). Dairy Microbiology–The Microbiology of Milk Products, Vol. 2, 2nd Edition. Elsevier Applied Science Publisher, London.

Rosenkranz, S., Knirel, D., Dietrich, H., Flesch, M., Erdmann, E. and Bohm, M. (2002). Inhibition of the PDGF receptor by red wine flavonoids provides a molecular explanation for the French paradox. Journal of the Federation of American Societies for Experimental Biology 16: 1958–1960.

Roubos van den Hil, P., Dalmas, E., Nout, R. and Abee, T. (2008). Inactivation of *Bacillus cereus* cells and spores by low molecular weight proteinaceous compounds extracted from soybean tempeh. The 21st International ICFM Symposium Evolving microbial food quality and safety Food Micro (2008), Aberdeen, pp. 107.

Rowland, I., Wiseman, H., Sanders, T., Adlercreutz, H. and Bowey, E. (1999). Metabolism of oestrogens and phytoestrogens: role of the gut microflora. Biochemical Society Transactions 27: 304–308.

Rozwandowicz, A. (2007). Tempeh fermentation as the process improving the nutritional quality of legume seeds. Ph.D. thesis, Technical University of Lodz, Łódź, pp. 69.

Shurtleff, W. and Aoyagi, A. (1979). The Book of Tempeh. Ten Speed Press, Berkeley, California, pp. 111.

Steinkraus, K.H. (1997). Classification of fermented foods: worldwide review of household fermentation techniques. Food Control 8(5-6): 311–317.

Steinkraus, K.H. (1996). Handbook of Indigenous Fermented Foods. Marcel Dekker Inc., New York, New York.

Stilling, B.R. and Hackler, L.R. (1965). Amino acid studies on the effect of fermentation time and heat-processing of tempeh. Journal of Food Science 30: 1043–1048.

Sugawara, E. and Sakurai, Y. (1999). Effect of media constituents on the formation by halophilic yeast of the 2 (or 5)-ethyl-5 (or 2)-methyl-4-hydroxy-3 (2H)-furanone aroma component specific to miso. Bioscience, Biotechnology, and Biochemistry 63: 749–752.

Sunesen, L.O. and Stahnke, L.H. (2003). Mould starter cultures for dry sausages-selection, application and effects. Meat Science 65: 935–948.

Suzuki, H., Sonne, H., Kawamura, K. and Ishihara, K. (2008). Liver injury due to 3-amino-1-methyl-5H-pyrido[4,3-*b*] indole (Trp-P-2) and its prevention by miso. Bioscience, Biotechnology, and Biochemistry 72(8): 2236–2238.

Takano, T. (2002). Anti-hypertensive activity of fermented dairy products containing biogenic Peptides. Antonie van Leeuwenhoek 82: 333–340.

Tanasupawat, S. and Visessanguan, W. (2008). Thai fermented foods. Microorganisms and Their Health Benefits. pp. 495–506. *In*: R. Farnworth (ed.). Handbook of Functional Fermented Foods. CRC Press, Boca Raton.

Tanasupawat, S., Shida, S., Okada, S. and Komagata, K. (2000). *Lactobacillus acidipiscis* sp. nov. and *Weissella thailandensis* sp. nov. isolated from fermented fish in Thailand. International Journal of Systematic and Evolutionary Microbiology 50: 1479–1485.

Tanasupawat, S., Okada, S. and Komagata, K. (1998). Lactic acid bacteria found in fermented fish in Thailand. The Journal of General and Applied Microbiology 44: 193–200.

Tanasupawat, S., Hashimoto, Y., Ezaki, T., Kozaki, M. and Komagata, K. (1992). *Staphylococcus piscifermentans* sp. nov. from fermented fish in Thailand. International Journal of Systematic Bacteriology 42: 577–581.

Tanasupawat, S., Hashimoto, Y., Ezaki, T., Kozaki, M. and Komagata, K. (1991). Identification of *Staphylococcus carnosus* strains from fermented fish and soy sauce mash. The Journal of General and Applied Microbiology 37: 479–494.

Tsugane, S., Tei, Y., Takahashi, T., Watanabe, S. and Sugano, K. (1994). Salty food intake and risk of *Helicobacter pylori* infection. Japanese Journal of Cancer Research 85: 474–478.

Wang, H.L., Ruttle, D.I. and Hesseltine, C.W. (1969). Antibacterial compound from a soybean product fermented by *Rhizopus oligosporus*. Proceedings of the Society for Experimental Biology and Medicine 131: 579–583.

Watanabe, H. (2013). Beneficial biological effects of miso with reference to radiation injury, Cancer and Hypertension. Journal of Toxicologic Pathology 26: 91–103.

Watanabe, H., Kashimoto, N., Kajimura, J. and Kamiya, K. (2006). A miso (Japanese soybean paste) diet conferred greater protection against hypertension than a sodium chloride diet in Dahl salt-sensitive rats. Hypertension Research 29: 731–738.

Wiesel, I., Rehm, H.J. and Bisping, B. (1997). Improvement of tempe fermentations by application of mixed cultures consisting of *Rhizopus* sp. and bacterial strains. Applied Microbiology and Biotechnology 47: 218–225.

www.lactalis.pl

Yamamoto, S., Sobue, T., Kobayashi, M., Sasaki, S. and Tsugane, S. (2003). Soy, isoflavones, and breast cancer risk in Japan. Journal of the National Cancer Institute 95: 906–913.

Yan, S. (1989). A socio-medical study of adult diseases related to the life style of Chinese in Japan. Nippon Eiseigaku Zasshi (Japanese Journal of Hygiene) 44: 877–886.

4

Legume-based Food Fermentation
Biochemical Aspects

Amit Kumar Rai[1] and Kumaraswamy Jeyaram[2],*

1. Introduction

Fermentation is an age-old practice of food preservation and improvement of food flavour, appearance, texture and nutritional properties. Legumes based fermented foods are very popular in Asian countries as well as large parts of Africa as the main source of protein (Reddy et al. 1983). There is a wide-ranging list of fermented legumes that is being collated from Asian countries. There is diversity in the art of preparation and microbiota involved, which reflects on flavour and nutritional properties. Legumes fermentation with different starter cultures including *Bacillus* spp., fungi and lactic acid bacteria (LAB) results in different types of products. The biological process of fermentation results in several biochemical changes by enzymes produced by microorganisms involved in fermentation (McCue and Shetty 2003, Rai and Jeyaram 2015). The fermentation process not only improves the organoleptic and nutritional properties, but also enhances health promoting properties (Zhang et al. 2006, Sanjukta et al. 2015) and

[1] Institute of Bioresources and Sustainable Development, Sikkim Centre, Tadong, Sikkim, India.
[2] Microbial Resources Division, Institute of Bioresources and Sustainable Development, Imphal, Manipur, India.
* Corresponding author: jeyaram.ibsd@nic.in

reduces antinutritional components (Egounlety and Aworh 2003, Osman 2011, Difo et al. 2014). The bioactive compounds are produced either by the transformation of legume components by the microbial enzymes or by the microbes during fermentation.

The recent research interest in the microbial transformation of food components during fermentation is becoming popular due to the enhancement of components responsible for health benefits (Vlieg et al. 2011). In the recent past, peptides and polyphenols in fermented legumes have attracted many researchers across the globe. These compounds have potentials in reducing the risk of several metabolic disorders (Zhang et al. 2006, Cho et al. 2011, Rai et al. 2015, Sanjukta et al. 2015). The fermentation process in legumes results in the formation of peptides and free polyphenols due to the hydrolysis of protein and bound polyphenols, respectively (Cho et al. 2011, Sanjukta et al. 2015). Fermentation also results in the production of bioactive compounds by the microbes involved in the fermentation. These compounds include antibacterial peptides, lipopeptides, surfactins, mucilage and fibrinolytic enzymes (Cao et al. 2009, Xiao-Hong et al. 2009, Singh et al. 2014). Fermented legumes have been reported to exhibit several health benefits such as antioxidant, antibacterial, antihypertensive, lipid lowering, anti-diabetic, immunomodulatory and anticancer properties (Hwang et al. 1997, Cao et al. 2009, Kwon et al. 2011, Sanjukta et al. 2015). This chapter focuses on biochemical changes that occur during legumes fermentation and their impact on several functional properties.

2. Traditional Legume-Based Fermented Foods

Leguminous seeds are being fermented in their traditional way of preparation in several countries. Among these, soybean is most commonly used because it is widely cultivated and is one of the richest sources of plant protein, along with several bioactive components, and *Bacillus* spp. are the key fermenting bacteria in many of these soybeans products (*natto, kinema, chungkookjang*); some by solely a fungus starter (*sufu, tempeh, douchi*) and in some cases by a mix (*doenjang*). The popular Asian fermented soybeans products are *natto, tofuyo, miso* in Japan; *douchi, doubanjiang, sufu* in China; *chungkookjang, doenjang, meju* in Korea; *tempeh* in Indonesia; *thua-nao* in Thailand; *kinema, hawaijar, tungrymbai* and *bekang* in India (Jeyaram et al. 2008, He and Chan 2013, Tamang et al. 2015). Fermented soymilk is also prepared by using LAB for the improvement of functional properties (Pyo et al. 2005, Wang et al. 2006a). The other legume seeds used for fermentation includes African locust bean (*dawadawa, iru, kinda, soumbala, uri*), Bengal gram (*dhokla, khaman*), black gram (*bhallae, maseura, vadai, papad, wari*), velvet bean (*tempe*), jack bean (*tempe*) and African oil bean (*ugba*) (Tamang et al. 2015). *Iru* or *dawadawa* is a fermented African locust bean (*Parkia biglobosa*) cotyledon consumed in South and Northern

Nigeria (Odunfa 1983, Achi 2005, Adewumi et al. 2013, 2014). *Okpehe* is fermented *Prosopis africana* seeds which are popularly consumed in Nigeria (Oguntoyinbo et al. 2007). *Soumbala* is an alkaline fermented African locust bean consumed in Burkina Faso and other countries of West Africa (Sarkar et al. 2002). Some other lesser known fermented legumes include fermented *Cathormion altissionum* seeds (*Oso*) (Jolaoso et al. 2014).

3. Enzymes Responsible for the Biochemical Changes

Legumes are a rich source of several anti-nutritional factors, macromolecules, and phytochemicals linked to these macromolecules. The microbes involved in legume fermentation produce many enzymes, which hydrolyse or transform the legume components. A wide range of microbes such as *Bacillus* spp., LAB, molds and yeasts involved in legume fermentation are potential candidates for hydrolytic enzyme production (Gibbs et al. 2004, Donkor et al. 2005, Pyo and Lee 2007, Sanjukta et al. 2015). *Bacillus* spp. have been reported to produce proteases, amylase, lipase, galactanase, galactosidase, fructofuranosidase and phytase (Achi 2005, Cho et al. 2011, Hur et al. 2014). Similarly, molds such as *Rhizopus* spp., *Mucor* spp. and *Aspergillus* spp. are well known for producing several industrially important enzymes such as proteases, amylase, cellulases, glucoamylase, β-glycosidase, xylanase, mannanase, pectinases, phytases, β-xylosidase (Achi 2005, Chancharoonpong et al. 2012, Hur et al. 2014). These enzymes produced by the starter culture during fermentation are responsible for the hydrolysis of complex macromolecules into their monomeric form (e.g., proteases, amylases, cellulose, lipases), degradation of antinutritional factors such as phytic acid (e.g., phytase) and degradation of bound polyphenols to free form (e.g., β-glucosidase) (Egounlety and Aworh 2003, McCue and Shetty 2003, Cho et al. 2011, Rai and Jeyaram 2015).

Production of the enzymes during legume fermentation has been correlated either with the growth of the starter culture or release of bioactive components produced during fermentation (Zhang et al. 2007, Cho et al. 2011, Chancharoonpong et al. 2012). Several researchers have suggested that the increase in free phenolics content during soybean fermentation is due to β-glucosidase production (Kim and Yoon 1999, Sohn et al. 2000, Moktan et al. 2008, Cho et al. 2011). In another study, McCue and Shetty (2003) have suggested the role of carbohydrate degrading enzymes (α-amylase, α- and β-glucosidase, and β-glucuronidase) on the increase in phenolic antioxidants. Normally, enzyme production during legume fermentation correlates with the growth of the starter culture. Chancharoonpong et al. (2012) have studied the activity of neutral protease, alkaline protease and amylase activities during *koji* fermentation and correlated with the growth of *Aspergillus oryzae*. Production of the specific enzyme during legume fermentation depends on the microbes associated with legume fermentation. During African locust bean fermentation, Odunfa (1985) detected protease and lipase activity but did not detect amylase activity. Similarly, *Rhizopus* species during soybean fermentation produces a variety of carbohydrate degrading enzymes such as α-D galactosidase, α-D glucosidase (Nout and Rombouts 1990) but lesser amount of amylase (Hesseltine 1965).

Protein degrading enzymes during fermentation are gaining more popularity due to the increasing interest in bioactive peptides. The enzymes produced by the starter culture decide the type and content of the hydrolyzed products present in the final product. Different strains of the same phylotype may also form a different type of peptides during fermentation (Sanjukta et al. 2015). Lipase produced during fermentation results in the breakdown of glycerides into fatty acids. Apart from biochemical changes during legume fermentation, there are enzymes in the finished fermented product that are held responsible for many health benefits. Many fibrinolytic enzymes with potential application in thrombolytic therapy are present in the fermented legume products (Rai and Jeyaram 2015). Many strains of *Bacillus* spp. and LAB with fibrinolytic activity were isolated from various fermented foods (Mine et al. 2005, Wang et al. 2006a, 2009). *Bacillus subtilis* associated with *natto*, *douchi* and *hawaijar* fermentation have shown to possess fibrinolytic enzymes with higher activity (Fujita et al. 1993, Peng et al. 2003, Singh et al. 2014). Consumption of fermented legumes having fibrinolytic enzymes can be a potential alternative tool for the management of cardiovascular diseases (Rai and Jeyaram 2015).

4. Biochemical Changes During Pre-Treatment

Biochemical changes in legumes start during pre-treatment steps of fermented legume production. Pre-treatment or traditional processing steps such as de-hulling, soaking, washing and cooking have been associated with a reduction of several antinutritional factors (Khattab and Arntfield 2009, Kalpanadevi and Mohan 2013). The pre-treatment steps significantly affected the concentration of oligosaccharides. A reduction of about 50% of raffinose and more than 60% of sucrose and stachyose was reported during pre-treatment steps (Egounlety and Aworh 2003). Among the pre-treatment steps, cooking was the most successful in reducing the activity of trypsin inhibitor. A reduction in the activity of trypsin inhibitor by 82.2%, 86.1% and 76.2% in soybean, cowpea and ground bean was witnessed after cooking for 30, 7 and 15 min, respectively (Egounlety and Aworh 2003). In another study, cooking and germination of chickpeas resulted in significant reduction in carbohydrate fractions, antinutritional factors, vitamins and minerals (El-Adawy 2002).

Recently, Sharma et al. (2015) have optimised traditional processing (soaking and cooking) of soybean to achieve a minimum level of antinutritional factors in fermented soybean (*kinema*) by response surface methodology. In their study, they found a maximum reduction of tannins, phytic acid and trypsin inhibitors when the seeds were soaked at a ratio of (1:10) at 10°C for 20 h. However, optimum cooking for reducing tannins, phytic acid and trypsin inhibitors attained in soybean seeds by pressure cooking (beans: water - 1:5) was at cooking pressure of 1.10 kg cm^{-2} for 20 min (Sharma et al. 2015). Apart from antinutritional factors, pre-treatment steps also affect the content of bioactive compounds. Processing steps such as soaking, boiling and steaming affects the total phenolic content and antioxidant activity of chickpea, yellow pea and lentil (Xu and Chang 2008). In another study, Prondanov et al. (2004) showed that cooking and soaking of legumes such as chickpeas, lentils and faba bean resulted in the loss of vitamins such as thiamine, riboflavin and niacin. Thus, biochemical changes during pre-treatment steps of legume fermentation not only reduce the antinutritional factors but may also reduce some of the beneficial nutrients.

5. Biochemical Changes During Legume Fermentation

There are several biochemical changes occurred during legume fermentation, which have a positive or negative impact on the final fermented product. These biochemical changes are due to the action of enzymes and metabolites produced by the starter culture on the macro- and micro-molecules in legumes (Fig. 1). The biochemical changes include hydrolysis of macromolecules, biotransformation of bioactive components, reduction of antinutritional factors and production of microbial compounds.

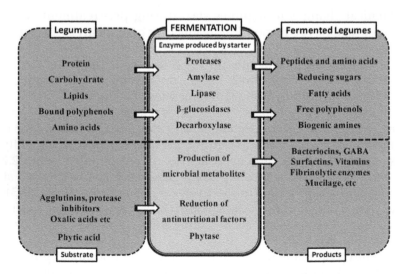

Figure 1. Biochemical changes associated with legume fermentation.

The changes in legume components during fermentation are discussed in the following section.

5.1 Proteins

Legumes are a rich source of proteins composed of essential amino acids, which are not only of nutritional importance but are the precursor of several biologically active peptides. Legume seeds accumulate large amounts of proteins (storage proteins) which could undergo proteolysis or hydrolysis to yield polypeptide and free amino acids that are physiologically relevant. Major storage proteins of legume seeds are oligomeric globulins that, according to their sedimentation coefficients, belong to the two groups of 7S and 11S proteins. The former are trimers of about 150 kDa, stabilized by hydrophobic interactions, electrostatic and hydrogen bonds while 11S proteins are hexamers of about 450 kDa (Carbonaro et al. 2015). Also, they contain some minor proteins/peptides such as amylase inhibitors, protease, lipoxygenase and lectins that are relevant to the nutritional and functional quality of the seed (Duranti 2006).

   Changes in legume proteins occur during fermentation by the action of proteases produced by the starter culture. The extent of hydrolysis and composition of free amino acids in the fermented products depends on the starter and legume variety used for fermentation (Sanjukta and Rai 2016). Increase in the free amino acid profile during fermentation has also been reported in *kinema* (Nikkuni et al. 1995), *thua nao* (Dajanta et al. 2011), *natto* (Nikkuni et al. 1995), *iru* (Odunfa 1985), *sufu* (Han et al. 2001) and *cheonggukjang* (Hong et al. 2008). Specific proteolytic enzymes produced by the starter culture strain leads to the generation of peptides with specific amino acid content and sequence. Changes in protein profile during African locust bean fermentation by *Bacillus subtilis* and *B. pumilus* strains for *soumbala* production was studied by Ouoba et al. (2003). They showed that the free amino acids profile of the final product was specific for each starter used for fermentation. Recently, Sanjukta et al. (2015) studied the effect of two proteolytic strains of *B. subtilis* (MTCC 1747 and MTCC 5480) and found a difference in free amino acid profile in the final product. In recent years, bioactive peptides resulting from hydrolysis of legume proteins during fermentation are gaining more popularity (Sanjukta and Rai 2016). Bioactive peptides are a chain of amino acids, which depending on their size and sequence of amino acids exerts several physiological effects.

5.2 Carbohydrate

Carbohydrate content in food legumes ranges from 22–68%, with starch being the major carbohydrate ranging from 22–45% (Hoover and Zhao

2003). Most of the legumes also contain nondigestible carbohydrates such as arabinogalactan, stachyose and raffinose (Odunfa et al. 1983). Microbes associated with legume fermentation have the ability to hydrolyze a broad range of carbohydrates. Hydrolysis of polysaccharides such as starch results in the formation of monosaccharides, which are readily digested by humans. Natural fermentation of lentil has shown that starch, sucrose, raffinose, stachyose and ciceritol content were reduced during fermentation (Vidal-Valverde et al. 1993). Similarly, stachyose content reduced by 91.5, 83.9, and 85.5%, respectively in cowpea, soybean and ground bean fermentation by *R. oligosporus* during *tempeh* production, whereas no significant change in raffinose content was observed. Galactose, glucose, fructose, melibiose and maltose content were found to increase during the first 30 and 36 hours of fermentation of cowpea and ground beans, respectively, which decreased on further fermentation (Egounlety and Aworh 2003). *Rhizopus* spp. have the ability to produce several carbohydrates degrading enzyme such as α-D galactosidase, α-D glucosidase and amylases, which helps in degradation of oligosaccharides and starch (Nout and Rombouts 1990, Egounlety and Aworh 2003). Apart from the changes in carbohydrate content in foods, the microbes associated with the fermentation process such as *B. subtilis*, *Lactobacillus* spp. also produce exopolysaccharides, which are also known to possess health benefits such as antioxidant and antitumour properties (Abdel-Fattah et al. 2012).

5.3 Lipids

Storage lipids in the legume seeds are the source of dietary fats, which are prone to changes during fermentation. These changes in the lipids components can determine the sensory properties and functionality of the final product. Total lipid content in legume seeds and their fatty acid composition can affect a range of quality attributes and health related benefits. Fermented soybeans products in many Asian countries are rich in linoleic acid (Sarkar et al. 1996) that is proven to possess protective effect on cardiovascular diseases. However, some of the reports suggest that triglycerides are hydrolyzed to free fatty acids during fermentation. In a recent study, Feng et al. (2014) have shown that triglyceride content during *koji* fermentation decreased significantly with the increase in free fatty acids, monoacyl glycerides and diacyl glycerides content. Similarly, increase in free fatty acids during fermentation was also established in *kinema* (Sarkar et al. 1996), *natto* (Kiuchi et al. 1976) and *tempeh* (Wagenknecht et al. 1961). Interestingly, the acid value was found to increase, and peroxide values found to reduce during *koji* (Feng et al. 2014) and chickpea fermentation (Paredes-López et al. 1991), which might be due to the preventing effect of antioxidants produced during the fermentation. *Tempeh* fermentation has shown to affect lipids composition due to hydrolysis of triglycerides into free fatty acids and glycerols (Nout and Keirs 2005). Also, the changes in lipids due to lipolysis during fermentation followed by secondary reactions of free fatty acids can result in the development of various oxidised products such as aldehydes, alcohols and ketones that are responsible for the flavour and sensorial properties of the fermented products (Visessangaun et al. 2004, Gambacorta et al. 2009, Feng et al. 2014).

5.4 Polyphenols

Polyphenols are one of the common bioactive components in legumes, which have been extensively studied in several legumes based fermented foods for their antioxidant activity. Polyphenols in legumes include compounds such as isoflavones, lignans, flavonols, and phenolic acids depending on the type of legumes. In unfermented legumes, these polyphenols are naturally present in glycosylated form (Cho et al. 2011), which prevents their absorption until it is hydrolyzed to a free form. Fermentation has been reported to increase the free polyphenols content in fermented legumes by the hydrolysis of bound polyphenols (Moktan et al. 2008, Oboh et al. 2008, Cho et al. 2011, Sanjukta et al. 2015). Due to the increase in free polyphenol content, the bioavailability of individual polyphenol increases, which further enhances the health benefits associated with polyphenols. Increase in polyphenols has been reported in fermented legume products such as *kinema* (Moktan et al. 2008), *cheonggukjang* (Cho et al. 2011), *tempe* (Nout and Kiers 2005), fermented African locust beans (Oboh et al. 2008), *thua nao* (Dajanta et al. 2013), fermented lentils (Torino et al. 2013) and fermented soy milk (Wang et al. 2006b).

Recently, Cho et al. (2011) have studied the conversion of individual polyphenols (isoflavones, phenolic acids and flavonols) during *cheonggukjang* fermentation with *B. subtilis*. Among several polyphenols, isoflavones have gained more popularity due to their potential health benefits. Twelve chemical forms of isoflavones are present in soybean. Their fermented forms include aglycones (daidzein, genistein, and glycitein) and glycosides (acetyl-, malonyl-, and β-glycosides). In their study, Cho et al. (2011) have shown that fermentation resulted in significant increase in the content of isoflavone aglycones (daidzein) accompanied by a decrease in isoflavone glucosides (daidzin). A similar report of a rise in aglycone form of isoflavone was reported in *meju* and *doenjang* fermentation (Kim and Yoon 1999). Fermentation has also shown to enhance phenolic acids such as gallic acid and flavanols such as catechin and epicatechin during soybean fermentation by *B. subtilis* (Cho et al. 2011). Thus, the polyphenol profile in the fermented product depends on the legume variety and starter used for fermentation.

5.5 Anti-nutritional Factors

Apart from bioactive components with health benefits, legumes are also a rich source of antinutritional factors. Legume fermentation not only enhances molecules responsible for health benefits but also reduces antinutritional factor presented in unfermented legumes. The antinutritional factors include phytic acids, saponins, α-galactoside, protease inhibitors, α-amylases inhibitors and lectins (Reddy et al. 1985, Urbano et al. 2000, Osman 2004, Udensi et al. 2007). Among all the antinutritional factors, phytic acid is of prime concern for nutrition and health of legume consumers. Fermentation of legumes with phytase producing starter cultures results in a reduction of phytic acid in the final products (Egounlety and Aworh 2003, Sharma et al. 2015). The phytic acid reduction is very much crucial because it reduces the bioavailability of essential minerals. *Tempeh* prepared by fermentations *Rhizopus oligosporus* of three locally grown legumes of West Africa (soybean, cowpea and ground bean) resulted in a reduction in phytic acid (Egounlety and Aworh 2003). In their study, Egounlety and Aworh (2003) have also shown elimination of trypsin inhibitors after 24 hours of cowpea and ground bean fermentation. In another study, feeding *tempe* to iron-deficient rats resulted in higher levels of iron in the liver in comparison to rats fed with unfermented soybean (Kasaoka et al. 1997). Traditionally fermented pearl millets for the preparation of *iohah* (fermented bread) resulted in significant reduction of phytic acid content, trypsin and amylase inhibitors (Osman 2004). In a recent study, Sharma et al. (2015) showed that fermentation of soybean to *kinema* resulted in 43.1 and 2.6% reduction in phytic acid content and trypsin inhibitory activity, respectively, over optimally cooked soybeans. Reduction of antinutritional factors is one of the important aspects during the production of fermented legume products, which starts during the pre-treatment steps and continues till the end of fermentation.

5.6 Biogenic Amines

Biogenic amines are ubiquitous components of protein rich fermented foods and are formed by decarboxylation of amino acids (Kim et al. 2012). Oral intake of biogenic amines may not cause adverse reaction till a particular level as the amine oxidases in the intestine detoxifies the compounds (Aksar and Treptow 1986). However, intoxication may happen if the capacity to metabolize biogenic amines is over saturated or impaired by specific inhibitors (Kim et al. 2012). Presence of biogenic amines in *doenjang* (Shukla et al. 2010), *natto* (Kim et al. 2012) *chunjang* (a black soybean paste), *Jajang* (a black soybean sauce) (Bai et al. 2013), *douche* (Tsai et al. 2007), *tempeh* (Nout et al. 1993) and *miso* (Kung et al. 2007) has been reported. The level of biogenic amines in fermented legumes depends on the type of legume used, microbial strains used and duration of the fermentation period.

Recently, Yang et al. (2014) have studied the biogenic amine content and composition of 30 fermented soybeans (*Aspergillus*-type *douchi*, *Mucor*-type *douchi* and *Aspergillus*-type *douchi*) and 33 fermented bean curd samples (gray *sufu*, red *sufu*, and white *sufu*). The total biogenic amines content in fermented soybean was reported in a relatively safe range in many samples by Yang et al. (2014). However, the concentration of tyramine, histamine, and β-phenethylamine was high enough in some of the fermented soybean products to cause a possible safety threat. Also among 30 samples analysed for biogenic amine content, eight samples were suggested to be unsafe (Yang et al. 2014). Analysis of total biogenic amines in fermented bean curd samples revealed that the content was more than 900 mg/kg in 19 white *sufu* samples. Biogenic amine content of 900 mg/kg has been determined to pose a safety hazard for the consumers. Consumption of these amines in a dose higher than 1000 mg kg^{-1} food leads to physiological disorders like a headache, nausea, rashes, brain haemorrhage, changes in blood pressure and abdominal cramps and flushing (Ladero et al. 2010, Shukla et al. 2010). Thus, selection of starter culture that result in low biogenic amines content during fermentation is ideal for legume fermentation.

5.7 Metabolites Produced by the Starter

Apart from hydrolysis and transformation of food components, several bioactive components produced by the starter cultures also improves the functional properties of fermented legumes. Microbes isolated from fermented legumes have shown to produce compounds, which exhibit antimicrobial, anticancer and immunomodulatory properties (Cao et al. 2009, Eom et al. 2014). Legume fermenting microbes have been reported to produce bacteriocins, lipopeptides and mucilaginous slime responsible for functional properties. Microbes associated with legume fermentation have been reported to produce peptides having antimicrobial and anticancer properties (Cao et al. 2009, Lee et al. 2012, Eom et al. 2014). A lipopeptide biosurfactant produced by *B. natto* TK-1 in *natto* have shown to exhibit antitumor activity (Cao et al. 2009). Production of γ-aminobutyric acid (GABA) has been found in legumes fermented with LAB and *B. subtilis* (Torino et al. 2013, Limon et al. 2015). The presence of GABA in fermented lentils (Torino et al. 2013), *tempeh* (Aoki et al. 2003) and fermented kidney bean (Limon et al. 2015) has been reported. GABA enriched fermented foods have gained interest due to their application in the treatment of hypertension.

6. Impact of the Biochemical Changes on Human Health

Fermentation improves the health benefits of legumes by hydrolyzing or transformation of bioactive components, which are already present in unfermented legumes. Legume-based fermented foods have been reported to possess several health benefits such as ACE-inhibitory, antioxidant, antimicrobial, antidiabetic and anticancer properties (Nout and Keirs 2005, Zhang et al. 2006, Moktan et al. 2008, Rai and Jeyaram 2015). Among fermented legumes, most of the studies on health benefits aspects have been done on fermented soybean products. The advantage of protein rich fermented food for their application as a health improving component is that they are economical in comparison to synthetic drugs available in the market, and they do not exhibit any side effects (Rai et al. 2015). The most promising components in fermented legumes, which have been exclusively studied for health benefits, are peptides and polyphenols (Table 1). Peptides and free polyphenols are the products of the biochemical changes mediated by the enzymes produced by the starter culture. Proteases and β-glucosidases are responsible for the formation of peptides and free polyphenols, respectively. In this chapter, components responsible for health benefits in fermented legumes are grouped as (i) polyphenols-mediated health benefits (ii) peptide-mediated health benefits.

6.1 Polyphenols-mediated Health Benefits

Fermentation has shown to enhance antioxidant properties of several legume fermented products due to the increase in free polyphenol contents. Polyphenols have been reported to play a beneficial role against diseases associated with oxidative stress such as cardiovascular, cancer, neurogenerative and inflammatory disorders (Sies 2010, Tomas-Barberan and Andres-Lacueva 2012). Transformation of isoflavones during soybean fermentation and their health benefits are shown in Table 2. Health benefits associated with polyphenols depend on their bioavailability. Fermentation improves the bioavailability of polyphenols by hydrolyzing the bound polyphenols to free form (Cho et al. 2011). Improvement of antioxidant properties due to increase in polyphenols content has been shown in several legume fermented foods such as *douche* fermented with *Aspergillus oryzae* (Wang et al. 2007), *kinema* fermented with *B. subtilis* (Moktan et al. 2008), *natto* fermented with *B. subtilis* (Iwai et al. 2002), *chungkookjang* by *B. subtilis* (Kwak et al. 2007), *tempeh* by *Rhizopus oligosporus* (Sheih et al. 2000), *miso* by *A. oryzae* and *Saccharomyces rouxii* (Esaki et al. 1997), fermented lentil by *Lactobacillus plantarum* (liquid fermentation) and *B. subtillis* (Torino et al. 2013), red bean fermented with *B. subtilis* and *L. bulgaricus* (Jhan et al. 2015), kidney bean fermented with *B. subtilis* and *L. plantarum* (Limon et al. 2015). Higher free polyphenols content in fermented soybean has been responsible factor for prevention of cancer (Jung et al. 2006), cardiovascular diseases (Larkin et al. 2008), diabetes (Ademiluyi et al. 2014) and antimicrobial against selected bacterial pathogens (Kiers et al. 2003, Kobayasi et al. 1992).

Table 1. Bioactive components and health benefits of selected fermented legumes.

Fermented legumes	Type of legume	Starter	Bioactive components	Functional properties	Reference
Natto	Soybean	*B. subtilis*	Isoflavone, peptides, Nattokinase	Antioxidant, Fibrinolytic properties	Iwai et al. 2002, Sumi et al. 1990
Tempeh	Soybean	*Rhizopus oligosporus*	Isoflavone, peptides	Anticancer, cholesterol lowering ability	Nout and Kiers 2005
Fermented kidney bean	Kidney bean	*L. plantarum* and *B. subtilis*	Anthocyanins, peptides, GABA, phenolic acids	Antioxidant, Antihypertensive	Limon et al. 2015
Chugkookjang	Soybean	*B. subtilis*	Polyphenols, peptides	Antioxidant, Antidiabetic	Kwak et al. 2007, Kwon et al. 2011
Douchi	Soybean	*Aspergillus* sp., *Mucor* sp.	Isoflavone, peptide	Antidiabetic, ACE inhibitory	Zhang et al. 2006, Wang et al. 2007
Fermented lentil	Lentil seeds	*L. plantarum* and *B. subtilis*	Isoflavone, peptide	Antioxidant, ACE inhibitory	Torino et al. 2013
Fermented Red bean	Red bean	*B. subtilis* and *L. delbrueckii*	Polyphenols, Nattokinase	Antioxidant, Fibrinolytic properties	Jhan et al. 2015
Kinema	Soybean	*B. subtilis*	Polyphenols	Antioxidant	Moktan et al. 2008
Fermented soybean	Black and yellow soybean	*B. subtilis* MTCC5480, *B. subtilis* MTCC1747	Peptides and polyphenols	Antioxidant	Sanjukta et al. 2015
Fermented soymilk	Soybean	Lactic acid bacteria	Polyphenols	Antioxidant	Wang et al. 2006b

ACE—angiotensin converting enzyme

Table 2. Transformation of isoflavones during soybean fermentation and their health benefits.

Glycosides

Compound	R^1	R^2	R^3
Daidzin	H	H	H
Genistin	OH	H	H
Glycitin	H	OCH$_3$	H
6"-O-Acetyldaidzin	H	H	CH$_3$
6"-O-Acetyl genistin	OH	H	CH$_3$
6"-O-Acetyl glycitin	H	OCH$_3$	CH$_3$
6"-O-Malonyldaidzin	H	H	CH$_2$COOH
6"-O-Malonylgenistin	OH	H	CH$_2$COOH
6"-O-Malonyl glycitin	H	OCH$_3$	CH$_2$COOH
6"-O-Succinyldaidzin	H	H	CH$_2$CH$_2$COOH
6"-O-Succinylgenistin	OH	H	CH$_2$CH$_2$COOH
6"-O-Succinyl glycitin	H	OCH$_3$	CH$_2$CH$_2$COOH

β-glucosidase

Bacillus spp., *Rhizopus* spp.

Aglycones

Compound	R^1	R^2
Daidzein	H	H
Geneistein	OH	H
Glycitein	H	OCH$_3$

Health benefits:

- High bioavailability of bioactive isoflavones,
- Antiproliferative activity on human cancer
- Prevention of mammary cancer
- Reduced risk of cardiovascular diseases
- Antidiabetic
- Anti mutagenic

(Dajanta et al. 2009, Anupongsanugool et al. 2005, Setchell 1998, Pan et al. 2000, He and Chen 2013, Liu et al. 2006, Park et al. 2003)

Hydroxylase

Aspergillus spp.

Aglycone derivatives

Human intestinal flora

Lactococcus

S-equol

Health benefits:

- Reducing menopause symptoms
- Prevent osteoporosis (Setchell et al. 2005, 2009)

8-hydroxy daidzein

8-hydroxy genestein

Health benefits:

- Antimutagenic activity,
- Depigmentation activities (Seo et al. 2013)

6.2 Peptide-mediated Health Benefits

As protein is one of the major component of legumes, their hydrolysis during fermentation lead to the formation of several peptides. Peptides are short chains of amino acids (2–20 amino acids), which depending on the size and amino acid sequence exert several beneficial physiological effect (Rai et al. 2015). Peptides in legume-based fermented foods have shown to exhibit several health benefits such as antioxidant, antihypertensive, anticancer and antidiabetic properties (Zhang et al. 2006, Kwon et al. 2011, Lee et al. 2012, Sanjukta et al. 2015). Bioactivities of the peptides formed during fermentative and enzymatic hydrolysis of soybean proteins are listed in Table 3. Most of the reports on health benefits due to peptides in fermented legume products are on antihypertensive properties. The antihypertensive effect due to peptides produced during food fermentation is by inhibition of angiotensin I-converting enzyme (ACE) (Rai et al. 2015). Several fermented soybean products such as *douche* (Zhang et al. 2006), *natto* (Okamoto et al. 1995), *sufu* (Iwamik and Buki 1986), *tempeh* (Gibbs et al. 2004) have shown to possess antihypertensive peptides. Apart from fermented soybean, the antihypertensive effect has been reported in fermented navy bean using LAB (Rui et al. 2015), fermented lentil with *L. plantarum* and *B. subtilis* (Torino et al. 2013). In addition, peptides have been suggested as a contributing factor for higher antioxidant properties in fermented legumes such as *natto* (Gibbs et al. 2004, Iwai et al. 2002), yellow and black soybean fermented with proteolytic *B. subtilis* (Sanjukta et al. 2015), bacterial *douche* (Fan et al.

Table 3. Bioactive peptides formed during hydrolysis of soybean proteins.

Bioactivity	Peptides	References
ACE inhibitory peptides	LVQGS	Rho et al. 2009
	VAHINVGK	Hernandez-Ledesma et al. 2004
	YVWK	
	DLP	Wu and Ding 2001
Hypocholesteromic	WGAPSL	Zhong et al. 2007
	LPYPR	Yoshikawa et al. 2003
	IAEK	Nagaoka et al. 2001
	LPYP	Kwon et al. 2002
Anticancer	X-MLPSYSPY	Kim et al. 2000
	SKWQHQQDSCRKQKQGVNL PCEKHIMEKIQGRGDDDDDDDDD	Galvez et al. 1997
Antihypertensive	HHL	Shin et al. 2001
	PGTAVFK	Kitts and Weiler 2003
Immunostimulation	MITLAIPVN	Yoshikawa et al. 2000
Antioxidant	LLPHH	Korhonen and Pihlanto 2003

2009), *Rhizopus* fermented *tempeh* (Wantabe et al. 2007), bacterial fermented kidney bean (Limon et al. 2015), *B. subtilis* and *L. plantarum* fermented lentil (Torino et al. 2013). In their study, Yang et al. (2012) have suggested that antidiabetic properties in fermented soybean are due to peptides along with isoflavones formed during fermentation. Bioactive peptides in several fermented legumes are yet to be explored, which may lead to the discovery of novel bioactive peptides.

7. Market Potential of Legume-Based Fermented Foods

Consumer preferences towards fermented legume products have increased due to the awareness of its health benefits. Increasing awareness and advanced research in the area of functional fermented legume products across the globe are the driving factors for its demand in the global market. In particular fermented soybean products have become more popular among U.S. consumers in recent years due to the recent studies on health benefits (He and Chen 2013). In the last few decades, Asia Pacific has been the dominating market for fermented soybean products followed by Europe and North America (Market and Market 2015). Some of the fermented soybean products are used as an alternative to meat and meat products (He and Chen 2013). Even fermented soymilk products such as soya yoghurt and cheese are gaining popularity due to the enhanced health-promoting components produced during soybean fermentation.

Among fermented legumes, most of the data available on annual production are for fermented soybean products. *Tempeh*, fermented soybean popular in Indonesia has an annual production of approximately 80,000 tonnes per year, and 14% of the Indonesian soybean production is used for *tempeh* production (Yakutsuka and Sasaki 1998). In a recent report by Codex Alimentarius Commission (CAC 2014), *natto* production had increased from 220,000 tons in 2011 to 225,000 tons in 2013. Increase in total production was reflected in the increase in total sale volume from 173 billion yen in 2011 to 196 billion yen in 2013 (CAC 2014). An annual production of *Sufu* products manufactured both commercially and domestically was estimated to over 300,000 metric tons in China (Han et al. 2001). In Korea, the growth rate of fermented soy sauce is about 15% per annum (Shurtleff and Aoyagi 2014). Most of the fermented soy sauce in Korea is produced by Kikkoman Foods, accounting for an annual production of about 19000 kilolitres (Shurtleff and Aoyagi 2014). In Japan, there are about 1600 *miso* manufacturing plants producing approximately 560,000 tons of *miso* annually, out of which 3000 tons of *miso* is exported (www.miyasaka-jozo.com). Recent advances in research on biochemical changes and health benefits of fermented legume products have affected the production and market potential of fermented legume products across the globe.

8. Conclusions

Legume fermentation results in a wide range of biochemical changes by the action of microbial enzymes on legume components and metabolites produced by the starter cultures. Legume fermentation not only improves the nutritional properties but also enhances bioactive components and reduces the antinutritional factors in raw legumes. There are possibilities of formation of novel bioactive compounds due to microbial transformation during the fermentation process. Research area on the characterisation of bioactive compounds such as peptides in fermented legume products and their physiological effects is only at its beginning. Till date, there are many fermented legume products produced by natural fermentation. Standardisation of fermentation methods with identified starter cultures can control the biochemical changes and production of beneficial metabolites. Fermentation with potential starter culture can also be applied for the utilisation of underutilized legumes by improving their health benefits. Such studies can lead not only in value addition of legumes having low economic values but also in the development of novel functional foods.

Keywords: *Bacillus*, bioactive peptides, fibrinolytic enzyme, angiotensin I converting enzyme (ACE), isoflavones, polyphenols, antioxidant activity, oligosaccharides, phytic acid, biogenic amines

References

Abdel-Fattah, A.M., Gamal-Eldeen, A.M., Helmy, W.A. and Esawy, M.A. (2012). Antitumor and antioxidant activities of levan and its derivative from the isolate *Bacillus subtilis* NRC1aza. Carbohydrate Polymers 89: 314–322.

Achi, A.O. (2005). Traditional fermented protein condiments in Nigeria. African Journal of Biotechnology 4: 1612–1621.

Ademiluyi, A.O., Oboh, G., Boligon, A.A. and Athayde, M.L. (2014). Effect of fermented soybean condiment supplemented diet on α-amylase and α-glucosidase activities in Streptozotocin induced diabetic rats. Journal of Functional Foods 9: 1–9.

Adewumi, G.A., Oguntoyinbo, F.A., Romi, W., Singh, T.A. and Jeyaram, K. (2014). Genome subtyping of autochthonous *Bacillus* species Isolated from Iru, a fermented *Parkia biglobosa* Seed. Food Biotechnology 28: 250–268.

Adewumi, G.A., Oguntoyinbo, F.A., Keisam, S., Romi, W. and Jeyaram, K. (2013). Combination of culture-independent and culture-dependent molecular methods for the determination of bacterial community of iru, a fermented Parkia biglobosa seeds. Frontiers in Microbiology 3: 436.

Anupongsanugool, E., Teekachunhatean, S., Rojanasthien, N., Pongsatha, S. and Sangdee, C. (2005). Pharmacokinetics of isoflavones, daidzein and genistein, after ingestion of soy beverage compared with soy extract capsules in postmenopausal Thai women. BMC Pharmacology and Toxicology 5(1): 2.

Aoki, H., Uda, I., Tagami, K., Furuya, Y., Endo, Y. and Fujimoto, K. (2003). The production of a new *tempeh*-like fermented soybean containing a high level of γ-amino butyric acid by anaerobic incubation with *Rhizopus*. Bioscience Biotechnology and Biochemistry 67: 1018–1023.

Askar, A. and Treptow, H. (1986). Biogenice amine in lebensmitteln. pp. 21–74. *In*: A. Askar and H. Treptow (eds.). Vorkommen, Bedeutung und Bestimmung. Stuttgart, Germany: Verlag Eugen Elmer.

Bai, X., Byun, B.Y. and Mah, J.H. (2013). Formation and destruction of biogenic amines in Chunjang (a black soybean paste) and Jajang (a black soybean sauce). Food Chemistry 141: 1026–1031.

CAC (2014). Codex Alimentarius Commission. Discussion paper on development of a regional standard for natto. ftp://ftp.fao.org/codex/Meetings/CCASIA/ccasia19/CRDs/as19_CRD10x.pdf.

Cao, X.H., Liao, Z.Y., Wang, C.L., Yang, W.Y. and Lu, M.F. (2009). Evaluation of a lipopeptide biosurfactant from *Bacillus natto* TK-1 as a potential source of anti-adhesive, antimicrobial and antitumor activities. Brazilian Journal of Microbiology 40: 373–379.

Carbonaro, M., Maselli, P. and Nucara, A. (2015). Structural aspects of legume proteins and nutraceutical properties. Food Research International 76: 19–30.

Chancharoonpong, C., Hsieh, P. and Sheu, S. (2012). Enzyme production and growth of *Aspergillus oryzae* S. on soybean *koji* fermentation. APCBEE Procedia 2: 57–61.

Cho, K.M., Lee, J.H., Yun, H.D., Ahn, B.Y., Kim, H. and Seo, W.T. (2011). Changes of phytochemical constituents (isoflavones, flavanols, phenolic acids) during cheonggukjang soybean fermentation using potential prebiotic *Bacillus subtilis* CS90. Journal of Food Composition and Analysis 24: 402–310.

Dajanta, K., Chukeatirote, E. and Apichartsrangkoon, A. (2011). Analysis and characterization of amino acid contents of *thua nao*, a traditionally fermented soybean food of Northern Thailand. International Food Research Journal 18: 595–599.

Dajanta, K., Chukeatirote, E., Apichartsrangkoon, A. and Frazier, R.A. (2009). Enhanced aglycone production of fermented soybean products by *Bacillus* species. Acta Biologica Szegediensis 53(2): 93–98.

Dajanta, K., Janpum, P. and Leksing, W. (2013). Antioxidant capacities, total phenolics and flavonoids in black and yellow soybeans fermented by *Bacillus subtilis*: A comparative study of Thai fermented soybeans (thua nao). International Food Research Journal 20: 3125–3132.

Difo, H.V., Onyike, E., Ameh, D.A., Ndidi, U.S. and Njoku, G.C. (2014). Chemical changes during open and controlled fermentation of cowpea (*Vigna unguiculata*) flour. International Journal of Food Nutrition and Safety 5: 1–10.

Donkor, O.N., Henriksson, A., Vasiljevic, T. and Shah, N.P. (2005). Probiotic strains as starter cultures improve angiotensin-converting enzyme inhibitory activity in soy yogurt. Journal of Food Science 70: 375–381.

Duranti, M. (2006). Grain legume proteins and nutraceutical properties. Fitoterapia 77: 67–82.

Egounlety, M. and Aworh, O.C. (2003). Effect of soaking, dehulling, cooking and fermentation with *Rhizopus oligosporus* on the oligosaccharides, trypsin inhibitor, phytic acid and tannins of soybean (*Glycine max* Merr.), cowpea (*Vigna unguiculata* L. Walp) and ground bean (*Macrotyloma geocarpa* Harms). Journal of Food Engineering 56: 249–254.

El-Adawy, T.A. (2002). Nutritional composition and antinutritional factors of chickpeas (*Cicer arietinum* L.) undergoing different cooking methods and germination. Plant Foods for Human Nutrition 57: 83–97.

Eom, J.S., Lee, S.Y. and Choi, H.S. (2014). *Bacillus subtilis* HJ18-4 from traditional fermented soybean food inhibits *Bacillus cereus* growth and toxin-related genes. Journal of Food Science 79: 2279–2287.

Esaki, H., Onozaki, H., Kawakishi, S. and Osawa, T. (1997). Antioxidant activity and isolation from soybeans fermented with *Aspergillus* spp. Journal of Agricultural and Food Chemistry 45: 2020–2024.

Fan, J., Zhang, Y., Chang, X., Saito, M. and Li, Z. (2009). Changes in the radical scavenging activity of bacterial-type *douchi*, a traditional fermented soybean product, during the primary fermentation process. Bioscience Biotechnology Biochemistry 73: 2749–2753.

Feng, Y., Chen, Z., Liu, N., Zhao, H., Cui, C. and Zhao, M. (2014). Changes in fatty acid composition and lipid profile during *koji* fermentation and their relationships with soy sauce flavour. Food Chemistry 158: 438–444.

Fujita, M., Nomura, K., Hong, K., Ito, Y., Asada, A. and Nishimuro, S. (1993). Purification and characterization of a strong fibrinolytic enzyme (nattokinase) in the vegetable cheese natto, a popular soybean fermented food in Japan. Biochemical and Biophysical Research Communications 197: 1340–1347.

Galvez, A.F., Revilleza, M.J.R. and De Lumen, B.O. (1997). A novel methionine-rich protein from soybean cotyledon: cloning and characterization of cDNA (accession no. AF005030). Plant Gene Register# PGR97-103. Plant Physiol 114: 1567–1569.

Gambacorta, G., Sinigaglia, M., Schena, A., Baiano, A., Lamacchia, C., Pati, S. and Notte, E.L. (2009). Changes in free fatty acid and diacylglycerol compounds in short ripening dry-cured sausage. Journal of Food Lipids 16: 1–18.

Gibbs, B.F., Zougman, A., Masse, R. and Mulligan, C. (2004). Production and characterization of bioactive peptides from soy hydrolysate and soy-fermented food. Food Research International 37: 123–131.

Han, B.Z., Rombouts, F.M. and Nout, M.J.R. (2001). A Chinese fermented soybean food. International Journal of Food Microbiology 65: 1–10.

He, F.J. and Chen, J.Q. (2013). Consumption of soybean, soy foods, soy isoflavones and breast cancer incidence: differences between Chinese women and women in Western countries and possible mechanisms. Food Science and Human Wellness 2(3): 146–161.

Hernández-Ledesma, B., Amigo, L., Ramos, M. and Recio, I. (2004). Angiotensin converting enzyme inhibitory activity in commercial fermented products. Formation of peptides under simulated gastrointestinal digestion. Journal of Agricultural and Food Chemistry 52(6): 1504–1510.

Hesseltine, C.W. (1965). A millennium of fungi, food and fermentation. Mycologia 57: 149–197.

Hong, J.Y., Kim, E.J., Shin, S.R., Kim, T.W., Lee, I.J. and Yoon, K.Y. (2008). Physicochemical properties of *cheonggukjang* containing Korean red ginseng and *Rubus coreanum*. Korean Journal of Food Preservation 15: 872–877.

Hoover, R. and Zhou, T. (2003). *In vitro* and *in vivo* hydrolysis of legume starches by amylase and resistant starch formation in legumes—a review. Carbohydrate Polymers 54: 401–417.

Hur, S.J., Lee, S.Y., Kim, Y.C., Choi, I. and Kim, G.B. (2014). Effect of fermentation on the antioxidant activity in plant-based foods. Food Chemistry 160: 346–356.

Hwang, J. (1997). Angiotensin I converting enzyme inhibitory effect of *doenjang* fermented by *B. subtilis* SCB-3 isolated from *meju*, Korean traditional food. Journal of the Korean Society Food Science & Nutrition 26: 776–783.

Iwai, K., Nakaya, N., Kawasaki, Y. and Matsue, H. (2002). Antioxidant function of Natto, A kind of fermented soybean: effect of LDL oxidation and lipid metabolism in cholesterol-fed rats. Journal of Agriculture and Food Chemistry 50: 3597–3610.

Iwamik, S.K. and F. Buki. (1986). Involvement of post-digestion hydropholic peptides in plasma cholesterol-lowing effect of dietary plant proteins. Agricultural and Biological Chemistry 50: 1217–1222.

Jeyaram, K., Singh, W.M., Premarani, T., Devi, A.R., Chanu, K.S., Talukdar, N.C. and Singh, M.R. (2008). Molecular identification of dominant microflora associated with 'Hawaijar'—A traditional fermented soybean (*Glycine max* (L.)) food of Manipur, India. International Journal of Food Microbiology 122: 259–268.

Jhan, J.K., Chang, W.F., Wang, P.M., Chou, S.T. and Chung, Y.C. (2015). Production of fermented red beans with multiple bioactivities using co-cultures of *Bacillus subtilis* and *Lactobacillus delbrueckii* subsp. *bulgaricus*. LWT—Food Science and Technology 63: 1281–1287.

Jolaoso, A.A., Ajayi, J.O., Ogunmuyiwa, S.I.O. and Albert, O.M. (2014). Functional, proximate and mineral composition of a lesser known fermented legume. Journal of Emerging Trends in Engineering and Applied Sciences 5: 129–134.

Jung, K.O., Park, S.Y. and Park, K.Y. (2006). Longer aging time increases the anticancer and antimetastatic properties of Doenjang. Nutrition 22: 539–545.

Kalpanadevi, V. and Mohan, V.R. (2013). Effect of processing on antinutrients and *in vitro* protein digestibility of the underutilized legume, *Vigna unguiculata* (L.) Walp subsp. *unguiculata*. LWT—Food Science and Technology 51: 455–461.

Kasaoka, S., Astuti, M., Uehara, M., Suzuki, K. and Goto, S. (1997). Effect of Indonesian fermented soybean *tempeh* on iron bioavailability and lipid peroxidation in anemic rats. Journal of Agriculture and Food Chemistry 45: 195–198.

Khattab, R.Y. and Arntfield, S.D. (2009). Nutritional quality of legume seeds as affected by some physical treatments 2. Antinutritional factors. LWT—Food Science and Technology 42: 1113–1118.

Kiers, J.L., Meijer, J.C., Nout, M.J.R., Rombouts, F.M., Nabuurs, M.J.A. and Van der Meulen, J. (2003). Effect of fermented soya beans on diarrhoea and feed efficiency in weaned piglets. Journal of Applied Microbiology 95: 545–552.

Kim, B., Byun, B.Y. and Mah, J.H. (2012). Biogenic amine formation and bacterial contribution in *Natto* products. Food Chemistry 135: 2005–2011.

Kim, J.S. and Yoon, S. (1999). Isoflavone contents and β-glucosidase activities in soybean, *meju* and *doenjang*. Korean Journal of Food Science Technology 31: 1405–1409.

Kim, S.E., Kim, H.H., Kim, J.Y., Kang, Y.I., Woo, H.J. and Lee, H.J. (2000). Anticancer activity of hydrophobic peptides from soy proteins. Biofactors 12(1-4): 151–156.

Kitts, D.D. and Weiler, K. (2003). Bioactive proteins and peptides from food sources. Applications of bioprocesses used in isolation and recovery. Current Pharmaceutical Design 9(16): 1309–1323.

Kiuchi, K., Ohta, T., Itoh, H., Takahayahsi, T. and Ebine, H. (1976). Studies on lipids of natto. Journal of Agriculture and Food Chemistry 24: 404–407.

Korhonen, H. and Pihlanto, A. (2003). Food-derived bioactive peptides-opportunities for designing future foods. Current Pharmaceutical Design, Vol. 9, No. 16, 1297–1308. ISSN: 1381–6128.

Kobayasi, S.Y., Okazaki, N. and Koseki, T. (1992). Purification and characterization of an antibiotic substance produced from *Rhizopus oligosporus* IFO 8631. Bioscience Biotechnology and Biochemistry 56: 94–98.

Kung, H.F., Tsai, Y.H. and Wei, C.I. (2007). Histamine and other biogenic amines and histamine-forming bacteria in *miso* products. Food Chemistry 101: 351–356.

Kwak, C.S., Lee, M.S. and Park, S.C. (2007). Higher antioxidant properties of *Chungkookjang*, a fermented soybean paste, may be due to increased aglycone and malonyl glycoside isoflavone during fermentation. Nutrition Research 27: 719–727.

Kwon, D.Y., Hong, S.M., Ahn, I.S., Kim, M.J., Yang, H.J. and Park, S. (2011). Isoflavonoids and peptides from *meju*, long-term fermented soybeans, increase insulin sensitivity and exert insulinotropic effects *in vitro*. Nutrition 27: 244–252.

Kwon, D.Y., Oh, S.W., Lee, J.S., Yang, H.J., Lee, S.H. and Lee, J.H. (2002). Amino acid substitution of hypocholesterolemic peptide originated from glycinin hydrolyzate. Food Science Biotechnology, Vol. 11, No. 3, 55–61. ISSN: 1226–7708.

Ladero, V., Calles-Enriquez, M., Fernandez, M. and Alvarez, M.A. (2010). Toxicological effects of dietary biogenic amines. Current Nutrition and Food Science 6: 145–156.

Larkin, T., Price, W.E. and Astheimer, L. (2008). The key importance of soy isoflavone bioavailability to understanding health benefits. Critical Reviews in Food Science and Nutrition 48: 538–552.

Lee, J.H., Namb, S.H., Seo, W.T., Yun, H.D., Hong, S.Y., Kim, M.K. and Cho, K.M. (2012). The production of surfactin during the fermentation of *cheonggukjang* by potential probiotic *Bacillus subtilis* CSY191 and the resultant growth suppression of MCF-7 human breast cancer cells. Food Chemistry 131: 1347–1354.

Limón, R.I., Penas, E., Torino, M.I., Martínez-Villaluenga, C., Dueñas, M. and Frias, J. (2015). Fermentation enhances the content of bioactive compounds in kidney bean extracts. Food Chemistry 172: 343–352.

Market and Market (2015). Soy Food Products Market: Trends and Global Forecasts 2012–2017. http://www.marketsandmarkets.com/Market-Reports/soybean-food-products-market-706.html.

McCue, P. and Shetty, K. (2003). Role of carbohydrate-cleaving enzymes in phenolic antioxidant mobilization from whole soybean fermented with *Rhizopus oligosporus*. Food Biotechnology 17: 27–37.

Mine, Y.A., Wong, H.K. and Jiang, B. (2005). Fibrinolytic enzymes in Asian traditional fermented foods. Food Research International 38: 243–250.

Miyasaka-jozo.com. (2015). 'Shinsyuichi | Miyasaka Brewery Website'. N.p., Web. 8 July 2015. http://www.miyasaka-jozo.com/english/miso/.

Moktan, B., Saha, J. and Sarkar, P.K. (2008). Antioxidant activities of soybean as affected by *Bacillus* fermentation to kinema. Food Research International 41: 586–593.

Nagaoka, S., Futamura, Y., Miwa, K., Awano, T., Yamauchi, K., Kanamaru, Y., Tadashi, K. and Kuwata, T. (2001). Identification of novel hypocholesterol peptides derived from bovine milk beta-lactoglobulin. Biochemistry Biophysics Research Communication. Vol. 28, No. 1, 11–17. ISSN: 0006-291X.

Nikkuni, S., Karki, T.B., Vilkhu, K.S., Suzuki, T., Shindoh, K., Suzuki, C. and Okada, N. (1995). Mineral and amino acid contents of *kinema*, a fermented soybean food prepared in Nepal. Food Science and Technology International 1: 107–111.

Nout, M.J.R. and Kiers, J.L. (2005). Tempe fermentation, innovation and functionality: update into the third millennium. Journal of Applied Microbiology 98: 789–805.

Nout, M.J.R. and Rombouts, F.M. (1990). Recent developments in *tempe* research. Journal of Applied Bacteriology 69: 609–633.

Nout, M.J.R., Ruikes, M.M.W., Bouwmeester, H.M. and Beljaars, P.R. (1993). Effect of processing conditions on the formation of biogenic amines and ethyl carbamate in soybean *tempe*. Journal of Food Safety 13: 293–303.

Oboh, G., Alabi, K.B. and Akindahunsi, A.A. (2008). Fermentation changes the nutritive values, polyphenol distribution, and antioxidant properties of *Parkia giglobosa* seeds (African locust beans). Food Biotechnology 22: 363–376.

Odunfa, S.A. (1985). Biochemical changes in fermenting African locust bean *(Parkia biglobom)* during '*iru*' fermentation. Journal of Food Technology 20: 295–303.

Odunfa, S.A. (1983). Carbohydrate changes in fermenting locust bean during *iru* preparation. Plant Foods and Human Nutrition 32: 1–10.

Oguntoyinbo, F.A., Sanni, A.I., Franz, C.M.A.P. and Holzapfel, W.H. (2007). *In vitro* selection and evaluation of Bacillus starter cultures for the production of Okpehe, a traditional African fermented condiment. International Journal of Food Microbiology 113: 208–218.

Okamoto, A., Hanagata, H., Kawamura, Y. and Yanagida, F. (1995). Anti-hypertensive substances in fermented soybean, Natto. Plant Foods for Human Nutrition 47: 39–47.

Osman, M.A. (2011). Effect of traditional fermentation process on the nutrient and antinutrient components of pearl millet during preparation of *lohoh*. Journal of the Saudi Society of Agriculture Science 10: 1–6.

Osman, M.A. (2004). Changes, in sorghum enzyme inhibitors, phytic acid, tannins, and *in vitro* protein digestibility occurring during Khamir (local bread) fermentation. Food Chemistry 88: 129–134.

Ouoba, L.I.I., Cantor, M.D., Diawara, B., Traore, A.S. and Jakobsen, M. (2003). Degradation of African locust bean oil by *Bacillus subtilis* and *Bacillus pumilus* isolated from soumbala, a fermented African locust bean condiment. Journal of Applied Microbiology 95: 868–873.

Pan, Y., Anthony, M., Watson, S. and Clarkson, T.B. (2000). Soy phytoestrogens improve radial arm maze performance in ovariectomized retired breeder rats and do not attenuate benefits of 17beta-estradiol treatment. Menopause 7: 230–235.

Paredes-López, O., González-Castañeda, J. and Cárabez-Trejo, A. (1991). Influence of solid substrate fermentation on the chemical composition of chickpea. Fermentation and Bioengineering 71: 58–62.

Park, K.Y., Jung, K.O., Rhee, S.H. and Choi, Y.H. (2003). Antimutagenic effects of doenjang (Korean fermented soypaste) and its active compounds. Mutation Research/Fundamental and Molecular Mechanisms of Mutagenesis 523, 43-53002E.

Peng, Y., Huang, Q., Zhang, R. and Zhang, Y. (2003). Purification and characterization of a fibrinolytic enzyme produced by *Bacillus amyloliquefaciens* DC-4 screened from *douchi*, a traditional Chinese soybean food. Comparative Biochemistry and Physiology-B Biochemistry and Molecular Biology 134: 45–52.

Prodanov, M., Sierra, I. and Vidal-Valverde, C. (2004). Influence of soaking and cooking on the thiamin, riboflavin and niacin contents of legumes. Food Chemistry 84: 271–277.

Pyo, Y.H., Lee, T.C. and Lee, Y.C. (2005). Enrichment of bioactive isoflavones in soymilk fermented with β-glucosidase-producing lactic acid bacteria. Food Research International 38: 5551–5559.

Pyo, Y.H. and Lee, T.C. (2007). The potential antioxidant capacity and angiotensin I-converting enzyme inhibitory activity of Monascus-fermented soybean extracts: evaluation of Monascus-fermented soybean extracts as multifunctional food additives. Journal of Food Science 72: 218–223.

Rai, A.K. and Jeyaram, K. (2015). Health benefits of functional proteins in fermented foods. pp. 455–476. *In*: J.P. Tamang (ed.). Health Benefits of Fermented Foods and Beverages. CRC Press, Taylor and Francis Group of USA.

Rai, A.K., Sanjukta, S. and Jeyaram, K. (2015). Production of Angiotensin I converting enzyme inhibitory (ACE-I) peptides during milk fermentation and its role in treatment of hypertension. Critical reviews in Food Science and Nutrition 10.1080/10408398.2015.1068736.

Reddy, N.R., Pierson, M.D., Sathe, S.K. and Salunkhe, D.K. (1985). Dry bean tannins—a review of nutritional implications. Journal of the American Oil Chemists Society 62: 541–549.

Reddy, N.R., Pierson, M.D., Sathe, S.K., Salunkhe, D.K. and Beuchat, L.R. (1983). Legume-based fermented foods: Their preparation and nutritional quality. CRC Critical Reviews in Food Science and Nutrition 17: 335–370.

Rho, S.J., Lee, J.S., Chung, Y.I., Kim, Y.W. and Lee, H.G. (2009). Purification and identification of an angiotensin I-converting enzyme inhibitory peptide from fermented soybean extract. Process Biochemistry 44(4): 490–493.

Rui, X., Wen, D., Li, W., Chen, X., Jiang, M. and Dong, M. (2015). Enrichment of ACE inhibitory peptides in navy bean (*Phaseolus vulgaris*) using lactic acid bacteria. Food and Function 6: 622–629.

Sanjukta, S., Rai, A.K., Ali, M.M., Jeyaram, K. and Talukdar, N.C. (2015). Enhancement of antioxidant properties of two soybean varieties of Sikkim Himalayan region by proteolytic *Bacillus subtilis* fermentation. Journal of Functional Foods 14: 650–658.

Sanjukta, S. and Rai, A.K. (2016). Production of bioactive peptides during soybean fermentation and their potential health benefits. Trends in Food Science and Technology 50: 1–10.

Sarkar, P.K., Hasenack, B. and Nout, M.J.R. (2002). Diversity and functionality of *Bacillus* and related genera isolated from spontaneously fermented soybeans (Indian *Kinema*) and locust beans (African Soumbala). International Journal of Food Microbiology 77: 175–186.

Sarkar, P.K., Jones, L.J., Gore, W., Craven, G.S. and Somerset, S.M. (1996). Changes in soya bean lipid profiles during *Kinema* production. Journal of the Science of Food and Agriculture 71: 321–328.

Seo, M.H., Kim, B.N., Kim, K.R., Lee, K.W., Lee, C.H. and Oh, D.K. (2013). Production of 8-hydroxydaidzein from soybean extract by *Aspergillus oryzae* KACC 40247. Bioscience, Biotechnology, and Biochemistry 77(6): 1245–1250.

Setchell, K.D. (1998). Phytoestrogens: the biochemistry, physiology, and implications for human health of soy isoflavones. The American Journal of Clinical Nutrition 68(6): 1333S–1346S.

Setchell, K.D., Zhao, X., Shoaf, S.E. and Ragland, K. (2009). The pharmacokinetics of S-(-) equol administered as SE5-OH tablets to healthy postmenopausal women. The Journal of Nutrition 139(11): 2037–2043.

Setchell, K.D., Clerici, C., Lephart, E.D., Cole, S.J., Heenan, C., Castellani, D. et al. (2005). S-equol, a potent ligand for estrogen receptor β, is the exclusive enantiomeric form of the soy isoflavone metabolite produced by human intestinal bacterial flora. The American Journal of Clinical Nutrition 81(5): 1072–1079.

Sharma, A., Kumari, S., Wongputtisin, P., Nout, M.J.R. and Sarkar, P.K. (2015). Optimization of soybean processing into *kinema*, a *Bacillus*-fermented alkaline food, with respect to a minimum level of antinutrients. Journal of Applied Microbiology 119: 162–176.

Sheih, I.C., Wu, H.Y., Lai, Y.J. and Lin, C.F. (2000). Preparation of high free radical scavenging tempeh by a newly isolated *Rhizopus* sp. R-69 from Indonesia. Food Science and Agricultural Chemistry 2: 35–40.

Shin, Z.I., Yu, R., Park, S.A., Chung, D.K., Ahn, C.W., Nam, H.S. et al. (2001). His-His-Leu, an angiotensin I converting enzyme inhibitory peptide derived from Korean soybean paste, exerts antihypertensive activity *in vivo*. Journal of Agricultural and Food Chemistry 49(6): 3004–3009.

Shukla, S., Park, H.K., Kim, J.K. and Kim, M. (2010). Determination of biogenic amines in Korean traditional fermented soybean paste. Food and Chemical Toxicology 48: 1191–1199.

Shurtleff, W. and Aoyagi, A. (2014). History of Soybeans and Soyfoods in Korea, and in Korean Cookbooks, Restaurants, and Korean Work with Soyfoods outside Korea, Soyainfo Center. http://www.soyinfocenter.com/books/174http://www.soyinfocenter.com/books/174.

Sies, H. (2010). Polyphenols and health: Update and perspectives. Archives of Biochemistry and Biophysics 501: 2–5.

Singh, T.A., Devi, K.R., Ahmed, G. and Jeyaram, K. (2014). Microbial and endogenous origin of fibrinolytic activity in traditional fermented foods of Northeast India. Food Research International 55: 356–362.

Sohn, M.Y., Seo, K.I., Lee, S.W., Choi, S.H. and Sung, N.J. (2000). Biological activities of *chungkugjang* prepared with black bean and changes in phytoestrogen content during fermentation. Korean Journal of Food Science and Technology 32: 936–941.

Sumi, H., Hamada, H., Nakanishi, K. and Hiratani, H. (1990). Enhancement of the fibrinolytic activity in plasma by oral administration of nattokinase. Acta Haematologica 84: 139–143.

Tamang, J.P., Thapa, N., Tamang, B., Rai, A. and Chettri, R. (2015). Microorganisms in fermented foods and beverages. pp. 1–110. In: J.P. Tamang (ed.). Health Benefits of Fermented Foods and Beverages. CRC Press, Taylor and Francis Group of USA.

Tomas-Barberan, F.A. and Andres-Lacueva, C. (2012). Polyphenols and health: Current state and progress. Journal of Agriculture and Food Chemistry 60: 8773–8775.

Torino, M.I., Limón, R.I., Martínez-Villaluenga, C., Mäkinen, S., Pihlanto, A., Vidal-Valverde, C. and Frias, J. (2013). Antioxidant and antihypertensive properties of liquid and solid state fermented lentils. Food Chemistry 136: 1030–1037.

Tsai, Y.H., Kung, H.F., Chang, S.C., Lee, T.M. and Wei, C.I. (2007). Histamine formation by histamine-forming bacteria in *douchi*, a Chinese traditional fermented soybean product. Food Chemistry 103: 1305–1311.

Udensi, E.A., Ekwu, F.C. and Isinguzo, J.N. (2007). Antinutrient factors of vegetable cowpea (sesquipedalis) seeds during thermal processing. Pakistan Journal of Nutrition 6: 194–197.

Urbano, G., Lopez-Jurado, M., Aranda, P., Vidal-Valverde, C., Tenorio, E. and Porres, J. (2000). The role of phytic acid in legumes: Antinutrient or beneficial function. Journal of Physiology and Biochemistry 56: 283–294.

Vidal-Valverde, C., Frias, J., Prodanov, M., Tabera, J., Ruiz, R. and Bacon, J. (1993). Effect of natural fermentation on carbohydrate, riboflavin and trypsin inhibitor activity of lentils. Zeitschrift für Lebensmittel-Untersuchung und-Forschung 197: 449–452.

Visessanguan, W., Benjakul, S., Riebroy, S., Yarchai, M. and Tapingkae, W. (2004). Changes in lipid composition and fatty acid profile of *Nham*, a Thai fermented pork sausage, during fermentation. Food Chemistry 94: 580–588.

Vlieg, J.E.T.H., Veiga, P., Zhang, C., Derrien, M. and Zhao, L. (2011). Impact of microbial transformation of food on health-from fermented foods to fermentation in the gastrointestinal tract. Current Opinion in Biotechnology 22: 211–219.

Wagenknecht, A.C., Mattick, L.R., Lewin, L.M., Hand, D.B. and Steinkraus, K.H. (1961). Changes in soybean lipids during tempeh fermentation. Journal of Food Science 26: 373–376.

Wang, Y.C., Yu, R.C. and Chou, C.C. (2006b). Antioxidative activities of soymilk fermented with lactic acid bacteria and bifidobacteria. Food Microbiology 23: 128–135.

Wang, C.T., Ji, B.P., Li, B., Nout, R., Li, P.L., Ji, H. and Chen, L.F. (2006a). Purification and characterization of a fibrinolytic enzyme of *Bacillus subtilis* DC33, isolated from Chinese traditional *Douchi*. The Journal of Industrial Microbiology and Biotechnology 33: 750–758.

Wang, L.J., Li, D. and Zou, L. (2007). Antioxidative activity of *douchi* (a Chinese traditional salt-fermented soybean food) extracts during its processing. International Journal of Food Properties 10: 1–12.

Wang, C., Du, M., Zeng, D., Kong, F., Zu, G. and Feng, Y. (2009). Purification and characterization of nattokinase from *Bacillus subtilis* Natto B-12. Journal of Agricultural and Food Chemistry 57: 9722–9729.

Watanabe, N., Fujimoto, K. and Aoki, H. (2007). Antioxidant activities of the water-soluble fraction in *tempeh*-like fermented soybean (GABA-*tempeh*). International Journal of Food Science and Nutrition 58: 577–587.

Wu, J. and Ding, X. (2001). Hypotensive and physiological effect of angiotensin converting enzyme inhibitory peptides derived from soy protein on spontaneously hypertensive rats. Journal of Agriculture and Food Chemistry 49: 501–506.

Xiao-Hong, C., Zhen-Yu, L., Chun-Ling, W., Wen-Yan, Y. and Mei-Fang, L. (2009). Evaluation of a lipopeptide biosurfactant from *Bacillus natto* TK-1 as a potential source of anti-adhesive, antimicrobial and antitumor activities. Brazilian Journal of Microbiology 40: 373–379.

Xu, B. and Chang, S.K.C. (2008). Effect of soaking, boiling, and steaming on total phenolic content and antioxidant activities of cool season food legumes. Food Chemistry 110: 1–13.

Yang, H.J., Kwon, D.Y., Kim, M.J., Kang, S. and Park, S. (2012). *Meju*, unsalted soybeans fermented with *Bacillus subtilis* and *Aspergilus oryzae*, potentiates insulinotropic actions and improves hepatic insulin sensitivity in diabetic rats. Nutrition and Meta Analysis 9: 37.

Yang, J., Ding, X., Qin, Y. and Zeng Y. (2014). Safety assessment of the biogenic amines in fermented soya beans and fermented bean curd. Journal of Agriculture and Food Chemistry 62: 7947–7954.

Yokotsuka, T. and Sasaki, M. (1998). Fermented protein foods in the Orient: *shoyu* and *miso* in Japan. Brian J.B. Wood. Microbiology of Fermented Foods Volume–1, Chapter–12, 351–415.

Yoshikawa, M., Takahashi, M. and Yang, S. (2003). Delta opioid peptides derived from plant proteins. Current Pharmaceutical Design 9(16): 1325–1330.

Yoshikawa, M., Fujita, H., Matoba, N., Takenaka, Y., Yamamoto, T., Yamauchi, R., Tsuruki, H. and Takahata, K. (2000). Bioactive peptides derived from food proteins preventing lifestyle-related diseases. Biofactors. Vol. 12, No. 1–4, 143–146. ISSN: 0951–6433.

Zhang, J.H., Tatsumi, E., Fan, J.F. and Li, L.T. (2007). Chemical components of *Aspergillus*-type *Douchi*, a Chinese traditional fermented soybean product, change during the fermentation process. International Journal of Food Science and Technology 42: 263–268.

Zhang, J.H., Tatsumi, E., Ding, C.H. and Li, L.T. (2006). Angiotensin I-converting enzyme inhibitory peptides in *douchi*, a Chinese traditional fermented soybean product. Food Chemistry 98: 551–557.

Zhong, F., Zhang, X., Ma, J. and Shoemaker, C.F. (2007). Fractionation and identification of a novel hypocholesterolemic peptide derived from soy protein Alcalase hydrolysates. Food Research International 40(6): 756–762.

5

Bioactive Molecules in Fermented Soybean Products and Their Potential Health Benefits

Samurailatpam Sanjukta,[1] *Amit Kumar Rai*[1,*]
and Dinabandhu Sahoo[2]

1. Introduction

Soybean (*Glycine max* (L.) *Merril*) is the most abundantly available source of plant protein, which also contains many bioactive components having positive implications on human health. Apart from bioactive compounds, soybean contain nutritional components such as carbohydrates, vitamins, lipids and minerals (Kim et al. 2006, Wang et al. 2008), and anti-nutritional factors like phytic acid, agglutinin, urease and trypsin inhibitors (Backer Ritt et al. 2004). Enhancement of beneficial compounds and reduction of anti-nutritional factors in soybeans is essential in order to exploit it as a functional food. Soybean proteins and polyphenols that are less bioavailable in native form are converted to highly bioavailable and functional form either by enzymatic hydrolysis using commercial enzymes

[1] Institute of Bioresources and Sustainable Development, Sikkim Centre, Tadong - 737102, Sikkim, India.
[2] Institute of Bioresources and Sustainable Development, Imphal - 795001, Manipur, India.
* Corresponding author: amitraikvs@gmail.com

or by fermentation. As enzymatic approach may not be economical for large-scale production therefore fermentation using microorganisms with specific enzyme (proteases, β-glucosidase and α-amylase) producing ability can be an economical and efficient alternative. Fermentation of soybean enhances its functional property by either hydrolysis of soy proteins and polyphenols (Sourabh et al. 2015, Dajanta et al. 2011, Kwak et al. 2012) or by the production of metabolites by microorganisms associated during fermentation (Cao et al. 2009, Lee et al. 2012). Fermentation not only enhances the functional property of soybean but it also reduces the toxic compounds such as phytic acid, agglutinin, proteinase-inhibitors and urease (Egounlety and Aworh 2003, Reddy and Pierson 1994). The bioactive compounds that are produced by the starter culture during fermentation are γ-amino butyric acid (GABA) (Watanabe et al. 2007), vitamins and provitamins (Denter et al. 1998), lipopeptides (Cao et al. 2009) and bacteriocins (Sanjukta and Rai 2016).

Fermented soybean products are consumed in many Asian and African countries and their popularity is increasing due to the recent research on health promoting properties. These fermented soybean products include *natto, tofuyo, miso* (Japan), *doenjang, cheonggukjang, meju, kanjang* (Korea), *douchi, sufu, doubanjiang* (China), *thua-nao* (Thialand), *tempeh* (Indonesia), *hawaijar, kinema, tungrymbai* (India) (Sanjukta and Rai 2016). Enormous *in vivo* and *in vitro* studies along with the epidemiological survey in human have shown that health promoting properties (nutritional and the therapeutic values) are higher in fermented soybean products in comparison to the unfermented form (Sanjukta and Rai 2016, Murekatete et al. 2012, Spector et al. 2003). The increase in health benefits in the fermented soybean products has been attributed to the increase in free isoflavones and bioactive peptides during the process (Cho et al. 2009). The bioactive molecules released during the process of soybean fermentation contribute to numerous therapeutic properties like antihypertensive (Zhang et al. 2006, Gibbs et al. 2004), antioxidant (Sanjukta et al. 2015, Watanabe et al. 2007), antidiabetic (Kwon et al. 2011), antitumor (Jung et al. 2006, Cao et al. 2009), immunomodulatory (Wagar et al. 2009), antimicrobial (Cao et al. 2009) and antithrombotic property (Park et al. 2012). The current book chapter gives detail information on the health promoting properties of the bioactive molecules present in fermented soybean products.

2. Production of Bioactive Molecules During Soybean Fermentation

In recent years food derived bioactive molecules have attracted many researchers for the development of functional foods having positive impact on human health. Raw soybean contain isoflavones (daidzin, genistin and glycitin) as glucosides form, phenolic acids and proteins, which on hydrolysis/transformation release molecules having higher bioactive properties (Watanabe et al. 2002). Health benefits in fermented soybean products are due to (i) bioactive molecules produced by the starter, and (ii) bioactive molecules produced during transformation/hydrolysis of raw soybeans by the enzymes produced by the starter. Many bioactive molecules such as aglycones, flavanols and phenolic acids are released during fermentation of soybean due to the increase in specific enzymatic activities during the process (Fig. 1). Most of the studies on health benefits due to consumption of fermented soybean have been attributed to free polyphenols and bioactive peptides produced during fermentation.

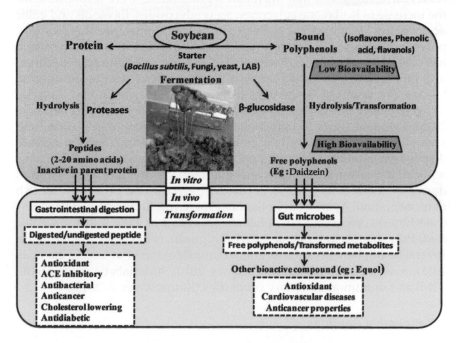

Figure 1. Production of bioactive molecules during soybean fermentation and their health benefits. LAB—lactic acid bacteria.

2.1 Polyphenols

Polyphenols are the most abundantly available bioactive components and are reported to possess antioxidant activity (Moktan et al. 2008, Sanjukta et al. 2015), antidiabetic property (Shukla et al. 2016), anticancer activity (Hirayama 1982, Zaid and El-Shenawy 2010) and ability to prevent cardiovascular diseases (Watanabe et al. 2002). The health benefits of any polyphenols depend on its bioavailability after its consumption. Bound polyphenols in the non-fermented soybean are hydrolyzed to free form by the enzymes produced by the microorganisms involved during fermentation (Fukutake et al. 1996). Thus, fermentation increases free polyphenol content, which results in its higher bioavailability and associated health benefits in comparison to unfermented soybean (Kwak et al. 2012, Moktan et al. 2008). In their study, Kano et al. (2006) have reported that isoflavone aglycones (free form) are absorbed 5–6 times higher than their corresponding glucosides (bound form) in Japanese men and women.

The increase in polyphenol content has been attributed to the increase in β-glucosidase activity of the starter culture (Cho et al. 2011, Hati et al. 2015, Kim and Yoon 1999, Sohn et al. 2000). In their study, Cho et al. (2011) have suggested that microbial β-glucosidase were responsible for the increase in isoflavone aglycones, acetylglycosides, flavanols and gallic acid during *cheonggukjang* fermentation. Isoflavones are the group of polyphenols present in soybean that possess estrogen like activity, which has several health benefits (Kwon et al. 2010). In another study, application of β-glucosidase producing *Lactobacillus rhamnosus* C6 strain resulted in biotransformation of isoflavones from glycones (genistin and daidzin) to aglycones (genistein and daidzein) in fermented soymilk (Hati et al. 2015). Different types of polyphenols such as aglycones, flavanols, and phenolic acids that enhanced the antioxidant property are found to increase during fermentation (Cho et al. 2009). In recent study, Cho et al. (2011) have reported increase in levels of alycones (daidzein, glycitein and genistein) and decreased levels of the glycosides (daidzin, glycitin and genistin) after 48 h of fermentation. Apart from isoflavones there was an increase in flavonols (catechin and epicatechin) and phenolic acids such as gallic acid and vanillic acid. Polyphenols in fermented soybean products have been reported for several health benefits such as antioxidant (Moktan et al. 2008, Cho et al. 2011), anticancer (Zaid and El-Shenawy 2010), antidiabetic (Kwon et al. 2010) and immunomodulatory properties (Karasawa et al. 2013).

2.2 Bioactive Peptides

Soy proteins on hydrolysis give biologically active peptides by the action of proteolytic enzymes produced by the microorganisms involved during fermentation (Rai and Jeyaram 2015, Sanjukta and Rai 2016). Glycinin (11S globulin) and β-conglycinin (7S globulin) are the major types of soy protein that constitute about 65–85% of the total soy proteins and are the precursors of most of the isolated bioactive peptides (Utsumi et al. 1997, Yang et al. 2000). During fermentation free amino acids and peptide content increases with increase in the activity of neutral and alkaline proteases produced by the starter culture (Chancharoonpong et al. 2012). Bioactive peptides are inactive within the sequence of parent protein and are produced by three different ways: (i) enzymatic hydrolysis using commercial proteases, (ii) fermentation using proteolytic starter, and (iii) gastrointestinal digestion (Rai et al. 2015). These peptides are mostly of 2–20 amino acids chain length, in which the amino acid compositions and sequence of amino acids determine the functionality. The functional specificity of the bioactive peptides in fermented soybean products depends on the specific microorganism and the soybean variety used for fermentation (Sanjukta et al. 2015, Rai and Kumar 2015). There are reports on the increased in the free amino acids contents in number of fermented soybean products such as *natto* (Nikkuni et al. 1995), *kinema* (Nikkuni et al. 1995), *thua nao* (Dajanta et al. 2011), *sufu* (Han et al. 2001) and *cheonggukjang* (Hong et al. 2008). The sequence or arrangement of amino acids in particular peptide decides its functionality exhibiting wide range of health promoting properties such as ACE inhibitory (Ibe et al. 2009), antioxidant (Ajibola et al. 2011), antimicrobial (Cao et al. 2009), antidiabetic (Yang et al. 2012) and anticancer properties (Lee et al. 2012). The detail updated information on bioactive peptides in different types of fermented soybean products is presented in Table 1.

3. Health Benefits of Bioactive Molecules in Fermented Soybean Products

Bioactive molecules produced during soybean fermentation have shown to possess several health promoting properties. Most of the research on health benefits of fermented soybean products have implicated that polyphenols and peptides are the key bioactive components responsible for the positive impact on human health. The health benefits exhibited by these molecules includes antioxidant, antihypertensive, antidiabetic, anticancer, immunomodulatory and antimicrobial properties (Kim et al. 2008, Karasawa et al. 2013, Sanjukta et al. 2015, Shin et al. 2001). The main objective of this section is to discuss the health benefits of fermented soybean products.

3.1 Antioxidant Property

Metabolic process during growth and maintenance of human cells and body can lead to the formation of by-product such as reactive oxygen species (ROS) that ultimately generates free radicals (Nimse et al. 2015). If the ROS production exceeds the capacity of individual antioxidant defense ability, it leads to oxidative stress followed by cell damage, ageing, diabetes and cardiovascular diseases (Nimse et al. 2015). Antioxidants in fermented soybeans prevent oxidation process either by free radicals scavenging, chelating metal-ions or by enhancing the production and activities of the antioxidant enzymes (Wang et al. 2008, Hu et al. 2004). Antioxidant activities

Table 1. Types of fermented soybean products and health benefits of soybean derived bioactive peptides.

Fermented soybean	Name of the starter	Health benefits	References
Bacillus fermented soybean products			
Fermented soybean	*Bacillus subtilis* MTCC5480, *B. subtilis* MTCC1747 *B. subtilis* SHZ	Antioxidant Antioxidant	Sanjukta et al. 2015 Yu et al. 2008
Natto	*B. natto* O9516 *B. natto* TK1	Antihypertensive Anticancer, Antimicrobial	Ibe et al. 2009 Cao et al. 2009
Cheonggukjang	*B. subtilis* CH-1023 *B. licheniformis* SCD 111067P	Antihypertensive Antidiabetic	Korhonen et al. 2003 Kwon et al. 2007
Doenjang	*B. subtilis* SCB 3	ACE-inhibitory	Hwang 1997
Douchi qu	*B. subtilis natto, B. subtilis* B1	ACE-inhibitory	Li et al. 2009
Lactic acid bacteria fermented soybean products			
Soy yogurt	*Lactobacillus delbrueckii* LB 1466, *Streptococcus thermophilus* St 1342, *Lb. acidophilus* LAFTI L10, *Bifidobacterium lactis* LAFTI B94, *Lb. paracasei* LAFTI L26	ACE-inhibitory	Donkor et al. 2005
Fermented Soymilk	*Lb. casei, Lb. acidophilus, Streptococcus thermophilus, Lb. bulgaricus, Bifidobacterium longum*	Antihypertensive	Tsai et al. 2006
Fermented Soymilk	*Lb. casei*	Antihypertensive	Bao and Chi 2016
Fermented Soymilk	*Lb. plantarum* C2	Antibacterial	Singh et al. 2015
Fungi fermented soybean products			
Douchi	*Aspergillus egypticus; Aspergillus oryzae*	Antihypertensive, Antioxidant	Zhang et al. 2006 Wang et al. 2008
Tempeh	*Rhizopus microspores*	Antioxidant, Antimicrobial	Gibbs et al. 2004 Moreno et al. 2002
Fermented soybean	*A. oryzae*	ACE inhibitory	Rho et al. 2009
Monascus fermented soybean	*Monascus pilosus* KFRI-1140	ACE inhibitory	Pyo and Lee 2007
Douchi qu	*A. oryzae, Mucor wutungkiao*	ACE inhibitory	Li et al. 2009
Shoyu	*Saccharomyces rouxii* and *Torulopsis versatilis*	ACE inhibitory	Kataoka 2005
Sufu	*Actinomucor taiwanensis*	ACE inhibitory	Wang et al. 2003
Mao-tofu	*Mucor micheli ex fries*	ACE inhibitory	Hang and Zhao 2012
Tofuyo	*A. oryzae* (as yellow *koji*), *Monascus* sp. (as red *kojis*)	Antihypertensive	Kuba et al. 2003, Yasuda et al. 1995

ACE – angiotensin converting enzyme

in fermented soybean products are mainly due to polyphenols (isoflavones, flavonoids, phenolic acids, etc.) and peptides (Sanjukta and Rai 2016). Apart from polyphenols and peptides there are compounds present in soybean at low concentration but can prevent oxidation that includes ascorbic acid, riboflavin, niacin, Vitamin B_6, and Vitamin B_{12}, β-carotene, carotenoids, ergosterol, α-, β-, γ- and δ-tocopherol (Denter et al. 1998). Among the health benefits of fermented soybean, most of the studies have been done on enhancement of antioxidant activity, therefore in the current chapter it has been grouped as (i) Polyphenol mediated antioxidant activity, and (ii) Peptide mediated antioxidant activity.

3.1.1 Polyphenol mediated antioxidant activity

Several authors have reported increase in antioxidant activity in fermented soybean products due to increase in free polyphenols (Kwak et al. 2007). Enhancement of antioxidant activity due to polyphenols has been reported in bacterial fermented soybean such as *kinema*, black soybean, *chungkookjang*, *thua nao* and fermented soya milk (Moktan et al. 2008, Kim et al. 2008, Sourabh et al. 2015). Antioxidant activities in methanolic extracts of *kinema* fermented with *Bacillus subtilis* increased due to increase in total polyphenol content, which was found to be 144% higher than non-fermented cooked soybean (Moktan et al. 2008). In another *Bacillus* fermented soybean product *chungkookjang*, it was found that increase in isoflavones (genistein and daidzein) was responsible for the enhancement of antioxidant activities during fermentation (Kim et al. 2008). They also concluded that the isoflavones released during fermentation inhibited lipid peroxidation, prevented oxidative DNA damages and stress. The methanolic extracts of *chungkookjang* prepared by fermentation of large black soybean with *Bacillus megaterium* SMY-212 also had higher antioxidant activities, which was due to the increase in concentration of total phenolic and isoflavone contents during fermentation (Shon et al. 2007). *In vitro* and *in vivo* studies of ethanolic extract of *chungkookjang* suggested that the increased isoflavones and the malonylglycosides could be the possible reason for higher antioxidant activity in Sprague-Dawley rats fed with high fats diet (Kwak et al. 2007). Methanolic extract of the *thua-nao* fermented with naturally occurring bacteria and *B. subtilis* TN51 resulted in higher antioxidant activity, which was correlated with increase in total phenolics released during fermentation (Dajanta et al. 2011).

Antioxidant activity in the methanolic extract depends on the type of soybean variety and starter used for fermentation. In their study, Dajanta et al. (2013) fermented black and yellow soybean with *B. subtilis* TN51 and found higher antioxidant activity and LDL peroxidation inhibiting property in fermented black soybean, which was due to higher phenolic and total flavonoids contents. Acetone extract of the black soybeans fermented with *B. subtilis* BCRC 14715 exhibited highest total phenolic and flavonoid contents and gave good scavenging and chelating properties (Juan and Chou 2010). Increase in antioxidant activity has also been reported during soymilk fermentation using β-glucosidase producing *Lactobacillus rhamnosus* CRL981 (Marazza et al. 2012). Further, it was also reported that mice fed with soymilk fermented with *Lb. rhamnosus* CRL981 had higher liver antioxidant enzyme activities in comparison to group fed with unfermented soymilk (Marazza et al. 2013).

There are several fungi fermented soybean products consumed in Asian countries that have been studied for enhancement of antioxidant activity (Esaki et al. 1997). Among thirty types of *Aspergillus* fermented soybean, soybean fermented with *Aspergillus satoi* exhibited strongest antioxidant activity in its methanolic extracts (Esaki et al. 1997). The compound showing highest antioxidant activity was identified as 2, 3-dihydroxybenzoic acid (a phenolic compound) by UV, EI-MS, and 1H- and 13C-NMR techniques. The ethanolic extracts of *Monascus*-fermented soybean, fermented for 15 d at 30°C showed enhancement of antioxidant activity by 5.2 to 7.4 fold as compared to unfermented soybean extracts (Pyo and Lee 2007). In *Monascus* fermentation soybean the factors responsible for the increase in antioxidant activity were found to be isoflavone aglycones and bioactive mevinolins derived during soybean fermentation. *Monascus* fermented soymilk also showed higher total phenolics isoflavones and Monacolin K content, which were resistant to homogenization and pasteurization, thus increasing the antioxidant property of the final product (Lim et al. 2010).

The isoflavones genistein and daidzein were also found to be responsible for the increase in antioxidant activity in *tempeh* extracted with 80% methanol (Murakami et al. 1984). HPLC analysis of the extracts confirmed the liberation of these isoflavones from its glycosides during fermentation. A novel antioxidant compound 3-hydroxyanthranilic acid was reported in the methanolic extract of *tempeh* fermented with *Rhizopus oligosporus* IFO-32002, 32003 (Esaki et al. 1997). Methanol extracts of soy *koji* fermented with *Aspergillus oryzae* exhibited good antioxidant activity, which on further analysis showed the presence of antioxidant compounds such as 6-hydroxydaidzein along with 8-hydroxydaidzein and 8-hydroxygenistein (Esaki et al. 1999). Soybean fermented with *Trichoderma harzianum* NBRI-1055 resulted in increase in free polyphenols and its water extracts exhibited antioxidant activities like free radical scavenging activity, hydroxyl and superoxide radicals scavenging properties, reducing power and metal ions chelating activity (Singh et al. 2010). HPLC analysis showed that the increase in antioxidant activity was due to the increase in free isoflavones, flavonoids and phenolic acids. Anthocyanins content and the total extractable phenolic compounds were increased in the methanolic extract of the black soybeans fermented with GRAS filamentous fungi (Lee et al. 2008). Total phenolic contents and antioxidant activities were increased in the soybean *koji* fermented with various GRAS filamentous fungi, including *Aspergillus sojae* BCRC 30103, *A. oryzae* BCRC 30222, *Aspergillus awamori*, *Actinomucor taiwanensis* and *Rhizopus* sp. (Lin et al. 2006). Among all the products, soybean *koji* fermented using *A. awamori* exhibited highest DPPH radicals scavenging activity, Fe^{2+} chelating ability and reducing power. In another study, Yao et al. (2010) have applied GRAS microorganisms, including *Aspergillus* sp., *Bacillus* sp. and yeast (*Issatchenkia orientalis*) for production of fermented black soybean at 30°C for 3 d and revealed that the antioxidant activities increased with the increased in the total phenolics and isoflavones content during fermentation. Symbiotic fermentation of soybean with LAB and yeast resulted in high antioxidant activity *in vitro*, also the antioxidant enzymes such as CAT, SOD, GPX were found to be increased in liver, which was concluded to be due to the high content of isoflavones, saponins and amino acids (Hu et al. 2004).

3.1.2 Peptide mediated antioxidant activity

As protein is one of the major components in soybean, there are more possibilities of production of antioxidant peptides during soybean fermentation (Sanjukta and Rai 2016). Antioxidant properties of peptides formed during fermentation depend on the types of amino acids present in the peptide. The amino acids reported to be present in antioxidant peptides includes Trp, His, Phe, Met, Gly, Leu, Ala, Tyr and Val (Guo et al. 2009, Ajibola et al. 2011, Sanjukta and Rai 2016). Functional groups of the amino acids such as imidazole in His, indolic in Trp and phenolic in Tyr are also one of the factors responsible for the antioxidant activity (Guo et al. 2009, Nam et al. 2008). In their study, Iwai et al. (2002), suggested that antioxidant activities including oxygen radicals scavenging activity and inhibition of plasma LDL oxidation of *Bacillus* fermented *natto* was due to the peptides content in the extract (Iwai et al. 2002). Antioxidant peptide isolated from *B. subtilis* SHZ fermented soybean exhibited significant antioxidant property by showing its ability to scavenge superoxide (62%) and hydroxyl (96%) radicals at concentration of 10 mg/ml (Yu et al. 2008). The fraction P1 obtained by RP-HPLC and LC-MS/MS, showed strongest scavenging and chelating activities. The aqueous extracts of *A. oryzae* fermented *douchi* were found to possess total isoflavones and peptides contents of 0.087% and 40.7%, respectively (Wang et al. 2008). It was reported that the antioxidant activity was found to be higher by the peptides in comparison to the isoflavones. *Douchi* fermented with *B. subtilis* B1, isolated from *douchi koji* resulted in higher degree of protein hydrolysis and production of smaller peptides enhancing the antioxidant activities (Fan et al. 2009). The water-soluble fraction of *Rhizopus* fermented *tempeh* (aerobic fermentation) gave higher antioxidant activity, which was found to be due to the increased free amino acids and peptides content during fermentation (Watanabe et al. 2007). Gibbs et al. (2004) have identified antioxidant peptides that were obtained by endoproteases digests (pronase, trypsin, Glu C protease, plasma proteases and kidney membrane proteases) in *R. oligosporus* NRRL 2710 fermented *tempeh*, which exhibited excellent antioxidant activities. In a recent report, Sanjukta et al. (2015) have reported the antioxidant activity of soybean varieties of Sikkim Himalayan region that were fermented with proteolytic *B. subtilis*. They also showed that activity in water extracts of soybean fermented with different strains of *B. subtilis* exhibited different trends on gastrointestinal digestion. As different strains are responsible for production of different types of peptide, there are possibilities of finding novel and potential antioxidant peptides.

3.2 Antihypertensive Property

Angiotensin I converting enzyme (ACE) is an important enzyme that is responsible for the conversion of angiotensin I to angiotensin II, which raises the blood pressure by acting directly on the blood vessels, sympathetic nerves, adrenal glands and also inactivates bradykinin, a potent vasodilator that can lead to hypertension and chronic cardiovascular diseases (Paul et al. 2006). In recent years, several protein rich fermented foods have been reported as a source of peptides, which are responsible for inhibition of ACE (Rai and Jeyaram 2015, Rai et al. 2015). In fermented soybean products ACE inhibitory or antihypertensive activities are mainly due to small peptides released by hydrolysis of soy proteins during fermentation. ACE inhibitory activity has been reported in fermented soybean products such as *natto, tofuyo, chunggugjang, douchi, tempeh, miso, sufu,* soy paste, soymilk, soy yogurt and soy sauce (Kuba et al. 2003, Tsai et al. 2006, Li et al. 2010, Ma et al. 2012).

ACE inhibitory activity of a peptide is due to the presence of hydrophobic amino acids (Try, Phe, Trp, Ala, Ile, Val, and Met) or positively charge amino acids (Arg and Lys) and Pro at the C terminal position of the ACE inhibitory peptides (Haque and Chand 2008, He et al. 2012, Rai et al. 2015). The ACE inhibitory activity in *tofuyo* extracts was reported by Kuba et al. (2003), in which two peptides (non-competitive inhibitors) were found to be the responsible factor for ACE inhibitory activity. The amino acid sequence Ile-Phe-Leu and Trp-Leu of the peptides were found in α- and β-subunits of β conglycinin and B-, B1A- and BX-subunits of glycinin and were resistant to gastrointestinal enzymes (pepsin, chymotrypsin and trypsin). Antihypertensive activity was reported in *B. subtilis* CH-1023 fermented *chunggugjang* by Korhonen et al. (2003). The effect of the ACE inhibitory peptide from *chunggugjang* was studied in 4 volunteers by Toshiro et al. (2004). After 2 hr of single dose administration of 20 g of *chunggugjang*, the systolic blood pressure of the volunteer reduced by 15 mmHg, whereas the diastolic blood pressure reduced by 8 mmHg. On further analysis, the peptide was found to be Lys-Pro (0.083 mg/100 g sample), which on purification exhibited ACE inhibitory activity of IC_{50} = 32.1 µM. The *in vitro* and *in vivo* study of *B. subtilis* natto O9516 fermented Japanese *natto* exhibited ACE inhibitory and antihypertensive properties, which on further study revealed that the activity was due to small peptide (Ibe et al. 2009). The peptide with ACE inhibitory activity *in vitro* was purified and orally administered to spontaneously hypertensive rats (SHR) single dose at different concentrations. Decreased in systolic blood pressure was observed within 5 hr of oral administration.

Several studies have shown that ACE inhibitory activity increased after fermentation, which proved that ACE inhibitors are produced during the process (Zhang et al. 2006, Hang and Zhao 2012). Study on ACE inhibitory activity of *douchi* fermented with *A. egyptiacus* revealed that the activity increased with the advancement in fermentation and the peptides obtained were pro-drug type or a mixture of pro-drug type and true inhibitor type peptides, which were resistant to gastrointestinal digestion (Zhang et al. 2006). The fraction with highest activity was found to possess Phe, Ile and Gly in the ratio of 1:2:5. Fungus fermented *sufu*, Chinese soybean cheese was studied for its ACE inhibitory activity with respect to time (Ma et al. 2012). There was 11.55–37.61% increase in ACE inhibitory activity during the process of fermentation to the ripening stage, which was correlated with the increased in peptide contents during the fermentation. Fermented soybean extract of soybean fermented at elevated temperature for better proteolytic hydrolysis was examined for ACE inhibitory activity (Rho et al. 2009). In the process, potent ACE inhibitory peptides was obtained with the amino acid sequence of Leu-Val-Gln-Gly-Ser by Edman degradation method that had good ACE inhibitory activity (IC_{50} –22 mg/ml), which was 66 fold higher in comparison to the control.

Comparative study of ACE inhibitory activity of different fermented soy products was done to know the potentiality in preventing hypertension and other cardiovascular related diseases. In comparison with *tempeh*, *tofuyo*, *miso* and *natto*, Chinese fermented soy paste exhibited lowest IC_{50} value of 0.012 mg/ml, which was due to specific peptides released during fermentation (Li et al. 2010). In another study, His-His-Leu, a tripeptide that exhibited excellent ACE inhibitory activity of IC_{50} value 2.2 µg/ml was isolated from Korean fermented soy paste, which was further synthesized and tested in SHR for its activity (Shin et al. 2001). In some cases fermented soybean seasoning (FSS) exhibited higher activity (IC_{50} = 454 µg/ml) than regular soya sauce (IC_{50} = 1620 µg/ml) (Nakahara et al. 2010). The activity

(IC$_{50}$ values) of ACE inhibitory peptides isolated from FSS were Val-Pro (480 µg/ml), Ala-Trp (10 µg/ml), Gly-Trp (30 µg/ml), Gly-Tyr (97 µg/ml), Ala-Tyr (48 µg/ml), Ser-Tyr (67 µg/ml), Ala-Phe (190 µg/ml), Ala-Ile (690 µg/ml) and Val-Gly (1100 µg/ml). There are some reports on ACE inhibitory activities exhibited by fermented soymilk and soy yogurt. ACE inhibitory property in soymilk fermented with *Lb. casei, Lb. acidophilus, S. thermophilus, Lb. bulgaricus, B. longum* was studied by Tsai et al. (2006). The study revealed that peptide content and free amino acid content were increased after 30 hr and gave the IC$_{50}$ value of 9.28 to 0.66 mg powder/ml. Another strain specific activity was observed in soymilk fermented with *E. faecium* strain resulted in production of peptides exhibiting high ACE inhibitory properties (Martinez-Villauenga et al. 2012). Good ACE inhibitory activities were observed in soy yogurt fermented with various mix cultures (Donkor et al. 2005). Thus, it can be concluded that fermented soybean products can be a potential source of peptides responsible for antihypertensive property.

3.3 Anticancer Property

Cancer is one of the major health problems globally and functional foods having cancer preventing ability have gained more interest among researchers. Recent scientific evidences have suggested that bioactive molecules in fermented soybean have the potential in reducing the risk of cancer. Polyphenol released in fermented soybean products have shown to possess anticancer effect and their efficiency depends on the microorganisms involved as a starter/co-starter (Jung et al. 2006). *Miso* fermented with *B. subtilis* and *B. subtilis* + *A. oryzae* exhibited increased isoflavone contents as compared to the non-fermented soybean, which exhibited anticancer property in human cancer cell lines HEPG2 (liver carcinoma), MCF7 (breast carcinoma), and HCT116 (colon carcinoma) (Zaid and El-Shenawy 2010). This study highlighted the importance of the microorganisms involved in the process of fermentation in exerting anticancer property. However, the impact of fermentation time was observed in *doenjang* that exhibited greater antimutagenic and anticancer activities as compared to *chungkukjang, miso* and non-fermented soybeans (Jung et al. 2006). The methanol extracts of *doenjang* fermented for 3, 6, and 24 mon were tested for prevention of tumor formation, natural killer cell activity (spleen), and glutathione *S*-transferase activity (liver) in sarcoma-180 cell–transplanted mice and found 2 to 3 fold increase in antitumor effects. The inhibitory effects were also studied on lung metastasis of colon 26-M3.1 cells in Balb/c mice from which it has been known that the 24 mon fermented *doenjang* showed 2 to 3 fold increase in antimetastatic effects. The antimutagenic and anticancer activities of *doenjang* correlated with the increased genistein contents. In relation with this study, Choi et al. (2001)

have studied the ability of the hexane fraction of *doenjang* on cell cycle progression in the human breast carcinoma MCF-7 cells. The treatment of *doenjang* hexane fraction resulted in induction of G1 phase arrest of the cell cycle.

There are studies on the anticancer or antitumor properties of fermented soybean products that are due to the low molecular weight peptides such as surfactins (Lee et al. 2012) and lipopeptides (Cao et al. 2009) released by the microorganisms during fermentation. Most of the anticancer/antitumor properties of the fermented soybean products are due to the short amino acid sequences that are mostly the surfactin like compounds or lipopeptides. *Choongkukjang* fermented with *B. subtilis* CSY191, isolated from *doenjang* produced a surfactin like compound that inhibits the growth of MCF-7 human breast cancer cells in a dose-dependent manner (IC_{50} of 10 µg/ml at 24 hr) however at 48 hr, the level of anticancer activity was increased from 2.6 to 5.1 fold (Lee et al. 2012). The same activity was observed in *doenjang* in which the surfactin-like compound when tested under artificial gastric conditions (pH 3.0 for 3 hr) could have survival of 58.3%. With further analysis, three isoforms of the surfactin-like compound were obtained whose amino acid sequences were identical (GLLVDLL). A lipopeptide biosurfactant produced by *B. natto* TK-1 exhibited antitumor property at dose dependent manner in K-562 and BEL-7402 cells with IC_{50} value of 19.1 mg/ml and 30.2 mg/ml, respectively (Cao et al. 2009). Fermented soy milk products are also reported for their positive effects on human breast carcinoma cell lines (MCF-7 cells) (Chang et al. 2002). Inhibition of tumor growth was observed when aqueous extract of fermented soy milk product was treated on MCF-7 cells.

3.4 Antimicrobial Property

Fermented soybean products exhibit antimicrobial activity and the compound responsible for antimicrobial effect are produced by the microorganisms (lactic acid bacteria and *Bacillus* spp.) associated during fermentation. Korean traditional fermented soybean paste, *kyeopjang* was found to possess antimicrobial peptide (AMP IC-1), which was produced by *B. subtilis* SCK-2 involved in the fermentation process (Yeo et al. 2011). Further studies showed that the strain exhibited a narrow antagonistic activity against *Bacillus cereus*, a food poisoning organism. Results on purification and characterization of AMP IC-1 showed that the peptide contains about 33 residues and 13 types of amino acids including Cys, Asp or Asn, Glu or Gln, Ser, Gly, Arg, Thr, Ala, Pro, Val, Ile, Leu, and Lys. *B. subtilis* SC-8, another potent microbe was isolated from Korean soy paste, *kyeopjang* that showed narrow antagonistic activity for the *B. cereus* group, which was due to the presence of the antibiotics biosynthesis coding genes in *B. subtilis* SC-8 genome (Yeo et al. 2012). Broad-spectrum antimicrobial activity was observed in *B. subtilis* HJ18-4 isolated from buckwheat *sokseongjang*, a traditional Korean fermented soybean food against *B. cereus* (Eom et al. 2014). Down-regulation of the *B. cereus* toxin related genes *groEL*, *nheA*, *nheC* and *entFM* was found to be the key factor for antibacterial activity of *B. subtilis* HJ18-4. It was discussed that the lipopeptides in *B. subtilis* strains could be the factor responsible for the antibacterial activity against *B. cereus*. Another lipopeptide biosurfactant produced by *Bacillus natto* TK-1 had anti-adhesive activity against many bacteria (Cao et al. 2009).

Tempeh exhibited good antimicrobial activity against Gram-positive pathogens such as *Listeria monocytogenes*, which was due to the partially purified bacteriocins (proteinaceous nature) from *E. faecium* LMG 19827 and *E. faecium* LMG 19828 (Moreno et al. 2002). It is reported that these bacteriocins (3.4 kDa for B1 bacteriocins, and 3.4 kDa & 5.8 kDa for B2 bacteriocins) were produced during late exponential/early stationary growth phase and were resistant to heat with bacteriostatic mode of action. Further, Yi et al. (1999) reported that *doenjang* exhibited antibacterial effects against bacterial pathogens such as *B. cereus*, *E. coil*, *L. monocytogenes* and *S. aureus*. Some bacteriocins with antimicrobial activity are temperature and pH dependent showing significant inhibition at optimum temperature when compared with uncontrolled condition (Onda et al. 2003). Antimicrobial activity in *miso*-paste was due to the bacteriocin produced by *Lactococcus* sp. GM005 which exhibited strong antibacterial effect against *Lactobacillus sakei* JCM1157T when maintained at 30°C at pH 6.0. Based on gel-filtration and tricine-SDS-PAGE analysis it was concluded that the bacteriocin was a tetrametric structure having molecular weight 9.6 kDa and 2.4 kDa, respectively with high proportion of hydrophobic amino acid residues and lanthionine. Zendo et al. (2005) have isolated bacteriocin produced by *Enterococcus mundtii* QU 2, which was stable at wide pH ranges and heat resistant showing inhibitory effect against indicator strains of *Enterococcus*, *Lactobacillus*, *Leuconostoc*, *Pediococcus* and *Listeria*. *B. subtilis* CSY191 isolated from *doenjang* produced a potential surfactin-like compound, which was partially resistant (58.3%) to artificial gastric conditions at pH 3 for 3 hr (Lee et al. 2012). On further analysis, three potential surfactin isoforms were found with protonated masses of m/z 1030.7, 1044.7, and 1058.71 with identical amino acids (GLLVDLL) and hydroxy fatty acids (13–15 C in length). Isoflavone in *doenjang* has also been reported for its antimicrobial effect against antibiotic resistant strains of *S. aureus* and *S. epidermidis* (Verdrengh et al. 2004). The antimicrobial property in fermented soybeans can be useful in preventing both spoilage causing bacteria and controlling antibiotic resistant pathogenic bacteria (Yi et al. 1999).

3.5 Antidiabetic Property

Fermented soybean products have shown to possess antidiabetic properties, which are mostly against Type II diabetes mellitus (Kwon et al. 2010). The antidiabetic property of fermented soybean products has been reported due to the presence of isoflavones and bioactive peptides (Kwon et al. 2011, Sanjukta and Rai 2016). The fermented soybean products, which has been reported for antidiabetic effects includes *meju, doenjang, kochujang,* and *chungkookjang* (Kwon et al. 2010, Yang et al. 2012). Soy isoflavones can have antidiabetes effect due to their structural similarities to estrogen, which are reported to have beneficial effect in prevention of Type II diabetes mellitus (Kwon et al. 2010).

Antidiabetic activity of water-soluble *touchi*-extract was studied by evaluating its efficiency in inhibiting α-glucosidase activity and eliciting anti-glycemic effects in genetically modified diabetic mice model KKAy (Fujita and Yamagami 2001). Eight weeks old male mice fed with *touchi* extract for 60 days had significantly lower fasting blood glucose levels of 6.68 ± 0.41 mmol/l in comparison with control group (8.75 ± 0.54 mmol/l). Traditionally fermented *meju* fermented for longer time of 20–60 d resulted in formation of bioactive molecules such as isoflavone aglycones and small peptides, which showed insulinotropic effect (Kwon et al. 2011). The methanol and water extracts prepared from 60 d fermented *meju* activated peroxisome proliferator activated receptor-γ in 3T3-L1 adipocytes. There was greater glucose stimulated insulin secretion capacity in Min6 insulinoma cells treated with water and methanol extracts of *meju* in comparison to cells treated with unfermented samples. *Chungkookjang* prepared with *Bacillus* spp. has also been reported to improve insulin secretion and sensitivity in type II diabetic rats (Yang et al. 2012).

Antidiabetic effect of *chungkukjang* was studied in 5 weeks old type II diabetic animal model (C57BL/KsJ-*db/db* mice) (Kim et al. 2008). Significant reduction of blood glucose and glycosylated hemoglobin level along with improvement in insulin tolerance was observed in mice supplemented with *chungkukjang* in comparison to control group. There was higher plasma insulin, pancreatic insulin and plasma glucagon level in mice supplemented with *chungkukjang*. It was concluded that the increase in physiological activity and enzymatic activity of the test mice was due to the aglycones isoflavone in fermented soybean. Soymilk fermented with β-glucosidase producing *Lb. rhamnosus* CRL-981 also showed antidiabetic effect (Marazza et al. 2013). Fermented soymilk when fed to streptozotocin induced diabetes animal model exhibited significant decrease in glucose levels, total cholesterol concentrations and increase antioxidant enzyme activities compared to animals fed with unfermented soymilk. These activities in animal models suggested that fermented soymilk enriched with isoflavone released during fermentation using β-glucosidase producing strain can be used for the treatment of diabetes. An antidiabetic effect of fermented soybean was also reported in streptozotocin induced diabetic rats (Ademiluyi et al. 2014). There was reduction in the blood glucose level and inhibitory effects on enzymatic activities (α-amylase and α-glucosidase) in streptozotocin induced diabetic rats fed with 10% fermented soybean for 14 d. It was concluded that the antidiabetes property of the fermented soybean was due to the effect of phytochemicals particularly polyphenols on starch digestion and activities of α-amylase and α-glucosidase.

Korean *doenjang* fermented with three fungal strains (*A. oryzae* J, *M. racemosus* 42 and *M. racemosus* 15) and bacterial strain *B. subtilis* TKSP 24 was studied for inhibition of α-glucosidase activities (Shukla et al. 2016). The fermented soybean products exhibited strong α-glucosidase inhibitory activities *in vitro* by 58.93–62.25% with IC_{50} values of 14.14–25.93. High phenolic content and enzymatic activity of the extract suggested that *doenjang* fermented with mix cultures could potentially prevent type II diabetes mellitus related disorders (Shukla et al. 2016). Inhibition of α-glucosidase and α-amylase activities along with reduction of postprandial glucose level by the methanolic extract of *B. subtilis* fermented soymilk exhibited the hypoglycemic effect in STZ-induced diabetic mice (Yi et al. 2009). The IC_{50} values of the extract for α-glucosidase and α-amylase were 0.77 and 0.94 mg/mL, respectively, which was correlated with the decrease of postprandial glucose level in diabetic mice. In a comparative study, *tempeh* having higher isoflavone aglycone has shown to exhibit higher antidiabetic effect in comparison to soymilk (Bintari et al. 2015). Research is needed on exploring peptides having antidiabetic effect as there are possibilities of getting novel peptides in fermented soybean possessing antidiabetic property.

3.6 Immunomodulatory Property

In recent years many food sources have been shown to possess immunomodulatory activities by boosting immune system (Soka et al. 2015) and exhibiting anti-allergic (Park et al. 2012), anti-inflammatory and anti-asthmatic properties (Bae et al. 2014). Immunomodulatory activities in fermented soybean products are either due to the bioactive molecules produced during hydrolysis or directly by the microorganism involved during fermentation (Sim et al. 2015). The larger amount of polyphenols (genistein and daidzein) and glutamine rich peptide in fermented non-salty soybean powder stimulated the cellular immune response in C3H/HeN mice (Karasawa et al. 2013). Antiasthmatic activity of *cheonggukjang* extract was studied in mouse model of ovalbumin induced asthma (C57BL/6 mice) by suppressing histamine release (Bae et al. 2014). Activity of *cheonggukjang* extract (70% ethanol) was studied after its intraperitoneal injection (100 mg/kg/d) in C57BL/6 mice followed by monitoring asthma related inflammations in lung tissues. Down-regulation of eosinophils and monocytes and suppression of histopathological changes in lungs responsible for allergic reactions were observed in *cheonggukjang* extract treated mice. It was concluded that *cheonggukjang* can be used as a dietary supplement for antiasthmatic effect for histamine mediated asthma.

Mixed culture fermented *tempeh* has also shown the potential of triggering immune response in Sprague-Dawley rats (Soka et al. 2015). There was significant increase in intestinal immunoglobulin A (IgA) gene expression in rats fed with *tempeh* supplemented diet in comparison to unfermented soybean. Soymilk fermented with *S. thermophilus* R0083 and *Lb. helveticus* R0052 were studied for its effect on modulation of tumor necrosis factor α (TNFα) induced gene expression in intestinal epithelial cells. There was down regulation of 33 pro-inflammatory TNFα induced genes in HT-29 intestinal epithelial cells treated with fermented soymilk (Lin et al. 2016). *B. subtilis* (*natto*) spores were also studied for immunomodulatory effect on murine macrophages (Xu et al. 2012). It was found that the spores stimulated macrophage activation by up-regulating activities of acid phosphatase and lactate dehydrogenase along with enhancement of immune function by increasing activity of inducible nitric oxide synthase activity and stimulating production of nitric oxide and cytokine. *Bacillus methylotrophicus* isolated from *doenjang*, a traditional fermented Korean soy paste showed good immunomodulatory property and ability to tolerate of gastrointestinal environment (Sim et al. 2015) for its application as a potential probiotic agent.

Anti-inflammatory effect of Korean traditional fermented soybean products *doenjang* and *cheonggukjang* was studied in 4 wk old Sprague–Dawley rats fed with high fat diet (Choi et al. 2011). From this study, it is known that *cheonggukjang* suppressed redox sensitive nuclear factor kB activation and modulated gene expression for inflammatory proteins including inducible nitric oxide synthase, vascular cell adhesion molecule-1 and cyclooxygenase-2. Soymilk fermented with *Lb. helveticus* R0052 or *Bifidobacterium longum* R0175 in combination with *S. thermophilus* ST5 also exhibited immunomodulatory activity in human monocyte model (U937) (Masotti et al. 2011). Even though there are reports on immunomodulatory effects of fermented soybean, more investigation is needed to get a clear idea on the mechanism of action of some of its bioactive metabolites.

4. Conclusions

Soybean being a protein rich food is also a storehouse of many nutritional and bioactive compounds providing various health benefits. Fermentation serves to be the most efficient and cost effective approach for improving the health benefits of soybean. The bioactive metabolites in fermented soybean include isoflavones aglycones, flavonoids, phenolic acids along with bioactive peptides, which are responsible for various health benefits such as antioxidant, antihypertensive, antitumor, antidiabetic, immunomodulatory and antimicrobial property. Production of bioactive molecules is dependent on the selection of starter culture at the strain level and optimization of fermentation condition for the production of bioactive metabolites without affecting the product quality. Research on the production and enhancement of the bioactive molecules in fermented soy products is trending towards the substitution of synthetic drugs having side effects. Therefore, proper engineering of the fermentation process and keen observation with respect to the criteria towards the generation of bioactive molecules can lead in development of product, which can substitute synthetic drugs.

Acknowledgements

The authors acknowledge the Department of Biotechnology, Government of India for the funding to carry out the research on fermented foods of Sikkim Himalayan region at Institute of Bioresources and Sustainable Development, Sikkim Centre, India.

Keywords: Soybean, Fermentation, bioactive molecules, polyphenols, peptides, isoflavones, health benefits

References

Ademiluyi, A.O., Oboh, G., Boligon, A.A. and Athayde, M.L. (2014). Effect of fermented soybean condiment supplemented diet on α-amylase and α-glucosidase activities in Streptozotocin-induced diabetic rats. Journal of Functional Foods 9: 1–9.

Ajibola, C.F., Fashakin, J.B., Fagbemi, T.N. and Aluko, R.E. (2011). Effect of peptide size on antioxidant properties of African yam bean seed (*Sphenostylis stenocarpa*) protein hydrolysate fractions. International Journal of Molecular Sciences 12: 6685–6702.

Becker-Ritt, A.B., Mulinari, F., Vasconcelos, I.M. and Carlini, C.R. (2004). Antinutritional and/or toxic factors in soybean (*Glycine max* (L.) Merril) seeds: comparison of different cultivars adapted to the southern region of Brazil. Journal of Science and Food Agriculture 84: 263–270.

Bae, M.J., Shin, H.S., See, H.J., Chai, O.H. and Shon, D.H. (2014). *Cheonggukjang* ethanol extracts inhibit a murine allergic asthma *via.* suppression of mast cell-dependent anaphylactic reactions. Journal of Medicinal Food 17(1): 142–149.

Bao, Z. and Chi, Y. (2016). *In Vitro* and *In Vivo* Assessment of Angiotensin-Converting Enzyme (ACE) Inhibitory Activity of Fermented Soybean Milk by *Lactobacillus casei* strains. Current Microbiology 73: 214–219.

Bintari, S.H., Putriningtyas, N.D., Nugraheni, K., Widyastiti, N.S., Dharmana, E. and Johan, A. (2015). Effect of *tempe* and soymilk on fasting blood glucose, insulin level and pancreatic beta cell expression (study on streptozotocin-induced diabetic rats). Pakistan Journal of Nutrition 14(4): 239–246.

Cao, X.H., Liao, Z.Y., Wang, C.L., Yang, W.Y. and Lu, M.F. (2009). Evaluation of a Lipopeptide biosurfactant from *Bacillus natto* TK-1 as a potential source of anti-adhesive, antimicrobial and antitumor activities. Brazilian Journal of Microbiology 40: 373–379.

Chancharoonpong, C., Hsieh, P. and Sheu, S. (2012). Enzyme production and growth of *Aspergillus oryzae* S. on soybean koji fermentation. Asia-Pacific Chemical Biology Environmental Engineering Society Conference Proceedings 2: 57–61.

Chang, W.H., Liu, J.J., Chen, C.H., Huang, T.S. and Lu, F.J. (2002). Growth inhibition and induction of apoptosis in MCF-7 breast cancer cells by fermented soy milk. Nutrition and Cancer 43(2): 214–226.

Cho, K.M., Lee, J.H., Yun, H.D., Ahn, B.Y., Kim, H. and Seo, W.T. (2011). Changes of phytochemical constituents (isoflavones, flavanols, and phenolic acids) during *cheonggukjang* soybeans fermentation using potential probiotics *Bacillus subtilis* CS90. Journal of Food Composition and Analysis 24: 402–410.

Cho, K.M., Hong, S.Y., Math, R.K., Lee, J.H., Kambiranda, D.M., Kim, J.M., Islam, S.M.A., Yun, M.G., Cho, J.J., Lim, W.J. and Yun, H.D. (2009). Biotransformation of phenolics (isoflavones, flavanols and phenolic acids) during the fermentation of *cheonggukjang* by *Bacillus pumilus* HY1. Food Chemistry 114: 413–419.

Choi, Y.H., Choi, B.T., Lee, W.H., Rhee, S.H. and Park, K.Y. (2001). *Doenjang* hexane fraction-induced G1 arrest is associated with the inhibition of pRB phosphorylation and induction of Cdk inhibitor p21 in human breast carcinoma MCF-7 cells. Oncology Reports 8(5): 1091–1096.

Choi, J., Kwon, S.H., Park, K.Y., Yu, B.P., Kim, N.D., Jung, J.H. and Chung, H.Y. (2011). The anti-inflammatory action of fermented soybean products in kidney of high-fat-fed rats. Journal of Medicinal Food 14(3): 232–239.

Dajanta, K., Janpum, P. and Leksing, W. (2013). Antioxidant capacities, total phenolics and flavonoids in black and yellow soybeans fermented by *Bacillus subtilis*: A comparative study of Thai fermented soybeans (*thua nao*). International Food Research Journal 20(6): 3125–3132.

Dajanta, K., Apichartsrangkoon, A. and Chukeatirote, E. (2011). Antioxidant properties and total phenolics of *thua nao* (a Thai fermented soybean) as affected by Bacillus-fermentation. Journal of Microbial and Biochemical Technology 3(4): 056–059.

Denter, J., Rehm, H.J and Bisping, B. (1998). Changes in the contents of fat-soluble vitamins and provitamins during *tempe* fermentation. International Journal of Food Microbiology 45(2): 129–134.

Donkor, O.N., Henriksson, A., Vasiljevic, T. and Shah, N.P. (2005). Probiotic strains as starter cultures improve angiotensin-converting enzyme inhibitory activity in soy yogurt. Journal of Food Science 70: 375–381.

Egounlety, M. and Aworh, O.C. (2003). Effect of soaking, dehulling, cooking and fermentation with *Rhizopus oligosporus* on the oligosaccharides, trypsin inhibitor, phytic acid and tannins of soybean (*Glycine max* Merr.), cowpea (*Vigna unguiculata* L. Walp) and groundbean (*Macrotyloma geocarpa* Harms). Journal of Food Engineering 56: 249–254.

Eom, J.S., Lee, S.Y. and Choi, H.S. (2014). *Bacillus subtilis* HJ18-4 from traditional fermented soybean food inhibits *Bacillus cereus* growth and toxin-related genes. Journal of Food Science 79: 2279–2287.

Esaki, H., Kawakishi, S., Morimitsu, Y. and Osawa, T. (1999). New potent antioxidative *o*-Dihydroxyisoflavones in fermented Japanese soybean products. Bioscience, Biotechnology and Biochemistry 63(9): 1999.

Esaki, H., Onozaki, H., Kawakishi, S. and Osawa, T. (1997). Antioxidant activity and isolation from soybeans fermented with *Aspergillus* spp. Journal of Agricultural and Food Chemistry 45: 2020–2024.

Fan, J., Zhang, Y., Chang, X., Saito, M. and Li, Z. (2009). Changes in the radical scavenging activity of bacterial-type *douchi*, a traditional fermented soybean product, during the primary fermentation process. Bioscience, Biotechnology and Biochemistry 73: 2749–2753.

Fujita, H. and Yamagami, T. (2001). Fermented soybean-derived *touchi*-extract with anti-diabetic effect *via*. α-glucosidase inhibitory action in a long-term administration study with KKAy mice. Life Sciences 70: 219–227.

Fukutake, M., Takahashi, M., Ishida, K., Kawamurai, H., Sugimura, T. and Wakabayashi, K. (1996). Quantification of Genistein and Genistin in Soybeans and Soybean Products Food and Chemical Toxicology 34: 457–461.

Gibbs, B.F., Zoygman, A., Masse, R. and Mulligan, C. (2004). Production and characterization of bioactive 643 peptides from soy hydrolysate and soy-fermented food. Food Research International 37: 123.

Guo, H., Kouzuma, Y. and Yonekura, M. (2009). Structures and properties of antioxidative peptides derived from royal jelly protein. Food Chemistry 113: 238–245.

Han, B.Z., Rombouts, F.M. and Nout, M.J.R. (2001). A Chinese fermented soybean food. International Journal of Food Microbiology 65: 1–10.

Hang, M. and Zhao, X.H. (2012). Fermentation time and ethanol/water-based solvent system impacted *in vitro* ACE-inhibitory activity of the extract of *mao-tofu* fermented by *Mucor* spp. Journal of Food 10: 137–143.

Haque, E. and Chand, R. (2008). Antihypertensive and antimicrobial bioactive peptides from milk protein. European Food Research Technology 227: 7–15.

Hati, S., Vij, S., Singh, B.P. and Mandal, S. (2015). Beta-glucosidase activity and bioconversion of isoflavones during fermentation of soymilk. Journal of Science of Food and Agriculture 95: 216–220.

He, R., Ma, H., Zhao, W., Qu, W., Zhao, J., Luo, L. and Zhu, W. (2012). Modeling the QSAR of ACE-inhibitory peptides with ANN and its applied illustration. International Journal of Peptides http://dx.doi.org/10.1155/2012/620609.

Hirayama, T. (1982). Relationship of soybean paste soup intake to gastric cancer risk. Nutrition and Cancer 3: 223–233.

Hong, J.Y., Kim, E.J., Shin, S.R., Kim, T.W., Lee, I.J. and Yoon, K.Y. (2008). Physicochemical properties of *cheonggukjang* containing Korean red ginseng and *Rubus coreanum*. Korean Journal of Food Preservation 15: 872–877.

Hu, C.C., Hsiao, C.H., Huang, S.Y., Fu, S.H., Lai, C.C., Hong, T.M., Chen, H.H. and Lu, F.J. (2004). Antioxidant activity of fermented soybean extract. Journal of Agricultural and Food Chemistry 52: 5735–5739.

Hwang, J. (1997). Angiotensin I converting enzyme inhibitory effect of *doenjang* fermented by *B. subtilis* SCB-3 isolated from *meju*, Korean traditional food. Journal of the Korean Society Food Science & Nutrition 26: 776–783.

Ibe, S. Yoshida, K., Kumada, K., Tsurushiin, S., Furusho, T. and Otobe, K. (2009). Antihypertensive effects of *natto*, a traditional Japanese fermented food, in spontaneously hypertensive rats. Food Science and Technology Research 15: 199–202.

Iwai, K., Nakaya, N., Kawasaki, Y. and Matsue, H. (2002). Antioxidative function of *natto*, a kind of fermented soybeans: effect on LDL oxidation and lipid metabolism in cholesterol-fed rats. Journal of Agricultural Food Chemistry 50: 3597–3601.

Juan, M.Y. and Chou, C.C. (2010). Enhancement of antioxidant activity, total phenolic and flavonoid content of black soybeans by solid state fermentation with *Bacillus subtilis* BCRC 14715. Food Microbiology 27: 586–591.

Jung, K.O., Park, S.Y. and Park, K.Y. (2006). Longer aging time increases the anticancer and antimetastatic properties of *doenjang*. Nutrition 22: 539–545.

Kano, M., Takayanagi, T., Harada, K., Sawada, S. and Ishikawa, F. (2006). Bioavailability of isoflavone after ingestion of soy beverages in healthy adults. The Journal of Nutrition 136: 2291–2296.

Karasawa, K., Sugiura, Y., Kojima, M., Uzuhashi, Y. and Otani, H. (2013). Fermented soybean powder with rice mold in the absence of salt stimulates the cellular immune system and suppresses the humoral immune response in mice. Journal of Nutritional Science and Vitaminology. (Tokyo) 59(6): 564–9.

Kataoka, S. (2005). Functional effects of Japanese style fermented soy sauce (*shoyu*) and its components. Journal of Bioscience and Bioengineering 100: 227–234.

Kim, J.S. and Yoon, S. (1999). Isoflavone contents and β-glucosidase activities in soybean, meju and *doenjang*. Korean Journal of Food Science Technology 31: 1405–1409.

Kim, S.L., Berhow, M.A., Kim, J.T., Chi, H.Y., Lee, S.J. and Chung, I.M. (2006). Evaluation of soyasaponin, isoflavone, protein, lipid, and free sugar accumulation in developing soybean seeds. Journal of Agricultural Food Chemistry 54: 10003–10010.

Kim, D.J., Jeong, Y.J., Kwon, J.H., Moon, K.D., Kim, H.J., Jeon, S.M., Lee, M.K., Park, Y.B. and Choi, M.S. (2008). Beneficial effect of *Chungkukjang* on regulating blood glucose and pancreatic β-cell functions in C75BL/KsJ-*db/db* mice. Journal of Medicinal Food 11(2): 215–223.

Korhonen, H. and Pihlanto, A. (2003). Food-derived bioactive peptides-opportunities for designing future foods. Current Pharmaceutical Design 9: 1297.

Kuba, M., Tanaka, K., Tawata, S., Takeda, Y. and Yasuda, M. (2003). Angiotensin I-converting enzyme inhibitory peptides isolated from *tofuyo* fermented soybean. Bioscience, Biotechnology and Biochemistry 67: 1278–1283.

Kwak, C.S., Lee, M.S. and Park, S.C. (2007). Higher antioxidant properties of Chungkookjang, a fermented soybean paste, may be due to increased aglycone and malonylglycoside isoflavone during fermentation. Nutrition Research 27: 719–727.

Kwak, C.S., Park, S.C. and Song, K.Y. (2012). *Doenjang*, a fermented soybean paste, decreased visceral fat accumulation and adipocyte size in rats fed with high fat diet more effectively than nonfermented soybeans. Journal of Medicinal Food 5(1): 1–9.

Kwon, D.Y., Jang, J.S., Hong, S.M., Lee, J.E., Sung, S.R. and Park, H.R. (2007). Long-term consumption of fermented soybean-derived *chungkookjang* enhances insulinotropic action unlike soybeans in 90% pancreatectomized diabetic rats. European Journal of Nutrition 46: 44–52.

Kwon, D.Y., Daily, J.W., Kim, H.J. and Park, S. (2010). Antidiabetic effects of fermented soybean products on type 2 diabetes. Nutrition Research 30: 1–13.

Kwon, D.Y., Hong, S.M., Ahn, I.S., Kim, M.J., Yang, H.J. and Park, H. (2011). Isoflavonoids and peptides from *meju*, long-term fermented soybeans, increase insulin sensitivity and exert insulinotropic effects *in vitro*. Nutrition 27: 244–252.

Lee, Y.L., Yang, J.H. and Mau, J.L. (2008). Antioxidant properties of water extracts from *Monascus* fermented soybeans. Food Chemistry 106: 1128–1137.

Lee, J.H., Nam, S.H., Seo, W.T., Yun, H.D., Hong, S.Y., Kim, M.K. and Cho, K.M. (2012). The production of surfactin during the fermentation of *cheonggukjang* by potential probiotic *Bacillus subtilis* CSY191 and the resultant growth suppression of MCF-7 human breast cancer cells. Food Chemistry 131: 1347–1354.

Li, F.J., Yin, L.J., Cheng, Y.Q., Yamaki, K., Fan, J.F. and Li, L.T. (2009). Comparison of angiotensin I-converting enzyme inhibitor activities of pre-fermented *douchi* (a Chinese traditional fermented soybean food) started with various cultures. International Journal of Food Engineering 5: 1556–3758.

Li, F.J., Yin, L.J., Cheng, Y.Q., Saito, M., Yamaki, K. and Li, L.T. (2010). Angiotensin I converting enzyme inhibitory activities of extracts from commercial Chinese style fermented soypaste. Japan Agricultural Research Quarterly 44: 167–172.

Lim, J.Y., Kim, J.J., Lee, D.S., Kim, G.H., Shim, J.Y., Lee, I. and Imm, J.Y. (2010). Physicochemical characteristics and production of whole soymilk from Monascus fermented soybeans. Food Chemistry 120: 255–260.

Lin, C., Wei, Y. and Chou, C. (2006). Enhanced antioxidative activity of soybean *koji* prepared with various filamentous fungi. Food Microbiology 23: 628–633.

Lin, Q., Mathieu, O., Tompkins, T.A., Buckley, N.D. and Green-Johnson, J.M. (2016). Modulation of the TNFα-induced gene expression profile of intestinal epithelial cells by soy fermented with lactic acid bacteria. Journal of Functional Foods 23: 400–411.

Ma, Y., Cheng, Y., Yin, L., Wang, J. and Li, L. (2012). Effects of processing and NaCl on angiotensin I-converting enzyme inhibitory activity and γ-aminobutyric acid content during *sufu* manufacturing. Food and Bioprocess Technology 6: 1782–1789.

Marazzaa, J.A., LeBlanca, J.G., de Gioria, G.S. and Garro, M.S. (2013). Soymilk fermented with *Lactobacillus rhamnosus* CRL981 ameliorates hyperglycemia, lipid profiles and increases antioxidant enzyme activities in diabetic mice. Journal of Functional Foods 5: 1848–1853.

Marazzaa, J.A., Nazarenob, M.A., de Gioria, G.S. and Garro, M.S. (2012). Enhancement of the antioxidant capacity of soymilk by fermentation with *Lactobacillus rhamnosus*. Journal of Functional Foods 4: 594–601.

Martinez-Villaluenga, C., Torino, M.I., Martin, V., Arroyo, R., Garcia-mora, P., Pedrola, I.E., Vidal-Valverde, C., Rodriguez, J.M. and Frias, J. (2012). Multifunctional properties of soy milk fermented by *Enterococcus faecium* strains isolated from raw soy milk. Journal of Agriculture and Food Chemistry 60: 10235–10244.

Masotti, A.I., Buckley, N., Champagne, C.P. and Green-Johnson, J. (2011). Immunomodulatory bioactivity of soy and milk ferments on monocyte and macrophage models. Food Research International 44: 2475–2481.

Moktan, B., Saha, J. and Sarkar, P.K. (2008). Antioxidant activities of soybean as affected by Bacillus fermentation to *kinema*. Food Research International 41: 586–593.

Moreno, M.R.F., Leisner, J.J., Tee, L.K., Ley, C., Radu, S., Rusul, G., Vancanneyt, M. and Vuyst, L.D. (2002). Microbial analysis of Malaysian *tempeh*, and characterization of two bacteriocins produced by isolates of *Enterococcus faecium*. Journal of Applied Microbiology 92: 147–157.

Murekatete, N., Hua, Y., Kong, X. and Zhang, C. (2012). Effects of fermentation on nutritional and functional properties of soybean, maize, and germinated sorghum composite flour. International Journal of Food Engineering 8(1): 6.

Murakami, H., Asakawa, T., Terao, J. and Matsushita, S. (1984). Antioxidative stability of tempeh and liberation of isoflavones by fermentation. Agricultural and Biological Chemistry 48: 2971–2975.

Nam, K.A., You, S.G. and Kim, S.M. (2008). Molecular and physical characteristics of squid (*Todarodes pacificus*) skin collagens and biological properties of their enzymatic hydrolysates. Journal of Food Science 73: 243–255.

Nakahara, T., Sano, A., Yamaguchi, H., Sugimoto, K., Chikata, H., Kinoshita, E. and Uchida, R. (2010). Antihypertensive effect of peptide-enriched soy sauce-like seasoning and identification of its angiotensin I-converting enzyme inhibitory substances. Journal of Agricultural and Food Chemistry 58: 821–827.

Nikkuni, S., Karki, T.B., Vilkhu, K.S., Suzuki, T., Shindoh, K., Suzuki, C. and Okada, N. (1995). Mineral and amino acid contents of *kinema*, a fermented soybean food prepared in Nepal. Food Science and Technology International 1: 107–111.

Nimse, S.B. and Pal, D. (2015). Free radicals, natural antioxidants, and their reaction mechanisms. Royal Society of Chemistry 5: 27986–28006.

Onda, T., Yanagida, F., Tsuji, M., Shinohara, T. and Yokotsuka, K. (2003). Production and purification of a bacteriocin peptide produced by Lactococcus sp. strain GM005, isolated from *Miso*-paste. International Journal of Food Microbiology 87: 153–159.

Park, D.K., Choi, W.S. and Park, H.J. (2012). Antiallergic activity of novel isoflavone methyl-glycosides from *Cordyceps militaris* grown on germinated soybeans in antigen-stimulated mast cells. Journal of Agricultural and Food Chemistry 60: 2309–2315.

Paul, M., Mehr, A.P. and Kreutz, R. (2006). Physiology of local renin-angiotensin systems. Physiological Reviews 86: 747–803.

Pyo, Y.H. and Lee, T.C. (2007). The potential antioxidant capacity and angiotensin I-converting enzyme inhibitory activity of Monascus-fermented soybean extracts: evaluation of Monascus-fermented soybean extracts as multifunctional food additives. Journal of Food Science 72(3): 218–223.

Rai, A.K. and Jeyaram, K. (2015). Health benefits of functional proteins in fermented foods. pp. 455–474. *In*: J.P. Tamang (ed.). Health Benefits of Fermented Foods and Beverages. CRC Press, Taylor and Francis Group of USA.

Rai, A.K., Sanjukta, S. and Jeyaram, K. (2015). Production of Angiotensin I converting enzyme inhibitory (ACE-I) peptides during milk fermentation and its role in treatment of hypertension. Critical Reviews in Food Science and Nutrition doi:10.1080/10408398.2015.1068736.

Rai, A.K. and Kumar, R. (2015). Potential of microbial bio-resources of Sikkim Himalayan region, ENVIS Bulletin 23: 99–105.

Reddy, N.R. and Pierson, M.D. (1994). Reduction in antinutritional and toxic components in plant foods by fermentation. Food Research International 27: 281.

Rho, S.J., Lee, J.S., Chung, Y.I., Kim, Y.W. and Lee, H.G. (2009). Purification and identification of an angiotensin I-converting enzyme inhibitory peptide from fermented soybean extract. Process Biochemistry 44: 490–493.

Sanjukta, S. and Rai, A.K. (2016). Production of bioactive peptides during soybean fermentation and their potential health benefits. Trends in Food Science and Technology 50: 1–10.

Sanjukta, S., Rai, A.K., Muhammed, A., Jeyaram, K. and Talukdar, N.C. (2015). Enhancement of antioxidant properties of two soybean varieties of Sikkim Himalayan region by proteolytic *Bacillus subtilis* fermentation. Journal of Functional Foods 14: 650–658.

Shin, Z.I., Yu, R., Park, S.A., Chung, D.K., Ahn, C.W., Nam, H.S., Kim, K.S. and Lee, H.J. (2001). His-His-Leu, an angiotensin I converting enzyme inhibitory peptide derived from Korean soybean paste, exerts antihypertensive activity *in vivo*. Journal of Agricultural Food Chemistry 49: 3004.

Sohn, M.Y., Seo, K.I., Lee, S.W., Choi, S.H. and Sung, N.J. (2000). Biological activities of chungkugjang prepared with black bean and changes in phytoestrogen content during fermentation. Korean Journal of Food Science and Technology 32: 936–941.

Shon, M.Y., Lee, J., Choi, J.H., Choi, S.Y., Nam, S.H., Seo, K.I., Lee, S.W., Sung, N.J. and Park, S.K. (2007). Antioxidant and free radical scavenging activity of methanol extract of *chungkukjang*. Journal of Food Composition and Analysis 20: 113–118.

Shukla, S., Park, J., Kim, D.H., Hong, S.Y., Lee, J.S. and Kim, M. (2016). Total phenolic content, antioxidant, tyrosinase and α-glucosidase inhibitory activities of water soluble extracts of noble starter culture *doenjang*, a Korean fermented soybean sauce variety. Food Control 59: 854–861.

Sim, I., Koh, J.H., Kim, D.J., Gu, S.H., Park, A. and Lim, Y.H. (2015). *In vitro* assessment of the gastrointestinal tolerance and immunomodulatory function of *Bacillus methylotrophicus* isolated from a traditional Korean fermented soybean food. Journal of Applied Microbiology 118: 718–726.

Singh, H.B., Singh, B.N., Singh, S.P. and Nautiyal, C.S. (2010). Solid-state cultivation of *Trichoderma harzianum* NBRI-1055 for modulating natural antioxidants in soybean seed matrix. Bioresource Technology 101: 6444–6453.

Singh, B.P., Vij, S., Hati, S., Singh, D., Kumari, P. and Minj, J. (2015). Antimicrobial activity of bioactive peptides derived from fermentation of soy milk by 'Lactobacillus plantarum' C2 against common foodborne pathogens. Journal of Fermented Foods 4: 77–85.

Soka, S., Suwanto, A., Sajuthi, D. and Rusmana, I. (2015). Impact of *tempeh* supplementation on mucosal immunoglobulin A in Sprague-Dawley rats. Food Science and Biotechnology 24(4): 1481–1486.

Sourabh, A., Rai, A.K., Chauhan, A., Jeyaram, K., Taweechotipatr, M., Panesar, P.S., Sharma, R., Panesar, R., Kanwar, S.S., Walia, S., Tanasupawat, S., Sood, S., Joshi, V.K., Bali, V., Chauhan, V. and Kumar, V. (2015). Health related issues and indigenous fermented products. pp 303–343. *In*: V.K. Joshi (ed.). Indigenous Fermented Foods of South Asia. Taylor and Francis Group of USA.

Spector, D., Anthony, M., Alexander, D. and Arab, L. (2003). Soy consumption and colorectal cancer. Nutrition and Cancer 47: 1–12.

Toshiro, M., Jae, Y.H., Sung, H.J., Seok, L.D. and Bok, K.H. (2004). Isolation of angiotensin I-converting enzyme inhibitory peptide from *chungkookjang*. Korean Journal of Microbiology 40: 355–358.

Tsai, J.S., Lin, Y.S., Pan, B.S. and Chen, T.J. (2006). Antihypertensive peptides and γ-aminobutyric acid from prozyme 6 facilitated lactic acid bacteria fermentation of soymilk. Process Biochemistry 41: 1282–1288.

Utsumi, S., Matsuma, Y. and Mori, T. (1997). Structure-function relationships of soy proteins. pp. 257–291. *In*: S. Damodaran and A. Paraf (eds.). Food Proteins and Their Applications. CRC press. New York: Dekker.

Verdrengh, M., Collins, L.V., Bergin, P. and Tarkowski, A. (2004). Phytoestrogen genistein as an anti-staphylococcal agent. Microbes and Infection 6: 86–92.

Wagar, L.E., Champagne, C.P., Buckley, N.D., Raymond, Y. and Green-Johnson, J.M. (2009). Immunomodulatory properties of fermented soy and dairy milks prepared with lactic acid bacteria. Journal of Food Science 74(8): 423–430.

Wang, L., Saito, M., Tatsumi, E. and Li, L. (2003). Antioxidative and angiotensin I converting enzyme inhibitory activities of *sufu* (fermented *tofu*) extracts. Japan Agricultural Research Quarterly 37(2): 129–132.

Wang, W., Neal, A.B., Mark, A.B. and Elvira, G.D.M. (2008). Conglycinins among sources of bioactives in hydrolysates of different soybean varieties that inhibit leukemia cells *in vitro*. Journal of Agricultural and Food Chemistry 56: 4012–4020.

Watanabe, N., Fujimoto, K. and Aoki, H. (2007). Antioxidant activities of the water-soluble fraction in *tempeh*-like fermented soybean (GABA-*tempeh*). International Journal of Food Science and Nutrition 58: 577–587.

Watanabe, S., Uesugi, S. and Kikuchi, Y. (2002). Isoflavones for prevention of cancer, cardiovascular diseases, gynecological problems and possible immune potentiation. Biomedicine and Pharmacotherapy 56(6): 302–312.

Xu, X., Huang, Q., Mao, Y., Cui, Z., Li, Y., Huang, Y., Rajput, I.R., Yu, D. and Li, W. (2012). Immunomodulatory effects of *Bacillus subtilis* (*natto*) B4 spores on murine macrophages. Microbiology and Immunology 56(12): 817–824.

Yang, J.H., Mau, J.L., Ko, P.T. and Huang, L.C. (2000). Antioxidant properties of fermented soybean broth. Food Chemistry 71: 249–54.

Yang, H.J., Kwon, D.Y., Kim, M.J., Kang, S., Kim, D.S. and Park, S. (2012). Jerusalem artichoke and *chungkookjang* additively improve insulin secretion and sensitivity in diabetic rats. Nutrition and Metabolism 9: 112.

Yasuda, M., Matsumoto, T., Sakaguchi, M. and Kinjyo, S. (1995). Production of *tofuyo* using the combination of red and yellow kojis. Nippon Shokuhin Kagaku Kogaku Kaishi (in Japanese) 42: 38–43.

Yao, Q., Xiao-nan, J. and Dong, P.H. (2010). Comparison of antioxidant activities in black soybean preparations fermented with various microorganisms. Agricultural Sciences in China 9: 1065–1071.

Yeo, I.C., Lee, N.K. and Hahm, Y.T. (2012). Genome Sequencing of *Bacillus subtilis* SC-8, Antagonistic to the *Bacillus cereus* group, isolated from traditional Korean fermented-soybean food. Journal of Bacteriology 194: 536–537.

Yeo, I.C., Lee, N.K., Cha, C.J. and Hahm, Y.T. (2011). Narrow antagonistic activity of antimicrobial peptide from *Bacillus subtilis* SCK-2 against *Bacillus cereus*. Journal of Bioscience and Bioengineering 112(4): 338–344.

Yi, N., Hwang, J.Y. and Han, J.S. (2009). Hypoglycemic effect of fermented soymilk extract in STZ-induced diabetic mice. Journal of Food Science and Nutrition 14(1): 8–13.

Yi, S., Yang, J., Jeong, J., Sung, C. and Oh, M. (1999). Antimicrobial activities of soybean paste extracts. Journal of the Korean Society Food Science and Nutrition 28(6): 1230–1238.

Yu, B., Zhao-Xin, L., Xiao-Mei, B., Feng-Xia, L. and Xian-Qing, H. (2008). Scavenging and anti-fatigue activity of fermented defatted soybean peptides. European Food Research and Technology 226: 415–421.

Zaid, A.A. and El-Shenawy, N.S. (2010). Effect of *miso* (A soybean fermented food) on some human cell lines; HEPG2, MCF7 and HCT116. Journal of American Science 6(12): 1274–1282.

Zendo, T., Eungruttanagorn, N., Fujioka, S., Tashiro, Y., Nomura, K., Sera, Y., Kobayashi, G., Nakayama, J., Ishizaki, A. and Sonomoto, K. (2005). Identification and production of a bacteriocin from *Enterococcus mundtii* QU 2 isolated from soybean. Journal of Applied Microbiology 99: 1181–1190.

Zhang, J.H., Tatsumi, E., Ding, C.H. and Li, L.T. (2006). Angiotensin I-converting enzyme inhibitory peptidesin *douchi*, a Chinese traditional fermented soybean product. Food Chemistry 98: 551–557.

6

Miso, the Traditional Fermented Soybean Paste of Japan

Ken-Ichi Kusumoto[1],* and *Amit Kumar Rai*[2]

1. Introduction

In Japan, there are various types of fermented foods that form an essential component of Japanese cuisine. Japanese traditional foods include a variety of fermented seasonings using filamentous fungi termed as 'koji mold', mainly *Aspergillus oryzae*. This includes soy sauce, mirin (sweet rice wine), su (rice vinegar), and miso (fermented soybean paste). Apart from *A. oryzae*, molds used in Japanese food industries for the production of koji include *A. sojae*, *A. tamarii*, *A. awamori*, *A. saitoi* and *A. kawachi* (Esaki et al. 1997). The uniqueness of these fermented foods is the utilization of highly active fungal enzymes for the degrading biological polymers, i.e., polysaccharides such as starch and pectin, protein and lipid in food matrices. Hydrolysis of macromolecules in cereals and legumes, results in release of monomer such as monosaccharide, free amino acids and free fatty acids. The resulting foods are a product of a wide range of biochemical

[1] Food Research Institute, National Agriculture and Food Research Organization, Tsukuba, Ibaraki 305-8642, Japan.
[2] Institute of Bioresources and Sustainable Development, Sikkim Centre, Tadong - 737102, Sikkim, India.
* Corresponding author: kusumoto@affrc.go.jp

reactions, which give sweet and *umami* tastes. The fermentation techniques followed are unique in their methodology and are traditionally maintained at household levels.

Among these fermented foods, miso is a paste-like half-solid food having sweet and salty and *umami* tastes, which has been gaining popularity worldwide. Miso is absolutely essential in Japanese dietary culture and is produced by Japanese traditional methodology as well as by modern cooking. Miso has been manufactured over a thousand years in Japan and is used to cook miso soup and side dishes as seasoning. Miso soup are still the staple dishes of Washoku (Japanese dietary culture), which was listed in Intangible Cultural Heritage of UNESCO on December 2013. Several health benefits from soybean derived bioactive metabolites as well as some of the microorganism present in miso during fermentation have been reported (Onda et al. 2003, Hirota et al. 2000, Wu et al. 2000). Bioactive compounds (peptides, isoflavones) formed or released by the action of fungal enzymes on soybean during miso production have shown to exhibit antioxidant, antidiabetic, reducing the risk of cancer and antihypertensive properties (Hirota et al. 2000, Kwon et al. 2010, Shimizu et al. 2015, Zaid and El-Shenawy 2010). The current chapter focuses on the detail of types of miso, processing methods, biochemical changes, flavor enhancement and improvement in health benefits during miso production.

2. Types of Miso

Raw materials used for miso fermentation mainly include soybean, salt, water, and rice or barley. Approximately, 400 kilo tons of miso was manufactured in Japan in the year 2015 (JFMMC 2015). The type of miso depends on the concentration of salt and type of koji (solid state fermented grain by koji mold) used for miso production. The type of koji depends on the substrate (rice, barley or only soybean) used for growing the koji mold. Among them, Rice-koji-type (kome miso; 'kome' means rice grain) is the most manufactured and used for miso fermentation, which accounts 80 percent of the total miso production. Apart from rice-koji, barley-koji-type miso (mugi miso; 'mugi' means barley grain) and soybean-koji-type miso (mame miso; 'mame' means soybean grain) are also manufactured. The main microorganism for miso fermentation is koji mold, which is used as conidia, and called as moyashi (fungal flower on the rice). These fungal conidia are manufactured and supplied by industries for maintaining fungal starter strains. Some of the miso manufacturers also use salt-tolerant yeasts and/or lactic acid bacteria. In the case of kome miso, conidia of koji mold are inoculated evenly on the husked, steam-cooked rice grain and are incubated under appropriate temperature and high humidity.

Kome miso is classified into 'sweet miso', 'sweet-salty miso', and 'salty miso' based on the ratio of koji and soybean used for production. Among them, salty miso is the most manufactured type of kome miso. Kome miso is also classified into 'white miso', light colored miso', and 'dark colored miso', according to its color. Another classification of Kome miso has been done based on its physical property (i) 'Tsubu-miso' (granular type miso) contain the grain of steamed soybean, and (ii) 'Koshi-miso' (ground type miso), which is very smooth texture as the fermented materials are grinded finely. 'Koji-miso' (koji type miso) contains the grain shape of rice koji. Mugi miso is also classified according to its color and taste whereas mame miso has no classification. Miso is mostly used in cooking miso soup with dashi (Japanese soup stock made from dried fish and/or dried kelp).

The reason behind the production of varieties of miso is due to the different tradition of dietary culture followed by the people living in different location of Japan. Salty miso is most widely preferred and manufactured in Japan (Fig. 1). Shinshu, a type of miso, which is named based on the location it comes (Shinshu' means the area of Nagano Prefecture). In contrast, mame miso (Fig. 1) is limited in the central area including Nagoya (Aichi Pref., Gifu Pref., and Mie Pref.). According to its characteristic flavor, mame miso is used in cooking miso soup and used in side dishes as a seasoning, like sauce for cutlet. In Kyoto City and its neighbor areas, people prefer white-sweet miso and light colored-sweet miso (Fig. 1, Saikyo miso, 'Saikyo' is one of the local areas of Kyoto). In any special occasion in this area, i.e., the New Year days, people often cook miso soup with rice cake ('mochi'), called

Shinshu miso   Sendai miso   Saikyo miso
(kome miso)   (kome miso)   (kome miso)

Mame miso   Mugi miso

Figure 1. The examples of miso.
Courtesy: Dr. Yoshiaki Kitamura (National Agriculture and Food Research
Organization, Japan).

as 'zouni'. Mugi miso as well as miso of the mixed type using both mugi miso and kome miso are manufactured and consumed mainly in 'Kyushu' area on the west side of Japan.

3. Processing of Miso

The processing of kome miso is described in this section. One of the important raw materials for manufacture of kome miso is soybean. Most of the soybean used in miso preparation are imported from abroad (around 90 percent), mainly from United States of America, Canada, and China. There are several points to select soybean as material for miso. The miso manufacturers prefer the large grain type of soybean with thin skin, high absorption of water, easy softening, light color (bright yellow) of the grain, and good fragrance after steaming. The type of soybean having high contents of total sugar and low lipid content is also preferred. Soybean is thoroughly washed to remove dust, soil and hull. It is important to polish the soybean grains and to remove husks, in order to give a better color to miso. Further, the grains are rinsed and soaked in water. This step is necessary to remove pigments or its precursors from the grain. Soaked soybean grains are then steamed and the cooked soybean is allowed to cool till room temperature. After cooling, the steamed grains are milled with chopper machine. The second raw material is rice, which is used as the solid media for koji mold. The quality of rice is not an important criterion as in case of rice wine production, but the rice used for miso production should be safe for consumption. Rice is treated to remove husk and hull, and then is polished. The salt used for miso preparation should have low contents of iron and copper. The quality of water for miso is same as drinking water, but it should have low content of iron, copper, manganese and calcium ions.

A. oryzae is a filamentous fungus, which is the main component of the *koji* starter. There are several strains of A. oryzae and each starter company provides the strains according to the needs of the miso manufacturer. Due to the history on application of A. oryzae in food industries, United States Food and Drug Administration (FDA) included A. oryzae in the list of generally regarded as safe (GRAS) organisms, which was also later supported by the World Health Organization (WHO) (Machida et al. 2002, Takahashi et al. 2002, Maeda et al. 2005). A. oryzae strain producing strong amylase activity is suitable for sweet *miso* and strong protease activity is usually suitable for salty miso production. Manufacturing of miso has several steps, e.g., koji making, treatment of soybean, material mixing, fermentation and aging, adjustment of product. The steps of koji making, boiling soybean, and material mixture are most important for the quality of the miso. The koji making process can especially directly affect the quality of the final product (Imai 2009).

During koji making, there are several steps for treatment of rice including polishing, rinsing, soaking, steaming and cooling. This also includes inoculation of *koji* mold, fermentation of *koji* mold, and finally, terminating the fermentation process by adding salt. The polished rice is subjected to soaking in water to let them absorb water completely. Then they are steamed and allowed to cool down to room temperature. Conidia of *koji* mold are inoculated and mixed with the steamed rice, and the whole mass is incubated at 30°C. Koji mold grows on the surface of the steamed rice increasing the temperature which finally reaches up to 40°C after few hours of inoculation. At this point, the mycelia of *koji* mold invade the rice grain and during the growth of koji mold, the enzyme activity increases. For sweet, white to bright yellow type of miso, the maximum temperature should be above 35°C to increase amylase activity. For salty, deep colored one, the temperature should be below 30°C to increase protease activity. Previously, *koji* making for miso production was performed with a multiple set of wooden trays called *'koji-buta'* (*koji* making tray) (Fig. 2). Nowadays, the manufacturers introduce automatically controlled *koji* making machine (Fig. 3).

After koji preparation, it is mixed with salt which is one third of the total amount of the substrate. The role of salt is to stop the growth of koji mold. The salted koji is mixed with the milled soybean, the rest of the salt, and water. Sometimes yeast (*Zygosaccharomyces rouxii*) and/or lactic acid bacteria (*Tetragenococcus halophilus*) is added into the mixture (Imai 2009). The mixture is packed in the barrels called miso daru, and they are kept at ambient temperature or in warm condition. The mixture is fermented for several months to sometimes one year or more. In case of sweet miso, the

Figure 2. Treatment of rice koji during fermentation directly by hand.
Courtesy: Mr. Takahiro Ogawa (Asahi Breweries Co. Ltd.).

Figure 3. Automatically controlled koji making machine.
Courtesy: Mr. Takahiro Ogawa (Asahi Breweries Co. Ltd.).

fermentation period is much shorter (one or two months). After the long period of fermentation, *Candida versatilis* and *C. etchellsii* are grown. They are concerned with fragrance production specific to well-fermented miso (Suezawa and Suzuki 2007).

4. Biochemistry of Miso

Soybean fermentation leads to a wide range of biochemical changes, which has a positive impact on the final product (Sanjukta and Rai 2016, Sourabh et al. 2015). These changes include production of enzymes, hydrolysis of macromolecules, degradation of antinutritional factors, transformation of bioactive metabolites, and production of flavor (Sourabh et al. 2015). Most of the changes during fermentation are due to the enzymes produced by *A. oryzae*, which includes acid, neutral and alkaline proteases, amylase, lipases, metallopeptidase, glutaminase, acid phosphatase and phytase (Liang et al. 2009, Shimizu et al. 1993, Chancharoonpong et al. 2012, Fujita et al. 2003). Apart from fermentation the biochemical changes also occur during ripening of miso. During ripening wide range of microorganisms are involved including molds, yeast and lactic acid bacteria, which helps in hydrolysis of soybean component (Kim et al. 2010, Onda et al. 2003). Increase in free amino acids during miso fermentation on soybean hydrolysis mediated by proteolytic enzymes has been reported by Nakadai and Nasuno (1977). It was also shown that both acid proteinase and acid carboxypeptidases produced by *A. oryzae* are responsible for increase in free amino acids. Peptides formed during soybean fermentation have been shown to exhibit several beneficial roles (Sanjukta and Rai 2016). Protein hydrolysis during *A. oryzae* fermentation has also been reported by Frias et al. (2008), which was correlated with reduction in immune-reactivity.

Lipids are important biomolecules, which are very prone to changes during fermentation and storage. Lipid analysis of rice-koji *miso* samples showed that triacylglycerols were gradually hydrolysed into free fatty acids, which further transformed into fatty acid ethyl esters (Yamabe et al. 2004). In another study, decrease in triglyceride content and increase in free fatty acids during fermentation was reported by Yoshida and Kajimoto (1972). They reported that linoleic acid content decreased in the triglyceride fraction, with the increase of linoleic acid in monoglycerides and diglycerides fractions. Isoflavones are one of the important bioactive molecules in soybean, which has been reported to possess several health benefits (Chiou and Cheng 2001, Sanjukta et al. 2015). In their study, Chiou and Cheng (2001) have shown that during *A. oryzae* fermentation, aglycosylated form of isoflavones (daidzein and genistein) increased resulting in the reduction in their glycosylated forms (Daidzin and geniatin). In another study, the isoflavone profile during rice-koji *miso* fermentation and aging process indicated that isoflavone glycosides decreased from 86.4% to 44.9% with the increase in isoflavone aglycones from 9.6% to 53.3% (Yamabe et al. 2007). Increase in free isoflavones in fermented soybean products during fermentation increases its bioavailability after consumption resulting in enhancement of its beneficial effect.

After fermentation there are several biochemical changes that take place during cooking or final product development. As mentioned above, miso is mostly used in miso soup with dashi, which is added during a cooking step at household level. Main ingredients for umami taste in dashi is considered to be nucleic acid monomer including inosinic acid derived from dried fish. Recently many manufacturers deliver miso with dashi (dashi-iri miso) as a final product. This type of miso makes the consumers omit the step of adding dashi into miso soup. Because of the easy use, the market of dashi-iri miso has become about one-fourth of the total miso markets. During the preparation of dashi-iri miso, fermented miso is heated to inactivate phosphatase, which is produced by koji mold during koji making process. Inadequate inactivation of this enzyme causes degradation of inosinic acid into inosine and phosphate, which has no taste. Therefore, the inactivation or removal of phosphatase is an essential step for preparation of dashi-iri miso, which is mostly done by heating miso. However, it often causes coloring of miso and some amount of flavor is lost. One of the solutions of this problem is a breeding of koji mold, which produces low amount of phosphatase (Marui et al. 2013).

5. Flavor Production in Miso

Initially, it was known that the fragrance of miso is derived from the ethyl esters and alcohol of low molecular weight. The taste of *miso* as seasonings is dependent on salt, amino acid, peptide, glucose and related oligosaccharide. Till date more than 200 flavor compounds have been identified to be associated with miso (Ohata et al. 2007). Glutamic acid is an amino acid, which is abundant in miso and is responsible for its umami taste (Kurihara 2009). The flavor compounds in miso are not only derived during fermentation, but also due to several reactions during ripening, especially the Maillard reaction (Inoue et al. 2016). Most of them are derived from the degradation of protein and polysaccharide by the function of koji enzymes (α-amylase, glucoamylase, isoamylase, proteinase, peptidase). It has been suggested that peptides and sugars, which are generated during miso ripening causes Maillard reaction contributing to the flavor of ripened miso (Ogasawara et al. 2006).

There are several important flavor compounds reported in miso, which are produced during fermentation. Study on aroma concentrates of raw miso types reported that the amounts of flavor compounds 4-hydroxy-2(*or* 5)-ethyl-5(*or* 2)-methyl-3(2H)-furanone (HEMF), methionol and 4-ethylguaiacol varied among types of miso prepared using different types of koji (Sugawara and Yonekura 1998). Among them, HEMF is a very important compound which gives a strong, sweet, and caramel like aroma in miso (Ohata et al. 2007). Lactic acid bacteria and yeast reported in miso have also been related in improving the flavor of miso (Suezawa and Suzuki 2007, Kim et al. 2010, Inoue et al. 2016). *Tetragenococcus halophilus* a halophilic strain, which is found to be associated during ripening of *miso*-paste play a significant role in the development of pleasing flavors and reduction of unpleasant flavors (Kim et al. 2010). *Zygosaccharomyces rouxii* is one of the predominant halotolerant yeast, which has been identified during spontaneous *miso* fermentation and is believed to be responsible for the distinct flavor development in *miso* (Kim et al. 2010). Production of HEMF specific to miso was promoted by the cultivation of Z. *rouxii* (Sugawara et al. 2007). In a recent study, Inoue et al. (2016) have shown that dimethyl trisulfide, a newly identified compound in cooked miso, contributed to the palatability and umami after taste. They also suggested that by controlling the concentration of dimethyl trisulphate by heating, the intensity of umami taste can be optimized in cooked soy miso.

6. Health Benefits of Miso

Fermented soybean products have shown to exhibit several health benefits due to the formation of bioactive molecules during fermentation (Sanjukta et al. 2015, Rai and Jeyaram 2015). Bioactive compounds in miso have also been suggested to exhibit several health benefits such as antioxidant, antidiabetic, anticancer and antihypertensive property (Esaki et al. 1997, Hirayama 1987, 1990, Kwon et al. 2010, Shimizu et al. 2015, Zaid and El-Shenawy 2010). In their study, Santiago et al. (1992) have shown that miso exhibited antioxidant property by scavenging free radicals such as DPPH, superoxide, hydroxyl, hydrogen and carbon-centered radicals, and suggested that the activity was due to isoflavones and saponins in miso. They also showed that miso inhibited formation of thiobarbituric acid reactive substances in rat cerebral cortex, thereby preventing damage caused by lipid peroxidation. Later, Hirota et al. (2000) isolated nine compounds from soybean miso exhibiting DPPH radical-scavenging activity. Among the antioxidant compounds, 8-hydroxygenistein, 8-hydroxydaidzein and syringic acid had higher DPPH radical scavenging activity, which was similar to that of α-tocopherol.

Four of the isolated isoflavones were examined for antiproliferative effect using three cancer cell lines. Among them, highest activity (IC_{50} = 5.2 μM) was observed by 8-Hydroxygenistein against human promyelocytic leukemia cells (HL-60). Later, Jung et al. (2006) showed that miso could prevent cancer by increasing natural killer cell activity in spleens and decreasing tumor formation. Enhancement in antioxidant activity in miso is dependent on *Aspergillus* sp. at strain level as soybean fermented with different strains resulted in difference in antioxidant activity (Esaki et al. 1997). Miso prepared with mixed culture starter (*A. oryzae* and *B. subtilis*) was more effective in reducing oxidative stress by stimulating endogenous antioxidants and had the ability to reduce lipid peroxidation (El-Shenawy et al. 2012). Miso fermented with *A. oryzae* and *B. subtilis* has also shown the ability to inhibit the proliferation of three human tumor cell lines—MCF7 (breast carcinoma), HEPG2 (liver carcinoma), and HCT116 (colon carcinoma) (Zaid and El-Shenawy 2010). It was suggested that the inhibitory effect in fermented soybean (miso) was due to their higher isoflavone content in comparison to unfermented soybean.

In another study, casein-added miso paste has been shown to possess peptides exhibiting angiotensin-converting enzyme inhibitory activity (Inoue et al. 2009). It was concluded that casein-added miso paste had tripeptides (Val-Pro-Pro and Ile-ProPro), which were already reported as antihypertensive agents in spontaneously hypertensive rats. Several studies have been done to find the relationship between consumption of miso and risk of stomach cancer (Hirayama 1982, Wu et al. 2000). Miso has also shown to be effective in suppression breast tumors in rats, lung tumors, and liver tumors in mice (Watanabe 2013). It was also reported that consumption of miso (3 times/day) reduces the risk and occurrence of breast cancer (Yamamoto et al. 2003). There was reduction and delay in appearance of mammary adenocarcinomas in dimethylbenz [a]anthracene

(DMBA) induced female Sprague-Dawley rats fed with miso supplemented diet (Baggot et al. 1990). The study suggested that the lower incidence of breast cancer in Japanese female population may be due to the consumption of miso.

In a recent study it was found that miso significantly activated the function of p-glucoprotein (p-gp) and CYP3A, which are important for drug bioavailability, resistance and interaction (Yu et al. 2014). Another recent study showed that salt induced hypertension rats, when treated with low dose of miso extracts decreased blood pressure, which was not mediated by increased sodium excretion from kidney (Shimizu et al. 2015). Apart from health benefits due to bioactive molecules in miso, lactic acid bacteria (e.g., *Lactobacillus* sp.) present in miso are known to have probiotic effect and ability to produce antibacterial compounds (Murook and Yamshita 2008, Onda et al. 2003).

7. Conclusions

Miso is a popular *A. oryzae* fermented soybean paste consumed in Japan which has a significant impact on human health. There are several varieties according to the type of preparations, which are specific to different locations in Japan. Apart from koji mold, lactic acid bacteria and yeast are present during miso fermentation and are responsible for improvement of flavor and health benefits. Health benefits in miso are due to the bioactive molecules formed during fermentation, which mainly includes peptides and isoflavones. The mechanism of action of peptides for the prevention and treatment of cardiovascular diseases need to be carried out.

Acknowledgements

The authors would like to sincerely thank Dr. Yoshiaki Kitamura for his kind offer of photographs of miso and Mr. Takahiro Ogawa for his kind offer of photographs of koji making.

Keywords: Koji, miso, *Aspergillus oryzae*, isoflavones, soybean paste

References

Baggot, J.E., Ha, T., Vaughn, W.H., Juliana, M.M., Hardin, J.M. and Grubbs, C.J. (1990). Effect of miso (Japanese soybean paste) and NaCl on DMBA-induced rat mammary tumors. Nutrition and Cancer 14: 103–109.
Chancharoonpong, C., Hsieh, P. and Sheu, S. (2012). Enzyme production and growth of *Aspergillus oryzae* S. on soybean koji fermentation. Asia-Pacific Chemical Biology Environmental Engineering Society Conference Proceedings 2: 57–61.
Chiou, A.Y.Y. and Cheng, S.L. (2001). Isoflavone transformation during soybean koji preparation and subsequent miso fermentation supplemented with ethanol and NaCl. Journal of Agriculture and Food Chemistry 49: 3656–3660.

El-Shenawy, N.S., Abu, Z.A. and Amin, G.A. (2012). Preparation of different types of miso with mixture of starters and their effects on endogenous antioxidant of liver and kidney of mice. Journal of Animal Physiology and Animal Nutrition 96: 102–110.

Esaki, H., Onazaki, H., Kawakishi, S. and Osawa, T. (1997). Antioxidant activity and isolation from soybeans fermented with *Aspergillus* spp. Journal of Agricultural and Food Chemistry 45: 2020–2024.

Frias, J., Song, Y.S., Martinez-Villaluenga, C., Gonzale de Mejia, E. and Vidal-Valverde, C. (2008). Immunoreactivity and amino acid content of fermented soybean products. Journal of Agricultural and Food Chemistry 56: 99–105.

Fujita, J., Shigeta, S., Yamane, Y., Fukuda, H., Kizaki, Y., Wakabayashi, S. and Ono, K. (2003). Production of two types of phytase from *Aspergillus oryzae* during industrial koji making. Journal of Bioscience and Bioengineering 95: 460–465.

Hirayama, T. (1990). Life-style and mortality. A large-scale census-based cohort study in Japan. *In*: J. Wahrendorf (ed.). Contributions to Epidemiology and Biostatistics, Vol. 6. Basel: Karger.

Hirayama, T. (1987). National burden of disease of urinary organs—An epidemiological consideration. Hinyokika Kiyo 33: 1550–1555.

Hirayama, T. (1982). Relationship of soybean paste soup intake to gastric cancer risk. Nutrition and Cancer 3: 223–233.

Hirota, A., Taki, S., Kawaii, S., and Yano, M. (2000). 1,1-diphenyl-2-picrylhydrazyl radical-scavenging compounds from soybean miso and antiproliferative activity of isoflavones from soybean miso towards the cancer cell lines. Bioscience, Biotechnology and Biochemistry 64: 1038–1040.

Imai, S. (2009). *Miso* as fermented food. pp. 28–35. *In*: Kanzume Gijutu Kenkyukai (ed.). Application of Microorganisms and Enzymes in Food Processing for Traditional Foods. Nippon Shokuryo Newspaper Corporation, Tokyo, Japan (In Japanese).

Inoue, K., Gotou, T., Kitajima, H., Mizuno, S., Nakazawa, T. and Yamamoto, N. (2009). Release of antihypertensive peptides in miso paste during its fermentation, by the addition of casein. Journal of Bioscience and Bioengineering 108: 111–115.

Inoue, Y., Kato, S., Saikusa, M., Suzuki, C., Otsubo, Y., Tanaka, Y., Watanbe, H. and Hayase, F. (2016). Analysis of the cooked aroma and odorants that contribute to umami aftertaste of soy miso (Japanese soybean paste). Food Chemistry 213: 521–528.

Jung, K.O., Park, S.Y. and Park, K.Y. (2006). Longer Aging Time Increases the Anticancer and Antimetastatic Properties of Doenjang. Nutrition 22: 539–545.

JFMMC (2015). Japan Federation of Miso Manufacturers Cooperatives, http://www.zenmi. jp/data/seisansyukka/2000-2015syuruibetusyukkaHP.pdf).

Kim, T.W., Lee, J.H., Park, M.H. and Kim, H.Y. (2010). Analysis of bacterial and fungal communities in Japanese-and Chinese-fermented soybean pastes using nested PCR-DGGE. Current Microbiology 60: 315–320.

Kurihara, K. (2009). Glutamate: from discovery as a food flavor to role as a basic taste (umami). The American Journal of Clinical Nutrition 90: 719S–722S.

Kwon, D.Y., Daily, J.W., Kim, H.J. and Park, S. (2010). Antidiabetic effects of fermented soybean products on type 2 diabetes. Nutrition Research 30: 1–13.

Liang, Y., Pan, L. and Lin, Y. (2009). Analysis of extracelular proteins of *Aspergillus oryzae* grown on soy sauce koji. Bioscience, Biotechnology and Biochemistry 73: 192–195.

Machida, M. (2002). Progress of *Aspergillus oryzae* genomics. Advances in Applied Microbiology 51: 81–106.

Maeda, H., Yamagata, Y., Abe, K., Hasegawa, F., Machida, M., Ishioka, R., Gomi, K. and Nakajima, T. (2005). Purification and characterization of a biodegradable plastic degrading enzyme from *Aspergillus oryzae*. Applied Microbiology Biotechnology 67: 778–788.

Marui, J., Tada, S., Fukuoka, M., Wagu, Y., Shiraishi, Y., Kitamoto, N., Sugimoto, T., Hattori, R., Suzuki, S. and Kusumoto, K. (2013). Reduction of the degradation activity of umami-enhancing purinic ribonucleotide supplement in miso by the targeted suppression of acid phosphatases in the *Aspergillus oryzae* starter culture. International Journal of Food Microbiology 166: 238–243.

Murooka, Y. and Yamshita, M. (2008). Traditional healthful fermented products of Japan. Journal of Industrial Microbiology and Biotechnology 35: 791–798.

Nakadai, T. and Nasuno, S. (1977) The action of acid proteinase from *Aspergillus oryzae* on soybean proteins. Agricultural and Biological Chemistry 41: 409–410.

Ogasawara, M., Yamada, Y. and Egi, M. (2006). Taste enhancer from the long-term ripening of *miso* (soybean paste). Food Chemistry 99: 736–741.

Ohata, M., Kohama, K., Morimitsu, Y., Kubota, K. and Sugawara, E. (2007). The Formation Mechanism by Yeast of 4-Hydroxy-2(or 5)-ethyl-5(or 2)-methyl-3(2H)-furanone in Miso. Bioscience, Biotechnology, and Biochemistry 71: 407–413.

Onda, T., Yanagida, F., Tsuji, M., Shinohara, T. and Yokotsuka, K. (2003). Production and purification of a bacteriocin peptide produced by *Lactococcus* sp. strain GM005, isolated from Miso-paste. International Journal of Food Microbiology 87: 153–159.

Rai, A.K. and Jeyaram, K. (2015). Health benefits of functional proteins in fermented foods. pp. 435–474. *In*: J.P. Tamang (ed.). Health Benefits of Fermented Foods and Beverages. CRC Press, Taylor and Francis Group of USA.

Sanjukta, S. and Rai, A.K. (2016). Production of bioactive peptides during soybean fermentation and their potential health benefits. Trends in Food Science and Technology 50: 1–10.

Sanjukta, S., Rai, A.K., Muhammed, A., Jeyaram, K. and Talukdar, N.C. (2015). Enhancement of antioxidant properties of two soybean varieties of Sikkim Himalayan region by proteolytic *Bacillus subtilis* fermentation. Journal of Functional Foods 14: 650–658.

Santiago, L.A., Hiramatsu, M. and Mori, A. (1992). Japanese soybean paste miso scavenges free radicals and inhibits lipid peroxidation. Journal of Nutrition Science and Vitaminology 38: 297–304.

Shimizu, N., Du, D.D., Sakuyama, H., Ito, Y., Sonoda, M., Kawakubo, K. and Uehara, Y. (2015). Continuous subcutaneous administration of miso extracts attenuates salt-induced hypertension in dahl salt-sensitive rats. Food and Nutrition Sciences 6: 693–702.

Shimizu, M. (1993). Purification and characterization of phytase and acid phosphatase produced by *Aspergillus oryzae* K1. Bioscience, Biotechnology, and Biochemistry 57: 1364–1365.

Sourabh, A., Rai, A.K., Chauhan, A., Jeyaram, K., Taweechotipatr, M., Panesar, P.S., Sharma, R., Panesar, R., Kanwar, S.S., Walia, S., Tanasupawat, S., Sood, S., Joshi, V.K., Bali, V., Chauhan, V. and Kumar, V. (2015). Health related issues and indigenous fermented products. pp. 303–343. *In*: V.K. Joshi (ed.). Indigenous Fermented Foods of South Asia. Taylor and Francis Group of USA.

Suezawa, Y. and Suzuki, M. (2007). Bioconversion of ferulic acid to 4-vinylguaiacol and 4-ethylguaiacol and of 4-vinylguaiacol to 4-ethylguaiacol by halotolerant yeasts belonging to the genus *Candida*. Bioscience, Biotechnology and Biochemistry 71: 1058–1062.

Sugawara, E., Ohata, M., Kanazawa, T., Kubota, K. and Sakurai, Y. (2007). Effects of the amino-carbonyl reaction of ribose and glycine on the formation of the 2(or 5)-ethyl-5 (or 2)-methyl-4-hyroxy-3(2H)-furanone aroma component specific to miso by halo-tolerant yeast. Bioscience Biotechnology and Biochemistry 71: 1761–1763.

Takahashi, T., Chang, P.K., Matsushima, K., Yu, J., Abe, K., Bhatnagar, D., Cleveland, T.E. and Koyama, Y. (2002). Nonfunctionality of *Aspergillus sojae aflR* in a strain of *Aspergillus parasiticus* with a disrupted *aflR* gene. Applied and Environmental Microbiology 68: 3737–3743.

Watanabe, H. (2013). Beneficial biological effects of miso with reference to radiation injury, cancer and hypertension. Journal of Toxicologic Pathology 26: 91–103.

Wu, A.H., Yang, D. and Pike, M.C. (2000). A meta-analysis of soyfoods and risk of stomach cancer: the problem of potential confounders. Cancer Epidemiology, Biomarkers and Prevention 9: 1051–1058.

Yamabe, S., Kobayashi-Hattori, K., Kaneko, K., Endo, H. and Takita, T. (2007). Effect of soybean varieties on the content and composition of isoflavone in rice-koji miso. Food Chemistry 100: 369–374.

Yamabe, S., Kaneko, K., Inoue, H. and Takita, T. (2004). Maturation of fermented rice-koji miso can be monitored by an increase in fatty acid ethyl ester. Bioscience Biotechnology Biochemistry 68: 250–252.

Yamamoto, S., Sobue, T., Kobayashi, M., Sasaki, S. and Tsugane, S. (2003). Soy, isoflavones, and breast cancer risk in Japan. Journal of the National Cancer Institute 95: 906–913.

Yoshida, H. and Kajimoto, G. (1972). Changes in lipid components during miso making studies on the lipids of fermented foodstuffs (Part 1). Journal of Japanese Society of Food and Nutrition 25: 415–421.

Yu, C. P., Hsieh, Y.W., Lin, S.P., Chi, Y.C., Hariharan, P., Chao, P.D.L. and Hou, Y.C. (2014). Potential modulation on P-glycoprotein and CYP3A by soymilk and miso: *in vivo* and *ex-vivo* studies. Food Chemistry 149: 25–30.

Zaid, A.A. and El-Shenawy, N.S. (2010). Effect of *miso* (A soybean fermented food) on some human cell lines; HEPG2, MCF7 and HCT116. Journal of American Science 6: 1274–1282.

Soy Sauce Fermentation

Shanna Liu

1. Introduction

Soy sauce is a product that is derived from soybean, wheat and other grain through microbial fermentation and enzymatic activity. During this process, proteins of soybean are degraded into amino acids, peptides and other water-soluble nitrogen content. Starches of wheat are decomposed into monosaccharides, disaccharides and polysaccharides. Reducing sugars and amino acids will induce reaction known as Maillard or non-enzymic browning reaction, providing typical dark brown color as well as aroma compounds (Kim and Lee 2008). Sugars are utilized by yeast and bacteria and changed to alcohol and organic acid, which are further converted to esters. Therefore, soy sauce production involves varieties of materials and microorganisms, leading to the extracted liquid from the aged mash juice containing amino acids, peptides, reducing sugars, polysaccharides, alcohol, aldehyde, esters, and organic acids. It is not only a traditional oriental condiment with special flavor, aroma and taste, but also functional food with bioactive components possessing anti-carcinogenic, anti-oxidation, anti-microbial, anti-platelet and immune-modulating activities (Tsuchiya et al. 1999, Gao et al. 2013).

College of Food Science and Bioengineering, Tianjin Agricultural University, 22 Jinjing Road, Tianjin, 300384, P.R. China.
Tianjin Engineering and Technology Research Center of Agricultural Products Processing, Tianjin 300384, P.R. China.
E-mail: shannaliu@tjau.edu.cn

Soy sauce is regarded as the invention of the Chinese and has more than 3,000 years of documented history (Wicklow et al. 2007, Gao et al. 2011). Zhouli, a historical book that was written in the Zhou Dynasty transcribes soy sauce as an indispensable seasoning agent, but made from animal materials. Until Han Dynasty, there was record of using soybean and wheat to produce sauce. A famous Chinese medical book, Qi Min Yao Shu, written in the Northern Wei Dynasty, documented the detailes of the sauce producing technology including microorganisms, time for manufacture and ratio of brine. After thousands of years in practice, Chinese people accumulated rich experience in soy sauce manufacture, and passed down this technology through generations. That explains the mountain of literature on soy sauce manufacture in China. Soy sauce fermentation was introduced into Japan about A.D. 700 years ago and gradually it spread to Southeast Asia and then to many other parts of the world. The process of soy sauce-making has been developed from household handcraft to the modern industry. Although the production process may differ according to the regions, the basic steps are summarized as koji production, brine fermentation and refining (Fig. 1).

Firstly, raw materials are moistening with water, cooked thoroughly to obtain the denatured proteins and starches in the materials. Secondly, the cooked materials are inoculated with starter mold (such as *Aspergillus oryzae*) at proper temperature, humidity and ventilation conditions to make koji. When the koji is ready, brine is added for fermentation at ambient temperature. The moromi mash juice is hydrolyzed by enzymes from the koji *via* soaking with brine. Lactic acid bacteria such as *Pediococcus halophilus* are selected for growth and production of lactic acid. With the decreasing of pH, alcohol fermentation occurs and aroma compounds form gradually. Finally the aged moromi is pressed for extraction of liquid product. Soy sauce is taken to pasteurization and blended in accordance with specialized standard. Finished soy sauce is produced after sedimentation or centrifugation, and then packaged (Hamada et al. 1989, Luh 1995).

2. Materials Used in Fermentation

The following materials are required for soy sauce fermentation.

2.1 Protein Source

Soybean (*Glycine max* L.) or defatted soybean is commonly used as protein material in soy sauce fermentation. Soybeans are rich in protein and oil content, which accounts for about 60% of dry soybeans by weight (Chen et al. 2012). As an excellent source of complete protein, soybeans contain all the eight essential amino acids particularly glutamic acid and aspartic

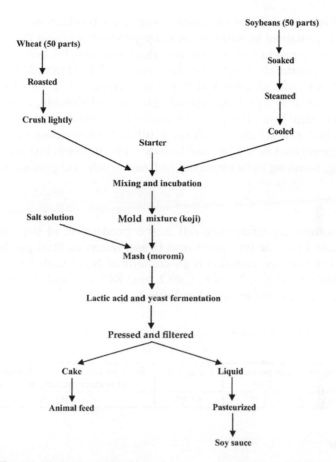

Figure 1. Production technology of soy sauce (Adapted from Luh 1995).

acid. Soybeans need to be cooked before consuming in order to destroy the trypsin inhibitors. The savory seasoning soy sauce is a representative of fermented soyfoods. Depending upon the area and specificity different materials may also be used, such as soybean meal, soybean cake, peanut cake and sunflower seedcake, the by-products of oil processing. For instance, the sheet-like soybean meal possesses higher total nitrogen content than the soybean. It is easy to cook, and presents a large surface for mold growth and enzyme accumulation. Meanwhile, the oil content is low in soybean meal, avoiding oil oxidation during storage.

2.2 Starch Source

Starch $(C_6H_{10}O_5)_n$ is a non-reducing, insoluble, white, tasteless and odorless powder which may be enzymatically degraded to sugars, contributing to the improvement of flavor and color in soy sauce production. This kind of material is contained in wheat bran as by-product of wheat flour milling, rice bran as by-product of rice milling, rice bran cake, broken rice, wheat, corn, sweet potato, etc. Sugars in wheat contain 70% starch, 2–3% dextrin, and 2–4% sucrose, glucose and fructose (Zhang 2009). Besides, wheat comprises about 10% protein, basically gliadin and glutenin, which are the source of umami taste. The wheat bran also has various vitamins, calcium and other minerals, which is the desired culture for *Aspergillus oryzae* growth and producing enzymes. The loose texture is conducive to koji-making and drenching, resulting in its increased utilization rate and production rate.

2.3 Salt

Salt is another important material in the production of soy sauce and plays a role in flavor formation and preservation of final product. Salt used in fermentation contains high content of NaCl with less impurity such as $MgSO_4$, $MgCl_2$, Na_2SO_4, $CaSO_4$ and KCl to avoid bitterness. The requirements for brine are listed in Table 1.

Table 1. Concentration, temperature and usage of brine (taking account of 100 kg material) (Li et al. 2009).

Baume degree (20°C)	Temperature (°C)	Usage (%)	Total amount of water (Taking account of mature moromi as 30%) (%)
About 13	40–45	62–65	92–95

2.4 Water

Usually 1 ton of soy sauce requires 6–7 tons of potable water for fermentation. Water source can be well water, tap water, clean water from river and lake.

3. Microbial Diversity

Soy sauce is rich in beneficial microorganism.

3.1 Mold

Aspergillus sojae or *A. oryzae* belongs to the variants of *Aspergillus* whose optimum temperature for growth is at 30–37°C and the enzymatic (amylase) activity will decrease when the temperature is higher than 37°C. *Aspergillus oryzae* is found to be dominant fungus in the koji and mash fermentation. It was shown that *Aspergillus parasiticus*, *Trichosporon ovoides* and *Trichosporon asahii* also appeared in the koji and at the early period of mash fermentation and disappeared thereafter (Wei et al. 2013). *Aspergillus* needs appropriate moisture and fresh air to grow especially during the growth period of hypha. Besides, the pH is controlled at 6.0 for better production of enzymes from microbial cells (Sandhya et al. 2005, Su and Lee 2001). The requirements for the strains in soy sauce fermentation include high activity of protease, amylase and glutaminase, no production of aflatoxin, fast growth with extensive cultivation conditions, being competitive to contaminating microorganisms, and no production of off-flavors (Matsushima et al. 2001).

3.2 Yeast

Halophilic yeasts play vital role in improving the quality of soy sauce moromi produces at the industrial scale (Cui et al. 2014). Common halophilic aromatic yeast including *Zygosaccharomyces roouxii*, *Torulopsis versatile*, *Candida versatilis* and *Candida etchellsii*, are used in the production of high-salt soy sauce (Wanakhachornkrai and Lertsiri 2003). For example, *Zygosaccharomyces roouxii* is able to grow at high content of sugar or salt. Even the saturated salt water does not inhibit its growth. The optimum temperature for growth is 28–30°C and the optimum pH is 4–5. *Zygosaccharomyces roouxii* may utilize glucose and maltose for fermentation, providing aroma components such as alcohol, ester, succinic acid, and furanone (Zhang 2009). *Torulposis mogii* is another genus of *Saccharomyces* with high tolerance to salt. It may also contribute to the flavor of soy sauce by producing 4-ethyl sores wood of phenol and phenethyl alcohol. *Zygosaccharomyces sojae*, a halophilic strain growing at the initial and middle stage of soy sauce mash fermentation is responsible for alcoholic fermentation and offering specific flavors.

3.3 Bacteria

It is reported that *Staphylococcus* and *Bacillus* were found in the whole fermentation process. *Kurthia* and *Klebsiella* appear during the koji step and fade away in the early stage of mash fermentation (Wei et al. 2013). Halophilic bacteria *Tetragenococcus halophilus*, *Lactobacillus acidipiscis*, *Lb. farciminis*, *Lb. pentosus*, and *Lb. plantarum* strains have been reported in Thai soy sauce (Tanasupawat et al. 2002). Lactic acid bacteria including *Pediococcus halaphilus*, *P. soyae*, *Tetracoccus soyae* and *Lb. plantarum* are typical bacteria involved in soy sauce fermentation. They are tolerant to salt and able to produce moderate amounts of lactic acid. Therefore, it may prevent low pH in the mash and give a soft taste for the final product (Song 2003).

4. Biochemical Properties

The following biochemical properties were observed during soy sauce fermentation.

4.1 Enzymes in Fermentation

The sources of enzymes during fermentation can be free enzymes added artificially, endogenic enzymes produced by microbial cells, and intrinsic enzymes in the soybean. Since the intrinsic enzymes in the soybean will be inactivated by the cooking process and there is no need to add free enzymes during traditional soy sauce fermentation, microbial enzymes are mainly discussed below.

4.1.1 Amylase

The α-amylases (EC 3.2.1.1), also called 1,4-α-D-glucan glucanohydrolase, break down starch chain into maltotriose, maltose, glucose or limit dextrin by acting on α-1,4 glycosidic bonds. The gelatinized starch granule decomposes into dextrin and small amount of sugar, leading to low viscosity. Further utilization of α-amylases results in total decomposition of dextrin into maltose or glucose (Li 2002a). The α-amylases are produced by microorganisms that include *Bacillus subtilis*, *A. niger*, *A. oryzae* and *Rhizopus*.

The glucoamylases (EC 3.2.1.3), also called glucan 1, 4-α-glucosidase hydrolyze the terminal 1, 4-linked-α-D-glucose residue successively from non-reducing ends of the starch chains with release of β-D-glucose. They can also act on α-1, 6 glycosidic bonds with relatively low rate and cause total degradation of amylopectin into glucose. Therefore, sugar's reducing capacity increases rapidly. The glucoamylases may weakly hydrolyze α-1, 3 carbon chains. Finally, starches are totally decomposed into glucose. This kind of enzymes can be produced by *A. niger*, *Rhizopus*, *Endomycopsis* and *Monascus* (Wu et al. 2003).

The α-amylases and glucoamylases work together contributing to glucose production from starch, providing carbon source for mold, yeast and other microorganisms, as well as basic substrates for ethanol fermentation, lactic acid and glutamic acid fermentation. Full saccharification is crucial to the quality of soy sauce such as better flavor and high solid content.

4.1.2 Protease

Protease (also named peptidase or proteinase) is any enzyme which catalyzes the hydrolysis of proteins. The peptide bonds which link amino acids together are splitted, generating small molecular peptones, peptides and amino acids. *Aspergilli* (*A. oryzae*, *A. flavus* and *A. terricola*) are rich sources of proteases, basically neutral and alkaline enzymes. The ways to degrade abundant proteins in soybeans are selection of strains with strong activity in protein hydrolysis, and destroying cell wall for better exposure of soybean protein from the texture.

4.1.3 Cellulase

Cellulases may decompose cellulose by hydrolyzing 1, 4-β-D-glycosidic linkages into monosaccharides or shorter polysaccharides and oligosaccharides. Cellulases produced by microorganisms such as *A. niger* and *Rhizopus* often occur as mixture of such enzymes containing Cx enzyme, C1 enzyme and β-1, 4-glucosidase. Cx enzyme is used for endo- and random-cutting of internal bonds to create reactive ends for the subsequent action of C1. C1 is responsible for the production of the cellobioce from the crystalline cellulose. Then the β-glucosidase may convert cellobiose into glucose (Chang et al. 1981). Recent study proved that β-glucosidase contributed to aroma enhancement by converting the aroma component from the combined state to the free state (Li 2002b).

4.1.4 Pectinase

Pectinase is known as an enzyme mixture of protopectinase, pectin methylesterase (PE), and pectinesterase, whose combined effect is the complete decomposition of pectin substance. The natural pectin changes to soluble pectin by protopectinase. Then pectin is catalyzed by pectin methylesterase and converts to pectic acid, which transforms to galacturonic acid by pectinesterase and enters into glucose metabolism. Molds are the main strains for producing pectinase such as *A. oryzae*, *A. wentii* and *A. niger*.

4.1.5 Lactate dehydrogenase

Lactate dehydrogenase (LDH, EC 1.1.1.27) catalyzes the conversion of pyruvate to lactate and back, as well as conversion of NADH to NAD$^+$ and back. LDH is produced intracellularly by a broad range of animal, plant and microorganisms and released after cell lysis (Krishnan et al. 2000). NADH pathway is rebuilt when the LDH activity is low, providing special flavor for the fermented food (Viana et al. 2005).

4.1.6 Alcohol dehydrogenase

Alcohol dehydrogenase (ADH) catalyzes conversion between alcohols and aldehydes or ketones with the reducing of NAD$^+$ to NADH. It is activated by combining with cofactor (non-protein components) including Zn^{2+} and NAD^+. The hydrolysate of cellulose and hemicellulose is rich in sugar, which is liable to produce alcohol during fermentation. The yeast may convert glucose to pyruvic acid by EMP pathway, and the acetaldehyde degraded from pyruvic acid by decarboxylase may change to ethanol with the help of ADH (Keating et al. 2004). Bacteria usually undergo ED pathway for the conversion of acetaldehyde to ethanol (Li and Zhang 1999).

4.1.7 Polyphenol oxidase

Polyphenol oxidase (PPO) catalyzes the o-hydroxylation of monophenol molecules to o-diphenols and further oxidation of o-diphenols to produce o-quinones. The polymerization of o-quinones and reaction with amino acid cause dark pigment, influencing the food color and quality. The depths of color of soy sauce partly depend on the PPO function since the hydrolyzed protein containing tyrosine is catalyzed by PPO on the condition of enough oxygen. This enzymatic reaction forms stronger color than non-enzymatic browning, especially during the late phase of mash making (Sun and Sun 2009).

4.1.8 Glutaminase

Glutaminase (EC 3.5.1.2) is a critical enzyme for the flavor of fermented foods, which converts L-glutamine to L-glutamic acid, increasing the amount of flavoring savory L-glutamic acid (Nandakumar et al. 2003). Glutamic acid is the major amino acid present in the fermented soy sauce and also contributes to the nutritional properties. When the materials are hydrolyzed by proteinase and peptidase, they are further catalyzed by glutaminase, from the koji mold (Weingand-Ziadé et al. 2003). The increased glutaminase activity leads to an increased amount of L-glutamic acid (Kim and Lee 2008). Glutaminase is sensitive to salt and markedly inhibited by high content of salt. Tolerance to salt with high glutaminase activity could be a desirable property for strain breeding.

4.2 Flavor Substances Formation

Flavor of soy sauce is a complicated and balanced system derived from aroma substances and taste substances, giving the product umami, salty, sweet and sour taste. Nearly 300 flavor substances have been isolated and identified in soy sauce, such as acids, alcohols, esters, phenols, furan, pyran, pyrazine, pyridine and sulfo-compounds. Although they are small in amount but their formation plays an important role in the characteristic flavor.

4.2.1 Proteolysis

During fermentation, protein material is cooked with the destruction of protein structure. Proteases of *A. oryzae* degrade the denatured protein into peptone, peptides and amino acids, and increase the content of water soluble nitrogen. With the help of aminopeptidase and carboxypeptidase, dissociative amino acids are formed, some of which are flavor substances. For instance, glutamic acid and aspartic acid provide delicious flavor. Glycine, alanine and tryptophan provide sweet flavor, and tyrosine gives bitter taste (Zong 2008). Soy sauce with good quality contains more than 1% total nitrogen (w/v), which are mainly composed of simple peptides and amino acids. According to China National Standard (GB18186-2000, fermented soy sauce), formaldehyde nitrogen is a decisive factor in evaluating the quality of soy sauce.

4.2.2 Saccharification

The α-amylases and the glucoamylases produced by *A. oryzae* during the koji cultivation generate fructose, maltose, pentose, dextrin as well as glucose, resulting in sweet flavor of soy sauce (Carlsen et al. 1996). Glucose can be utilized by yeast and bacteria and converts to compounds of low molecular weight containing ethanol, aldehyde, acetic acid and lactic acid. These materials are ingredients of soy sauce and also substrates for production of flavor components. Besides, reducing sugars obtained from starch saccharification are used in Maillard reaction, an important color reaction in food chemistry.

4.3 Steatolysis

Lipase produced by *A. oryzae* takes advantage of the raw material for formation of glycerol and fatty acids. The fatty acids can be degraded or oxidized to different kinds of short chain fatty acids, monounsaturated fatty acids, polyunsaturated fatty acid, alkane and olefin, which may be oxidized or cyclized to aldehydes, ketone, heterocyclic compounds, lactone and alkylfurans. The yeast fermented ethanol takes effect with fatty acids to form esters, the origins of fragrant aroma (Sun et al. 2009).

4.4 Organic Acid Production

Bacteria in soy sauce fermentation convert sugars to organic acid including lactic acid, acetic acid and succinic acid, which is the source of sourness. Furthermore, organic acid can be transformed into aldehydes with special fragment.

4.5 Ethanol Production

The yeasts make use of glucose for ethanol which produces esters with fatty acid. Esters are aroma compounds contributing to the rich and soft base of the flavor.

5. Technology in Commercialization

The various technologies adopted for soy sauce fermentation are discussed in this section.

5.1 Production Process

Fermented soy sauce using bean or defatted soybean, wheat or bran as raw material is attributed to microbial community succession and metabolic regulation. Meanwhile, brine concentration is a critical factor in defining soy sauce production technology.

5.1.1 Spontaneous liquid-state fermentation

Spontaneous liquid-state fermentation is an ancient traditional technology of China. The brine was brought to the liquid-state fermentation in the spring season and kept in the open environment for a long time. With the increasing ambient temperature, water evaporates and leads to concentrated mash juice. The open environment facilitates participation of wild-type microorganisms including lactic acid bacteria and yeast. The fermentation attains high level in summer and turns to mature stage in winter. In the southern part of China with relatively high average temperature, soy sauce products using this technology can be found. Since the starter is originated from nature and fermented for a long period, the soy sauce paste mash presents bronzing color and rich taste. But the product quality is unstable because of impure microbial cultures and improper production process control and specifications (Murooka and Yamshita 2008).

5.1.2 Low salt solid-state fermentation (LSSF)

LSSF is the most popular technology nowadays which accounts for majority of soy sauce production in China. It was invented in 1960s to fulfill the increasing demand for soy sauce. Manufacturing procedure of LSSF is illustrated in Fig. 2. Based on traditional fermentation, 12–13 baume degree of brine were prepared for inhibition of contaminating microorganism and yet having no influence on enzymatic activities. The starter is manufactured at 50–60°C and cultured at 40–55°C, which eliminated contaminated microorganisms growth and enzymes function at low temperatures. The main change is protein and starch hydrolysis instead of alcohol fermentation. Only half of the amino acid production rate can be reached, making the flavor of the final product inferior to that from high salt dilute-state fermentation (HSDF). However, study has revealed that the LSSF moromi had the higher total concentration of volatile compounds than HSDF, particularly 2, 3-butanediol and 2, 6-dimethylpyrazine (Zheng et al. 2013). Some physicochemical indexes of LSSF soy sauce in different grades are summarized in Table 2.

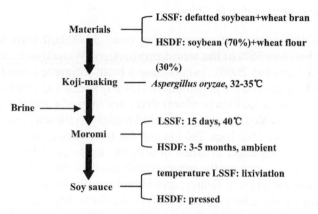

Figure 2. Manufacturing procedure of LSSF and HSDF. Adapted from Zheng et al. (2013).

Table 2. Physicochemical indexes of LSSF soy sauce (g/100 ml) (Li et al. 2009).

Index	Special grade	First-grade	Second-grade	Third-grade
Soluble saltless solid	≥20.0	≥18.0	≥15.0	≥10
Total nitrogen	≥1.6	≥1.4	≥1.2	≥0.8
Amino nitrogen	≥0.8	≥0.7	≥0.6	≥0.4

5.1.3 High salt diluted-state fermentation (HSDF)

Raw materials containing soybean and wheat are cooked before cultivation for koji-making and mixing with 20% brine. The fermentation process is conducted in a sealed fermentor or fermenting vat, which is able to control the temperature at different stages. The initial low starting temperature provides a chance for suitable microorganisms and enzymes to take effect. Then during the prolonged fermentation period the diverse substances are sufficiently fermented. The flavoring yeast (mainly *Zygosaccharomyces rouxii* and *Torulopsis*) and lactic acid bacteria propagating in the moromi improve the production of alcohol and other aroma-producing substance. HSDF is beneficial to higher amino acid production rate than LSSF, resulting in attractive flavor formation. Physicochemical indexes of HSDF soy sauce in different grades are summarized in Table 3.

Table 3. Physicochemical indexes of HSDF soy sauce (g/100 ml) (Li et al. 2009).

Index	First-grade	Second-grade	Third-grade
Soluble saltless solid	≥1.3	≥1.0	≥8
Total nitrogen	≥1.3	≥1.0	≥0.7
Amino nitrogen	≥0.7	≥0.55	≥0.4

5.2 Development Status

5.2.1 Development status in China

Nowadays, the output of soy sauce is over 5,000,000 tons in China, accounting for over 55% of the world production (Wanakhachornkrai and Lertsiri 2003, Yan et al. 2008). As consumers from different areas may have different dietary habits, the requirements for soy sauce are diverse, leading to the appearance of various products that vary in grade and classification. At present, large-scale factories are located mainly in the southern part and include brands like Hai Tian, Tao Da and Jia Le. Among them, Hai Tian's soy sauce maintains leads in terms of output and sales volume. Factories in Guangdong province are the representatives of the natural fermentation method, while factories in Beijing and Hebei province are inclined to use high salt liquid state fermentation. Overall, these high grade products account for small part of the whole market, and the most people consume medium- and low-end products of LSSF with different leading brands.

Studies have been done on the equipment, technical parameters and new product development. For instance, the models of total nitrogen dissolving speed and quantity have been set up for evaluating main factors influencing total nitrogen quantity (Gao and Shen 1994). Ultrafiltration and extrusion technology is adopted in raw material processing for simplification of technology, decreasing factory area, reducing fermentation period and increasing utilization rate.

5.2.2 Development status in other countries

The techniques of producing soy sauce in Japan came from China and the sauce has a great sales volume in Japan. Soy sauce production experiences long period of high salt liquid state fermentation, usually six months to one year (Yang and Li 2001, Lu 2002). Raw materials are selected on the basis of local resources. For example, the fish sauce and garlic soy sauce are produced from local cheap materials. The characteristics of Japanese soy sauce are low pH, high content of amino acids, organic acids, and reducing sugars, as well as rich aroma. Japan has established its position in high grade soy production (Table 4) in the international market of with US being the

Table 4. Classification and constitution of Japanese soy sauce (g/100 ml) (Li et al. 2009).

Classification	Color	pH	Baume degree	NaCl	Total nitrogen	Methyl nitrogen	Reducing sugar	Ethanol
Koikuchi	Dark brown	4.7	22.5	17.6	1.55	0.88	3.8	2.2
Usukuchi	Light brown	7.8	22.8	19.2	1.17	0.70	5.5	0.6
Taman	Dark brown	4.8	29.9	19.0	2.55	1.05	5.3	0.1
Saishikomi	Dark brown	4.8	26.9	18.6	2.39	1.11	7.5	Trace
Shiro	Yellow to brown	4.6	26.9	19.0	0.50	0.24	20.2	Trace

country of largest export volume. Korean soy sauce products are similar to the Japanese ones and are mainly consumed locally.

Studies of soy sauce production focus on strain improvement, fermentation optimization, modern solid-state fermentation and bio-reactor, etc. Although immobilized cell technology, membrane technology and cell fusion technology have been explored in soy sauce production, strain fermentation still dominates in application till date (Zhang et al. 2008). During recent years, addition of enzyme such as γ-glutamyl transpeptidase in fermentation has been tested to improve the delicious taste of soy sauce in Japan (Kijima and Suzuki 2007). Soy sauce with light color or colorless have also been developed as new products.

6. Perspectives and Conclusion

The daily consumption of soy sauce as condiment is taken for granted by large populations, especially in Asia where soy sauce has become an indispensable part of dietary culture. With the increasing living standard of Chinese people and spread of dishes that use soy sauce in their preparation, the market is displaying a further space for growth. For example, US, Canada and Australia are emerging countries for soy sauce consumption. As a seasoning that has been used since ancient times, soy sauce's traditional production technology has been enhanced with the introduction of modern technology. With the recent findings of bioactive compounds, soy sauce has proven to be competitive in the functional food market. Intensive theoretical study combined with modern fermentation technology will guide a broad prospect for soy sauce industry.

Keywords: Soy sauce, fermentation, *Aspergillus*, technology, production, enzyme

References

Carlsen, M., Nielsen, J. and Villadsen, J. (1996). Kinetic studies of acid-inactivation of α-amylase from *A. oryzae*. Chemical Engineering Science 51(1): 37–43.

Chang, M.M., Chou, T.Y.C. and Tsao, G.T. (1981). Structure, pretreatment and hydrolysis of cellulose. Bioenergy 20: 15-42.

Chen, K.I., Erh, M.H., Su, N.W., Liu, W.H., Chou, C.C. and Cheng, K.C. (2012). Soyfoods and soybean products: from traditional use to modern applications. Applied Microbiology and Biotechnology 96(1): 9–22.

Cui, R.Y., Zheng, J., Wu, C.D. and Zhou, R.Q. (2014). Effect of different halophilic microbial fermentation patterns on the volatile compound profiles and sensory properties of soy sauce moromi. European Food Research and Technology 239(2): 321–331.

Gao, C. and Shen, L. (1994). Theoretical analysis and total nitrogen solubility model of low-salt soy sauce fermentation process. China Condiment (in Chinese) 12(12): 10–12.

Gao, X.L., Cui, C., Ren, J.Y., Zhao, H.F., Zhao, Q.Z. and Zhao, M.M. (2011). Changes in the chemical composition of traditional Chinese-type soy sauce at different stages of manufacture and its relation to taste. International Journal of Food Science & Technology 46(2): 243–249.

Gao, X., Sun, P. and Li, J. (2013). Characterization and formation mechanism of proteins in the secondary precipitate of soy sauce. European Food Research and Technology 237(4): 647–654.

Hamada, T., Ishiyama, T. and Motai, H. (1989). Continuous fermentation of soy sauce by immobilized cells of *Zygosaccharomyces rouxii* in an airlift reactor. Applied and Microbiology Biotechnology 31(4): 346–350.

Keating, J.D., Robinson, J., Bothast, R.J., Saddler, J.N. and Mansfield, S.D. (2004). Characterization of a unique ethanologenic yeast capable of fermenting galactose. Enzyme and Microbial Technology 35(2): 242–253.

Kijima, K. and Suzuki, H. (2007). Improving the umami taste of soy sauce by the addition of bacterial γ-glutamyltranspeptidase as a glutaminase to the fermentation mixture. Enzyme and Microbial Technology 41(1): 80–84.

Kim, J.-S. and Lee, Y.-S. (2008). A study of chemical characteristics of soy sauce and mixed soy sauce: chemical characteristics of soy sauce. European Food Research and Technology 227(3): 933–944.

Krishnan, S., Gowda, L.R. and Karanth, N.G. (2000). Studies on lactate dehydrogenase of *Lactobacillus plantarum* spp. involved in lactic acid biosynthesis using permeabilized cells. Process Biochemistry 35(10): 1191–1198.

Li, L. (2002a). Soybean processing and utilization. Chemical Industry Press, Beijing (in Chinese) 219.

Li, Y. (2002b). Research progress on β-glucosidase. Journal of Anhui Agricultural University (in Chinese) 29(4): 42l–425.

Li, X., Wang, D., Li, P., Wang, Y., Xiu, K. and Zhang, L. (2009). Preservation and processing technology of soybean and beans. China Agriculture Press, Beijing (in Chinese) 82–92.

Li, Y. and Zhang, Z. (1999). Overview of fermentation industry. China Light Industry Press, Bejiing (in Chinese) 93–95.

Lu, Z. (2002). Classification and main components of soy sauce. China Condiment (in Chinese) 2: 45–46.

Luh, B.S. (1995). Industrial production of soy sauce. Journal of Industrial Microbiology 14(6): 467–471.

Matsushima, K., Yashiro, K., Hanya, Y., Abe, K., Yabe, K. and Hamasaki, T. (2001). Absence of aflatoxin biosynthesis in koji mold (*Aspergillus sojae*). Applied Microbiology and Biotechnology 55(6): 771–776.

Murooka, Y. and Yamshita, M. (2008). Traditional healthful fermented products of Japan. Journal of Industrial Microbiology & Biotechnology 35(8): 791–798.

Nandakumar, R., Yoshimune, K., Wakayamam, M. and Moriguchi, M. (2003). Microbial glutaminase: biochemistry, molecular approaches and applications in the food industry. Journal of Molecular Catalysis B-Enzymatic 23(2): 87–100.

Sandhya, C., Sumantha, A., Szakacs, G. and Pandey, A. (2005). Comparative evaluation of neutral protease production by *Aspergillus oryzae* in submerged and solid-state fermentation. Process Biochemistry 40(8): 2689–2694.

Song, G. (2003). New condiment production and application. China Light Industry Press. Beijing 1–58.

Su, N.W. and Lee, M.H. (2001). Purification and characterization of a novel salt-tolerant protease from *Aspergillus* sp. FC-10, a soy sauce koji mold. Journal of Industrial Microbiology and Biotechnology 26(4): 253–258.

Sun, C., Li, D. and Sun, L. (2009). The function of enzymes in naturally fermented soybean paste and soybean sauce products. China Food Additives (in Chinese) 03: 164–169.

Tanasupawat, S., Thongsanit, J., Okada, S. and Komagata, K. (2002). Lactic acid bacteria isolated from soy sauce mash in Thailand. Journal of General and Applied Microbiology 48(4): 201–209.

Tsuchiya, H., Sato, M. and Watanabe, I. (1999). Antiplatelet activity of soy sauce as functional seasoning. Journal of Agricultural & Food Chemistry 47(10): 4167–4174.

Viana, R., Yebra, M.J., Galán, J.L., Monedero, V. and Pérez-Martínez, G. (2005). Pleiotropic effects of lactate dehydrogenase inactivation in *Lactobacillus casei*. Research in Microbiology 156(5): 641–649.

Wanakhachornkrai, P. and Lertsiri, S. (2003). Comparison of determination method for volatile compounds in Thai soy sauce. Food Chemistry 83(4): 619–629.

Wei, Q., Wang, H., Chen, Z., Lv, Z., Xie, Y. and Lu, F. (2013). Profiling of dynamic changes in the microbial community during the soy sauce fermentation process. Applied Microbiology and Biotechnology 97(20): 9111–9119.

Weingand-Ziadé, A., Gerber-Décombaz, C. and Affolter, M. (2003). Functional characterization of a salt-and thermotolerant glutaminase from *Lactobacillus rhamnosus*. Enzyme and Microbial Technology 32(7): 862–867.

Wicklow, D.T., Mcalpin, C.E. and Yeoh, Q.L. (2007). Diversity of *Aspergillus oryzae*, genotypes (RFLP) isolated from traditional soy sauce production within Malaysia and Southeast Asia. Mycoscience 48(6): 373–380.

Wu, J., Wang, P. and Li, X. (2003). Progress and trend of glucoamylase. Nature Magazine (in Chinese) 25(3): 161–163.

Yan, L.J., Zhang, Y.F., Tao, W.Y., Wang, L.P. and Wu, S.F. (2008). Rapid determination of volatile flavor components in soy sauce using head space solid-phase micro-extraction and gas chromatography-mass spectrometry. Chinese Journal of Chromatography (in Chinese) 26(3): 285–291.

Yang, J. and Li, W. (2001). Discussion for improving the quality of soy sauce production in China. China Condiment (in Chinese) 5: 9–11.

Zhang, H., Jiang, Y. and Chen, M. (2008). Application of multi-strain inoculation in koji-making and fermentation in soy sauce production. China Brewing (in Chinese) 17: 1–3.

Zhang, Y. (2009). Study on optimizing enzyme system by using multi-strains during koji making and improving quality of soy sauce. Ph.D. thesis, Jiangnan University (in Chinese) 1–2.

Zheng, J., Wu, C.-D., Huang, J., Zhou, R.-Q. and Liao, X.-P. (2013). Analysis of volatile compounds in Chinese soy sauces moromi cultured by different fermentation processes. Food Science and Biotechnology 22(3): 605–612.

Zong, W., Leng, Y., Ge, D., Zhu, X. and Dai, C. (2008). Analysis of amino acids and comparison among different brands of soy sauce. Food & Machinery (in Chinese) 24(4): 105–107.

8

Rice-based Fermented Foods and Beverages

Functional and Nutraceutical Properties

Keshab C. Mondal,[1,]* *Kuntal Ghosh,*[2] *Bhabatosh Mitra,*[3]
Saswati Parua[4] *and Pradeep K. DasMohapatra*[1]

1. Introduction

Fermented foods practice is an integral part of age-old wisdom from ancient human civilization. Over the generations, this pioneering practice of food fermentation has expanded and improved to preserve and fortify the available food resources for sustained nutrition to meet nutritional needs. Ancient people adopted different preservation methods to store excess foods of plant and animal origin, particularly those that were seasonal and perishable. Based on archaeological evidence, it is thought that fermented foods probably originated during 7,000 to 8,000 B.C. in the areas of Indus Valley (Roy 1997, Samanta et al. 2011). It is evident from the Harappan civilization that people used different clay pots for preparing fermented foods and drinks (Barnett 2003). Fermented milk products, alcoholic beverages from fruits and cereal grains, and leavened breads were very

---
[1] Department of Microbiology, Vidyasagar University, Midnapore – 721102, India.
[2] Department of Food Science and Technology, College of Agriculture and Life Sciences, Chonbuk National University, Jeonju, Jeollabuk-do 561-756, Korea.
[3] Department of Biotechnology, Fakir Mohan University, Balasore, Odisha, India.
[4] Department of Physiology, Bajkul Milani Mahavidyalaya, Purba Medinipur, India.
* Corresponding author: mondalkc@gmail.com

popular during the early civilization in the Middle East and in the Indus Valley and later among the Egyptians, Greeks, and Romans. During the middle ages, the varieties of fermented foods and drinks were more popular and depended upon the availability of the raw materials, environmental conditions and the taste of the people.

The health-beneficial effects of fermented food were first advocated as far back as 76 A.D. by the Roman historian Pliny, who mentioned the use of fermented milk for treating gastrointestinal infections. Around 1850, the great French chemist, Pasteur, identified the particular role of microbes that initiate and continue the fermentation process. In the early 1900s, Tissier, a French pediatrician, proposed that bifidobacteria in food could be effective in preventing infections in infants (Tissier 1906, Grill et al. 1995). The positive role played by some selected bacteria is attributed to Eli Metchnikoff (FAO/WHO 2001). He suggested that "the dependence of the intestinal microbes on the food makes it possible to adopt measures to modify the flora in our bodies and to replace the harmful microbes by useful microbes" (Metchnikoff 1907).

The phrase "Let food be the medicine and medicine be the food", coined by Hippocrates over 2500 years ago is receiving a lot of interest today as food scientists and consumers realize the many health benefits of fermented foods. Campbell-Platt (1994) has defined fermented foods as those foods which have been subjected to the action of microorganisms or enzymes so that desirable biochemical changes cause significant quality improvement. Fermented foods are marketed globally as healthy foods, functional foods, therapeutic foods, nutraceutical based foods or bio-foods. The term 'functional food' was first coined in Japan and defined as 'Food products fortified with special constituents that possess advantageous physiological effects' (Kwak and Jukes 2001). Generally functional food enriched with nutraceuticals that either promoting health or decreasing the risk of diseases. The term "Nutraceuticals" which is a combination of nutrition and pharmaceutical was first coined by Stephen De Felice in 1989 (Romano et al. 2012). According to him, nutraceutical can be defined as "any substance that may be considered a food or part of a food and provides medical or health benefits, including the prevention and treatment of disease." Fermented foods have unique functional properties that impart some health benefits to consumers due to presence of functional microorganisms, which concentrate and enhance nutraceuticals (probiotics, prebiotics, phytochemicals, vitamins, essential amino acid and fatty acid, etc.), destroying undesirable components (mycotoxin, anti-nutritional factors), protecting food deteriorations (preventing growth of food borne pathogenic bacteria and fungi), ameliorating sensory qualities (taste, aroma, texture, consistency, and appearance of the food) and fortifying food with diverse types of bioactive compounds that have profound health beneficial effects (anti-microbial, anti-oxidant, anti-degenerative, gastroprotective, etc.) (Ray et al. 2016).

The variety of fermented foods prepared by mankind ranging from milk to alcohol, soybeans to cereals, vegetables to bamboo, meat to fish, and alcoholic beverages to non-alcoholic beverages depends upon locally available bioresources (Tamang et al. 2016b). According to Nout and Mortarjemi (1997), fermented foods are typically unique and vary according to regions due to the variation in climate, social patterns, consumption practices and most importantly the availability of raw materials. Asian and African countries have been carrying out lactic acid fermentation of cereals for the production of varieties of solids, porridge, beverages and gruels type food products. Asian countries are the major producer of rice and rice based fermented beverages which are very popular among the people. Different types of rice based fermented foods and beverages are regularly prepared and consumed in this area and these are synchronized in the social culture of native people. A global interest on rice and its fermented product is increasing due to their calorific value, unique quality, and high acceptability (Ray et al. 2016). Moreover, during the last decade, consumer lifestyle choices (vegetarianism, veganism, etc.), an increase in adverse reaction to dairy based food products (lactose intolerance, malabsorption, and allergies), and diet-related non-communicable diseases (cancer, cardiovascular disease, elevated blood pressure, diabetes, allergic, renal diseases, etc.) have led to an increasing demand for nondairy-based functional foods (Zannini et al. 2011). Emerging evidences supported that cereal based diet can only ameliorate the diseases/symptoms like coeliac disease, wheat allergy, gluten sensitive, lactose Intolerance, Cows' milk protein allergy or intolerance, Irritable Bowel Syndrome, cardio-vascular diseases and high blood pressure, etc., which are principally induced by gluten/wheat/lactose/dairy containing diets (Waters et al. 2015). Prevalence of hypolactasia, including lactose intolerance, is approximately 70% worldwide with increased occurrence in certain populations like Northern Europeans 2–15%, American Whites 6–22%, Central Europeans 9–23%, Blacks 60–80%, and Asians 95–100% (Lomer et al. 2008). Fermented cereal-based foods have a huge potential to fill this gap in the consumer market acting as potential vehicles for the delivery of nutraceuticals such as probiotics, prebiotics, dietary fiber, phytochemicals, anti-oxidants, minerals, vitamins, etc. (Ghosh 2016). Considering these beneficial effects, the grain-based fermented foods have now become more popular than conventional dairy-based products. The market of nondairy probiotic beverages is expanding, with a projected annual growth rate of 15% (between 2013 and 2015). It is predicted that the market will reach €65 billion by the year 2016 and of this, dairy-based produce accounts for approximately 43% of the market (Ghosh 2016).

2. Rice—The Principal Substrate

Rice (*Oryza sativa*) is a major food crop for the people of the world in general and Asians in particular; nearly 90% of the world's rice is produced and consumed in this region (Gross and Zhao 2014). Furthermore, rice-based diet provides two thirds of the calories for nearly 2.4 billion people in Asia. Almost all cultivated varieties of rice belong to a single species, *Oryza sativa* with about 120,000 varieties. China is the major producer of rice and the present production is 205 million MTonnes. India is the second largest producer and consumer of rice in the world. Rice production (around 4,000 varieties) crossed the mark of 155 million MTonnes in 2011–2012 accounting for 22.81% of global production in that year (Muthayya et al. 2014). It contains ~80% carbohydrates, 6–8% protein, ~3% fat and energy (1460–1560 KJ). White rice is a good source of magnesium, phosphorus, manganese, selenium, iron, folic acid, thiamine and niacin. It is low in fiber and its fat content is primarily omega-6 fatty acids, which are considered pro-inflammatory. The mineral content, starch quality, glycemic index, and anti-oxidant activity have made rice unique among cereals. In Asian sub-continent rice is considered as grain of life and has historically been perceived as hypoallergenic and antihypertensive. Considering its energy content, digestibility and health beneficial, ancient Indian texts aptly wrote in praise of this food grain, "Rice is vitality, rice is vigor too, and rice indeed is the means of fulfillment of all ends in life. All Gods, demons and human beings subsist on rice" (Amudha et al. 2011).

Rice is also used in different forms such as flour, paste, laja (parched rice), boiled, flattened, fried rice, and dried, and also as sprouted seedlings (Amudha et al. 2011). Due to it being the most available and common food resource in the Asian continent, preparation of different types of fermented foods and beverages from rice is a regular practice since time immemorial and which play an integral role in social, rituals, and festivals of ethnic people. Native people mix cereals, pseudocereals, and legumes for preparation of variety of rice based fermented foods to improve sensory and nutritive quality (Ray et al. 2016).

3. Microbial Association During Rice Based Food Fermentation

Traditional fermented foods are the consortia of synergistic microorganisms including multi-strain and multi-species, which modify the substrates biochemically and organoleptically into edible products that are culturally and socially acceptable to the consumers (Tamang et al. 2016a). Starch matrix in rice supports the growth of specific group of microbes mainly mold, yeasts, and lactic acid bacteria (LAB) and these groups of microbes play the major role for its enzymatic biotransformation. Wet processing and swelling of their starch granules upon cooking favours the growth and establishment of microbes which generally have starch hydrolytic abilities (Waters et al. 2015, Nout 2009). These are mainly derived from nature (natural or spontaneous) or from the addition of a starter culture (controlled fermentation by monoculture or multi-culture).

The mold mainly *Mucor, Rhizopus, Amylomyces*, and *Aspergillus* enzymatically (mostly amylases, proteases, etc.) decomposed the initial complex substrate into simple form and create anaerobic conditions that favour the growth of LAB, yeast and other organisms. Among the LAB, *Enterococcus, Lactococcus, Lactobacillus, Leuconostoc, Pediococcus, Streptococcus*, and *Weissella* are common bacteria associated with cereal fermentations. Besides, *Bacillus* sp. and *Bifidobacterium* sp. also participate in this fermentation (Waters et al. 2015, Gupta and Abu-Ghannam 2012, Tamang et al. 2016a). These group of bacteria liberate a versatile group of health beneficial metabolites including lactic acid, other organic acid, vitamins, enzymes, peptides and some of them also show probiotic potentialities (Lamsala and Faubionb 2009). Food originated probiotic bacteria are becoming popular because these are generally regarded as safe (GRAS), biochemically active, secretes different bioactive molecules, have no need of additional carrier molecules for their stability and do not show pathogenicity (Gupta and Abu-Ghannam 2012). Additionally, they can tolerate many environmental stresses, such as acidity, oxygen, and heat in respect to intestinal origin probiotic bacteria. Native strains of *Saccharomyces cerevisiae* are the principal yeast of most cereal fermentations. But other non-*Saccharomyces* yeasts are also significant in many cereal fermentations including *Candida, Debaryomyces, Hansenula, Kazachstania, Pichia, Trichosporon*, and *Yarrowia*. Yeast utilizes the sugar and produce alcohol and other aromatic metabolites that changes aroma and flavor (Aidoo et al. 2006, Law et al. 2011, Limtong et al. 2002). A list of common rice based fermented food and their associated microbes are depicted in Table 1.

4. Source of Microbes and Process of Fermentation

Rice based fermented foods and beverages in different countries are mostly prepared in the household or in cottage industries and follow traditional practice. Functional or non-functional participatory microbes in fermented foods are principally derived from two sources, from nature or starter. In case of natural (spontaneous) fermentation, the indigenous microbiota originate from uncooked substrates, utensils, containers, earthen pots, and environment (Law et al. 2011, Tamang et al. 2016b). Addition of starter or inoculums for preparation of fermented foods is also an ancient culture. It is believed that the usage of the dry form starters originated from China (Limtong et al. 2002). Starter culture is added or mixed in proper ratio

Table 1. Some common rice based fermented foods and beverages and their associated microbes.

Name of fermented food and beverage	Country	Substrate	Associated microbes	References
Bhaati jaanr	India, Nepal	Rice	*Mucor circinelloides, Rhizopus chinensis, Rhizopus stolonifer, Saccharomycopsis fibuligera, Pichia anomala, S. cerevisiae, Pd. pentosaceus, Lb. bifermentans*	Tamang and Thapa 2006
Chongju	Korea	Rice	*S. cerevisiae*, yeast, LAB	Rhee et al. 2003
Haria	India	Rice	*S. cerevisiae, lactic acid bacteria like Lb. fermentum, Bifidobacterium* sp., mould	Ghosh et al. 2014, Ghosh et al. 2015a, Ghosh et al. 2015b
Judima	India	Rice	*Pd. pentosaceous, B. circulans, B. catarosporous, B. pumilus, B. firmus, Debaryomyces hansenii, S. cerevisiae*	Arjun et al. 2014, Chakrabarty et al. 2010
Khaomak	Thailand	Rice	*Rhizopus, Mucor, Saccharomyces, Hansenula, Amylomyces rouxii*	Rhee et al. 2011, Yodtheon et al. 2010
Kodo ko jaanr	India, Nepal, Bhutan, Tibet	seeds of finger millets	*Saccharomycopsis fibuligera, Pichia anomala, S. cerevisiae, Candida glabrata, Pd. pentosaceus, Lb. bifermentans, Lb. fermentum*	Dutta et al. 2012, Thapa and Tamang 2006
Po: ro Apong	India	Rice	*S. cerevisiae, Hanseniaspora* sp., *Kloeckera* sp., *Pischia* sp., and *Candida* sp.	Kardong et al. 2012
Sake	Japan	Rice	*S. cerevisiae, Aspergillus oryzae*, LAB like *L. mesenteroides var sake* and *Lb. saké*, and other bacteria	Selhub et al. 2014, Uno et al. 2009
Takju	Korea	Rice, wheat	*Lb. paracasei, Lb. arizonensis, Lb. plantarum, Lb. harbinensis, Lb. parabuchneri, Lb. brevis, Lb. hilgardii, S. cerevisiae*	Aidoo et al. 2006, Bae et al. 2010, Jin et al. 2008, Lee et al. 1996, Park and Lee 2002
Tapai	Southeast Asia	Rice	*S. cerevisiae, C. krusei, C. pelliculosa, C. guillermondii, C. magnolia, Rhodotorula glutinis, Lb. brevis, Lb. plantarum, Lb. collinoides, Pediococcus* sp., Moulds	Chiang et al. 2006, De Angelis et al. 2007, Foo et al. 2001, Lilly and Stillwell 1965

Table 1. contd....

...Table 1. contd.

Name of fermented food and beverage	Country	Substrate	Associated microbes	References
Tape ketan	Southeast Asia	Rice or cassava	*S. cerevisiae, Hansenula anomala, Rhizopus oryzae, Chlamydomucor oryzae, Mucor, Endomycopsis fibuliger*	Ardhana and Fleet 1989, Cook et al. 1991, Cronk et al. 1977, Steinkraus 1997
Yellow wine	China	Rice with wheat Qu	*S. cerevisiae*	Shen et al. 2011
Zutho	India	Rice	*S. cerevisiae, Rhizopus* sp.	Das and Deka 2012, Tamang et al. 2012, Teramoto et al. 2002
Idli	India, Sri Lanka, Malaysia, Singapore	Rice and black gram dhal (*Phaseolus mungo*) (4:1)	*L. mesenteroides, Lb. delbrueckii, Lb. fermenti, Lc. lactis, Streptococcus faecalis, S. cerevisiae, Pd. cerevisiae, Debaryomyces hansenii, Hansenula anomala, Torulopsis candida, Trichosporon beigelii*	Blandino et al. 2003, Satish et al. 2013, Moktan et al. 2011
Dosa	India, Sri Lanka, Malaysia, Singapore	Rice and black gram dhal (*Phaseolus mungo*)	*L. mesenteroides, Streptococcus faecalis, Torulopsis candida, Lb. fermentum, B. amyloliquefaciens, Lb. lactis, L. delbruckii, and Lb. plantarum, S. cerevisiae, Debaryomyces hansenii, Trichosporon beigelli, Torulopsis sp., Trichosporon pullulans*	Blandino et al. 2003, Satish Kumar et al. 2013, Gupta and Tiwari 2014
Ang-kak	China, Taiwan, Thailand, Philippines	Red rice	*Monascus purpureus*	Steinkraus 1996
Khamak (Kao-mak)	Thailand	Glutinous rice	*Rhizopus* sp., *Mucor* sp., *Penicillium* sp., *Aspergillus* sp., *Endomycopsis* sp., *Hansenula* sp., *Saccharomyces* sp.	Alexandraki et al. 2013
Lao-chao	China	Rice	*Rhizopus oryzae, Rhizopus chinensis, Chlamydomucor oryzae, Sacchromycopsis* sp.	Blandino et al. 2003

Table 1. contd....

...Table 1. contd.

Name of fermented food and beverage	Country	Substrate	Associated microbes	References
Puto	Philippines	Rice	L. mesenteroides, Enterococcus faecalis, Pd. pentosaceus, Yeasts	Steinkraus 2004
Selroti	India, Nepal, Bhutan	Rice and wheat flour	Lactic acid bacteria like L. mesenteroides, Enterococcus faecium, Pd. entosaceus, and Lb. curvatus, S. cerevisiae, S. kluyveri, Debaryomyces hansenii, Pichia burtonii, Zygosaccharomyces rouxii	Yonzan and Tamang 2010
Pak-gard-dong	Thailand	Leafy vegetable, salt, boiled rice	Lb. plantarum, Lb. brevis, Pd. cerevisiae	Phithakpol et al. 1995
Mifen	China and South Asia	Rice dust	Lb. plantarum, Lb. curvatus, Lb. brevis, Lb. acidophilus, Lb. fermentum, Lb. delbruckii. delb, Lb. cellobiosus, Lb. lactics and Lb. helveticus, S. cerevisiae, C. tropicalis, C. parapsilosis, Pichia etchellsii/ carsonii	Lu et al. 2008

B.: Bacillus; C: Candida; Lb.: Lactobacillus; Lc.: Lactococcus; L.: Leuconostoc; Pd.: Pediococcus; S: Saccharomyces

prior to fermentation of the substrate. Starter culture(s) contain functional microorganism(s) which participate in synergistic fermentation. Amylolytic starters are commonly known as "Chinese yeast cake" in the Western world and are used in the form of starchy tablets. Hesseltine et al. (1988) stated that amylolytic starters in East and South Asia are typically a combination of fungus, yeast and bacteria. Mixed cultures were used for fermentation of carbohydrate-rich substrate such as starchy crops and cereals instead of sequential fermentation in the amylolytic starter, in which molds are able to degrade starch and yeasts and LAB continue the fermentation process and bio-embolden the rice with varieties of health beneficial components (Hesseltine et al. 1988, Aidoo et al. 2006). In India, a unique and typical starter used by the ethnic people for preparation of rice based beverage or wine, consists of plant residues instead of old culture. Specific parts of the plant are dried and bound with rice dust to form ball shaped structures, then sun dried and preserve for several months. The endophytic microbes in the plant parts act as functional microbes for mixed culture and multistage fermentation of rice, which is not possible by using old ferments (Ghosh et al. 2014). Apart from this, herbal products are good sources of therapeutic and preservative phytochemicals that add extra health beneficial potentialities to the rice-based fermented products. The occurrences of dominant microflora of the popular starter cultures have been widely studied. These starters or inoculums are homemade or commercially produced and are mostly used for preparation of rice based beverages and given different names in the Asian region (Table 2).

Preparation of rice based fermented products are vary according to locations and traditional practices. Generally, rice fermentation can be categorized into submerged and solid state process. Submerged process involving saccharification and liquefaction of rice occurs in presence of plenty of free water to liberate sugar and be converted to ethanol by yeast (Law et al. 2011, Ray and Sivakumar 2009). Preparation of sake and rice wine in East Asian countries are the examples of this type fermentation. Solid state fermentation is the common practice for varieties of rice based product preparation. General procedure for preparation of rice based beverages includes washing, soaking, steaming of rice grain and then it being air dried under sunlight. Thereafter mixed with starter powder and put into earthen pot or heaped on leaf surface (Law et al. 2011). Fermentation involves a three-step non-synchronized process. Initially, amylolytic molds are grown, which contribute to the saccharification and liquefaction of rice grain. Thereby, rice is decomposed and creates an anaerobic environment. This condition favours the growth of LAB and *Bifidobacterium*, and apart from lactic and acetic acid, they also produce different hydrolytic enzymes (active on both nutrients and antinutrients) and various metabolites. Later, alcohol-producing yeast (*Saccharomyces cerevisiae*) is grown, which enriches the ferment with vitamins, amino acids, etc., and aids specific flavor and aroma (Ghosh et al. 2015a, Ray et al. 2016). For preparation of rice or psuedocereal based snack or cake (*pitha*) preparation, the batter (semi-solid condition) of rice and other cereal (pulses) are subjected to microbial fermentation (spontaneous) that leads to the formation of carbon dioxide and other gases inside and makes the food spongy (Ray et al. 2016). The degree of fermentation depends on time, which again determines food taste, texture, appearance, aroma, and so on.

5. Evolution of Bioactive Metabolites During Course of Rice Fermentation

Being the largest choice of food ingredient globally, rice has been found to contain >5,000 small metabolites (Kind et al. 2009, Heuberger et al. 2010). There is great evidence that indicates to the bioenrichment of rice with different bioactive metabolites occurring during fermentation. The changes

Table 2. Uses of specific starter for preparation of rice based beverages.

Name of fermented beverage	Substrate	Starter material	Reported from	References
Apong (white)	Boiled rice	Ipoh (yeast)	Arunachal Pradesh, India	Kardong et al. 2012, Tiwari and Mahanta 2007
Ennog (black)	Boiled rice + burnt rice husk	Ipoh (yeast)	Arunachal Pradesh, India	Tiwari and Mahanta 2007
Zutho	Whole sprouted rice + unhulled sprouted rice	Piazu	Nagaland, India	Das et al. 2012, Teramoto et al. 2002
Tai Ahom	Rice (Bora variety)	Vekurpitha (yeast + leaves of some ethnomedicinal plants)	Assam, India	Saikia et al. 2007
Sujen	Rice + husk + medicinal plants + charcoal	*Mod pitha* (natural starter)	Assam, India	Deori et al. 2007
Bhaati Jaanr	Glutinous rice	*Marcha* (mixed culture of yeasts, filamentous moulds, Lactic acid bacteria	Eastern Himalayan regions of Nepal, India and Bhutan	Tamang and Thapa 2006
Tape Ketan	Glutinous rice	*Ragi*	Indonesia	Steinkraus 1995
Miso and Shoyu	Mold rice, soybean	*Aspergillus oryzae*, *Saccharomyces rouxii*	Japan, China, Malaysia and Thailand	Steinkraus 1995
Yellow wine	Glutinous rice with wheat Qu	*Saccharomyces cerevisiae*	China	Shen et al. 2011
Haria	Boiled rice	*Bakhar* (herbal based)	India	Ghosh et al. 2014, Ghosh et al. 2015a, Ghosh et al. 2015b
Makgeolli	Boiled rice	*Nuruk*	Korea	Jung et al. 2012

of metabolite composition in the final rice-based fermented products are related with the participating microbiota and nature of fermentation (Ryan et al. 2011). After fermentation, rice becomes enriched with various metabolites, some notable examples like phenolics (monophenols and polyphenols), flavons (mono-C-glycosides, malonylated O-hexosides, O-glycosides), vitamin E (tocopherols and tocotrienols), phytosterols, linolenic acid, anthocyanins, proanthocyanidins, γ-oryzanol, etc. The bioavailability of phenolics in rice is related with scavenging of reactive oxygen radicals (antioxidant activity), which is a mechanism of anticancer activity (Thompson and Thompson 2009, Henderson et al. 2012). Phutthaphadoong et al. (2009) noticed that the fermented brown rice and rice bran gained some chemopreventive potentialities, as it can suppress the carcinogenesis of the colon, liver, stomach, bladder, and esophagus. Rice bran fermented with *Saccharomyces boulardii* has the potential to reduce the growth of human B lymphomas. Its lipid soluble metabolites, like tocopherol and tocotrienols, are known to induce an anti-inflammatory activity. During fermentation, rice beer (*haria*) accumulated different maltooligosaccharides (G5-G2), such as maltotetrose (G4, 20.1 mg/g), maltotriose (G3, 28.1 mg/g), and maltose (G2). These are low in calories, inhibit the growth of intestinal pathogens, and are very nutritious for infants and aging people (Ghosh et al. 2015a). A number of pyranose derivatives like 2,3,4,5-tetra-O-acetyl-1-deoxy-β-D-glucopyranose, β-D-mannopyranose pentaacetate, β-D-galactopyranose pentaacetate, and 1,2,3,6-tetra-O-acetyl-4-O-formyl-D glucopyranose are also accumulated, which have profound immune-stimulatory, antioxidant, and antimutagenic activities (Ghosh et al. 2015a). In addition, a number of oligosaccharides, phenolics, and flavonoids in the rice beer show significant free radical scavenging activities, which can potentially reduce the risk of cardiovascular and other degenerative diseases (Ghosh 2016). Health promoting t-resveratro and phenolic acids such as gallic acid were found present in rice wine. Rice wine enriched with D-glucose, ethyl α-D-glucoside, glycerol, organic acids and amino acids and suggested to be a potential effective agent for the prevention and treatment of UV-induced skin aging (Seo et al. 2009).

6. Functional Properties of Rice Based Fermented Foods and Beverages

In most of the Asian countries, rice is fermented either by using mixed culture(s) into alcoholic beverages, or by natural fermentation into leavened product (Steinkraus 1994). Apart from basic nutrients, microbial association enriches the rice with different nutraceuticals, destroying the undesirable components (phytic acid, tannic acid, trypsin inhibitors, etc.) and contaminants (mycotoxins, antinutrients, heavy metals, etc.), and increasing the shelf-life of food (preventing the growth of pathogenic bacteria and fungi by evolving organic acids and proteinaceous substances) (Gupta and Abu-Ghannam 2012, Waters et al. 2015). During fermentation, the content of proteins and essential amino acids (particularly lysine content) are increased, that are helpful in preventing protein-energy malnutrition (Nout 2009). It helps to reduce the content of reactive saccharides (raffinose, stachyose, and verbascose, which can cause flatulence, diarrhoea, and indigestion), enriches the pool of vitamins (B vitamins), minerals (zinc and iron), phenolics, dietary fibers, phytochemicals, therapeutic components, etc. and increases the overall quality, digestibility, flavor, and aroma of the food (Nout 2009, Tamang et al. 2016b). Edible microbes specially LAB and Bifidobacteria exhibited extraordinary benefits to maintain the healthy composition of intestinal microbiota, improve immunity for protection from the diseases and to preserve the extra-intestinal host's physiology. From this point of view, fermented cereal is designated as 'naturally fortified functional food'.

The amassing health beneficial effects of rice based fermented food products is due to exaggerated functional properties including promulgation of health beneficial probiotic microorganisms as well as prebiotic non-digestible carbohydrates (such as β-glucan, arabinoxylan, galacto- and fructo-oligosaccharides and resistant starch) and thus gives it a status of being a synbiotic food (Ghosh et al. 2015a,b). Several curative properties attributed to probiotics, including production of antibacterial substances, improvement of intestinal barrier function, Immune modulation, degradation of intestinal carcinogenesis, reducing serum cholesterol, alleviation of lactose intolerance symptoms, improvement of specific enzymatic activities and nutritional enhancements. Profound growth of multispecies and multistrain of the so called probiotic microbes like LAB, Bifidobacteria and yeast in rice make this a health beneficial probiotic food (Stanton et al. 2005, Lamsala and Faubionb 2009). Probiotics derived short chain fatty acids (SCFA) especially butyrate has been identified as the target molecule that improves gastrointestinal as well as overall health of the individual. Apart from this, starchy cereal based substance can improve viability of probiotic organisms during storage and support its viability by making a barrier from acid and bile rich harsh environment in the intestine. The prebiotic ingredients in this type of food also facilitate the growth of probiotic organisms and kept it metabolically active in colon (Waters et al. 2015).

Lactic acid bacterial consortia in cereal based fermented foods showed broad spectrum antibacterial activities (diastatic power) by producing acid, H_2O_2 and proteinaceous bacteriocin like metabolites that make a strong barrier for growth of food borne pathogens like enterobacteriaceae and other gram-positive and gram-negative bacteria and fungi. These strains also exhibited anti-mycotoxigenic, mycotoxin-binding and proteolytic (neutralize toxic peptides and release flavor-contributing amino acids) activities (Kivanc et al. 2011, Waters et al. 2015), therefore, are helpful in biopreservation of fermented foods and their extended shelf-life.

Several genera of molds and LAB in fermented rice liberate various carbohydrases such as α-amylase, amyloglucosidase, maltase, invertase, pectinase, ß-galactosidase, cellulase, hemi-cellulase; and proteases, which predigested the food components, facilitated their rapid assimilation through the intestine as well as energy extraction (Tamang et al. 2016b). Proteolytic byproducts such as amino acids (lysine particularly, which is deficient in rice) and peptides (antibacterial, antifungal, antitoxin, other function) are also bioenriched in the fermented rice. Lorri and Svanberg (1993) mentioned that bulk properties of traditional starch-based weaning foods in developing countries are major causes of the high prevalence of malnutrition. Fermented cereal beverage can alleviate this problem through increased nutrition levels. The energy density of the lactic acid-fermented gruel is about 1.2 kcal/g as compared with 0.4 kcal/g in non-fermented gruel, with a three-fold increase in nutrient density to improve the nutritional status of malnourished infants and children (Lorri and Svanberg 1993). Phytase and polyphenol hydrolases can diminish the level of antinutrients in rice and simultaneously elevated the level of minerals (limited content of zinc and iron in rice) and phenolic compounds, which exerted antioxidant activities (Ghosh 2016). Dietary antioxidants are supposed to be protecting cells and tissues against damaged caused by reactive oxygen species (ROS). From *head to tail* there are many infectious and non-infectious diseases which are mediated by ROS and probiotic mediated cereal based functional foods are one of the preventive measure (Birben et al. 2012).

During fermentation of rice with mixed culture of microorganisms, the levels of micronutrients are enormously elevated. LAB have the capability to enrich fermented cereal products synthesizing water-soluble vitamins such as vitamin C and those included in the B-group (folates, riboflavin, and vitamin B12) among others (Waters et al. 2015). Ghosh (2016) described that in rice beer the content of folic acid, thiamine, pyridoxine and ascorbic acid (vitamin C) were improved more than 5 folds, 7 folds, 400 folds and 500 folds, respectively than unfermented rice. This represents another mechanism of LAB-mediated nutritional amelioration of rice based beverages.

7. Some Rice Based Fermented Foods and Beverages that are Popular Worldwide

Rice and other cereals are mainly used for preparation of versatile traditional fermented food dishes. However, the proportions and preparation process for all these fermented foods vary from place to place depending upon ancestral knowledge, bioavailability of substrate and culture of the society. The behavioral pattern of each fermented food also depends upon the participation of functional microbes. The climate, societal ritual and many other factors are related to the habitat of each microbe. For these reasons the taste, aroma, texture and other physico-chemical appearance of each traditional fermented food are different.

According to the physical appearance, rice based fermented food products can be differentiated into snack/cake, glutinous/paste and beverage/wine type. The information like raw materials, associated microbes, preparation process, nutritional details and health impacts of some popular and indicative rice-based fermented foods have been described in Table 3.

8. Conclusion

In contrast to the dairy based products, there is mass consumption of fermented rice products in Asian countries, due to their affordability and nutritional properties, and because these are also gluten free, lactose-free, vegetarian friendly, low cholesterol/fat containing healthy food. Rice has some inherent nutritive limitations. The co-fermentation of rice with other cereals, leguminous seeds, or herbs can improve the amino acid and mineral profiles, and therapeutic potentialities of food, on account of complementary actions. This type of mixed cereal fermented foods have a significant and growing global market as nutraceuticals rich functional foods, and have now became very popular in Europe, Japan and Australia. The standardization of process parameters and the adaptation of newer technologies for fermentation of rice by mixing with other cereals, and considering acceptability by the end users, are very essential for popularization of each regional food. Newly developed fermented rice-based food product must address markets globally including, high-nutrition markets (developing countries), lifestyle choice consumers (vegetarian, vegan, low-fat, low-salt, low-calorie), food-related non-communicable disease sufferers (cardiovascular disease, diabetes, kidney diseases and even for cancer), and green label consumers (Western countries).

Acknowledgement

Authors gratefully acknowledge the sponsoring agencies (SEED Division, DST; UGC, Govt. of India and DST, Govt. of West Bengal) for their financial support for conducting research in this field.

Keywords: Rice, fermented foods, microflora, functional components, health benefits

Table 3. Traditional preparation, nutraceuticals and functional properties of popular rice based fermented foods.

Fermented food	Traditional processes	Functional components	Health beneficial impacts	References
Idli	White polished rice and black gram dhal are washed and soaked in drinking water overnight separately. Soaked ingredients are ground separately. Slurries of rice and black gram dhal are mixed to form a thick batter. The batter is kept at room temperature overnight for fermentation. Then, the fermented batter is placed into a concaved *idli* pan for steaming, finally yielding the savoury spongy cake *idli*.	*Idli* contains approximately 3.4% protein, 20.3% carbohydrate, 70% moisture, 1% verbacose, 0.2% stachyose and raffinose. Fermentation increases level of amylase, protenase, total acids, batter volume, soluble solids, essential amino acids (lysine, cystine and metheionine), non-protein nitrogen, soluble vitamins (folate, vitamin A, B_1, B_2 and B_{12}) content, with reduction in antinutrientphytic acid.	*Idli* is generally regarded as anti-obesity and weight loosing diet. Useful to reduce the risk of cardiovascular diseases, high blood pressure and stroke. This is used as a dietary supplement to treat children suffering from protein calorie malnutrition and kwashiorkor. The micronutrients like iron, zinc, folate and calcium prevent anaemia and facilitate the oxygenation of blood and nourishment of the muscle and bone. The carbohydrate as well as dietary fibre content promotes healthy digestion and formation of bulky stool.	Blandino et al. 2003, Satish et al. 2013, Moktan et al. 2011
Dosa	The raw ingredients are soaked and grinded separately with added water to get smooth consistent batter. The batter is allowed to ferment overnight at room temperature. The colloid fermented batter is spreaded in the form of a thin layer on a flat heated plate which is smeared with a little edible oil. Within a few minutes, a circular semisoft to crisp *Dosa* is formed.	*Dosa* fermentation increases in amount of total acids, total volume, total solids, non-protein nitrogen, free amino acids, amylase, proteinases, vitamins B_1 and B_2, folic acid, amino nitrogen, formation of diols, antimicrobial and antioxidant substances. Antinutrients are reduced and enhances the bio-accessibility of zinc and iron.	Appropriate vegan diet for individuals with wheat allergies or gluten intolerance. Low glycemic load and glycemic index of *Dosa* helps to fight pre and post-diabetic condition. It offers adequate energy for prolonged physical endurance. Some people believes that *dosa* has medicinal property and can be used to increase fertility, weight of foetus, breast milk. It is also considered as a remedy for rheumatism and neural disorders.	Blandino et al. 2003, Satish Kumar et al. 2013, Gupta and Tiwari 2014

Dhokla	Thick batter prepared from the raw materials is allowed to ferment overnight at room temperature. The fermented batter is poured into the greased tray and placed in the steamer in open condition. The spongy formed cake with quite high water content is *Dhokla*.	Each serving *dhokla* (213 g) contains 384 cal, 59 g of carbohydrate, 6.6 g of free sugars, 10.6 g of dietary fiber, 11.7 g of protein, 11.8 g of fat, 89 mg of sodium, 551 mg potassium, adequate amount of calcium, iron, vitamin A and C, and is free from cholesterol.	*Dhokla* has low glycemic index (34.96) and is very useful for diabetic patients. It helps to reduce blood cholesterol and body weight and protects from cardiovascular diseases.	Steinkraus 1996, Satish Kumar et al. 2013, Aidoo et al. 2006, Moktan et al. 2011, Roy et al. 2009
Uttapam	Batter prepared from the soaked raw materials were fermented at room temperature for 5–6 h. The fermented batter is spreaded over buttery/greased pan into a round shape. Toppings like chopped vegetables, paneer, capsicum and onion are added over the flat batter and cooked in low flame.	Uttapam is a zero trans-fat food having approximately 160 Cal per 50 g serving, 0.4 g of fat, 34 g carbohydrate, 3.0 g dietary fiber, 5.0 g of protein with source calcium, ferrous, vitamin A and C.	Being cholesterol-free food item, *uttapam* is a prescribed food for high sugar and cholesterol patients. It is easily digestible and can reduce body weight and prevent obesity.	Saraniya and Jeevaratnam 2014
Selroti	Traditionally, batter is prepared using rice and wheat flour, sugar, butter or fresh cream and spices. It is kept for fermentation at room temperature for 3–4 h. Next, it is deeply fried in ring shape (golden brown in colour) yielding spongy (bread like) and pretzel like food item.	An average serving of 260 g *Selroti* gives approximately 694 Cal, 138.0 g carbohydrates, 8.4 g proteins, 14.8 g fat, 42.0 g sugars and 2.68 g dietary fibers. Minerals like sodium, potassium, iron, calcium and vitamin A and C are also present in *Selroti*. This is free from gluten and trans-fat.	*Selroti* is generally offered for good health and a recommended diet for protecting dyslipidaemia and cardio-metabolic risks.	Satish et al. 2013, Yonzan and Tamang 2010

Table 3. contd.

Table 3. contd....

Fermented food	Traditional processes	Functional components	Health beneficial impacts	References
Adai and Vada	The traditional process starts with the batter preparation using rice and lentils. After fermentation, the batter is fried with some edible oil in flattened donut shape. Sometimes the batter is mixed with spices like red chillis, fresh pepper corns, cumin, ginger, pieces, green chilies', fresh coriander leaves, coconut pieces, etc., to give an intense flavor to the *Adai* and *Vada*.	These are low calorie protein and iron rich foods. This have 197 Cal, 505 g fat, 20 mg sodium, 350 mg potassium, 39.6 g total carbohydrates, 6.5 g dietary fibers, 1.7 g sugars and 7.6 g proteins. It contains 2% vitamin C, 3% calcium and 14% iron.	This is considered as healthy snack as it is rich with proteins, iron and dietary fibers. This is recommended for kids as well as women for the health benefits.	Blandino et al. 2003, Chavan and Kadam 1989
Sour Rice	Cooked rice is cooled down to room temperature and adequate water is added to it. This watery rice is allowed to ferment overnight at room temperature. The fermented rice with water is consumed along with cooked vegetables/others stuffs.	The microbes by fermentation produce copious amounts of vitamin B complex, and vitamin K. About 100 g of sour rice contains 73.91 mg iron, 303 mg sodium, 839 mg potassium and 850 mg calcium.	Fermented sour rice is a high energy rich, body rehydrating food. It controls the bowel movement as well as fiber release in the stool, prevents constipation. The fermented rice restores healthy intestinal flora and can prevent gastrointestinal ailments like duodenal ulcer, infectious ulcerative colitis, Crohn's disease, irritable bowel syndrome (IBS), celiac disease, candida infection, etc.	Blandino et al. 2003, Tamang 2012, Ray and Swain 2013, Choi et al. 2014
Chitou/ Appam	Parboiled rice and urad dahl paste are mixed to prepare the *Appam* batter which is fermented overnight. The fermented batter is next mixed	Chitou/Appam can give 138.8 Cal, 3.7 g total fat, 0.1 g MUFA and PUFA, 31.7 mg sodium, 13.5 mg potassium, 23.2 g total	This is a healthy, simple and easy digestible, nutritionally enriched food.	Roy et al. 2007, Ray and Swain 2013

	with sugar, grated coconut or other seasonings. The final mixture is poured into earthen mold, covered and fried in low heat to obtain traditional *Chitou* and *Appam*.	carbohydrate, 1.1 g dietary fiber, 2.1 g protein, with vitamin A, B-complex, calcium, folate, iron, niacin, riboflavin, and thiamine.		
Chakuli pitha (rice cake)	The coat free black gram (*Phaseolus mungo*) and rice flour batter are kept for 10–12 h and then fried over a hot greased pan.	Not reported	Energy rich delicious food, it is digest easily and suitable in the ailing persons, pre and post-natal women and children.	Roy et al. 2007
Enduri Pitha	Fermented batter of black gram and rice flour are filled in a folded leaves of turmeric (*Curcuma longa* L.) and then cooked over steam.	Not reported	Helps in strengthening the immune system, fighting against worms and different infections common in winter season.	Roy et al. 2007
Podo Pitha	The fermented (2–4 h) batter of rice and black gram mixed with different sweeteners and coconut, then wrapped with banana (*Musa paradisiaca* L.) or sal (*Shorearobusta*) leaves and roasted in an oven or earthen oven with help of charcoal to bake in low but continuous heat for 5–10 h. The pitha is cut into pieces and served.	Not reported	This is tasty and energy rich food containing much carbohydrate and free sugars.	Roy et al. 2007
Haria	Para boiled rice is mixed with *Bakhar* (1:100) and transferred to an earthen pot and kept in dark room for 3–5 days for fermentation. Diluted with water and sieved to get the Haria.	pH-3.61, alcohol-3–4% (v/v), altooligosaccharides, pyranose sugar derivatives, vitamins, minerals and phenolic rich beverage.	Protects from gastrointestinal ailments like dysentery, diarrhoea, amebiosis, acidity, vomiting. It exerted significant level of antioxidant activity.	Ghosh et al. 2014, Ghosh et al. 2015a,b

Table 3. contd.

Table 3. contd....

Fermented food	Traditional processes	Functional components	Health beneficial impacts	References
Apong	Ash of paddy husk and straws is mixed with cooked glutinous rice, *Epop* is mixed (1:30) and transferred into earthen pot. Allowed to ferment for 20 days at 30–35°C. The ferment is filtered to get clear brownish filtrate *Apong*.	pH-4.06, lactic acid-0.5%, carbohydrate-46 mg/ml, reducing sugar 3 mg/ml, protein-1.05 mg/ml, free amino acids-2.43 mg/ml, ethanol-7.52%–18.5%, amylase-2.4 U/ml.	It is nutricious, energy-rich refreshing drink having antimicrobial, antioxidant and other age preventing effects. Apong is also helpful in preventing formation of kidney stones.	Das et al. 2012, Kardong et al. 2012
Judima	Powdered *Humao* is mixed (1:100) with air dried boiled rice and left to ferment in room temperature. After 3–4 days, the mixture is transferred to *khulu* (a triangle shaped bamboo cone) and the fermented juice judima is collected by back sloping.	pH-4.4, titratable acidity-0.45%, carbohydrate-32 mg/ml, protein-0.97 mg/ml, free amino acids-3.21 mg/ml. ethanol-16% (v/v), trace elements like Cu, Cr, Mn, Fe, K, Na, Se.	*Judima* has anti-inflammatory, anti-allergic, anti-oxidant, anti-bacterial, anti-fungal, anti-spasmodic, hepatoprotective, hypolimidemic, neuroprotective, hypotensive, anti-aging and anti-diabetic potentialities.	Chakrabarty et al. 2014, Arjun et al. 2014
Zutho	Cooled rice porridge mixed with grist and poured into earthen jar. Allowed to ferment for 2–3 days. *Zutho* is formed and drank directly.	pH-3.6, acidity-5.1 ml, reducing sugar-6.3 mg/ml, total sugar-39.7 mg/ml, ethanol-5%.	Zutho boosts immune system, lowers insulin level of blood, prevents loss of appetite, lowers bad cholesterol, assists in healing wound and prevents infection.	Das et al. 2012, Teramoto et al. 2002
Bhaati Jaanr	Powdered *marcha* (2%) is mixed with cooked air dried glutinous rice and kept in a vessel for 1–2 days at room temperature for 2–8 days. A thick paste is made by stirring the ferments and consumed directly as *Bhatti Jaanr*.	pH-3.5, alcohol-5.9%, ash-1.7%, protein-9.5%, fat-2.0%, crude fiber-1.5%, carbohydrate-86.9%, food value (per 100 g)-404.1 kcal, trace elements like Ca, K, P, Fe, Mg, Mn, Zn are also present.	It is consumed as a staple food or mild alcoholic sweet beverage. It is recommended for ailing persons and post-natal women to regain their strength.	Tamang and Thapa 2006

			Selhub et al. 2014, Uno et al. 2009, Zhu and Trampe (2013)
		Sake is the national drink in Japan. It has many pharmacological attributes including antioxidant, antiaging, anti-inflammatory, anticarcinogenic, and chemopreventive activity. It also contain isomaltooligosaccharides that enhances the growth of *Bifidobacteria* and *Lactobacilli*.	
Sake	Steamed and cooled polished rice mixed with *koji*, water, and yeast starter for fermentation at low temperature for 20–30 days.	It contain high quantity of ethanol (15–20%), ferulic acid (0.1–6.6 µg/ml), isomaltooligosaccharides and other nutraceuticals.	
Takju	Steamed and cooled rice is mixed traditional starter *nuruk*, fermented, and the resulting liquid is aged in a bottle.	Contain alcohol, minerals (Cu, Ca, Mn, Na, P, Zn, Fe, and Mg), high amount of sugars (30–90 µg/ml), and polyphenols (100–250 µg/ml).	Aidoo et al. 2006, Bae et al. 2010, Jin et al. 2008, Lee et al. 1996, Park and Lee 2002
			Takju exhibits various biological effects, including antioxidant activity, immunomodulating activity, and anticomplementary activity due to the presence of bioactive compounds such as polyphenols, polysaccharides, and polysaccharide–peptide complexes.
Tapai	Soft and sticky boiled rice mixed with traditional amylolytic starter. Cassava tubers are also widely used materials for the tapai production. The inoculated substrate can placed in wide mouth glass jars or in banana leaves (to obtain pleasant aroma). Fermentation continued for 24–48 hours at ambient temperatures (25–30°C).	pH 3.4–4.0, titratable acidity 0.86, contain protein (8.7%), fat (0.29%), ash 0.5%, and crude fiber (0.56%). Tapai contained essential amino acid, lysine and thiamine.	Chiang et al. 2006, De Angelis et al. 2007, Foo et al. 2001, Lilly and Stillwell 1965, Chiang et al. 2006
			This fiber rich foods are recommended to prevent several diseases and regulate the energy intake and satiety, which would decrease obesity. Participating LAB produce bacteriocin which can kill the pathogens.

Table 3. contd.

Table 3. contd....

Fermented food	Traditional processes	Functional components	Health beneficial impacts	References
Tape ketan	Procedure is as like tape but need prolonged incubation.	Contain 30.7–39.3 % of total soluble solid, pH ranges from 4.2 to 4.5, soluble crude protein 13.8–18.4% (w/w), titratable acidity 5.74–8.11, ethanol ~ 8.5%, v/v (varies according to the duration of incubation).	Energy rich and healthy beverage. Ethanol content provides calories and prevents food borne pathogen. Protein (7–16%) and lysine content are also significantly high.	Ardhana and Fleet 1989, Cook et al. 1991, Cronk et al. 1977, Steinkraus 1997
Khaomak	Its preparation procedure is also same as tape but substrate is only steamed glutinous rice.	Contain fat (3.72 gm/100 gm DM), protein (8.17 g/100 g DM), dietary fibre (4.01 gm/100 gm DM), carbohydrate (74.09 gm/100 gm DM), amylose (8.9%), and give high calories (362.55 cal/100 gm DM). It also contain phenolics (665.16 gm/100 gm) and pigmented molecule anthocyanin (256.61 gm/100 gm).	Showed broad antimicrobial and antioxidant activities.	Rhee et al. 2011, Yodtheon et al. 2010
Rice wine	Steamed rice mixed (1:100 ratio) with starter culture (composed of *Rhizopus* and *Sacharomycopsis* spp.) and incubated in a try for 2 days. Then transferred into a sealed jar and kept for 2 weeks. The fermented mass is squeezed using cheese cloth to collect alcoholic juice. Pasteurized juice was allowed to stand for one to three months in dark, cool place.	The ethanol concentration of the solid state fermented rice wines was in the range of 0.001–0.01% (w/v). Other parameters like pH (4.5–5.5), protein (0.45–0.99% w/w), ash (0.10–0.30% w/w), total solid (1.72–14.34% w/w), glucose (0.41–0.79%w/v). Also contained t-resveratro and phenolic acids such as gallic acid, ethyl α-D-glucoside, glycerol, organic acids and amino acids.	Antioxidant and potential effective agent for the prevention and treatment of UV-induced skin aging.	Law et al. 2011

Yellow wine	Chinese rice wine is typically fermented from glutinous rice with 'wheat Qu' as a saccharifying agent and pure cultured yeast (*Saccharomyces cerevisiae*) as a fermentation starter. The material is then incubated at 28–30°C for 48 h in a special room, and dried at 45°C until the humidity is lower than 12% (w/w).	Contain ethanol, sugar, amino acids, polyphenols.	It is therapeutically active in the prevention of cancer and cardiovascular disease. It shows strong antioxidant activity. It reduces the risk of coronary disease.	Li et al. 2010, Que et al. 2006
Rice noodles [*Mifen* in China, *Khanom Jeen* in Thailand, *Mohingar* in Myanmar, *Khao Pen* in Laos and *Banh Da* in Vietnam]	The rice grains are statically fermented naturally, without a starter, for 3–6 days, then wet-milled, steamed, and extruded into rice noodles.	pH 4.0, organic acids (lactic acid and acetic acid), free reducing sugars 3.5–4.0g%, less starchy, contain abundant amount of enzymes, vitamins and other nutrients.	Healthy food than any other non-fermented noodles.	Lu et al. 2008

References

Aidoo, K.E., Rob Nout, M. and Sarkar, P.K. (2006). Occurrence and function of yeasts in Asian indigenous fermented foods. FEMS Yeast Research 6: 30–39.

Alexandraki, V., Tsakalidou, E., Papadimitriou, K. and Holzapfel, W.H. (2013). Status and trends of the Conservation and Sustainable use of Microorganisms in Food Processes. Commission on Genetic Resources for Food and Agriculture. FAO Background Study Paper No. 65.

Amudha, K., Sakthivel, N. and Mohamed, Y.M. (2011). Rice-a novel food with medicinal value. Agriculture Review 32: 222–227.

Ardhana, M.M. and Fleet, G.H. (1989). The microbial ecology of tape ketan fermentation. International Journal of Food Microbiology 9: 157–165.

Arjun, J., Verma, A. and Prasad, S. (2014). Method of preparation and biochemical analysis of local tribal wine Judima: an indigenous alcohol used by Dimasa tribe of North Cachhar Hills district of Assam, India. International Food Research Journal 21: 463–470.

Bae, S.H., Jung, E.Y., Kim, S.Y., Shin, K.S. and Suh, H.J. (2010). Antioxidant and immuno-modulating activities of Korean traditional rice wine, Takju. Journal of Food Biochemistry 34(s1): 233–248.

Barnett, J.A. (2003). Beginnings of microbiology and biochemistry: the contribution of yeast research. Microbiology 149: 557–567.

Birben, E., Sahiner, U.M., Sackesen, C., Erzurum, S. and Kalayci, O. (2012). Oxidative stress and antioxidant defense. WAO Journal 5: 9–19.

Blandino, A., Al-Asceri, M.E., Pandiella, S.S., Cantero, D. and Webb, C. (2003). Cereal based fermented foods and beverages. Food Research International 36: 527–543.

Campbell-Platt, G. (1994). Fermented foods—a world perspective. Food Research International 27: 253–257.

Chakrabarty, J., Sharma, G. and Tamang, J.P.T. (2010). Substrate utilisation in traditional fermentation technology practiced by tribes of North Cachar Hills district of Assam. Assam University Journal of Science and Technology 4: 66–72.

Chavan, J.K. and Kadam, S.S. (1989). Nutritional improvement of cereals by fermentation. Critical Reviews in Food Science and Nutrition 28: 349–400.

Chiang, Y.W., Chye, F.Y. and Mohd Ismail, A. (2006). Microbial diversity and proximate composition of tapai, a Sabah's fermented beverage. Malaysian Journal of Microbiology 2: 1–6.

Choi, J.S., Kim, J.W., Cho, H.R., Kim, K.Y., Lee, J.K., Ku, S.K. and Sohn, J.H. (2014). Laxative effects of fermented rice extract (FRe) in normal rats. Toxicology and Environmental Health Science 6: 155–163.

Cook, P., Owens, J. and Campbell-Platt, G. (1991). Fungal growth during rice tape fermentation. Letters in Applied Microbiology 13: 123–125.

Cronk, T., Steinkraus, K., Hackler, L. and Mattick, L. (1977). Indonesian tape ketan fermentation. Applied and Environmental Microbiology 33: 1067–1073.

Das, A. and Deka, S. (2012). Fermented foods and beverages of the North-East India. International Food Research Journal 19: 377–392.

Das, A., Raychaudhuri, U. and Chakraborty, R. (2012). Cereal based functional food of Indian subcontinent: a review. Journal of Food Science and Technology 49: 665–672.

De Angelis, M., Rizzello, C.G., Alfonsi, G., Arnault, P., Cappelle, S., Di Cagno, R. and Gobbetti, M. (2007). Use of sourdough lactobacilli and oat fibre to decrease the glycaemic index of white wheat bread. British Journal of Nutrition 98: 1196–1205.

Deori, C., Begum, S.S. and Mao, A. (2007). Ethnobotany of sujen, a local rice beer of Deori tribe of Assam. Indian Journal of Traditional Knowledge 6(1): 121–125.

Dutta, A., Das, D. and Goyal, A. (2012). Purification and characterization of fructan and fructansucrase from *Lactobacillus fermentum* AKJ15 isolated from Kodo ko jaanr, a fermented beverage from north-eastern Himalayas. International Journal of Food Sciences and Nutrition 63: 216–224.

Foo, H., Lim, Y. and Rusul, G. (2001). Isolation of bacteriocin producing lactic acid bacteria from Malaysian fermented food, Tapai. Proceeding of the 11th World Congress of Food Sciences and Technology, Seoul, Korea.

Ghosh, K. (2016). Profiling of nutrients and lactic acid bacteria in a rice-based fermented beverage. Ph.D. Thesis, Vidyasagar University, Midnapore, India.

Ghosh, K., Ray, M., Adak, A., Halder, S.K., Das, A., Jana, A., Parua, S., Vágvölgyi, C., Das Mohapatra, P.K., Pati, B.R. and Mondal, K.C. (2015b). Role of probiotic *Lactobacillus fermentum* KKL1 in the preparation of a rice based fermented beverage. Bioresource Technology 188: 161–168.

Ghosh, K., Ray, M., Adak, A., Dey, P., Halder, S.K., Das, A., Jana, A., Parua, S., Das Mohapatra, P.K., Pati, B.R. and Mondal, K.C. (2015a). Microbial, saccharifying and antioxidant properties of an Indian rice based fermented beverage. Food Chemistry 168: 196–202.

Ghosh, K., Maity, C., Adak, A., Halder, S.K., Jana, A., Das, A., Parua, S., Das Mohapatra, P.K., Pati, B.R. and Mondal, K.C. (2014). Ethnic Preparation of Haria, a Rice-Based Fermented Beverage, in the Province of Lateritic West Bengal, India. Ethnobotany Research and Applications 12: 39–49.

Grill, J.P., Manginot-Durr, C., Schneider, F. and Ballongueet, J. (1995). Bifidobacteria and probiotic effects: action of *Bifidobacterium* species on conjugated bile salts. Current Microbiology 31: 23–27.

Gross, B.L. and Zhao, Z. (2014). Archaeological and genetic insights into the origins of domesticated rice. Proceeding of the National Academy of Science 111: 6190–6197.

Gupta, A. and Tiwari, S.K. (2014). Probiotic potential of *Lactobacillus plantarum* LD1 isolated from batter of dosa, a south Indian fermented food. Probiotics and Antimicrobial Proteins 6: 73–81.

Gupta, S. and Abu-Ghannam, N. (2012). Probiotic fermentation of plant based products: Possibilities and opportunities. Critical Reviews in Food Science and Nutrition 52: 183–199.

Henderson, A.J., Ollila, C.A., Kumar, A., Borresen, E.C., Raina, K., Agarwal, R. and Ryan, E.P. (2012). Chemopreventive properties of dietary rice bran: Current status and future prospects. Advances in Nutrition 3: 643–653.

Hesseltine, C.W., Rogers, R. and Winarno, F.G. (1988). Microbiological studies on amylolytic oriental fermentation starters. Mycopathologia 101: 141–155.

Heuberger, A.L., Lewis, M.R., Chen, M.H., Brick, M.A., Leach, J.E. and Ryan, E.P. (2010). Metabolomic and functional genomic analyses reveal varietal differences in bioactive compounds of cooked rice. PLoS One 5: e12915.

Jin, J., Kim, S.-Y., Jin, Q., Eom, H.-J. and Han, N.S. (2008). Diversity analysis of lactic acid bacteria in Takju, Korean rice wine. Journal of Microbiology and Biotechnology 18: 1678–1682.

Joint FAO/WHO Expert Consultation on Evaluation of Health and Nutritional Properties of Probiotics in Food Including Powder Milk with Live Lactic Acid Bacteria, October, 2001.

Jung, M.J., Nam, Y.D., Roh, S.W. and Bae, J.W. (2012). Unexpected convergence of fungal and bacterial communities during fermentation of traditional Korean alcoholic beverages inoculated with various natural starters. Food Microbiology 30: 112–123.

Kardong, D., Deori, K., Sood, K., Yadav, R., Bora, T. and Gogoi, B. (2012). Evaluation of nutritional and biochemical aspects of Po: ro apong_ (Saimod)-A home made alcoholic rice beverage of Mising tribe of Assam, India. Indian Journal of Traditional Knowledge 11: 499–504.

Kind, T., Scholz, M. and Fiehn, O. (2009). How large is the metabolome? A critical analysis of data exchange practices in chemistry. PLoS One 4: e5440.

Kivanc, M., Yilmaz, M. and Cakir, E. (2011). Isolation and identification of lactic acid bacteria from boza, and their microbial activity against several reporter strains. Turkish Journal of Biology 35: 313–324.

Kwak, N.S. and Jukes, D.J. (2001). Functional foods. Part 1: the development of a regulatory concept. Food Control 12: 99–107.

Lamsala, B.P. and Faubionb, J.M. (2009). The beneficial use of cereal and cereal components in probiotic foods. Food Reviews International 25: 103–114.

Law, S.V., Abu Bakar, F., Mat Hashim, D. and Abdul Hamid, A. (2011). Popular fermented foods and beverages in Southeast Asia. International Food Research Journal 18: 475–484.

Lee, J.-S., Lee, T.-S., Noh, B.-S. and Park, S.O. (1996). Quality characteristics of mash of takju prepared by different raw materials. Korean Journal of Food Science and Technology 28: 330–336.

Li, B.B., Zeng, J.H., Liu, X.-q., Zhuge, Q. and Yu, Y.F. (2010). Study on quantitative relationships between amino acids and sensory taste of yellow rice wine [J]. Liquor-Making Science & Technology 10: 004.

Lilly, D.M. and Stillwell, R.H. (1965). Probiotics: growth-promoting factors produced by microorganisms. Science 147(3659): 747–748.

Limtong, S., Sintara, S., Suwannarit, P. and Lotong, N. (2002). Yeast diversity in Thai traditional alcoholic starter. Kasetsart Journal of Natural Sciences 36: 149–158.

Lomer, M.C.E., Parkes, G.C. and Sanderson, J.D. (2008). Lactose intolerance in clinical practice –myths and realities. Alimentary Pharmacol. Therapeutics 27: 93–103.

Lorri, W. and Svanberg, U. (1993). Lactic acid-fermented cereal gruels: Viscosity and flour concentration. International Journal of Food Science and Nutrition 44: 207–213.

Lu, Z.H., Peng, H.H., Cao, W., Tatsumi, E. and Li, L.T. (2008). Isolation, characterization and identification of lactic acid bacteria and yeasts from sour Mifen, a traditional fermented rice noodle from China. Journal of Applied Microbiology 105: 893–903.

Metchnikoff, E. (1907). The prolongation of life. *In*: Optimistic Studies. William Heinemann, London, UK.

Moktan, B., Roy, A. and Sarkar, P.K. (2011). Antioxidant activities of cereal-legume mixed batters as influenced by process parameters during preparation of dhokla and idli, traditional steamed pancakes. International Journal of Food Science and Nutrition 62: 360–369.

Muthayya, S., Sugimoto, J.D., Montgomery, S. and Maberly, G.F. (2014). An overview of global rice production, supply, trade, and consumption. Annals of the New York Academy of Science 1324: 7–14.

Nout, M.J.R. (2009). Rich nutrition from the poorest-cereal fermentations in Africa and Asia. Food Microbiol 26: 685–692.

Nout, M.J.R. and Motarjemi, Y. (1997). Assessment of fermentation as a household technology for improving food safety: A joint FAO/WHO workshop. Food Control 8(5-6): 221–226.

Park, C.-S. and Lee, T.S. (2002). Quality characteristics of Takju prepared by wheat flour Nuruks. Korean Journal of Food Science and Technology 34: 296–302.

Phithakpol, B., Varanyanond,W., Reungmaneepaitoon, S. and Wood, H. (1995). The Traditional Fermented Foods of Thailand. Kuala Lumpur: ASEAN Food Handling Bureau.

Phutthaphadoong, S., Yamada, Y., Hirata, A., Tomita, H., Taguchi, A., Hara, A., Limtrakul, P.N., Iwasaki, T., Kobayashi, H. and Mori, H. (2009). Chemopreventive effects of fermented brown rice and rice bran against 4-(methylnitrosamino)-1-(3-pyridyl)-1-butanone-induced lung tumorigenesis in female A/J mice. Oncology Reports 21: 321–327.

Que, F., Mao, L., Zhu, C. and Xie, G. (2006). Antioxidant properties of Chinese yellow wine, its concentrate and volatiles. LWT—Food Science and Technology 39: 111–117.

Ray, M., Ghosh, K., Singh, S. and Mondal, K.C. (2016). Folk to functional: An explorative overview of rice-based fermented foods and beverages in India. Journal of Ethnic Foods 3: 5–18.

Ray, R.C. and Sivakumar, S.S. (2009). Traditional and novel fermented foods and beverages from tropical root and tuber crops: review. International Journal of Food Science and Technology 44: 1073–1087.

Ray, R.C. and Swain, M.R. (2013). Indigenous fermented foods and beverages of Odisha, India: an overview. *In*: V.K. Joshi (ed.). Indigenous Fermented Foods of South Asia. USA: CRC Press.

Rhee, S.J., Lee, C.-Y.J., Kim, K.K. and Lee, C.H. (2003). Comparison of the traditional (Samhaeju) and industrial (Chongju) rice wine brewing in Korea. Food Science and Biotechnology 12: 242–247.

Rhee, S.J., Lee, J.-E. and Lee, C.H. (2011). Importance of lactic acid bacteria in Asian fermented foods. Microbial Cell Factories 10(1): S5.

Romano, M., Vitaglione, P., Sellitto, S. and D'Argenio, G. (2012). Nutraceuticals for protection and healing of gastrointestinal mucosa. Current Medicinal Chemistry 19: 109–117.

Roy, A., Bijoy, M. and Prabir, S.K. (2009). Survival and growth of foodborne bacterial pathogens in fermentating batter of dhokla. Journal of Food Science and Technology 46: 132–135.

Roy, A., Moktan, B. and Sarkar, P.K. (2007). Traditional technology in preparing legume based fermented foods of Orissa. Indian Journal of Traditional Knowledge 6: 12–16.

Roy, M. (1997). Fermentation technology. pp. 437–447. In: A.K. Bag (ed.). History of Technology of India, The Indian National Science Academy, New Delhi.

Ryan, E.P., Heuberger, A.L., Weir, T.L., Barnett, B., Broeckling, C.D. and Prenni, J.E. (2011). Rice bran fermented with *Saccharomyces boulardii* generates novel metabolite profiles with bioactivity. Journal of Agricultural Food Chemistry 59: 1862–1870.

Saikia, B., Tag, H. and Das, A. (2007). Ethnobotany of foods and beverages among the rural farmers of Tai Ahom of North Lakhimpur district, Asom. Indian Journal of Traditional Knowledge 6(1): 126–132.

Samanta, A.K., Kolte, A.P., Senani, S., Sridhar, M. and Jayapal, N. (2011). Prebiotics in ancient Indian diets. Current Science 101: 43–46.

Saraniya, A. and Jeevaratnam, K. (2014). Purification and mode of action of antilisterial bacteriocins produced by *Lactobacillus pentosus* SJ65 isolated from uttapam batter. Journal of Food Biochemistry 38: 612–619.

Satish, R.K., Kanmani, P., Yuvaraj, N., Paari, K.A., Pattukumar, V. and Arul, V. (2013). Traditional Indian fermented foods: a rich source of lactic acid bacteria. International Journal of Food Science and Nutrition 64: 415–428.

Selhub, E.M., Logan, A.C. and Bested, A.C. (2014). Fermented foods, microbiota, and mental health: ancient practice meets nutritional psychiatry. Journal of Physiological Anthropology 33: 2.

Seo, M.Y., Chung, S.Y., Choi, W.K., Seo, Y.K., Jung, S.H., Park, J.M., Seo, M.J., Park, J.K., Kim, J.W. and Park, C.S. (2009). Anti-aging effect of rice wine in cultured human fibroblasts and keratinocytes. Journal of Bioscience and Bioengineering 107: 266–271.

Shen, F., Ying, Y., Li, B., Zheng, Y. and Zhuge, Q. (2011). Multivariate classification of rice wines according to ageing time and brand based on amino acid profiles. Food Chemistry 129: 565–569.

Stanton, C., Ross, R.P., Fitzgerald, G.F. and Sinderen, D.V. (2005). Fermented functional foods based on probiotics and their biogenic metabolites. Current Opinion in Biotechnology 16: 198–203.

Steinkraus, K.H. (1997). Bio-enrichment: production of vitamins in fermented foods. pp. 603–621. In: Microbiology of Fermented Foods, Springer.

Steinkraus, K. (1995). Handbook of Indigenous Fermented Foods, revised and expanded. CRC Press.

Steinkraus, K. (1994). Nutritional significance of fermented foods. Food Research International 27: 259–267.

Steinkraus, K.H. (2004). Industrialization of Indigenous Fermented Foods. NewYork, NY: Marcel Dekker, Inc.

Steinkraus, K.H. (1996). Hand book of Indigenous Fermented Food, 2nd Edn. New York, NY: Marcel Dekker, Inc.

Tamang, J.P., Shin, D.H., Jung, S.J. and Chae, S.W. (2016b). Functional properties of microorganisms in fermented foods. Frontiers in Microbiology [doi: 10.3389/fmicb.2016.00578].

Tamang, J.P. and Thapa, S. (2006). Fermentation dynamics during production of Bhaati jaanr, a traditional fermented rice beverage of the Eastern Himalayas. Food Biotechnology 20(3): 251–261.

Tamang, J.P., Tamang, N., Thapa, S., Dewan, S., Tamang, B., Yonzan, H., Rai, A.K., Chettri, R., Chakrabarty, J. and Kharel, N. (2012). Microorganisms and nutritional value of ethnic fermented foods and alcoholic beverages of North East India. Indian Journal of Traditional Knowledge 11: 7–25.

Tamang, J.P., Watanabe, K. and Holzapfel, W.H. (2016a). Review: Diversity of microorganisms in global fermented foods and beverages. Frontiers in Microbiology [doi: 10.3389/fmicb.2016.00377].

Teramoto, Y., Yoshida, S. and Ueda, S. (2002). Characteristics of a rice beer (zutho) and a yeast isolated from the fermented product in Nagaland, India. World Journal of Microbiology and Biotechnology 18: 813–816.

Thapa, S. and Tamang, J. (2006). Microbiological and physico-chemical changes during fermentation of Kodo ko Jaanr, a traditional alcoholic beverage of the Darjeeling hills and Sikkim. Indian Journal of Microbiology 46: 333.

Thompson, M.D. and Thompson, H.J. (2009). Biomedical agriculture: a systematic approach to food crop improvement for chronic disease prevention. Advances in Agronomy 102: 1–54.

Tissier, H. (1906). Traitement des infections intestinales par la me' thode de la flore bacte' rienne de l'intestin. CR Soc Biol 60: 359–361. [English translation: Treatment of intestinal infections by the commensal flora method].

Tiwari, S. and Mahanta, D. (2007). Ethnological observations fermented food products of certain tribes of Arunachal Pradesh. Indian Journal of Traditional Knowledge 6: 106–110.

Uno, T., Itoh, A., Miyamoto, T., Kubo, M., Kanamaru, K., Yamagata, H., Yasufuku, Y. and Imaishi, H. (2009). Ferulic acid production in the brewing of pice wine (Sake). Journal of the Institute of Brewing 115: 116–121.

Waters, D., Maucha, A., Coffeyb, A., Arendt, E. and Zannini, E. (2015). Lactic acid bacteria as a cell factory for the delivery of functional biomolecules and ingredients in cereal-based beverages: a review. Critical Reviews in Food Science and Nutrition 55: 503–520.

Yodtheon, J., Chantarapanont, W. and Rimkeeree, H. (2010). Screening pure cultures by testing starch hydrolysis and alcohol fermentation for developing sweeten rice (Khaomak) beverage. Proceedings of the 48th Kasetsart University Annual Conference, Kasetsart, 3–5 March, 2010. Subject: Agro-Industry. Kasetsart University.

Yonzan, H. and Tamang, J.P. (2010). Microbiology and nutritional value of selroti, an ethnic fermented cereal food of the Himalayas. Food Biotechnology 24: 227–247.

Zannini, E., Pontonio, E., Waters, D. and Arendt, E. (2011). Applications of microbial fermentations for production of gluten-free products and perspectives. Applied Microbiology and Biotechnology, pp. 1–13.

Zhu, Y. and Trampe, J. (2013). Koji—where East meets West in fermentation. Biotechnology Advances 31: 1448–1457.

9

Maize (*Zea mays* L. subsp. *mays*) Fermentation

Lila Lubianka Domínguez-Ramírez, Gloria Díaz-Ruiz and Carmen Wacher*

1. Introduction: Brief History of Maize

Maize (/'meɪz/) from the Spanish *maíz*, originally from *taino* word *maisi* or *maiji*, is a large grain plant also known as corn. According to its scientific classification it belongs to the *Plantae* Kingdom, phylum *Magnoliophyta*, class *Liliopsida*, order *Liopsidia*, *Poaceae* family (as wheat, barley, rye and grass, perennial or annual grasses), it is from genus *Zea* and at species level is known as *Zea mays*. This specie has four subspecies: *Z. mays* subsp. *huehuetenangensis* that can be found in Western Guatemalan highlands, *Z. mays* subsp. *mexicana* endemic in the Central and Northern Mexican Highlands, *Z. mays* subsp. *parviglumis*, these tree species are teosintes (*teocinte*, *teocintle* or *teosintle*), taller than common grasses and are wild relatives of maize, and the fourth is the only domesticated member of *Zea* genus: *Z. mays* subsp. *mays*, the variety cultivated nowadays as corn (from now on, maize).

Z. mays subsp. *parviglumis* is the closest modern relative of maize; genetically this teosinte is almost like the "wild" ancestor of maize with evidence at morphological, phylogenetic, eco-geographic, biochemical

Departamento de Alimentos y Biotecnología, Facultad de Química, UNAM, 04510 Mexico City, Mexico.
  E-mail: lila.lubianka@yahoo.com.mx; gloriadr@unam.mx
* Corresponding author: wacher@unam.mx

and molecular levels (Doebley 1990, Dorweiler et al. 1993, Buckler and Holtsford 1996, Wang et al. 1999, Matsuoka et al. 2002). Maize arose from an annual teosinte species sharing a common ancestor with Z. *mays* subsp. *parviglumis*. There is also evidence that an ancestral native teosinte from Central Balsas River Valley (on the Pacific slopes of the Mexican states of Guerrero and Michoacán) was the ancestor (Piperno and Flannery 2001). The archeological evidence recovered from caves at these geographic sites (Xihuatoxtla, Guerrero) are micro-remains (phytoliths and starch grains recovered from stone tools related directly with cultivation and for cooking) of maize still teosinte-like dated as early as 8,700 years ago (4250 B.C.), being evidence of an early domestication of teosinte to maize (Piperno et al. 2009, Ranere et al. 2009). This hypothesis is well accepted and this ancestor is known as Balsas teosinte. This information is consistent with genetic evidence (molecular clocks) that estimates maize divergence from teosinte almost 9,000 years ago (Matsuoka et al. 2002).

More archeological evidences tracing maize distribution through Mesoamerica are phytolites and pollen found at San Andrés, Tabasco, Mexico, dated as 7,300 years old, also associated with stone tools and ceramics of maize agriculture (Pohl et al. 2007). These are older than macrofossils (cobs, pollen grains and sediments) of primitive maize found at Guilá Naquitz Cave in Oaxaca, Mexico, dated as 6,200 years old (Benz 2001, Piperno and Flannery 2001) and with those, at once thought to be the ancient evidence of first stages of maize domestication (MacNeish 1985), found in the Valley of Tehuacán, Puebla, Mexico dated *ca.* 6,000 years (Mangelsdorf et al. 1964, 1967) but, at this site, specimens dated *ca.* 4,700 years were also found (Long et al. 1989). All together, these data suggest that Mexico is the place of origin and domestication of maize with evidence of its introduction into South America (Andean Regions) from about 2,200 to 1,900 years ago (Staller and Thompson 2002) and its dispersion into North America no later than 2,100 years ago (Merrill et al. 2009).

Biological and archeological evidence shows that the evolution of maize has been a process that not only was natural selection but a complex combination of biological and cultural pressure. This was guided by humans who selected the best seeds according to their needs (cultural process), for an increased productivity or for a better growth under specific conditions (soil, weather, etc.). The seeds from the best plants that were able to develop under those specific conditions (biological pressure guided by ecological factors) were selected. These procedures brought a wide variety in maize.

One of the first proposals for maize classification was made by Sturtevant (1883), which is still in use, in which he divided maize into six types being different by the structure of the kernel (phenotypes): dent, flint, floury (soft), pop, sweet and pod (Smoesmith 1918, McCann 2001).

Dent corn has the horny endosperm mainly at the edges of the kernel and the soft endosperm in the center and at the crown. Crops have from medium to large size with ears having a large number of grains and large diameter, requiring a medium to long season to mature. The grains are long, wedge-shaped, closely set on the cob with the yellow varieties being very common.

Flint has a soft and starchy endosperm in the center of the grain and surrounded by the horny endosperm. The grain shrinks uniformly in maturing so no indentation occurs. The ear has a flinty, glossy appearance with yellow or white color. This variety is adapted to Northern States of USA and Canada.

Soft (floury) variety is characterized by the absence of the horny endosperm, the entire interior of the grain, aside from the germ, being composed of a soft, starchy endosperm and it needs a long growing season to mature. The shape of the ear is similar to the flinty variety but is larger in diameter. There are crops in white, blue and black colors being the most common and the appearance of the ear is not as glossy as in flint variety.

Pop variety has a horny and hard endosperm this gives the popping characteristic (turning inside out and enlarging the white mass) upon being heated. It has two groups according to its shape: the *rice* one, whose grains are longer and very sharp at the crown and the *pearl* group, this has smooth and rounded grains with a more compact arrangement on the cob. It is only for human consumption.

Sweet corn endosperm is translucent with horny look and wrinkled texture of mature kernels. Some of its larger varieties are produced for forage and as soiling crops. Before maturing it is consumed as vegetable more than as grain. Pod variety is characterized because each kernel is enclosed in a pod or with a husk surrounding each kernel.

Classification by races is the other way to consider maize diversity (Fig. 1).

It is important to think of a race as a maize population with more similarities among its members than with other population members (Benz 1997). This means more genetic heritage shared and, therefore, a direct relationship. For example, The National Commission for the Knowledge and Use of the Biodiversity (CONABIO, Mexico) has a record of 64 races from Mexico, all of them fitting in seven groups because they share similar shape and are cultivated in the same geographic region (http://www.biodiversidad.gob.mx/usos/maices/razas2012.html; 13 May 2016).

This huge diversity is reflected in the complex maize genome. On the sequenced Z. *mays* subsp. *mays* B73 there are 32,000 genes in 10 chromosomes forming a genome of 2,300 million base pairs (nucleotides) (Schnable et al. 2009); the human genome is composed of 20,000 genes organized into 23 chromosomes, representing a genome of 2,900 million base pairs. Even

Figure 1. Different Mexican maize races exhibited at the International Festival of Cacao and Chocolate at the National Museum of Popular Cultures, Coyoacan, Mexico City, 2014. All of them of a different color, size and shape of the corn ear. The right hand photo is a close-up of specimens from 3 races, from left to right: Apachito, endemic in Chihuahua; Vandeño from Michoacán and Conic corn ear from San Luis Potosí.

so, comparing this genome with the one obtained for *Z. mays* L. spp. 2233 (a Mexican pop variety), the last one shows a compact genome but with high gene density being 25% shorter than the reference B73 (Corona et al. 2006). So, this complexity in maize varieties is not a surprise.

The existence of many races and varieties of maize and with the further engineering of its evolution by men; its long culinary history is not a surprise either. This culinary history begins with the maize domestication and is evident from the existence of stone tools for grinding maize like hand stones and milling stones that are archeological finds from Central Balsas and Tabasco (Pohl et al. 2007, Piperno et al. 2009, Ranere et al. 2009). At Salinas la Blanca, Guatemala, evidence of ceramics with lime adhesions have been associated with the use of nixtamalization (heat and alkali treatment of maize) and were dated from 1,000 to 800 B.C. Also, paintings on ancient building walls show some of these traditions.

In Mesoamerica, maize is the staple food and is not only used as food but also as medicine and at rituals in ceremonial context according to ancient traditions orally transmitted from one generation to another. Also, fermented and non-fermented maize products are important as part of these traditions and a contribution to the culinary richness.

In Africa maize is also a staple food. In the last decades of the 20th Century, maize represented 50% of the calorific contribution in the local diet (McCann 2001). The Europeans took maize to Africa during the beginning of its colonial period, and they have developed their own traditional fermented maize products since then. Most of African fermented products are prepared at home level but some of them have been scaled up to industrial level.

2. Fermentation

The history of fermented foods and beverages goes hand in hand with human development. Fermentations contribute to the enrichment of the human diet not only by preserving food and making it available for long periods of time, but also bringing new flavors (sweet, sour, alcoholic), aromas (buttery, fruity, aged), and textures (smooth, spreadable) to food. However it can also enhance substrate digestibility and at the nutritional level, increase the protein content and improve its quality. Vitamins are also produced or made accessible. It also helps to detoxify food, during soaking or cooking of raw materials, mycotoxins or phytates may be reduced or destroyed and also during fermentation, the levels of minerals are increased in the final product, most of the time those that are frequently lacking in the diet, especially in children (Steinkraus 1997).

Food modifications during fermentation are the result of the growth of microorganisms and their metabolism but mainly their enzymes activities such as amylases, proteases and lipases, which hydrolyze polysaccharides, proteins and lipids in the foods. Fermentable carbohydrates are converted into different proportions of ethanol, organic acids, carbon dioxide and water; if proteins and lipids are used new flavors and aromas (such as fruity, buttery, aged) could be developed by the production of amines from proteins or esters, and aldehydes or ketones from lipids. If new sensorial attributes are unpleasant, as those produced by cadaverine or putrescine odors, or if a toxic is present in food, it could be considered as spoiled and/ or a potential health risk.

Most of the contributions of fermentation in food safety are associated with the development of the fermentative microbiota which are mostly stable during the whole process, and then these microorganisms occupy the niche limiting or inhibiting the growth of less desirable microorganisms by direct substrate competition and also by the production of antimicrobial compounds which act against them. Lactic acid bacteria (LAB) produce metabolites with antimicrobial activity such as: organic acids, carbon dioxide, hydrogen peroxide and bacteriocins (Steinkraus 1983, Daeschel 1989). For example, the first bacteriocin described and used in food is nisin, produced by *Lactococcus lactis* subsp. *lactis* (Riley and Wertz 2002). Some enterococci used for cheese ripening and also member of the *Steptococcus bovis/Streptococcus equinus* Complex (SBSEC) produce bacteriocins or bacteriocin like substances with activity against other LAB, spoilage bacteria or even pathogenic bacteria (Pieterse et al. 2008, DeVuyst and Tsakalidou 2008, Rashid et al. 2009, Jans et al. 2012). More known are those produced by *Lactobacillus* genus than those produced by streptococci, i.e., lactocine produced by *Lactobacillus sakei*, brevicine by *Lb. brevis*, curvacine by *Lb. curvatus* orplantaricine produced by *Lb. plantarum* (Monroy et al. 2009).

From the fermentations classification brought in by Steinkraus (1997, 2002), lactic acid fermentation and alcoholic fermentations are among the most important for maize products.

Lactic acid fermentations are generally considered as safe. The main microbiota is composed of LAB: *Lb. plantarum, Lb. delbrueckii, Lb. fermentum, Lb. reuteri, Lactococcus lactis* subsp. *lactis, Lc. lactis* subsp. *cremoris, Leuconostoc mesenteroides, Pediococcus acidilactici,* etc., all of them characterized as Gram-positive rod or cocci, non motile, with low G-C% content, being anaerobic-aerotolerant or microaerophilic bacteria, they are negative for catalase and oxidase. LAB are homo-fermentative when the main fermentation product is lactic acid and they are hetero-fermentative if other than lactic acid carbon dioxide, ethanol and acetic acid or others are also produced.

In alcoholic fermentations ethanol and carbon dioxide are produced and act as antimicrobial substances that inhibit other microorganisms, therefore the products obtained are generally safe. Yeasts such as: *Saccharomyces cerevisiae, Candida guilliermondii, C. kefyr,* are responsible for these fermentations but a bacterium, *Zymomonas mobilis,* is an exception. Substrates used in this kind of fermentations are: sugar cane juice, fruit juice, diluted honey or germinated cereal grains (malting). If starchy raw materials like maize kernels are used germination alone is not enough to get fermentable carbohydrates making the addition of hydrolyzed starch or an enzyme source with amylase activity necessary (such as human saliva in Mexican *tesgüino* or traditional Andean *chicha*). In other cases, microbiota could provide amylases as was previously mentioned. For instance, amylolytic LAB have the main role during *pozol* fermentation (Díaz-Ruiz et al. 2003).

Traditional maize fermented products considered in this chapter are examples of lactic acid fermentation or a combination of alcoholic and lactic acid fermentations from the growth of LAB and/or yeasts. Food safety of these fermented products is relevant because most of the time they are made at household level, under unhygienic conditions or using not purified water. Furthermore, the nutritional value as well as their capacity to prevent or to improve the health of consumers is appreciated nowadays. This is an opportunity to study them to understand the way they are prepared, the microbial ecology, their contribution to fermentation process and also their action in human health. This would allow the optimization of their preparation methods and their microbiological quality for assuring safety.

3. Maize Pretreatments and Fermentation

Before fermentation, maize grains undergo pre-treatments that include one or more of the following steps:

3.1 Soaking in Water

It is important to soak the crops or kernels when they are dry. This helps to clean them up by removing floating materials and also to moisturize and soften them to make grinding easier. Also, steeping in water promotes the activation of grains enzymes like amylases, proteases and phosphatases (Nche et al. 1996, Eyzaguirre et al. 2006); however, some loss of soluble materials, such as carbohydrates, may occur.

3.2 Percolation

Separation of water from the grains is used to remove the remaining unwanted materials and to get clean kernels for the grinding step.

3.3 Grinding

In Mexico a hand mill (a manual metal mill) is used but in some villages the traditional *metate* (a hand mill stone, Fig. 2) is still used. In Africa, mechanical grinding mills, millstones or mortar and pestle are used. Grinding can be

Figure 2. *Metate*, it is a mill stone still used these days in some rural areas for grinding maize or other dried food products. Ethnographic Room: The *nahuas*, Anthropology National Museum, Mexico City, 2016.

with less or more water according to whether the needed end product is going to be a solid (e.g., dough) or liquid (thin pap like or flour in water suspension) fermentation. Particle size and product homogeneity could be different in each case.

3.4 Kneading

This is necessary to keep together the ground kernels and to shape them, for example, into dough balls for solid state fermentation. These are allowed to ferment naturally or they can be cooked before fermentation.

3.5 Addition of Enzymes or Other Materials

Plant materials called "catalysts" are used as enzyme sources that enhance microbial development or to bring more flavor or fragrance (*tesgüino, tejuino, chicha, mahewu*) or to promote alcoholic fermentation (*tesgüino*) (Lappe and Ulloa 1989). Examples of catalysts used to make *tesgüino* are: barks such as *batari* or *kakwara* (*Randia echinocarpa, R. watsoni* and *R. laevigata*), leaves of *roninowa* (*Stevia serrata*) or *rojisuwi* (*Chimaphila maculata*), stems of *basiawi* (*Bromus arizonicus*), roots of *gotoko, otoko* or *whole gotoko* (*Phaseolus metcalfei* and *Plumbagi scandans*) (Taboada et al. 1977, Lorence-Quiñones et al. 1999).

According to the ancient tradition, human saliva is added to *tesgüino* and to *chicha*, acting as catalyst. Saliva is source of amylases, important to convert starch in to fermentable sugars. In rural regions of Peruvian North Coast, after grinding, pieces of the mash are taken, chewed and then returned to the preparation before dilution in water and fermentation to make the traditional *chicha* (Nicholson 1960, Hayashida 2008). Otherwise, to make *mahewu*, millet malt is added before fermentation. Malting releases amylases from the grains (Gadaga et al. 1999, Odunfa et al. 2001). For *tejuino* fermentation, sugar cane juice is added, increasing non-starch polysaccharide, mono- and d-isaccharides, as glucose, fructose and sucrose for microbial growth, possibly enhancing yeast development (Guerra-Flores et al. 2009).

3.6 Germination

To prepare *tesgüino* and *chicha*, kernels are germinated (malting) before grinding. Germination activates amylases and other enzymes, important for starch hydrolysis. After being steeped in water, kernels are filtered and then are left to germinate under pine leaves and tree branches for *tesgüino* (Lappe and Ulloa 1989) while grains to prepare *chicha* are covered with maize leaves, tree leaves, cloths or plastics (Nicholson 1960, Hayashida 2008).

3.7 Natural Fermentation or Inoculation

It is assumed that fermentation conditions were the strategies used by people from antiquity to reproduce a fermented food, as then, in most cases, a natural fermentation occurs. The procedures have been passed on from generation to generation, so that it is known that if you use a certain substrate and follow the traditional procedure, you obtain the right product. Usually complex microbiotas participate and the better fit microorganisms prevail. In microbial ecology, molecular microbial ecology methods are now being used and the structure and function of fermented food microbiotas have been described, as also the interactions among microorganisms and the metabolic routes in the fermenting food (Ampe et al. 1999, Ampe and Miambi 2000, Ben Omar and Ampe 2000).

In some cases, to improve fermentation, a sample of old fermented product is added as inoculum. This is called backslopping and is used, for example, in the elaboration of *togwa* (Mugula et al. 2003, Molin 2008).

3.8 Vessels and Wrappings

It is important to highlight special issues that help to make each fermentation step unique. These contribute to build different microbial communities and therefore the final product will have unique characteristics. For example, when the substrate is placed in a closed vessel, anaerobic or micro-aerophilic conditions will be obtained and LAB which can be in small numbers initially, outgrow the rest of the microbiota, but in some cases yeasts take over and then an alcoholic fermentation occurs.

The dough balls of *pozol* are wrapped in banana tree leaves (Cañas et al. 1993a,b) and for *kenkey* they are wrapped in maize leaves or banana tree leaves (Bartle 2007) to preserve humidity and protect the dough during fermentation. These conditions contribute to maintain a high water activity (a_w), which promotes bacterial growth. For *tesgüino, chicha* and *togwa*, special pottery is used for their liquid state fermentation (helping to maintain constant temperature, humidity and dark conditions); in the case of *tesgüino*, clay pots special for this purpose are used and, traditionally, they are never washed (possibly contributing the inoculum) (Lappe and Ulloa 1989). Gourd bottles, plastic containers or clay pottery are used for *togwa* (Mugula et al. 2003, Molin 2008) and ceramic or clay sealed pots serve to contain *chicha* during fermentation at least for 24 h (Vallejo et al. 2013). These clay containers also help to maintain a constant temperature and humidity during fermentation and possibly they contribute with flavor. Most of times, these are the basic steps to prepare maize for fermentation but in some cases other steps as *nixtamalization* are used.

3.9 Nixtamalization

This process has been used since pre-Hispanic times (as indicated by the existing Mexican tradition and by the archeological evidence, like the pottery at Salinas la Blanca, Guatemala, with remains of lime associated with the use of nixtamalization and dated from 1,000 to 800 B.C.) to prepare maize for different foods such as: *tortilla* (flat pancakes used to accompany most meals and one of the main foods for Mexicans), *tamales* (steamed nixtamal dough containing lard, savory or sweet ingredients, beaten to make them fluffy and wrapped in banana or maize leaves), *atole* (beverage made with ground nixtamal in water or milk which is heated until it gets the texture of a thin porridge). Nixtamalized kernels are known as *nixtamal* and its dough is called *masa*.

The steps involved in the process are described in Fig. 3. Clean kernels are cooked in lime water using two spoons (from 10 to 20 g) of lime per kg of maize (to get a water solution of 1 to 2% w/v) boiling them for 30 to 55 min (Illesca 1943, Rangel-Meza et al. 2004, Yañez 2005). After boiling, kernels can be left in the same lime water for 5 to 15 h (overnight) (Illesca

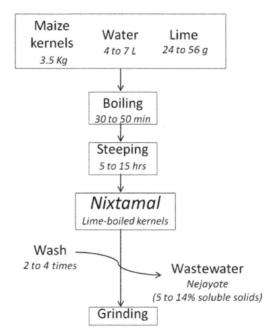

Figure 3. Steps involved in *nixtamalization* process to get maize dough with special characteristics to prepare *tortilla* or other Mexican dishes based on maize dough (Modified from: Rangel-Meza et al. 2004).

1943, Salinas 2003, Rangel-Meza et al. 2004, Yañez 2005) to get *nixtamal* (treated kernels) and then washed or they can be washed immediately with cold water, from 2–4 times, also rubbing with the hands to remove the husk and clean up the lime (Rangel-Meza et al. 2004).

Nutritional and sensorial features change during nixtamalization. This process promotes physical and chemical changes, having an interesting impact on the microbiota that develops during fermentation.

Nixtamalization gives special features to the final dough. Possibly, it was devised to remove the husks, which happens because hemicelluloses are solubilized in alkaline solutions, so after the treatment they can be removed by rubbing the grains with the hands. Grains get softer and ready for grinding; starch gelatinizes partially (15% of total starch) losing crystallinity and increasing its solubility (Salinas 2003). A different, particular and desirable flavor is acquired, with changes also in moisture, adhesivity and cohesivity due to this treatment (Ruiz-Gutierrez et al. 2012). Nixtamalization brings about a better quality for handling the dough for tortillas and other Mexican dishes with special rheological characteristics, like its plasticity. Texture changes are associated with partial starch gelatinization and retrogradation by cooking and soaking. During nixtamalization pH increases almost to 12.5 (Salinas 2003) and the pH value registered for nixtamalization wastewater, named *nejayote*, is *ca.* 11.0; although after washing, the dough's pH is close to neutrality, from 7.5 to 6.8 (Nuraida et al. 1995, Cabra 2000, Díaz-Ruiz et al. 2003); moisture content compared with uncooked kernel is higher (from 28.27 to 36.75% in uncooked kernels against 39.77 to 45.77% in nixtamal) and also this process brings a yellow color to the final product, whose intensity depends on lime concentration and boiling time (Sefa-Dedeh et al. 2004). Niacin is released making it available (Sefa-Dedeh 2003, Rosado et al. 2005), so that people who consume this dough do not suffer from pellagra; calcium concentration also increases, almost 400% (Fernández-Muñoz et al. 2004). Other minerals can also be present depending on the quality of lime (Pflugfelder et al. 1988).

Because this process helps to remove the pericarp, it also affects the aleuron, the tip cap and the kernels' germ, as these are reservoirs of lipids, a decrease from 2 to 6% of total lipids, apart from saponification and solubilization of fatty acids (Martínez-Bustos et al. 2001) takes place. Raw fiber decreases from 13 to 6% in the dough, because of the loss of pericarp (Acevedo et al. 1990), but soluble fiber increases from 0.88 to 1.31% (Saldaña and Brown 1984). Thiamine, riboflavin and carotenoid concentrations are reduced (Sefa-Dedeh et al. 2004).

4. Four Methods to Prepare Traditional Maize Fermented Foods

Hereby, considering the order of these steps for the traditional fermented maize products reviewed here, four basic variants of the process can be observed. Briefly these are:

1) Foods made with dried crops or kernels. These have been described as *kenkey, mawè, ogi* and also *tejuino* and *pozol* (being different only because of the first *nixtamalization* step). Dried crops or kernels are soaked in water for some hours or a few days. Then, they are drained, wet milled and kneaded to get dough that is allowed to ferment naturally.

2) A variation of this process is described for *kenkey* where the complete maize ears are soaked in water till they acquire a smooth texture, allowing an overnight fermentation before being wet milled.

3) Foods made with maize flour as raw material, which is the case of *mahewu, togwa* and *uji.* Water is added to the flour to get a pap consistency and is allowed to spontaneously ferment and then it is cooked or cooked at first and then allowed to ferment.

4) Foods using tender maize, e.g., *atole agrio.* Literally sour gruel (Villahermosa, Tabasco, Mexico), maize is mashed to get a dough and this is allowed to ferment (solid state fermentation) or, after mashing the mash is blended with water and is allowed to ferment (liquid fermentation). After fermentation *atole agrio* is cooked.

Foods that use germinated grains, as *tesgüino* (from Northern Mexico) and *chicha* (from Peru). In both cases these alcoholic beverages use kernels that are germinated (malting) under especial conditions (high moisture and without light) to stimulate enzyme action and then they are wet ground. In the case of the *tesgüino*, maize gruel is obtained by boiling in water ground non-germinated grains and the germinated mash is added to produce the sugars that will be the substrate for the fermentation.

For *chicha*, germinated kernels are dried and then ground to get flour, then, water is added and the mixture is boiled, afterwards the fermentation takes place.

5. Traditional Maize Fermented Foods

It is assumed that the main role of traditional fermented food was their contribution to increase diversity in the diet, as can be observed from the different kinds of fermented products that have been prepared from maize. According to the four variants described above, a description of some African and Latin American traditional fermented foods processes is presented. A brief description of them is in Table 1.

Table 1. Brief comparison of traditional fermented maize products from Africa and Latin America.

Product	Origin	Food category	Functional microorganism	Reference
Made with dried crops or kernels soaked in water				
African				
Kenkey	West Africa Ghana, Ivory Coast, Togo	Fermented maize dough, bread-like, steam cooked	LAB and yeasts	McKay and Baldwin 1990 Hounhouigan et al. 1993 Olsen et al. 1995 Hayford et al. 1999 Bartle 2007
Mawè	South African and some West African countries	Fermented (uncooked) maize dough	LAB and yeasts	Hounhouigan et al. 1993 Hounhouigan et al. 1994 Greppi et al. 2013
Ogi	West Africa Benin, Nigeria	Sour pap consumed as weaning food or porridge	LAB and yeasts	Beuchat 1987 Agati et al. 1998 Ampe and Miambi 2000 Blandino 2003 Omemu et al. 2007 Molin 2008 Oguntoyinbo et al. 2011 Greppi et al. 2013
Mexican				
Pozol	Southeastern Mexico	Fermented nixtamalized maize dough dissolved in water, is consumed as refreshing beverage	LAB and yeasts	Nuraida 1998 Ampe et al. 1999b Ben Omar et al. 2000 Díaz-Ruiz et al. 2003 López 2006
Tejuino	Northwestern Mexico	Nixtamalized maize dough used to get a fermented thin pap consumed as beverage	LAB and yeasts	Guerra-Flores et al. 2009
Made with maize flour as raw material				
African				
Mahewu	South and West Africa Some parts of Arabian Gulf	Fermented thick porridge consumed as beverage	LAB	Chavan and Kadam 1989 Steinkraus et al. 1993 Matshenka et al. 2013

Table 1. contd....

Table 1. contd.

Product	Origin	Food category	Functional microorganism	Reference
Togwa	East Africa	Fermented diluted porridge consumed as beverage	LAB and yeasts	Mugula et al. 2003 Molin 2008
Uji	Eastern Africa Kenya, Uganda, Tanzania	Gruel like consumed as creamy soup, weaning food or beverage	LAB	Masha et al. 1998 Uvere and Awada 2013
Using tender maize				
Mexican				
Atole agrio	Southeastern State Tabasco	Beverage thin gruel like	LAB	Valderrama 2012
Using germinated grains				
Mexican				
Tesgüino	North and Northeastern Mexico	Alcoholic beverage, thin gruel like	Yeast and LAB	Lappe and Ulloa 1989 Lorence-Quiñones et al. 1999 Herrera-Suárez 2003
Andean				
Chicha	Andean South American countries Peru, Ecuador, Brazil, Bolivia, Colombia, Argentina	Alcoholic beverage	Yeasts and LAB	Nicholson 1960 Escobar 1993 Lorence-Quiñones et al. 1999 Hayashida 2008 Vallejo et al. 2013

LAB: Lactic acid bacteria

5.1 Foods Made with Dried Crops or Kernels Soaked in Water

African *kenkey, mawè* and *ogi,* and the Mexican traditional fermented *pozol* and *tejuino* (that differ from the African foods in a first step of nixtamalization) need to moisturize dried grains. The way to prepare them is described briefly in Fig. 4.

5.1.1 Kenkey

It is fermented maize dough consumed in West Africa: Ghana, Eastern Ivory Coast and some parts of Togo (McKay and Baldwin 1990, Bartle 2007), in which a cooked bread-like product (Hayford et al. 1999) is obtained. White maize kernels are soaked in water from 1 to 2 days, then water is removed and the grains are wet ground, more water can be added and it is allowed

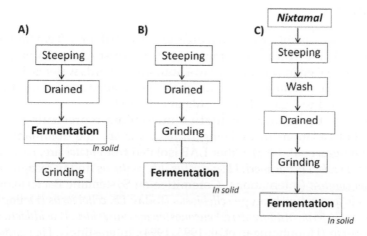

Figure 4. Processes involved in the preparation of traditional fermented maize doughs and paps: (A) *kenkey,* (B) *kenkey, mawè* and *ogi,* (C) *nixtamal* maize dough for *pozol* and *tejuino. Nixtamal,* treated kernels, can be steeped in lime water overnight and then washed or immediately after nixtamalization they can be washed with cold water.

to naturally ferment from 2 to 4 days (Hounhouigan et al. 1993). Another way to make it is using complete maize ears, steeping them in water till they have a smooth texture and then they are allowed to ferment overnight and finally drained and wet ground (McKay and Baldwin 1990). In both ways, the fermented dough obtained is steam cooked using maize leaves as containers, or as some people do at Cape Cost in Ghana, using banana tree leaves (Bartle 2007). Also, two types of *kenkey* can be mentioned: *ga-kenkey* and *fanti-kenkey,* being different because *ga-kenkey's* dough is salted before fermentation and is wrapped in maize leaves, for *fanti-kenkey* plantain leaves are used to wrap it (Amoa-Awua et al. 2007).

LAB are the main microbiota in *kenkey*: *Lactobacillus plantarum, Lb. brevis, Lb. fermentum/reuteri, Pediococcus acidilactici* and *P. pentosaceus* were identified (Hounhouigan et al. 1993, Olsen et al. 1995). It seems like *Lb. fermentum* and *Lb. reuteri* dominate the fermentation (Olsen et al. 1995) particularly the first one (Hayford et al. 1999). In this product, fermentation process increases protein availability making 20% more available than in raw maize (Bartle 2007).

Some yeast and molds have been found too. The main yeasts are *Candida krusei* and *Saccharomyces cerevisiae,* found in the raw material, during steeping and fermentation (Halm et al. 1993, Jespersen et al. 1994, Obiri-Danso 1994). Other yeasts are: *Thrichosporon* sp., *Kluyveromyces* sp., *Debaryomyces* sp., *C. tropicalis* and *C. kefyr*; molds like: *Penicillium citrinum, Fusarium subglutinans, Aspergillus flavus, A. parasiticus* and *A. wentii* were also detected (Hounhouigan et al. 1993, Jespersen et al. 1994, Obiri-Danso 1994, Olsen et al. 1995, Blandino 2003).

5.1.2 Mawè

This is another kind of fermented maize (uncooked) dough. It is a traditional food from South Africa and some places from West Africa. To prepare it, kernels are soaked in water from 2 to 4 hours, afterwards water is discarded and kernels are wet milled to get dough that is kneaded and left to naturally ferment from 1 to 3 days (Hounhouigan et al. 1993).

LAB are the main fermentative microbiota representing 94% of the total of isolations made being *Lactobacillus* sp., the dominant genus (Hounhouigan et al. 1993). Other LAB isolated from *mawè* are: *Lactococcus lactis*, *Lb. brevis*, *Lb. buchneri*, *Lb. curvatus*, *Lb. confusus* (today recognized as *Weissella confusa* (International Committee on Systematic Bacteriology of 2000)), *Lb. fermentum* biotype *cellobiosus* (today *Lb. cellobiosus* (Dellaglio et al. 2004)), *Lb. fermentum/reuteri*, *Leuconostoc mesenteroides*, *P. acidilactici* and *P. pentosaseus* (Hounhouigan et al. 1993, 1994). Interestingly, *Lb. plantarum*, common in other maize products and vegetable fermentations having a dominant role, was not found in *mawè*. Yeasts were found too, the main at the beginning of the fermentation were: *Candida krusei*, *C. kefyr*, *C. glabrata* and *Kluyveromyces marxianus*; then, after 72 h, a succession was observed changing the dominance to *S. cerevisiae* followed by *C. krusei* and *C. glabrata* (Hounhouigan et al. 1994, Greppi et al. 2013).

5.1.3 Ogi

It is traditional in Nigeria, Benin and other West African countries. *Ogi* is a sour flavored pap made with maize, sorghum or millet commonly used as weaning food or as a daily porridge (Beuchat 1987, Blandino 2003, Molin 2008). To make this pap, the selected grains are soaked in water from 1 to 3 days, and then they are wet ground and sieved to get a filtrate-flour like suspension that is allowed to ferment from 1 to 3 days (Beuchat 1987, Steinkraus 1998, Molin 2008, Oguntoyinbo and Narbad 2012). Then, when the mash is soured enough, it is decanted to make dough cakes which can be wrapped in leaves to be sold or used directly mixing with water and boiling to get gruel or a porridge (Beuchat 1987, Molin 2008). This cooking step contributes to gelatinize the starch (Oguntoyinbo and Narbad 2012) modifying the texture and the viscosity of the final pap, making it convenient as a weaning food.

During steeping in water some molds have been found, but the fermentative microbiota developed in the flour suspension are mainly LAB dominated by *Lb. plantarum* followed by *Lb. fermentum*, *P. pentosaceus* and *Streptococcus gallolyticus* subsp. *macedonicus* but also non-LAB such as members of *Bacillus* and *Staphylococcus* genera have been isolated too (Agati et al. 1998, Ampe and Miambi 2000, Oguntoyinbo et al. 2011, Oguntoyinbo and Narbad 2012). Some isolates found in *ogi* were identified as related to *Lb. amylolyticus, Lb. delbrueckii* subsp. *bulgaricus, Lactococcus lactis* subsp. *lactis* (Oguntoyinbo et al. 2011). Fast development of yeasts at the start of fermentation has also been observed, *S. cerevisiae, Pichia membranaefaciens, Candida* spp. and *Geotrichum* spp. seem to be dominant (Beuchat 1987, Omemu et al. 2007). *Candida mycoderma, C. tropicalis, Geotrichum candidum, G. fermentas, Rhodotorula graminis* have been found; *Cephalosporium* sp., *Fusarium* sp., *Aspergillus* sp. and *Penicillium* sp. were additionally identified (Blandino 2003, Omemu et al. 2007).

5.1.4 Pozol

In the Mayan territories in Southeast Mexico (Tabasco, Campeche, Yucatan, Quintana Roo, Oaxaca, Chiapas and also Guatemala and Honduras), *pozol* is another traditional fermented beverage made with maize, which is locally very popular. The process to prepare the traditional *pozol* from Chiapas Highlands was described by Cañas et al. (1993a,b) having two main varieties: the indigenous and the *ladino* or *mestizo* processes.

According to this, after nixtamalization, kernels are washed with fresh water and they are soaked overnight (indigenous pozol) or boiled in fresh water a second time for 3 to 8 hours (mestizo pozol). After straining, kernels are hand milled (with manual metal mill or with electric mill, less commonly with *metate*) and the dough obtained is made into balls by hand-made and wrapped with banana tree leaves. Inside this envelope fermentation occurs for some hours to some days (4 or 5 days) and sometimes a month or more at room temperature. To prepare the beverage, the fermented dough balls are dispersed in fresh water in order to obtain an acid, non alcoholic and refreshing beverage. Other ingredients can be added to the drink: sugar, honey, cocoa beans (if they are ground together with maize dough before its fermentation, the drink being called *chorote*), dry chili powder, salt or even lemon.

LAB are the first to grow and have the main role in *pozol*, representing 90 to 97% of the total fermenting microbiota (Ampe et al. 1999), but being a natural fermentation, this niche is complex and other bacteria, yeasts and molds are also present (Ulloa et al. 1987, Wacher et al. 1993, Nuraida et al. 1995, Wacher et al. 2000). Those LAB identified as part of *pozols* native microbiota are: *Lactobacillus fermentum, Lb. plantarum, Lactococcus lactis, Leuconostoc citreum, Ln. mesenteroides, Streptococcus infantarius* and *Weissella confusa* (Nuraida 1988, Ampe et al. 1999, Ben Omar and Ampe 2000, Olivares-Illana et al. 2002, Díaz-Ruiz et al. 2003, López 2006). *W. cibaria* and *W. paramesenteroides* have been found too (Paz 2014).

Another isolation and identification was of nitrogen fixing *Agrobacterium azotophilum* from pozol has been described. This bacterium shows antimicrobial activity (Ulloa and Herrera 1972) and was later designated as *Bacillus* sp. strain CS93, possibly a variant of *Bacillus subtilis* (Ray et al. 2000). Yeasts such as *Candida* sp., grew to 10^{10} CFU/g of fresh dough in 8 days, *Geotrichum candidum* and *Trichosporon cutaneum* were also identified with a growth of 10^8 CFU/g, also in 8 days (Ulloa 1974); molds like *Fusarium moniliforme* which was isolated from maize grains, *Trichoderma viride* and *Monilia sitophila* isolated from nixtamal were found too; e.g., *Cladosporium cladosporioides* and *Monilia sitophila*, reaching to 10^{10} CFU/g (Ulloa 1974). Pathogenic or potentially pathogenic molds such as *Candida parapsilosis*, *Trichosporon cutaneum, Geotrichum candidum, Aspergillus flavus* were found only at the beginning of fermentation (Ulloa 1974).

Besides its importance in nutrition as a food, it is known from tradition that *pozol* has medicinal properties too in its use to control diarrhea and reduce fever. In the ancient Mayan legends Hanincol (maize-field meal) is a ceremony in which gods are asked to heal those who had injured them and *pozol* is used as a consecrated beverage. Strains of *Streptococcus* sp., *Weissella* sp. and *Leuconostoc* sp. have been found to be potential probiotics (Rodríguez et al. 2011) and some *Weisella* strains produce bacteriocin like substances with activity against other LAB, spoilage bacteria and some pathogens (Paz 2014). Also, is known that *Bacillus* sp. strain CS93 produce antimicrobial compounds: bacilysin, chlorotetaine and iturin A (Phister et al. 2004). This seems to explain its medical use to control diarrhea and intestinal infections.

5.1.5 Tejuino

This beverage is not as well documented as the other traditional fermented products reviewed here, but is quite important in the communities from Northwestern Mexico (Sonora, Chihuahua, Coahuila, Jalisco, Colima, Nayarit and also Oaxaca at southeast Mexico). It is locally consumed during festivals and sport competitions and is a non-alcoholic *mestizo* variant of *tesgüino* but *nixtamal* is used to prepare it. After washing and grinding lime-boiled kernels, sugar cane juice and water are added to get a thin pap which is filtered and left to naturally ferment, also some old *tejuino* can be added as inoculum source; fermentation occurs for 2 to 3 days (Guerra-Flores et al. 2009). To drink this refreshing and bittersweet beverage some lemon juice, salt and chili powder can be added.

During fermentation molds and yeasts as *Aspergillus* sp., *Penicillum* sp., *Candida guillermondii* and *Saccharomyces* sp. have been found, also LAB like *Lb. delbrueckii, Lb. acidophilus* and *Leuconostoc mesenteroides* (Guerra-Flores et al. 2009).

5.2 Made with Maize Flour as Raw Material

Maize flour is the main raw material to prepare some traditional fermented foods such as the African *mahewu*, *togwa* and *uji*. A diagram describing briefly the process involved in their preparation is shown in Fig. 5.

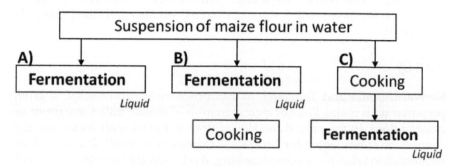

Figure 5. Brief description and comparison of the processes involved in preparing traditional fermented products based on maize flour: (A) *mahewu*, (B) *uji*, (C) *togwa* and *uji*. All of them are beverages and liquid fermentations.

5.2.1 Mahewu

This is also known as *amahewu* or *magou* and it is a non-alcoholic beverage consumed in South and West Africa and some parts of the Arabian Gulf (Chavan and Kadam 1989).

The process to prepare it can begin with maize porridge mixed with water or with *sadza* (thick porridge made with maize flour in water and cooked); sorghum, wheat flour and malted millet could also be added in household process and the fermentation happens at room temperature (Gadaga et al. 1999, Odunfa et al. 2001). At industrial level, *Lb. delbrueckii* is added as starter culture (Steinkraus 2002).

LAB are the main microbiota of which some genera have been identified; *Lactococcus lactis* subsp. *lactis* is the dominant (Steinkraus et al. 1993); *Lactobacillus* spp., *Lactococcus raffinolacticus*, *Ln. mesenteroides* and *P. dextrinicus* were identified as also *Bacillus subtilis*, *B. megaterium*, *B. coagulans* and *S. oralis* (Matsheka et al. 2013).

5.2.2 Togwa

An African beverage, *togwa* can be used as weaning food and is consumed in Eastern African countries. *Sadza*, water and old *togwa* are mixed and kept in gourd bottles, plastic containers or in clay pottery to ferment during 9 to 24 hours (Mugula et al. 2003, Molin 2008).

In Tanzanian *togwa* the main microbiota is composed of LAB and in lesser numbers by yeasts. Some isolated bacteria in order of domiance are: *Lb. plantarum, Lb. brevis, Lb. cellobiosus, Lb. fermentum, Weissella confusa* and *P. pentosaceus*, yeasts like *C. tropicalis, C. pelliculosa, P. kudriavzevii* (basonym, *Issatchenkia orientalis*) and *S. cerevisiae* have been also identified (Mugula et al. 2003).

5.2.3 Uji

Kenya, Uganda and Tanzania, in Eastern Africa, produce *uji*, a gruel prepared with maize flour or a combination of maize, millet, sorghum or cassava flour. In the procedure, sieved flour is mixed with water and the suspension is fermented for 1 to 3 days (Masha et al. 1998). If sieved-flour fermentation takes place before cooking, it is boiled afterwards. At the end, sugar or salt can be added and it can be consumed as weaning food, creamy soup or drunk as non-alcoholic beverage (Masha et al. 1998).

Eventhough there is not enough information about its microbial ecology, some LAB were isolated from *uji* and were identified as: *Lb. fermentum, Lb. cellobiosus* and *Lb. buchneri* (Uvere and Awada 2013).

5.3 Foods Using Tender Maize

There are a variety of procedures to prepare Mexican *atole agrio*, which is a sour gruel. It can be made with fermented maize grains or the gruel can be made first and then fermented. Mature or tender maize can be used and other ingredients such as some plants can be added.

5.3.1 Atole agrio de dobla

It is different from other fermented maize products because tender maize is used to prepare it, so it is not necessary to moisturize the kernels to get the dough. Besides, the fermentation can be developed in solid or in liquid state (Valderrama 2012) as in Fig. 6.

In Mexican Southeastern state of Tabasco, before maize harvesting the stalks are folded to protect maize ears and preserve their humidity and tenderness till the harvest, this maize is called *maíz de dobla*. Atoleagrio is a traditional beverage consumed usually from May to September, during the harvest time. Also, as a tradition, women consume it after childbirth and during breastfeeding because they believe that its consumption enhances

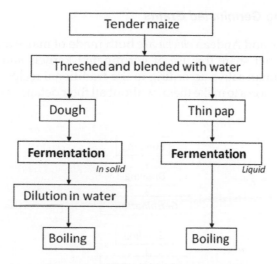

Figure 6. *Atole agrio*, traditional fermented thin gruel like beverage from Mexico uses tender crops called *maíz de dobla* as raw material. This diagram shows the process to get this product by solid or liquid fermentation.

milk production (Valderrama 2012). It is a sour, refreshing, non-alcoholic drink.

Maíz de dobla is threshed, kneaded into balls and left to ferment in solid or blended with water (in a *masa*/water relation of 1:1 w/v) and left to ferment at room temperature (Valderrama 2012). For the solid state fermentation, the shaped balls are transferred into a plastic bowl, covered with a thin fabric and they are left to naturally ferment for 12 hours, overnight, at room temperature (31 to 32°C). Afterward, water is added to get a thin pap-like beverage which is boiled for almost 10 minutes, less or more according to the desired consistency. For the liquid fermentation, more water is added after grinding to get the thin pap and it is left to ferment into plastic containers for 6 hours at room temperature, then it is boiled for 6 minutes or for the necessary time until it gets thick.

Yeasts such as *Candida intermedia* (mainly), *Pichia kudriavzevii*, *Clavispora lusitaniae* (showing probiotic potential) and *C. parapsilosis* were found in *atole agrio* (Manrique et al. 2016). Because of unhygienic conditions or manipulation during the process some members of the *Enterobacteriaceae* family have been reported such as: *Serratia marcescens*, *Klebsiella pneumoniae*, *Enterobacter cloacae* and *Morganella morganii* subsp. *sibonii*, also bacteria from *Pseudomonadaceae* family like: *Pseudomonas aeruginosa* and *P. leuteola* (Esquivel et al. 2013), but their concentrations decrease along with the fermentation; besides, coliforms diminish as fermentation goes on.

5.4 Foods Using Germinated Grains

Mexican *tesgüino* and Andean *chicha* are both made of maize and are beer-like beverages that share the use of germinated kernels and the use of catalysts like human saliva according to the traditional recipe. A diagram that shows the process to make them, without all their details, is represented in Fig. 7.

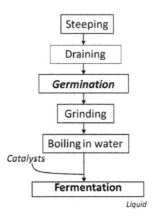

Figure 7. General description of the procedure to prepare two traditional fermented beverages, Mexican *tesgüino* and the South American *chicha*. Both products use, before dough fermentation, catalysts like: barks, roots, fruits, human saliva (according to the traditional recipe), etc.

5.4.1 Tesgüino

Indigenous people from some North and North-western states of Mexico consumed a traditional fermented beverage described as a maize beer-like, *tesgüino*, an alcoholic drink with an acidic taste and viscous consistency, being not at all like a beer but more like an alcoholic *atole*. This beverage is prepared by the *yaquis* and *pimas* of Sonora, *tarahumaras* (or *raramuri*) from Chihuahua, *guarijíos* from Sonora and Chihuahua, *tepehuanos* from Durango and *huicholes* from Nayarit and Jalisco (Ulloa et al. 1987, Herrera-Suárez 2003).

According to the ethnic group's practises, preparation and (available) substrates used can vary, as does its name; e.g., *tesgüino* prepared with maize is also known as *batari*. In the *tarahumara* community this beverage has a ceremonial role in the religious context, shamanic rituals, funeral ceremonies, during sporting festivals and in community festivities called *tesgüinadas* in which all the village participants make economic and political decisions. Or as a communitarian gift for the work they have done for the town (Ulloa et al. 1987, Lappe and Ulloa 1989). As it is an alcoholic drink, only diluted with water it can be given to children and nurslings (Herrera-Suárez 2003) and provides energy to Tarahumara runners, who are known for their stamina.

To obtain *tesgüino*, after soaking the kernels several days, they are drained and prepared to germinate in the dark, putting them on a thin layer and covering with pine tree leaves and branches. After germination, grains are milled and boiled with water till a yellowish thin pap (*atole*) is obtained. When the beverage is at room temperature, catalysts are added (Taboada et al. 1977, Lorence-Quiñones et al. 1999).

Fermentation takes place in special containers, clay pots that are solely for this purpose and which are never washed, they are known as *ollas tesgüineras*, these are recognized as the main source of fermentative inoculum (Lappe and Ulloa 1989).

During the first fermentation stages a lactic fermentation occurs, then an alcoholic/lactic fermentation and at the final stages an alcoholic/acetic fermentation happens. Hetero-fermentative LAB and yeasts such as *Brettanomyces intermedius* may be responsible for the last one (Ulloa et al. 1987, Lappe and Ulloa 1989, Herrera-Suárez 2003). Alcoholic fermentation is promoted by *enzymes* that cause the mash to be effervescent when added to it. The LAB detected belong to the genera: *Lactobacillus, Leuconostoc, Pediococcus* and *Streptococcus* (Lappe and Ulloa 1989); yeasts detected during the first fermentation stages are: *B. intermedius, Cryptococcus albidus* and *Pichia membranaefaciens* and during the whole process of fermentation: *C. guilliermondii, Hansenula anomala, Saccharomyces cerevisiae* and *S. kluyveri* were found (Lappe and Ulloa 1989, Herrera-Suárez 2003). The final product is a sour and alcoholic, thick, effervescent beer-like product with an ethanol concentration of almost 3.73% (Lorence-Quiñones et al. 1999).

5.4.2 Chicha

From Peru and the Andean regions of Ecuador, Brazil, Bolivia, Colombia and Argentina, *chicha* is an alcoholic, clear, effervescent beverage described as with a cider taste made with maize or/and chickpea, wheat, barley, quinoa, fruits and sesame (Nicholson 1960, Hayashida 2008, Vallejo et al. 2013). This is a traditional fermented product with pre-Hispanic history, using indigenous knowledge with a few changes in its preparation since then. Some authors suggest the early role of maize in Andean regions as part of ceremonies and rituals as a fermented intoxicant beverage and there is archeological evidence of the consumption of *chicha* at Coastal Ecuador dating from 4,150 to 3,850 years ago that support this view (Burger and van der Merwe 1990, Staller and Thompson 2002, Staller 2003).

The process to get *chicha* may vary according to local traditions (or even the family tradition) and whether it is going to be drunk as refreshing beverage or during rituals or traditional festivities. At first, kernels are steeped in water from 8 hours to 4 days (Nicholson 1960, Hayashida 2008, Vallejo et al. 2013). Water is strained and kernels are allowed to germinate from 1 or 2 days to one week putting them together over a flat surface, covering them with maize leaves, tree leaves, cloth or plastic to maintain humidity and temperature (Nicholson 1960). Afterwards grains are covered with *jute* (vegetal textile fiber) from 1 to 12 days (Hayashida 2008) and seeds are dried under the sun from 1 to 2 days (Nicholson 1960). Dried seeds are used to get a flour, then water is added and the mixture is boiled from 3 to 10 hours adding water to maintain the initial volume and stirring frequently; afterward, when this mash is at room temperature, inner pot's walls are rubbed with mash (Hayashida 2008). This is done instead of the ancient practice of adding chewed mash (Nicholson 1960, Hayashida 2008) (saliva acts as catalyst bringing amylases to the fermentation). Once again, the drink is cooked few hours and left to cool down before adding sugar (also as catalyst as sugar cane juice for *tejuino*) and then is filtered through a cotton fabric to finally allow to naturally ferment for 1 or 2 days at room temperature (Nicholson 1960, Hayashida 2008). Fermentation can be done in ceramic or sealed clay pots (Vallejo et al. 2013).

Yeasts and LAB have been identified in the drink. The main microbiota are yeasts: *S. cerevisiae* and *S. apiculata*, *S. pastorianus* have also been found (Escobar 1993, Vallejo et al. 2013), and from the LAB group: *Lactococcus lactis*, *Lb. plantarum*, *Lb. brevis*, *Lb. casei* and *Leuconostoc mesenteroides*, *E. faecium*, *E. durans*, *E. hirae*, *P. acidilactici* and *W. viridescens* (Escobar 1993, Lorence-Quiñones et al. 1999, Elizaquível et al. 2015).

6. Similarities and Differences among Traditional Fermented Maize Products

In spite of the distance and for all the cultural differences between African and Mesoamerican countries that produce fermented maize products, there are similarities among them. Most pretreatments are used in both places to prepare maize for fermentation. Soaking and percolation are common treatments in foods as African *kenkey*, *mawè* and *ogi* and also in Mexican *pozol* and *tejuino*; dough is made in African *kenkey* and for Mexican *pozol*; and germination is a common procedure to prepare maize beers and Mexican *tesgüino* and Andean *chicha*. However, nixtamalization seems to be a Mesoamerican technology without an African counterpart process.

Maize pretreatments contribute to defining the main microbiota that guide the process because they imply physical and chemical changes that act as selective agents or as selective substrates for microorganisms. The low pH of most maize fermented products is selective for LAB, these and their bacteriocins further select other LAB species or strains, for example,

Lb. plantarum, which is one of the most acid resistant LAB species. Pathogenic bacteria are also killed by the same antimicrobial strategies but acid resistant *E. coli* pathogenic groups may survive (Sainz et al. 2001). *Streptococcus* spp. is the most frequently associated LAB in *pozol*, in contrast to *Lactobacillus* spp. in other fermented maize products. This is possibly associated to the nixtamalization process, as members of this genus are known to survive alkaline pH and thermal treatments (Domínguez-Ramírez and Wacher 2014).

If all are maize fermented products, the question arises as to why some are acidic while others follow an alcoholic fermentation. The addition of enzymes as saliva or as a germinated mash containing enzymes to hydrolyze the substrates to produce simpler sugars that the microorganisms are able to use, seems to be the reason for following the alcoholic instead of the lactic fermentation. In some cases, as in *tesgüino*, this is enhanced with the use of *catalysts*.

Furthermore, pretreatments contribute with the introduction of microorganisms not only by hand manipulation but also from the use of special containers, as clay pots, and even from the practice of backslopping as inoculum source.

In spite of particular differences among the preparation of traditional fermented maize products (as nixtamalization for example), they share not only most of the pretreatments but also some members of their native microbiota. Table 2 shows the microorganisms that have been found in each maize product. For example, members of *Lactobacillus* genus are common in all the fermented maize products reviewed in this chapter, particularly *Lb. fermentum* and *Lb. plantarum*. If not both at least one of them is present during fermentation of these traditional maize food products in both, African and Latin American. These products use dried corn or kernels and it is necessary to soak them in water for several hours where a first "fermentation" occurs decreasing the initial pH value of the substrate prior to the main fermentation. This acidic pH is a selective factor for acid tolerant LAB like member of *Lactobacillus* genus (*Lb. fermentum* and *Lb. plantarum* which have a minimal growth pH value of 3.0 and 3.4 respectively [Giraud et al. 1991, Le Blanc et al. 2004]), which can develop during the first stages of the main fermentation. This could be a reason to explain how nixtamalized fermented maize doughs (where lactobacilli are not the dominant) are microbiologically different from the others, where in fermentation initial pH value can be neutral or slightly alkaline and possibly contain higher pH microenvironments.

Members of *Pediococcus* genus, as *P. acidilactici* and *P. pentosaceus*, are present in almost all the African products reviewed, except for *uji*, and also in the Mexican *tesgüino*; they do not dominate in nixtamalized fermented maize dough. There are some species from this genus that produce diacetyl as end product giving a buttery or butterscotch aroma to the final product.

Table 2. Bacteria and yeasts found in traditional fermented maize products from Africa and Latin America.

Product	LAB	Enterobacteriaceae	Other microorganisms	Yeasts	Reference
Made with dried crops or kernels soaked in water					
African					
Kenkey	*Lactobacillus fermentum* *Lb. brevis* *Lb. plantarum* *Lb. reuteri* *Pediococcus acidilactici* *P. pentosaceus*		*Aspergillus flavus* *A. parasiticus* *A. wentii* *Fusarium subglutinans* *Penicillium citrium*	*Candida kefyr* *C. krusei* *C. tropicalis* *Debaromyces* sp. *Kluyveromyces* sp. *Trichosporon* sp. *Saccharomyces cerevisiae*	Halm et al. 1993 Hounhouigan et al. 1993 Jespersen et al. 1994 Obiri-Danso 1994 Olsen et al. 1995 Hayford et al. 1999 Blandino 2003
Mawè	*Lb. brevis* *Lb. buchneri* *Lb. cellobiosus* *Lb. curvatus* *Lb. fermentum* *Lb. reuteri* *Lactococcus lactis* subsp. *lactis* *Leuconostoc mesenteroides* *P. acidilactici* *P. pentosaceus* *Weissella confusa*	*E. cloacae* *Escherichia coli*		*C. kefyr* *C. krusei* *C. glabrata* *Kluyveromyces maexianus* *S. cerevisiae*	Hounhouigan et al. 1993 Hounhouigan et al. 1994 Greppi et al. 2013

Ogi	*Lb. plantarum* *Lb. fermentum* *Lb. delbrueckii* *Lb. amylolyticus* Latococcus lactis subsp. lactis *P. pentosaceus* *S. gallolyticus* subsp. *macedonicus*	*Enterobacter* sp.	*Corynebacterium* sp. *Bacillus* sp. *Staphylococcus* sp. *Aspergillus* sp. *Cephalosporium* sp. *Fusarium* sp. *Penicillium* sp.	*C. mycoderma* *C. tropicalis* *Geotrichum candidum* *G. fermentas* *Pichia membranaefaciens* *Rhodotorula graminis* *S. cerevisiae*	Beuchat 1987 Agati et al. 1998 Ampe and Miambi 2000 Blandino 2003 Omemu et al. 2007 Oguntoyinbo et al. 2011 Greppi et al. 2013

Made with dried crops or kernels soaked in water

Mexican

Pozol	*Lactococcus lactis* subsp. *lactis* *Lb. fermentum* *Lb. plantarum* *Leuconostoc citreum* *Leuconostoc mesenteroides* *Streptococcus infantarius* *Weissella cibaria* *W. confusa* *W. paramesenteroides*	*Enterobacteriaceae*	*Bacillus* sp. *Aspergillus flavus* *Cladosporium cladosporioides* *Fusarium moniliforme* *Monilia sitophila* *Trichoderma viride*	*Candida* sp. *C. parasilopsis* *G. candidum* *T. cutaneum*	Ulloa and Herrera-Suárez 1972 Ulloa 1974 Nuraida 1988 Wacher 1995 Ampe et al. 1999b Ben Omar et al. 2000 Ray et al. 2000 Olivares-Illana et al. 2002 Díaz-Ruiz et al. 2003 López 2006 Paz 2014
Tejuino	*Lb. acidophilus* *Lb. delbrueckii* *L. mesenteroides*		*Aspergillus* sp. *Penicillium* sp.	*C. guilliermondii* *Saccharomyces* sp.	Guerra-Flores et al. 2009

Table 2. contd....

Table 2. contd.

Product	LAB	Enterobacteriaceae	Other microorganisms	Yeasts	Reference
Made with maize flour as raw material					
African					
Mahewu	*Lactobacillus* spp. *Lactococcus lactis* subsp. *lactis* *Lactococcus reffinolacticus* *L. mesenteroides* *P. dextrinicus* *S. oralis*		*B. coagulans* *B. megaterium* *B. subtilis* *Clostridium* spp.	*	Steinkraus et al. 1993 Matshenka et al. 2013
Togwa	*Lb. plantarum* *Lb. brevis* *Lb. cellobiosus* *Lb. fermentum* *P. pentosaceus* *W. confusa*			*C. tropicalis* *C. pelliculosa* *P. kudriavzevii*** *S. cerevisiae*	Mugula et al. 2003
Uji	*Lb. buchneri* *Lb. cellobiosus* *Lb. fermentum*	Coliforms		*	Masha et al. 1998 Uvere and Awada 2013
Using tender maize					
Mexican					
Atole agrio	*Lb. plantarum* *Lb. delbrueckii* *Lactococcus lactis* subsp. *lactis*	*Enterobacter cloacae* *Klebsiella pneumonia* *Morganella morganii* *Serratia marcescens*	*Pseudomonas aeruginosa* *Ps. Leuteola*	*C. intermedia* *C. parasilosis* *Clavispora lusitaniae* *P. kudriavzevii*	Valderrama 2012 Esquivel 2013 Manrique et al. 2016

Using germinated grains				
Mexican				
Tesgüino	*Lactobacillus* sp.	*B. megaterium*	*Brettanomyces intermedius*	Lappe and Ulloa 1989
	Leuconostoc sp.		*C. guilliermondii*	Lorence-Quiñones et al. 1999
	Pediococcus sp.		*Cryptococcus albidus*	Herrera-Suárez 2003
	Streptococcus sp.		*Hansenula anomala*	
			P. membranaefaciens	
			S. cerevisiae	
			S. kluyveri	
Andean				
Chicha	*Lactococcus lactis*	*Acetobacter* sp.	*S. apiculata*	Escobar 1993
	Lb. plantarum		*S. cerevisiae*	Lorence-Quiñones et al. 1999
	Lb. brevis		*S. pastorianus*	Vallejo et al. 2013
	Lb. casei			Elizaquível et al. 2015
	Leuconostoc sp.			
	L. mesenteroides			
	Enterococcus faecium			
	E. durans			
	E. hirae			
	P. acidilactici			
	W. viridescens			

LAB: Lactic acid bacteria

* Yeasts are present but species or genus are not given

** *Pichia kudriavzevii* (basonym: *Issatchenkia orientalis*)

Other LAB common in these products are members of *Leuconostoc* genus like *Leuconostoc mesenteroides*, present in nixtamalized fermented maize doughs (*pozol, tejuino*) and in maize beers (*tesgüino, chicha*). Also, *Leuconostoc mesenteroides* is used as starter culture in some dairy products and bread fermentations (Server-Busson et al. 1999). Some strains of this species can change the viscosity of the fermented product because of their ability to produce polysaccharides (Tallgren et al. 1999).

Other interesting LAB found are: *S. gallolyticus* subsp. *macedonicus*, identified in *ogi*, but this is not common in plant-based fermentations but in cheese ripening; *S. infantarius* subsp. *infantarius* found in *pozol* is not common either in these kinds of fermentations, it has been associated with mammalian environments and also found in fermented dairy products (Tsakalidou et al. 1998, Abdelgadir et al. 2008, Jans et al. 2012, 2013). Less common is the description of members of *Weissella* genus in these maize based products, such as *W. cibaria, W. paramesenteroides* or as *W. confusa* which is a non-amylolytic LAB also considered *Lactobacillus*-like or *Leuconostoc*-like bacteria. Little is known about these LAB and their contribution to the fermentation process, however, it has been reported that some strains show antimicrobial activity against *Staphylococcus aureus, Streptococcus agalactiae, E. coli* ATCC 25922 and *Klebsiella pneumoniae* possibly by the production of bacteriocin or bacteriocin-like substances (Serna-Cock et al. 2010, 2012).

Because lactic acid fermentations occur in most maize fermentations, LAB have been studied exhaustively. This highlights the need to consider yeasts in future approaches because less is known about their contribution, not only as health promoters but also to learn more about their impact on substrate modifications and their contribution towards the flavors and aromas of the end product. The maize fermented products included in this chapter share in common mainly members of *Saccharomyces* and *Candida* genera.

7. Attempts to Industrialize Traditional Fermented Maize Food Production

More information about microbial ecology (yeast role and microbial succession, etc.) is needed because most of the fermentations occur spontaneously and naturally, giving most of the time products with variable quality even representing a health risk. This is the reason why it is important to determine which microorganisms are dominant and could direct natural fermentations.

There have been serious approaches to escalate the production of some African traditional fermented maize products. One of these efforts is to obtain starter cultures to standardize the fermentation and enhance food safety.

Other example is escalation of *mahewu* production for which *Lb. brevis* and *Lb. bulgaricus* var. *delbrueckii* have been studied as starter cultures reducing fermentation times to hours (instead of days) (Holzapfel and Taljaard 2004, McMaster et al. 2005), but even they were not identified as part of native microbiota of it if the end product is well accepted.

Once a strain is characterized and if the substrate is similar, analogies can be drawn among traditional fermented products, so that it would be possible to design starter cultures to standardize the fermentation and improve food quality using health promoting strains or combining strains that enhance the growth of each other (LAB and yeasts), perhaps reducing fermentation times or directing the fermentation to produce certain end products to improve the nutritional value and sensorial quality of food.

Likewise, it is important to use Good Manufacturing Practices (GMP) and Hazard Analysis and Control of Critical Points (HACCP). They are effective in managing the process quality to assure toxicological and microbiological safety of the end product and to standardize its production toaccord with local laws. One example of the implementation of these programs is documented by Amoa-Awua et al. (2007), who developed a semi-commercial *kenkey* production plant in Ghana, monitoring the process to upgrade the plant, improving GMF and then implementing them by auditing and verifying the HACCP.

8. Conclusions

The role of maize and fermented maize foods as staple foods for Mesoamerican and some African countries is highlighted. African and Mesoamerican (Latin American) fermented maize products share pretreatment processes, except for nixtamalization which is only Mesoamerican; as well as LAB species from *Lactobacillus* genus like *Lb. plantarum, Lb. brevis* and/or *Lb. fermentum* (Table 2). Lactic acid fermentation occurs in all of these traditional foods, LAB have been studied exhaustively because of this, but it is important to mention the need to consider yeasts in future approaches because less is known about their contribution to the fermentation and on the end product. The need to improve its nutritional value is also important as fermentations help to enrich this value and also our diets. Traditional fermented maize products are getting more attention thanks to this and are becoming more attractive for researchers. Also, as they are produced mainly at the home level, so that their quality and safety is not always appropriate, improving and standardizing their production is an interesting research field. It is important to consider the contribution of maize pretreatments as those cause selection of the microbiota that is going to develop during fermentation. Then, given the pretreatments that the particular product needs, some considerations have to be made if the product is going to be scaled up. If the main microbiota is known, a starter culture could be designed to inoculate the pasteurized substrate or the natural one, expecting the inoculum to drive the fermentation. Altogether, these efforts are being made to understand the role of the microbiota, its structure and diversity in this kind of fermented maize products in an attempt to assure their safety and good quality and even to improve it. The need to complete this knowledge is important to produce them under controlled conditions, bringing the opportunity to produce them in rural areas having a good quality and to extend their consumption further than the local. This gives us the opportunity to rediscover our ancient traditions and consider their benefits at the nutritional level as well as by adding such interesting flavors and textures to our diets. Traditional fermented products are a valuable source of new knowledge too because they can provide novel probiotic bacteria or prebiotic compounds, even sources of bacteriocins or bacteriocin-like substances with potential use in the food industry.

Acknowledgements

We are grateful to CONACyT project CB-2009-131615, PAPIIT/UNAM project IN218714 and project PAIP 5000-9099 from Facultad de Química, UNAM. Domínguez-Ramírez, L.L. thanks to CONACyT for grant 344938.

Keywords: Maize, fermented maize foods, traditional foods, lactic acid bacteria, yeasts, lactic acid fermentation, alcoholic fermentation

References

Abdelgadir, W., Nielsen, D.S., Hamad, S. and Jakobsen, M. (2008). A traditional Sudanese fermented camel's milk product, *Gariss*, as a habitat of *Streptococcus infantarius* subsp. *infantarius*. International Journal of Food Microbiology 127: 215–219.
Acevedo, E. and Bressani, R. (1990). Contenido de fibra dietética de alimentos Centroamericanos. Guatemala. Archivos Latinoamericanos de Nutrición 40: 439–451.
Agati, V., Guyot, J.-P., Morlon-Guyot, J., Talamond, P. and Hounhouigan, J. (1998). Isolation and characterization of new amylolytic strains of *Lactobacillus fermentum* from fermented maize doughs (mawè and ogi) from Benin. Journal of Applied Microbiology 85: 512–520.
Ampe, F., Ben Omar, N., Moizan, C., Wacher, C. and Guyot, J.P. (1999). Polyphasic study of the spatial distribution of microorganisms in mexican pozol, a fermented maize dough, demonstrates the need for cultivation independent methods to investigate traditional fermentations. Journal of Applied Microbiology 65: 5464–5473.
Ampe, F. and Miambi, E. (2000). Cluster analysis, richness and biodiversity indexes derived from denaturing gradient gel electrophoresis fingerprints of bacterial communities demonstrate that traditional maize fermentations are driven by transformation process. International Journal of Food Microbiology 60: 91–97.
Amoa-Awua, W.K., Ngunjiri, P., Anlobe, J., Kpodo, K., Halm, M., Hayford, A.E. and Jakobsen, M. (2007). The effect of applying GMP and HACCP to traditional food processing at a semi-commercial kenkey production plant in Ghana. Food Control 18: 1449–1457.

Bartle, P. (2007). *Kwasi Bruni.* Los europeos y el maíz. Estudios sobre los Akan de África Occidental. M. L. Sada (trad.)www.cec.vcn.bc.ca/mpfc/index.htm; http://cec.vcn.bc.ca/rdi/kw-bruns.htm; 13 May 2016.

Ben Omar, N. and Ampe, F. (2000). Microbial community dynamics during production of the Mexican fermented maize dough Pozol. Applied and Environmental Microbiology 66: 3664–3673.

Benz, B.F. (1997). Diversidad y distribución prehispánica del maíz mexicano. Ed. Raíces/Instituto Nacional de Antropología e Historia. Arqueología Mexicana 5: 16–23.

Benz, B.F. (2001). Archaeological evidence of teosinte domestication from Guilá Naquitz, Oaxaca. PNAS 98: 2104–2106.

Beuchat, L.R. (1987). Traditional fermented food products. pp. 294–295. *In*: L.R. Beuchat (ed.). Food and Beverage Mycology, Van Nostrand Reinhold, New York.

Blandino, A., Al-Aseeria, M.E., Pandiella, S.S., Cantero, D. and Weeb, C. (2003). Cereal-based fermented foods and beverages. Food Research International 36: 527–543.

Buckler, E.S. IV and Holtsford, T.P. (1996). *Zea* systematic: Ribosomal ITS evidence. Molecular Biology and Evolution 13: 612–622.

Burger, R.L. and van der Merwe, N.J. (1990). Maize and the origin of highland chavin civilization. American Antropologyst 92: 85–95.

Cabra, V.C. (2000). Estudio cinético comparativo de la fermentación de masa de pozol elaborada en dos condiciones higiénicas distintas. Dissertation, Facultad de Química, UNAM, Ciudad de México, México.

Cañas, A., Barzana, E., Owens, J. and Wacher, C. (1993a). La elaboración de pozol en Los Altos de Chiapas. Ciencias 44: 219–229.

Cañas, A., Barzana, E., Owens, J. and Wacher, C. (1993b). Estudio de la variabilidad en los métodos de producción de pozol en los Altos de Chiapas. Alimentos fermentados indígenas de México. UNAM, Ciudad de México, México, pp. 69–74.

Chavan, J.K. and Kadam, S.S. (1989). Critical reviews in food science and nutrition. Food Science 28: 348–400.

Corona, G., Jiménez, G., Hernández, G., Vega, J., Fernández, A., Vielle, J., Martínez de la Vega, O., Hernández, A. and Herrera, L. (2006). Genoma del maíz mexicano *Zea mays* L. ssp. 2233 por el método de Sanger. XXVI Congreso Nacional de la Sociedad Mexicana de Bioquímica. Guanajuato, Guanajuato, México.

Daeschel, M.A. (1989). Antimicrobial substances from lactic acid bacteria for use as food preservatives. Food Biotechnology 43: 164–167.

Dellaglio, F., Torriani, S. and Felis, G.E. (2004). Reclassification of *Lactobacillus cellobiosus* Rogosa et al., 1953 as a later synonym of *Lactobacillus fermentum* Beijerinck 1901. International Journal of Systematic and Evolutionary Microbiology 54: 809–812.

De Vuyst, L. and Tsakalidou, E. (2008). *Streptococcus macedonicus*, a multi-functional and promising species for dairy fermentations. International Dairy Journal 18: 476–485.

Díaz-Ruiz, G., Guyot, J.P., Ruiz-Teran, F., Morlon-Guyot, J. and Wacher, C. (2003). Microbial and physiological characterization of weak amylolytic but fast growing lactic acid bacteria: a functional role in supporting microbial diversity in Pozol. Applied and Environmental Microbology 69: 4367–4374.

Doebley, J. (1990). Molecular evidence and evolution of maize. Economic Botany 44: 6–27.

Domínguez-Ramírez, L.L. and Wacher, C. (2014). *Streptococcus infantarius* subsp. *infantarius* persistence in Pozol from Villahermosa, Tabasco. 6th Food Science Biotechnology and Safety Meeting, AMECA. Nuevo Leon, Monterrey, México.

Dorweiler, J., Stec, A., Kermicle, J. and Doebley, J. (1993). Teosinte glume architecture 1: A Genetic Locus Controlling a Key Step in Maize Evolution. Science 262: 233–235.

Elizaquível, P., Pérez-Cataluña, A., Yépez, A., Aristimuño, C., Jiménez, E., Cocconcelli, P.S., Vignolo, G. and Aznar, R. (2015). Pyrosequencing vs. culture-dependent approaches to analyze lactic acid bacteria associated to chicha, a traditional maize-based fermented beverage from Northwestern Argentina. International Journal of Food Microbiology 198: 9–18.

Escobar, A., Gardner, A. and Steinkraus, K.H. (1993). South American fermented maize chicha. pp. 402–406. *In*: K.H. Steinkraus (ed.). Handbook of Indigenous Fermented Foods. Marcel Dekker, New York.

Esquivel, A.K., Espinosa, J., Centurión, D., Flores, M.T., Wacher, C., Reyes-Duarte, D. and Díaz-Ruiz, G. (2013). Study of Enterobacteria present in atole agrio from Villahermosa, Tabasco. XV Congreso Nacional de Biotecnología y Bioingeniería. Cancún, Quintana Roo, México.

Eyzaguirre, R Z., Nienaltowska, K., de Jong, L.E.Q., Hasenack, B.B.E. and Nout, M.J.R. (2006). Effect of food processing of pearl millet (*Pennisetum glaucum*) IKMP-5 on the level of phenolics, phytate, iron and zinc. Journal of the Science of Food and Agriculture 86: 1391–1398.

Fernández-Muñoz, J.L., Rojas-Molina, I., González-Dávalos, M.L., Leal, M., Valtierra, M.E., San Martín-Martínez, E. and Rodríguez, M.E. (2004). Study of calcium diffusion in components of maize kernels during traditional nixtamalization process. Cereal Chemistry Journal 81: 65–69.

Gadaga, T.H., Mutukumira, A.N., Narvhus, J.A. and Feresu, S.B. (1999). A review of traditional fermented foods and beverages of Zimbabwe. International Journal of Food Microbiology 53: 1–11.

Giraud, E., Brauman, A., Keleke, S., Lelong, B. and Raimbault, M. (1991). Isolation and physiological study of an amylolytic strain of *Lactobacillus plantarum*. Applied Microbiology and Biotechnology 36: 379–383.

Greppi, A., Rantsiou, K., Padonou, W., Hounhouigan, J., Jespersen, L., Jakobsen, M. and Cocolin, L. (2013). Determination of yeast diversity in ogi, mawè, gowé and tchoukoutou by using culture-dependent and -independent methods. International Journal of Food Microbiology 165: 84–88.

Guerra-Flores, S., Solís-Pacheco, J., Camarillo-Miranda, A., Reyes-Blanco, M., Gónzalez, E. and Aguilar-Uscanga, B. (2009). Aislamiento e Identificación de Microorganismos Nativos en la Fermentación de Tejuino Artesanal. XIII Congreso Nacional de Biotecnología y Bioingeniería y VII Simposio Internacional de Producción de Alcoholes y Levaduras. Acapulco, Guerrero, México.

Halm, M., Lillie, A., Sorensen, A.K. and Jakobsen, M. (1993). Microbiological and aromatic characteristics of fermented maize dough for kenkey production in Ghana. International Journal Food Microbiology 19: 135–143.

Hayashida, F.M. (2008). Ancient beer and modern brewers: Ethnoarchaeological observations of *chicha* production in two regions of the North Coast of Peru. Journal of Anthropological Archaeology 27: 161–174.

Hayford, A.E., Petersen, A., Vogensen, F.K. and Jakobsen, M. (1999). Use of Conserved Randomly Amplified Polymorphic DNA (RAPD) Fragments and RAPD Pattern for Characterization of *Lactobacillus fermentum* in Ghanaian Fermented Maize Dough. Applied and Environmental Microbiology 65: 3213–3221.

Herrera-Suárez, T. (2003). Impresiones de un Breve Recorrido de la Memoria a Través de más de Medio Siglo en la UNAM. Forjadores de la Ciencia en la UNAM, Ciclo de conferencias "Mi vida en la Ciencia". Coordinación de la Investigación Científica, UNAM, Ciudad de México, México, pp. 7–13.

Holzapfel, W.H. and Taljaard, J.L. (2004). Industrialization of mageu fermentation South Africa. pp. 363–407. *In*: K.H. Steinkraus (ed.). Industrialization of Indigenous Fermented Foods. Marcel Dekker, New York.

Hounhouigan, D.J., Nout, M.J.R., Nago, C.M., Houben, J.H. and Rombouts, F.M. (1993). Characterization and frequency distribution of species of lactic acid bacteria involved in the processing of maw, a fermented maize dough from Benin. International Journal of Food Microbiology 18: 279–287.

Hounhouigan, D.J., Nout, M.J.R., Nago, C.M., Houben, J.H. and Rombouts, F.M. (1994). Microbiological changes in mawè during natural fermentation. World Journal of Microbiology and Biotechnology 10: 410–413.

http://www.scielo.org.ve/scielo.php?script=sci_arttext&pid=S0004-06222003000200011; 13 May 2016.

Illesca, R. (1943). La teoría química de la formación del Nixtamal. Revista de la Sociedad Mexicana de Historia Natural 4: 129–136.

International Committee on Systematic Bacteriology (2000). Subcommittee on the taxonomy of *Bifidobacterium, Lactobacillus* and related organism. International Journal of Systematic and Evolutionary Microbiology 50: 1391–1392.

Jans, C., Bugnard, J., Njage, P.M.K., Lacroix, C. and Maile, L. (2012). Lactic acid bacteria diversity of African raw and fermented camel milk products reveals a highly competitive, potentially health-threatening predominant microflora. Food Science Technology 47: 371–379.

Jans, C., Kaindi, D.W.M., Böck, D., Njage, P.M.K., Kouamé-Sina, S.M., Bonfoh, B., Lacroix, C. and Meile, L. (2013). Prevalence and comparison of *Streptococcus infantarius* subsp. *infantarius* and *S. gallolyticus* subsp. *macedonicus* in raw and fermented dairy products from East and West Africa. International Journal of Food Microbiology 167: 186–195.

Jespersen, L., Halm, M., Kpodo, K. and Jakobsen, M. (1994). Significance of yeast and moulds occurring in maize dough fermentation for 'kenkey' production. International Journal of Food Microbiology 24: 239–248.

Lappe, P. and Ulloa, M. (1989). Estudios Étnicos, Microbianos y Químicos del Tesgüino Tarahumara. UNAM, Ciudad de México, México, pp. 6–41, 51–60.

Le Blanc, J.G., Garro, M.S. and de Giori, G.S. (2004). Effect of pH on *Lactobacillus fermentum* growth, raffinose removal, α-galactosidase activity and fermentation products. Applied Microbiology and Biotechnology 65: 119–123.

Long, A., Benz, B.F., Donahue, D.J., Jull, A.J.T. and Toolin, L.J. (1989). First direct AMS dates on early maize from Tehuacan, Mexico. Radiocarbon 31: 1035–1040.

López, G. (2006). Estudio de la variabilidad microbiológica del pozol mediante PCR-DGGE con el gen *rpoB*.M.S. Thesis, Facultad de Química, UNAM, Ciudad de México, México.

Lorence-Quiñones, A., Wacher-Rodarte, C. and Quintero-Ramírez, R. (1999). Cereal fermentation in Latin American Countries. *In*: N.F. Haard, S.A. Odunfa, C.-H. Lee, R. Quintero-Ramírez, A. Lorence-Quiñones and C. Wacher-Rodarte (eds.). Fermented Cereals A Global Perspective. FAO, Rome. http://www.fao.org/docrep/x2184e/x2184e10.htm; 13 May 2016.

MacNeish, R.S. (1985). The archaeological record on the problem of the domestication of corn. Science 143: 171–178.

Mangelsdorf, P.C., MacNeish, R.S. and Galinat, W.C. (1964). Domestication of corn. Science 143: 538–545.

Mangelsdorf, P.C., MacNeish, R.S. and Galinat, W.C. (1967). Prehistoric wild and cultivated maize. pp. 178–200. *In*: D.S. Byers (ed.). The Prehistory of the Tehuacan Valley, Vol. I: Environment and Subsistence. The University of Texas Press, London.

Manrique, R., Díaz-Ruiz, G. and Wacher, C. (2016). Probiotic potential of yeasts isolated from *atole agrio* from Villahermosa, Tabasco. 7th International Symposium on Probiotics. Ciudad de México, México.

Martínez-Bustos, F., Martínez-Flores, H.E., San Martín-Martínez, E., Sánchez-Sinencio, F., Chang, Y.K., Barrera-Arellano, D. and Rios, E. (2001). Effect of the components of maize on the quality of masa and tortillas during the traditional nixtamalization process. Journal of the Science of Food and Agriculture 81: 1455–1462.

Masha, G.G.K., Ipsen, R., Petersen, M.A. and Jakobsen, M. (1998). Microbiological, rheological and aromatic characteristics of fermented *Uji* (an East Africa Sour Porridge). World Journal of Microbiology & Biotechnology 14: 451–456.

Matsheka, M.I., Magwamba, C.C., Mpuchane, S. and Gashe, B.A. (2013). Biogenic amine producing bacteria associated with three different commercially fermented beverages in Botswana. African Journal of Microbiology Research 7: 342–350.

Matsuoka, Y., Vigouroux, Y., Goodman, M.M., Sanchez, J., Buckler, E. and Doebley, J. (2002). A single domestication for maize shown by multilocus microsatellite genotyping. PNAS 99: 6080–6084.

McCann, J. (2001). Maize and Grace. History, Corn, and Society. Africa's New Landscape, 1500–1999. Comparative Study of Society and History 43: 246–272.

McKay, L.L. and Baldwin, K.A. (1990). Applications for biotechnology: present and future improvements in lactic acid bacteria. FEMS Microbiology Reviews 7: 3–14.

McMaster, L.D., Kokott, S.A., Reid, S.J. and Abratt, V.R. (2005). Use of traditional African fermented beverages as delivery vehicles for *Bifidobacterium lactis* DSM 10140. International Journal of Food Microbiology 102: 231–237.

Merrill, W.L., Hard, R.J., Mabry, J.B., Fritz, G.C., Adams, K.R., Ronery, J.R. and MacWilliams, A.C. (2009). The diffusion of maize to the Southwestern United States and its impact. PNAS 106: 21019–21026.

Molin, G. (2008). *Lactobacillus plantarum*. The Role in Foods and in Human Health. pp. 363–365. *In*: E.R. Farnworth (ed.). Handbook of Fermented Functional Foods. CRC Press, Boca Raton, Florida.

Monroy, M. del C., Castro, L., Fernández, F.J. and Mayorga, L. (2009). Revisión Bibliográfica: Bacteriocinas producidas por bacterias probióticas. ContactoS 73: 63–72.

Mugula, J.K., Nnko, S.A.M., Narvhus, J.A. and Sørhaug, T. (2003). Microbiological and fermentation characteristics of togwa, a Tanzanian fermented food. International Journal of Food Microbiology 80: 187–199.

Nche, P.F., Odamtten, G.T., Nout, M.J.R. and Rombouts, F.M. (1996). Soaking of maize determines the quality of aflata for kenkey production. Journal of Cereal Science 24: 291–297.

Nicholson, E. (1960). Chicha Maize Types and Chicha Manufacture in Peru, Economic Botany 14: 290–291.

Nuraida, L. (1988). Studies on microorganisms isolated from pozol, a Mexican corn fermented maize dough. M.S. Thesis. Faculty of Agriculture and Food, Department of Food Science and Technology. University of Reading, Berkshire, UK.

Nuraida, L., Wacher, C. and Owens, J.D. (1995). Microbiology of pozol, a Mexican fermented maize dough. World Journal of Microbiology and Biotechnology 11: 567–571.

Obiri-Danso, K. (1994). Microbiological studies on corn dough fermentation. Cereal Chemistry 71: 186–188.

Odunfa, S.A., Adeniran, S.A., Teniola, O.D. and Nordstrom, J. (2001). Evaluation of lysine and methionine production in some *Lactobacilli* and yeasts from ogi. International Journal of Food Microbiology 63: 159–163.

Oguntoyinbo, F.A., Tourlomousis, P., Gasson, M.J. and Narbad, A. (2011). Analysis of bacterial communities of traditional fermented West African cereal foods using culture independent methods. International Journal of Food Microbiology 145: 205–210.

Oguntoyinbo, F.A. and Narbad, A. (2012). Molecular characterization of lactic acid bacteria and *in situ* amylase expression during traditional fermentation of cereal foods. Food Microbiology 31: 254–262.

Olivares-Illana, V., Wacher-Rodarte, C., Le Borgne, S. and López-Mungía, A. (2002). Characterization of a novel cell-associated levansucrase from a *Leuconostoc citreum* strain isolated from pozol, a fermented corn beverage of Mayan Origin. Journal of Industrial Microbiology and Biotechnology 28: 112–117.

Olsen, A., Halm, M. and Jakobsen, M. (1995). The antimicrobial activity of lactic acid bacteria from fermented maize (kenkey) and their interactions during fermentation. Journal of Applied Bacteriology 79: 506–512.

Omemu, A.M., Oyewole, O.B. and Bankole, M.O. (2007). Significance of yeasts in the fermentation of maize for *ogi* production. Food Microbiology 24: 571–576.

Paz, C.A. (2014). Estudio del efecto antimicrobiano de bacterias del género *Weissella* aisladas del pozol. Dissertation, Facultad de Química, UNAM, Ciudad de México, México.

Phister, T.G., O'Sullivan, D.J. and McKay, L.L. (2004). Identification of bacilysin, chlorotetaine, and iturin a produced by *Bacillus* sp. strain CS93 isolated from Pozol, a Mexican fermented maize dough. Applied and Environmental Microbiology 70: 631–634.

Pieterse, R., Todorov, S.D. and Dicks, L.M. (2008). Bacteriocin ST91KM, produced by *Streptococcus gallolyticus* subsp. *macedonicus* ST91KM, is a narrow-spectrum peptide active against bacteria associated with mastitis in dairy cattle. Canadian Journal of Microbiology 54: 525–531.

Piperno, D. and Flannery, K.V. (2001). The earliest archaeological maize (*Zea mays* L.) from highland Mexico: new accelerator mass spectrometry dates and their implications. Proceeding of the National Academy of Science 98: 2101–2103.

Piperno, D.R., Ranere, A.J., Holst, I., Iriarte, J. and Dickau, R. (2009). Starch grain and phytolith evidence for early ninth millennium B.P. maize from Central Balsas River Valley, Mexico. PNAS 106: 5019–5024.

Pflugfelder, R.L., Rooney, L.W. and Waniska, R.D. (1988). Fraction and Composition of Commercial Corn Masa. Cereal Chemistry 65: 262–266.

Pohl, M.E.D., Piperno, D.R., Pope, K.O. and Jones, J.G. (2007). Microfossil evidence for pre-Columbian maize dispersals in the neotropics from San Andrés, Tabasco, Mexico. PNAS 104: 6870–6875.

Ranere, A.J., Piperno, D.R., Holst, I., Dickau, R. and Iriarte, J. (2009). The cultural and chronological context of early Holocene maize and squash domestication in the Central Balsas River Valley, Mexico. PNAS 106: 5014–5018.

Rangel-Meza, E., Muñoz-Orozco, A., Vázquez-Carrillo, G., Cuevas-Sánchez, J., Merino-Castillo, J. and Miranda-Coli, S. (2004). Nixtamalización, elaboración y calidad de tortilla de maíz de Ecatlán, Puebla, México. Agrociencia 38: 53–61.

Rashid, M.H.U., Togo, K., Ueda, M. and Miyamoto, T. (2009). Characterization of bacteriocin produced by *Streptococcus bovis* J2 40-2 isolated from traditional fermented milk "Dahi". Animal Science Journal 80: 70–78.

Ray, P., Sánchez, C., O'Sullivan, D.J. and McKay, L.L. (2000). Classification of a bacteria isolate, from Pozol, exhibiting antimicrobial activity against several Gram-Positive and Gram-Negative bacteria, yeasts and molds. Journal of Food Protection 63: 1123–1132.

Riley, M.A. and Wertz, J.E. (2002). Bacteriocins: Evolution, ecology, and application. Annual Review of Microbiology 56: 117–137.

Rodríguez, A., Villalva, B., Flores, M.T., Sainz, T., Eslava, C., Díaz-Ruiz, G. and Wacher, M. del C. (2011). Estudio del potencial probiótico de bacterias ácido lácticas aisladas del pozol. XIV Congreso Nacional de Biotecnología y Bioingeniería. Querétaro, Querétaro, México.

Rosado, J.L., Cassís, L., Solano, L. and Duarte-Velázquez, M.A. (2005). Nutrient addition to corn masa flour: Effect on corn flour stability, nutrient loss and acceptability of fortified corn tortillas. Food and Nutrition Bulletin 26: 266–272.

Ruiz-Gutierrez, M.G., Quintero-Ramos, A., Meléndez-Pizarro, C.O., Talamás-Abbud, R., Barnard, J., Márquez-Meléndez, R. and Lardizábal-Gutiérrez, D. (2012). Nixtamalization in two steps with different calcium salts and their relationship with chemical, textures and thermal properties in masa and tortilla. Journal of Food Process Engineering 35: 772–783.

Sainz, T., Wacher, C., Espinoza, J., Centurión, D., Navarro, A., Molina, A., Inzunza, A., Cravioto, A. and Eslava, C. (2001). Survival and characterization of *Escherichia coli* strains in a typical Mexican acid-fermented food. International Journal of Food Microbiology 71: 169–176.

Saldaña, G. and Brown, H.E. (1984). Nutritional composition of corn and flour tortillas. Journal of Food Science 49: 1202–1203.

Salinas, Y., Herrera, J.A., Castillo, J. and Pérez, P. (2003). Cambios físico-químicos del almidón durante la nixtamalización del maíz en variedades con diferente dureza de grano. ALAN Archivos Latinoamericanos de Nutrición 53.

Schnable, P.S. et al. (2009). The B73 Maize genome: complexity, diversity, and dynamics. Science 326: 1112–1115.

Sefa-Dedeh, S., Cornelius, B. and Afoakwa, E.O. (2003). Effect of fermentation on the quality characteristics of nixtamalized corn. Food Research International 36: 57–64.

Sefa-Dedeh, S., Cornelius, B., Sakyi-Dawson, E. and Ohene-Afoakwa, E. (2004). Effect of nixtamalization on the chemical and functional properties of maize. Food Chemistry 86: 317–324.

Serna-Cock, L., Valencia, L.J. and Campos, R. (2010). Cinética de fermentación y acción antimicrobiana de *Weissella confusa* contra *Staphylococcus aureus* y *Streptococcus agalactiae*. Revista de la Facultad de Ingeniería de la Universidad de Antioquia 55: 55–65.

Serna-Cock, L., Rubiano-Duque, L.F., Loaiza-Castillo, N.B. and Enríquez-Valencia, C.E. (2012). *Weissella confusa* como un agente protector en la inocuidad alimentaria contra patógenos Gram negativos. Revista Alimentos Hoy 21: 102–114.

Server-Busson, C., Foucaud, C. and Leveau, J.-Y. (1999). Selection of dairy *Leuconostoc* isolates for important technological properties. Journal of Dairy Research 66: 245–56.

Smoesmith, V.M. (1918). The study of corn. Orange Judd Company, New York, pp. 11–17.

Staller, J.E. (2003). An Examination of the palaeobotanical and chronological evidence for an early introduction of maize (*Zea mays* L.) into South America: A response to pearsall. Journal of Archeological Science 30: 373–380.

Staller, J.E. and Thompson, R.G. (2002). A multidisciplinary approach to understanding the initial introduction of maize into coastal Ecuador. Journal of Archaeological Science 29: 33–50.

Steinkraus, K.H. (1983). Lactic acid fermentation in the production of food from vegetables, cereals and legumes. Antonie van Leeuwenhoek 49: 337–348.

Steinkraus, K.H., Ayres, R., Olek, A. and Farr, D. (1993). Biochemistry of *Saccharomyces*. pp. 517–519. *In*: K.H. Steinkraus (ed.). Handbook of indigenous fermented foods. Marcel Dekker, New York.

Steinkraus, K.H. (1997). Classification of fermented foods: worldwide review of household fermentation techniques. Food Control 8: 311–317.

Steinkraus, K.H. (1998). Bio-enrichment: production of vitamins in fermented foods. pp. 603–619. *In*: Wood (ed.). Microbiology of Fermented Foods. Blackie Academic and Professional, London.

Steinkraus, K.H. (2002). Fermentations in world food processing. Comprehensive Reviews in Food Science and Food Safety 1: 23–32.

Sturtevant, E.L. (1883). Proposed classification of corn varieties. Pacifica Rural Press 26: 402.

Taboada, J., Ulloa, M. and Herrera, T. (1977). Studies on tesgüino fermented maize beverage consumed in northern and central Mexico. Symposium on Indigenous Fermented Foods. Bangkok, Thailand.

Tallgren, A.H., Airaksinen, U., von Weissenberg, R., Ojamo, H., Kuusisto, J. and Leisola, M. (1999). Exopolysaccharide-producing bacteria from sugar beets. Applied and Environmental Microbiology 65: 862–64.

Tsakalidou, E., Zoidou, E., Pot, B., Wassill, L., Ludwig, W., Devriese, L.A., Kalantzopoulos, G., Schleifer, K.H. and Kersters, K. (1998). Identification of streptococci from Greek Kasseri cheese and description of *Streptococcus macedonicus* sp. nov. International Journal of Systematic Bacteriology 48: 519–527.

Ulloa, M. and Herrera, T. (1972). Descripción de dos especies nuevas de bacterias aisladas de pozol: *Agrobacterium azotophilum* y *Achromobacter pozolis*. Revista Latinoamericana de Microbiología 15: 199–202.

Ulloa, M. (1974). Mycofloral succession in pozol from Tabasco, Mexico. Boletin de la Sociedad Mexicana de Micologia 8: 17–48.

Ulloa, M., Herrera-Suárez, T. and Lappe, P. (1987). Pozol. pp. 13–20. *In*: Fermentaciones tradicionales indígenas de México Vol. 16. Instituto Nacional Indigenista, Ciudad de México.

Uvere, P.O. and Awada, L.H. (2013). Complementary local foods for infants in developing countries. pp. 75–93. *In*: R.R. Watson, G. Grinble, V.R. Preedy and S. Zibadi (eds.). Nutrition in Infancy Vol. 1. Human Press/Springer, New York.

Valderrama, A. (2012). Diversidad de Bacterias Lácticas del Atole Agrio de Villahermosa Tabasco. Dissertation, Facultad de Química, UNAM, Ciudad de México, México.

Vallejo, J.A., Miranda, P., Flores-Félix, J.D., Sánchez-Juanes, F., Ageitos, J.M., González-Buitrago, J.M., Velázquez, E. and Villa, T.G. (2013). Atypical yeasts identified as *Saccharomyces cerevisiae* by MALDI-TOF MS and gene sequencing are the main responsible of fermentation of chicha, a traditional beverage from Peru. Systematic and Applied Microbiology 36: 260–264.

Wacher, C., Cañas, A., Cook, E., Barzana, E. and Owen, J. (1993). Sources of microorganisms in pozol a traditional Mexican fermented maize dough. World Journal of Microbiology and Biotechnology 9: 269–274.

Wacher, C., Cañas, A., Barzana, E., Lappe, P., Ulloa, M. and Owens, J.D. (2000). Microbiology of Indian and Mestizo pozol fermentations. Food Microbiology 17: 251–256.

Wang, R.-L., Stec, A., Hey, J., Lukes, L. and Doebley, J. (1999). The limits of selection during maize domestication. Nature (London) 398: 236–239.

www.cic-tic.unam.mx/cic/mas_cic/publicaciones/download/forjadores/Teofilo_Herrera_Suarez.pdf; 23 January 2016.

Yañez, Y. (2005). Nixtamalización por extrusión de las fracciones del grano de maíz para la obtención de harinas instantáneas. M.S. Thesis, IPN, Ciudad de México, México.

10

Bread Fortification
Health and Nutritional Benefits

Lucia Padalino, Amalia Conte and
Matteo Alessandro Del Nobile *

1. Introduction on Enriched Bread

Bread is a staple food consumed all over the world. It is mainly made of wheat flour (*Triticum aestivum* or *Triticum durum*), water, salt and yeasts by a series of processes involving mixing, kneading, proofing, shaping and baking (Dewettinck et al. 2008). Bread has an important role in human nutrition. Generally, wheat bread is considered to be a good source of energy and irreplaceable nutrients for human body. This is especially true for the products made from wholegrain or high-yield flour types. Bread prepared from refined flour is nutritionally much poorer and does not adequately meet the requirements for many macro- or micro-nutrients (Škrbić and Filipčev 2008, Rosell et al. 2015). It has been reported that bread made from refined flour has low micro-nutrient content (Isserliyska et al. 2001). Wheat protein also lacks the balance of essential amino acids lysine, threonine and valine. Therefore, there have been many on going investigations on enhancing the nutritive value of bakery products to fulfill the expanding demands of modern dietary habits, considering the products' protein, mineral, vitamin and/or fiber contents. It was observed that the additions of some flours

Dip. Scienze Agrarie, degli Alimenti e dell'Ambiente, Università di Foggia, via Napoli 25, 71122 Foggia.
 E-mail: lucia.padalino@unifg.it; amalia.conte@unifg.it
* Corresponding author: matteo.delnobile@unifg.it

having valuable nutrient profile improved the bread properties (e.g., the protein quality and quantity, dietary fibre content, and unsaturated fats profile), extended the shelf life, made crumb softer and lowered firming rate (Raffo et al. 2003). Bakery products supplemented with various nutritious, protective or ballast substances, have been gaining popularity worldwide. Mixed grain, wholegrain breads and related products are even considered as functional foods because they are convenient vehicles for important nutrients and phytochemicals. A variety of wheat flour substitutions have been tried in bread formulations to improve the nutritional quality: for instance, pseudocereal grains such as buckwheat, amaranth and quinoa are rich in a wide range compounds, e.g., flavonoids, phenolic acids, trace elements, fatty acids and vitamins with known effects on human health (Gorinstein et al. 2008, Kalinova and Dadakova 2009, Li and Zhang 2001, Tomotake et al. 2007); legumes such as soybean are a good resource of vegetable proteins, fat, lysine, and other biologically active components (isoflavones) that may be effective in reducing the risk of coronary heart diseases and several cancers (Trainer and Holden 1999, Hasler 2002, Murphy et al. 1999); vegetables are good sources of functional substances such as dietary fibre, carotenoids, polyphenolics, tocopherols, vitamins C, minerals, organic acids and others (Schieber et al. 2001a). Also, gluten-free cereal foods are frequently made using refined gluten-free flour or starch and are generally not enriched or fortified (Thompson et al. 1999). As a result, many gluten-free cereal foods do not contain the same levels of B-vitamins, iron and fibre as their gluten-containing counterparts (Thompson et al. 1999, 2000). A need to improve their nutritional quality has been raised by many medical and nutritional experts (Kupper et al. 2005, Thompson et al. 2005). It is worth noting that the enrichment of wheat flour with non-conventional ingredients results in alteration of both rheological and sensory properties of final breads. Specifically, the manufacture of bread with non-wheat flours presents considerable technological difficulty because their proteins lack the ability to form the necessary gluten network for holding the gas produced during the fermentation (Gallagher et al. 2003, Arendt et al. 2002). Gluten is an essential structure-building protein that provides viscoelasticity to the dough, good gas-holding ability and good crumb structure of the resulting baked product (Gallagher et al. 2004). Some of the most important approaches developed to mimic the properties of gluten in gluten-free bakery products involve the use of gums, hydrocolloids and protein-based ingredients (Arendt et al. 2002). In general, acceptable bread with non-traditional ingredients can also be obtained with appropriate formulation and process optimization. More details can be found in the following paragraphs where specific types of fortified bread are taken into account.

2. Bread Enriched With Pseudo-Cereals and Legumes

Pseudocereals are not often used to produce bread but they can be useful in diet therapy of celiac disease (Thompson 2001). Homemade breads containing pseudocereals could be an important part of traditional type food, i.e., "slow food", which is now becoming more popular. In particular, increasing consumption of this kind of bread in our daily menu can improve the nutrient profiles of our diet because it is known that the pseudocereals' grains can be excellent sources of nutrients (Lin et al. 2009, Pasko et al. 2008). Besides being important energy sources due to their starch content, amaranth, quinoa and buckwheat provide good-quality proteins, dietary fibres and lipids rich in unsaturated fats (Alvarez-Jubete et al. 2009) and show antioxidant, anti-nflammatory, and anticarcinogenic activities (Lin et al. 2008). Moreover, they contain adequate levels of important micronutrients, such as minerals and vitamins and significant amounts of other bioactive components, such as saponins, phytosterols, squalene, fagopyritols and polyphenols (Berghofer et al. 2002, Taylor et al. 2002, Wijngaard et al. 2006). As result, pseudocereals can provide beneficial health effects (Christa and Sorel-Smietana 2008, Martirosyan et al. 2007), therefore bread with addition of buckwheat, amaranth or quinoa flour, as a staple product could diversify ordinary (daily) model of nutrition. The amino acid profile of the proteins of amaranth is comparable to that of egg, and the nutritional quality of the proteins of quinoa is comparable to that of caseins (Schoenlechner et al. 2008). Specifically, pseudocereals flours show a balanced amino acid spectrum with high methionine and lysine contents and for this reason are superior to common cereals (Aubrecht and Biacs 2001, Drzewiecki et al. 2003, Gorinstein et al. 2002, Peiretti et al. 2013). Compared to cereals, quinoa has a higher concentration of fat with elevated levels of unsaturated fatty acids and phospholipids that, due to the presence of vitamin E, remain stable during storage (Ng et al. 2007). A growing number of studies have investigated the application of pseudocereals in the production of gluten-free products such as bread. The nutritional properties and baking characteristics of amaranth, quinoa and buckwheat have been assessed in gluten-free matrices (Alvarez-Jubete et al. 2010), achieving breads with superior nutritional features and acceptable sensory scores. Gambus et al. (2002) studied the feasibility of amaranth as an alternative gluten-free ingredient to improve the nutritional quality of gluten-free breads. In a more recent study (Kiskini et al. 2007), amaranth-based gluten-free bread fortified with iron was successfully formulated. Buck-wheat has also been investigated as a composite flour in the development of high-quality gluten-free breads in two recent studies with promising results (Moore et al. 2004, Renzetti et al. 2008). In other recent studies, both nutritional properties and baking characteristics of amaranth, quinoa and buckwheat have been assessed (Alvarez-Jubete et al. 2009, 2010). The authors found that the replacement of potato starch with a pseudocereal flour resulted in gluten-free breads with an increased content of important nutrients such as protein, fiber, calcium, iron and vitamin E. The resultant bread also had a significantly higher content of polyphenol compounds and antioxidant activity. Buckwheat and quinoa bread volumes were found increased in comparison to the control and all the pseudocereal-containing breads were characterized by a desirable significant softer crumb structure (Arendt et al. 2002).

In wheat flour matrices some studies demonstrated the feasibility of partial/low replacement of wheat flour with pseudocereals for processing baked goods (Tosi et al. 2002, Schoenlechner et al. 2008, Angioloni and Collar 2011a,b). The use of a blend of buckwheat, amaranth, chickpea and quinoa flours subjected to sourdough fermentation by selected γ-aminobutyric acid (GABA) producing strains, allowed the manufacture of a bread enriched with GABA. This finding should be considered as a promising possibility for enhancing nutritional, functional, sensory, and technological properties of bread. The addition of quinoa and/or buckwheat seeds (at levels of 30 and 40%) previously subjected to hydrothermal process, resulted in a valuable effect on the nutritive value of breads (Demin et al. 2013).

Pseudocereals flour exhibits higher qualitative and quantitative lipid profiles than wheat flours (Hager et al. 2012). Lipids have a significant effect on the quality and texture of baked goods because of their ability to associate with proteins, due to their amphipathic nature (hydrophilic and hydrophobic groups present), and with starch, forming inclusion complexes (Goesaert et al. 2005). The high levels of lipids present in amaranth and quinoa flours may have implications in relation with both crumb structure and crumb texture. In breadmaking applications, protein and starch lipid binding in wheat flour and bread systems have been reported to correlate with loaf volume, crumb structure, softness and/or texture of bread (Collar et al. 2001). Alvarez-Jubete et al. (2010a) reported that the gluten-free breads containing buckwheat or quinoa flour had a significant higher volume in comparison with gluten-free control bread (100% rice flour) attributed to the presence of natural emulsifiers as monoglycerid in the pseudocereal flours. In fact, the use of emulsifers (monoglycerides) in baking has been shown to have a softening effect on bread crumb (Hoseney 1998). Specifically, the emulsifers can form complexes with amylose, thus limiting starch swelling during baking and leaching of amylose into solution (Hoseney 1998, Belitz et al. 2004). As a result, fewer entanglements between starch granules and amylose in solution take place, leading to breads with a softer crumb structure (Schober et al. 2009). The pseudocereal containing breads were characterised by a significant softer crumb texture effect that was attributed to the presence of natural emulsifiers in the pseudocereal flours Alvarez-Jubete et al. (2010a). In addition, the high level of polar lipids in pseudocereal seeds may have functionality as gas cell stabilising agents during bread making. In particular, the lipids present in the pseudocereal flours may act as surface-active agents and thus contribute to gas cell stabilisation prior to starch gelatinisation (Alvarez-Jubete et al. 2010b). Park et al. (2005) found that the substitution of 7.5 to 10% quinoa flour for hard-type wheat flour significantly increased the loaf volume of bread, but more than 15% substitution distinctly decreased it. A combination of microbial lipase and 15% substitution of quinoa flour improved the bread quality with a loaf volume distinctly higher than that of the control (100% wheat flour) and also suppressed the staling of bread during storage. In other studies, the incorporation of up to 30% (w/w) of amaranth in the dough improved the score of loaf color, 10% of substitution had positive effect on specific volume, whereas taste, aroma, texture, and overall acceptability of bread were not influenced (Mlakar et al. 2008). Analysis of sensory quality results of breads showed that addition of pseudocereals flour (especially buckwheat and quinoa flours) to wheat flour may increase acceptable quality attributes such as taste, colour and odour. Quinoa breads that deserved overall sensory acceptability with no significant differences with respect to the breads were made from the blends wheat: quinoa (50:50) (Valcárcel-Yamani et al. 2012). Also replacement of wheat flour by up to 50% of amaranth led to coloured breads with good sensory perception (Valcárcel-Yamani et al. 2012). The flavor of the wheat bread enriched with amaranth flour was described as nutty and was preferred over the flavor of the white bread control. These observations suggest that addition of buckwheat flour into bread can improve not only nutritional value but also sensory properties of bread. In fact, the addition of buckwheat flour up to 15% or 30% levels to wheat flour satisfactory improved bread properties and attributes such as colour, odour and taste. In particular, sample with 30% buckwheat flour had an interesting and natural taste and was crusty as more than 40% of testers declared such opinion. Results suggest that 30% of buckwheat flour improved the sensory value of bread and reduced attributes (i.e., gummy), which could decrease value of bread. Thus, the bread supplemented with 30% buckwheat flour recorded the highest sensory profile as also reported by Wronkowska et al. (2008).

The legumes are generally good sources of slow-release carbohydrates and are rich in proteins (18–25%); soybean is unique in containing about 35–43% proteins. Incorporation of legumes into cereal-based foods deserves a special attention since high levels of lysine in legumes complement lysine deficiencies in cereal-based diets. With protein contents double than that of other cereal crops, legumes are an economical, environmentally sustainable protein source with low starch bioavailability, and high resistant starch content, to potentially improve the nutritional value of breads (Patterson et al. 2010), in line with the current suitable dietary trends (Miller-Jones 2009). The incorporation of significant amounts of legumes has been successfully achieved in producing cereal-based goods with no strict gluten-related requirements, such as in biscuit (Serrem et al. 2011), pasta (Jayasena et al. 2008) and cake (Gómez et al. 2008). Conversely, in bread-making applications, the lack of gluten proteins to meet dough viscoelastic and fermentative restrictions has generally constrained the incorporation of substantial amounts of legumes (15% of wheat flour replacement) into wheat dough systems to achieve nutritional and health endorsing effects. A general increase in crumb hardness and crumb grain heterogeneity and roughness concomitantly with a slight decrease in specific volume was observed in wheat breads with increased content of legume flour. High levels of legumes as chickpea, lentil and soybean incorporated into baked products without any structuring agent are cost effective and nutritionally advantageous although technologically very challenging (Angioloni and Collar 2012). However, high-legume-wheat blends appear as an efficient strategy to obtain sensorially accepted and nutritionally enhanced bread with no dramatic technological impairment when structuring agents (gluten/hydrocolloids) are incorporated.

3. Bread Enriched With Fruit and Vegetable Flours, Seed Oil, Fiber and By-Products

Functional foods, such as products with pro- and prebiotics, dietary fibre, low fat, etc., are a new category of promising foods with many properties (i.e., low cholesterol, antioxidant, anti-ageing, anticancer, etc.) that have rendered products quite appealing (Arvanitoyannis et al. 2005). Despite the growing interest to issues related to health and diet, a decline in sales of fruit and vegetables has been recorded in recent years. In this regard, to meet consumer demands an increasing number of foods enriched or fortified with plant products have been introduced; this may be an alternative route to the consumption of minor plant components that may have health benefits (Schieber et al. 2001b, Mildner-Szkudlarz and Bajerska 2016). Moreover, the use of functional ingredients arising from natural sources is preferable compared with the use of synthetic additives, which are less appreciated by consumers. According to nutritionists, at least five servings a day should be consumed, paying particular attention to the choice of green and yellow vegetables and citrus fruits (Heimendinger and Chapelsky 1996). Indeed, several clinical studies indicate that consumption of adequate amounts of fruits and vegetables provides beneficial effects on the organism by exerting a protective role against cardiovascular disease, cancer, respiratory (asthma and bronchitis), cataracts and constipation (Strinmetz and Potter 1996, Van poppel and Van den Berg 1997). The beneficial effects arising from the consumption of foods of plant origin are mainly attributed to the presence of organic micro-nutrients such as carotenoids, polyphenolics, tocopherols, vitamins C, minerals, organic acids and others (Schieber et al. 2001a). Being that bread is a food consumed every day, it can be considered as a carrier of nutraceutical substances that have a beneficial role on consumer health (Almana and Mahmoud 1994). The incorporation of vegetables flours as ingredients to the bread may affect the processing properties and the quality of the final product, so the use of structuring agents may be considered a valuable aid to improve bread quality (Guarda et al. 2004).

From the screening of vegetable flours carried out by Mastromatteo et al. (2012) it was found that durum wheat bread with 10% yellow pepper flour had high sensory quality in terms of taste and appearance. On the contrary, durum wheat bread with the highest amount of yellow pepper flour recorded a low overall quality score with respect to the control bread (100% durum wheat). In fact, the bread had an unpleasant taste and a low loaf volume with a compact crumb characterised by the presence of few small bubbles that caused product unacceptability. Others vegetables that can be used in the breadmaking are potatoes (*Solanum tuberosum* L.). They contain good quality edible grade protein, dietary fiber, several minerals and trace elements, essential vitamins and little or negligible fat (Misra and Kulshrestha 2003). Potato flour has become the most viable value-added product due to the versatility as a thickener and color or flavor improver. In bread making, potato can be blended with wheat flour in different ways, currently as starch, thermo-treated flour and native flour. In general, the suitable substitution level of wheat flour with potato flour is lower than 10%, because higher levels produced unacceptable bread in terms of loaf volume, flavor, and texture (Greene and Bovell Benjamin 2004).

The addition of dietary fibre to bakery products increases dietary fibre intake, prolongs freshness due to its capacity to retain water and thus reduces economic losses (Elleuch et al. 2011, Kohajdová et al. 2011). Traditionally, the fibre components used as functional ingredients are obtained from the cereal industry (Fuentes-Alventosa et al. 2009, González-Centeno et al. 2010, Górecka et al. 2009). However, by-products derived from various vegetables as potatoes, carrots (Chantaro et al. 2008, Shyamala and Jamana 2010), red beets (Shyamala and Jamana 2010), onions (Benítez et al. 2011) and tomatos (Navarro-González et al. 2011) are also good sources of dietary fibres. In general, the residues derived from vegetable processing contain a higher soluble dietary fibres content, present a better insoluble/soluble dietary fibres ratio, and also have better functional properties than those obtained from cereal processing (González-Centeno et al. 2010). Moreover, these residues are inexpensive and available in large quantities (Chantaro et al. 2008, Shyamala and Jamuna 2010), exhibit a lower caloric content, and, often, include other interesting compounds such as antioxidants (Chantaro et al. 2008, González-Centeno et al. 2010) which might provide additional health benefits (González-Centeno et al. 2010). Initial studies on the introduction of potato peels to bread date back to the 1970s and 1980s when concerns about low levels of dietary fibre in the American diet and related health risk emerged. The addition of peels obtained from various processes (abrasion, steam, caustic, and manual peeling) to wheat flours at levels of 5, 10 and 15% resulted in breads of acceptable sensory quality (Toma et al. 1979). It appears that potato peel can be included in bread for the purpose of increasing dietary fiber without unduly sacrificing quality and acceptability. However, a musty aroma in the breads containing peel possibly influenced their flavor and overall rating scores, roughly in proportion to the amount of peel they contained (Orr et al. 1982). Carlson et al. (1981) reported that tomato seeds, an abundant waste of tomato processing, contain high amounts of crude fat and protein. He also showed that tomato seed protein is especially high in lysine, the limiting amino acid of cereal products, and that the supplementation of wheat flour bread at the 10 and 20% levels increased lysine content by 40.2 and 69%, respectively. Canella et al. (1979) reported that tomato seeds, which make up 55% of canning waste, contain 28.4–31.0% protein and 36.0–37.9% fat. Yaseen et al. (1991) reported that wheat breads baked with the addition of 1 or 2% tomato seed flour to the wheat flour were equal in quality to the control wheat bread, but greater additions resulted in crumb darkening. Brodowski et al. (1980) reported that tomato seed protein contains approximately 13% more lysine than soy protein, which would allow it to be used in fortifying foods low in lysine content. Yaseen et al. (1991) studied the effects of adding tomato seed meal to wheat flour on the gas production of dough and the organoleptic characteristics of bread. The overall quality of bread with tomato seed flour is correlated with the amount of addition. The 5 and 10% recipes received a fancy grade in quality measurements while the 15% recipe received an extra standard grade in quality. The crust and crumb of the breads fortified with 5% tomato seed flour were golden; at higher levels of supplementation they were darker. Mokhnacheva et al. (1975) reported that wheat breads baked with 1 or 2% tomato seed flour added to wheat flour were equal in quality to control wheat bread, but greater additions resulted in crumb darkening. Taste score decreased as the level of tomato seed flour increased. A slightly bitter taste at 10% or greater replacement level may be due to a steroid compound found in crushed tomato seed (Zagibalov et al. 1985).

Also carrot pomace powder can be considered as suitable ingredient for wheat bread supplementation due to high content of fibres and low energetic value (Kohajdova et al. 2012). The incorporation of carrot pomace powder significantly influences taste and odour of wheat bread. It has been primarily attributed to terpenoids and sugars, which are mainly responsible for the carrot flavour of carrot pomace powder (Jones 2009). The results of the study by Kohajdova et al. (2012) show that blend flours which contained up to 3% of carrot pomace powder could be incorporated in the formulation to produce wheat bread with acceptable quality. Furthermore, it was found that addition of higher levels (5 and 10%) of carrot pomace powder to wheat flour negatively affected rheological parameters of wheat dough and qualitative and sensory properties of wheat bread. In particular, overall acceptability of wheat bread with higher content of carrot pomace powder (5 and 10%) was markedly decreased because it negatively affected taste, odour, colour and hardness of final products (Kohajdova et al. 2012). The hardening effect observed after addition of dietary fibre results from the dilution of gluten content (Sivam et al. 2010) and also due to the thickening of the walls surrounding the air bubbles in the crumb (Gómez et al. 2003). Increasing of carrot pomace powder level resulted in appreciable darker crust colour of wheat bread. Incorporation of carrot pomace powder also allowed increasing the intensity of orange colour of crumb due to carotenoids (ß-carotene) (Jones 2009).

The enrichment of cereal-based products with seeds oil is also very interesting. As result various bread types enriched with combinations of whole oilseed are being readily accepted by consumers due to their high content of polyunsaturated fatty acids, vegetable protein, phosphorus, iron, magnesium, vitamin E, niacin, folate and phytoestrogens. For example, sunflower seeds contain around 20% protein, high levels of potassium (710 mg/100 g) and magnesium (390 mg/100 g) and are especially rich in polyunsaturated fatty acids (approximately 31.0%) in comparison with other oilseeds (Food Standards Agency Institute of Food Research 2002). Mainly, due to its superior nutritional quality and relatively low amount of anti-nutritional factors, sunflower seed has great potential to be, not only a protein source, but also a valuable supplement in human diet, providing considerable amounts of antioxidants, minerals and unsaturated fatty acids. It was reported that sensorially acceptable and nutritionally improved bread can be made with the addition of as much as 16% of sunflower seed (Škrbić and Filipčev 2008). Also, Nadeem et al. (2010) observed that the scores for overall crust and crumb color, flavor and taste of breads increased with dehulled seeds supplementation. Specifically, the breads prepared with dehulled seeds of up to 14% level of supplementation in wheat flour were found to be acceptable with respect to all sensory attributes. The incorporation of the oil seeds caused decrease in bread firmness and reduction in volume due to the decrease in gluten proteins and increase in dietary fiber contents (Nadeem et al. 2010). On the other hand, the addition of the sunflower seeds flour significantly improves the chemical composition (ash, fat, fiber and protein) of samples bread as compared to the control (100% wheat flour).

Pumpkin seeds could also be successfully utilized as good source of edible protein (320 g/kg) and oil (450 g/kg). Moreover, they exhibit unique functional properties (high water and fat absorption, as well as good emulsification properties) and high lysine content, which suggest the ability to be incorporated in bakery products. El-Soukkary (2001) did not observe significant differences between the control bread (wheat flour) and the bread fortified with pumpkin seeds, as regard to crust and crumb color, flavor, odor and overall acceptability. Addition of pumpkin seed to wheat flour improved the nutritional quality of bread. In particular, the ash contents and also the macro- and micro-mineral contents of fortified breads, especially with Ca, K, P, Mg, Cu, Fe and Mn were increased compared with the control and to other oilseeds (peanuts, sunflowers and rapeseed meal). Makinde and Akinoso (2014) evaluated the effect of sesame flour supplementation on physical, nutritional and sensory quality of bread. In particular, the addition of black sesame seeds flour to wheat flour with different incorporation levels (5–20%) in bread blend led to significant increase in nutritive value but with inferior quality with respect to crumb, crust color and bread volume with respect to the 100% wheat bread (Makinde et al. 2013). It was concluded that a substitution of 5% sesame flour into wheat flour gave the bread with the best overall quality acceptability.

Recently, mushrooms mycelia have become attractive as functional foods and as source of bioactive components as lovastatin, γ-aminobutyric acid and ergothioneine (Wasser and Weis 1999, Tseng et al. 2008). However, mycelia are not used directly as food but could be used as food flavouring materials and food ingredients. Therefore, the incorporation of mycelium into the formulation of different conventional foods to provide beneficial effects is of great interest. The addition of the mushrooms mycelia at 5% negatively influenced the loaf volume and colour. In fact, the wheat bread enriched with mushrooms mycelia was smaller in loaf volume and moderately coloured (browned) with respect to the bread control. The colour change of mycelium-supplemented bread might be related to the original mycelial pigments and the oxidation of phenolic compounds of mycelia during baking. Besides, the Maillard reaction might occur during baking since mycelium contained more soluble sugars and free amino acids than wheat flour (Ulziijargal 2009, Chen 2009). Overall, mushroom mycelium could be incorporated into bread to provide beneficial health effects. However, the bread formulation could be modified and the process of bread-making could be manipulated to improve consumer acceptability of mycelium-supplemented bread and to maintain contents of the functional components after baking.

Fruit flours have also good prospect as raw materials in baked product manufacture. The introduction of fruit flours into wheat flour for bread making may lead to some changes in chemical composition, modifications of dough rheology, physical and sensory property of bread (Olaoye et al. 2006).

Date pits could be regarded as an excellent source of food ingredients with interesting technological functionality that could also be used in food as an important source of dietary fiber (Besbes et al. 2009, Bouaziz et al. 2010). The bread fortified with date pits powder has improving effects on the quality and nutritive values of pan bread, as well as it has a hypoglycemic effect and could have a protective effect against diabetes complications as well as improvement of lipid profile. However, Najafi (2011) showed that bread containing 10% date seed had higher dietary fiber content and similar sensory properties to the wheat bran control, but lower color, flavor, odor, and overall acceptability sensory scores. Replacing up to 10% and 15% of date pits with wheat flour reduced bread volume, as was also confirmed by Kawka et al. (1999) and Reda (2006). The crust and crumb colors were significantly different in bread containing 5% and 10% compared to the control bread, while, in pan bread containing 15% date pits there was a significant increase than that of the control (El-Porai et al. 2013).

Mase et al. (2013) examined the processing of bread in which a proportion of the wheat flour (25%) was replaced with unripe banana or apple flour. They observed that bread made with these fruits were lower in specific loaf volume than the conventional wheat bread. The addition of fruit increased the dietary fiber content and the antioxidative activity. Unripe plantain has also been found to have medium glycemic index (Oboh and Erema 2010). Thus, diets based on unripe plantain have been recommended for diabetics. On the contrary, the sensory evaluation of unripe fruits bread by the panel was not favorable in terms of color, flavor and taste (Mase et al. 2013). Specifically, the crumb and crust color of the breads with unripe banana flour was browner than the control bread. The apple flour bread had a green color and a sour taste. The green color originates from the green color of peel and sour taste is from the organic acid in the flesh. If unripe fruit flours are used in bread-making, further additives are necessary to improve the bread palatability.

Chestnut flour may also be used in gluten-free flour breads due to its nutritional and health benefits. Chestnut flour contains high quality proteins with essential amino acids (4–7%), relatively high amount of sugar (20–32%), starch (50–60%), dietary fiber (4–10%), and low amount of fat (2–4%). It also contains vitamin E, vitamin B group, potassium, phosphorous, and magnesium (Sacchetti et al. 2004, Chenlo et al. 2007). Although chestnut flour has good nutritional quality and aroma, it may yield inferior quality of baked products with low volume and unacceptable dark color. These defects may be caused by the inadequate starch gelatinization, high amount of sugar and fiber. Demirkesen et al. (2010) found that when bread was prepared using only chestnut flour, the hardest structure with the lowest volume was observed because of the rigid and compact structure of the fibrous chestnut flour dough. The increase in the chestnut flour content decreased the loaf volume but increased the hardness of breads. Relatively high sugar content of chestnut flour may also hinder or reduce the starch gelatinization during baking, leading to low specific volume and firm texture of bread. Sugars are known to delay starch gelatinization by reducing the water activity of the system and stabilizing the amorphous regions of the starch granule by interacting with starch chains (Sumnu et al. 2000). Thus, breads cannot entrap the gas bubbles leading to lower volume and harder structure. Breads prepared by using only chestnut flour had lower flavor score as compared to breads made with chestnut/rice flour ratio 30/70 (Demirkesen et al. 2010). This may be due to the off-flavor formation as a result of Maillard reactions. In addition, increasing sugar content triggered Maillard and caramelization reactions resulting in undesirable dark color.

Obiegbuna et al. (2013) reported that the use of date palm fruit pulp meal as sugary agent to replace granulated sugar in bread production improved the nutritional value of bread. The overall organoleptic quality was not affected even at 100% replacement. In particular, the preference for crumb color increased with increasing replacement of granulated sugar with date palm fruit pulp meal.

Peng et al. (2010) investigated the impact of grape seed extract on bread quality. Specifically, the incorporation of grape seed extract caused favourable change in colour of bread without causing significant changes in other sensory properties as compared to the sample control (100% wheat flour). Moreover, the sample bread enriched with grape seed extract showed an increase in total antioxidant content with respect to the bread control. It is an abundant source of catechins and proanthocyanidins with a strong antioxidant and free radical scavenging activity (Liang et al. 2004, Wu et al. 2005). In addition, it shows other biological effects such as the inhibition of platelet aggregation, anti-inflammation and anti-ulcer activity (Saito et al. 1998, Vitseva et al. 2005).

4. Role of Hydrocolloids in Bread Enriched With Non-Conventional Ingredients

As reported above, the substitution of wheat with other flours progressively reduce bread quality. This has been attributed to reduced flour strength and gas retention capacity due to lessening gluten content, thereby reducing bread volume and sensory appeal of most baked composite bread. To improve the quality of wheatless bread, gluten may be to some extent replaced by natural or synthetic raw materials, which can significantly swell in water and form the structural equivalent of the gluten network in wheat dough. Food hydrocolloids (or gums) are high-molecular weight hydrophilic biopolymers used as functional ingredients in the food industry (Collar 2016, Collar et al. 1998, 1999). The term hydrocolloids embraces all polysaccharides extracted from plants, seaweed and microbial sources, as well as gums derived from plant exudates, and modified biopolymers prepared by chemical treatment of cellulose (Dickinson 2003). Hydrocolloids and their mixtures impact rheology of dough, as well as its baking properties and final bread texture. Gluten-free bread based on starch is less tasty than traditional bread and has a high staling tendency. The crumb which after baking is wet and sticking together, on the next day becomes dry, rough and crumbly. Because homemade gluten-free bread is prepared for several days, it is very important to prevent sufficient organoleptic quality during storage (Ylimaki et al. 1991, Malcolmson et al. 1993, Gambuś et al. 2001, Sanchez et al. 2002). In particular, the hydrocolloids are added to bakery products to control water absorption and consequently dough rheology (Mandala et al. 2007), improving their shelf life by keeping the moisture content constant and retarding the staling (Davidou et al. 1996, Rojas et al. 1999, Collar et al. 1999). The specific action of these polymers is expressed particularly in improving the structure and viscoelastic properties of dough, slow down the retrogradation of amylose, facilitate the absorption and retention of water, extend shelf life and give softness, smoothness and stability over time. The reason for this is the water retaining, leading to higher moisture content in the final baked product and consequently, retrogradation of starch and bread firming is reduced (Sharadanant and Khan 2003). Many parameters of gluten-free bread depend on the amount and type of non-starch hydrocolloids used as gluten replacers, as this determines interactions between them and starch, which is the main component of dough. There are reports on the interactions between starch and other polysaccharide hydrocolloids, such as pectin, guar gum and xanthan gum (Funami et al. 2005, Eidam et al. 1995). In some studies dealing with the quality of gluten-free bread, the synergistic action of guar gum and pectin in the mixture with corn starch was reported (Gambuś et al. 2001). Various hydrocolloids such as hydroxy-propyl-methyl-cellulose, methyl-cellulose, carboxymethyl-cellulose, locust bean gum, guar gum, xanthan, pectin, and β-glucan have been studied for the quality improvement gluten-free bread (Lazaridou et al. 2007).

The cellulose derivatives (carboxymethyl-cellulose and hydroxy-propyl-methyl-cellulose) are obtained by chemical modification of cellulose, which ensures their uniform properties, in opposition to the hydrocolloids from natural sources that have a high variability (Guarda et al. 2004). Moreover, despite the presence of hydrophobic groups in the chain, this polymer partially maintains the hydrophilic properties of cellulose (Sarkar and Walker 1995, Bárcenas and Rosell 2005). Their use in bread-making improves the quality (loaf volume, moisture content, crumb texture, and sensorial properties) and retard the crumb hardening and amylopectin retrogradation (Armero and Collar 1996, Rosell et al. 2001, Guarda et al. 2004, Bárcenas and Rosell 2005). This network strengthens the gas cells of dough (in the initial stages of baking) which expand during baking and consequently the gas losses are reduced and bread volume improved (Dziezak 1991, Haque et al. 1993, Sarkar and Walker 1995, Bárcenas and Rosell 2005). McCarthy et al. (2005) also reported a slight decrease in gluten-free bread volume when increasing the addition level of hydrocolloid. However, Gujral and Rosell (2004) obtained higher volume when the addition level of hydroxy-propyl-methyl-cellulose to rice flour increased by 4%. The hydrocolloid has the ability to change from solution to gel during heating, thus forming a thermostable network that shields the bread dough from volume and moisture content losses during baking (Bell 1990, Dziezak 1991). In addition, when the temperature rises during baking, the hydroxy-propyl-methyl-cellulose forms gels through the interaction of the hydrocolloid chains, thus obtaining a temporary network (Haque et al. 1993, Sarkar and Walker 1995) which strengthens the dough expansion and protects the dough against volume loss. This gel also acts as a barrier against the loss of moisture but it disappears at cooling; therefore, it provides better texture and softness without conferring any adverse effect on the palatability of final product (Bell 1990, Bárcenas et al. 2004). The hydroxy-propyl-methyl-cellulose network formed during baking could act as a barrier for gas diffusion, decreasing the water vapor losses (Bell 1990, Dziezak 1991) and increasing the final moisture content of the loaf (Bárcenas and Rosell 2005). The improving effect of this hydrocolloid on the sensory quality of bread could be explained by its influence on the crumb texture yielding softer crumbs (Bárcenas and Rosell 2005). Crumb hardening is a complex process caused by simultaneous phenomena (Gray and BeMiller 2003). One of the most important phenomenon is amylopectin retrogradation; however, amylose recrystallization (Hug-Iten et al. 1999), moisture content loss (He and Hoseney 1990), interactions between starch and gluten (Martin et al. 1991) and moisture redistribution also contribute to bread hardening (Bárcenas and Rosell 2007). Microstructure analysis revealed a possible interaction between hydroxy-propyl-methyl-cellulose and other bread constituents (Bárcenas and Rosell 2004, 2005), as well as the ability of this hydrocolloid to interact with the effective water present in the system (Sarkar and Walker 1995, Schiraldi et al. 1996).

Carboxymethylcellulose (CMC) is a cellulose derivative with carboxymethyl groups (-CH2-COOH) bound to some of the hydroxyl groups present in the glucopyranose monomers that form the cellulose backbone. CMC is used in baked goods mostly to retain moisture, improve the body or mouth-feel of the products (Chinachoti 1995), control rheological properties of dough, improve the volume and structural uniformity of baked products, or increase the shelf life of cereal products (Dziezak 1991, Chinachoti 1995, Gimeno et al. 2004). In addition, CMC improves the volume yield of certain dough as a result of the viscosity drop during baking. CMC is also used in combination with other stabilizers, such as pectin or locust bean gum. Eduardo et al. (2014) found that the addition of CMC to composite cassava-maize-wheat bread soave improved bread quality parameters such as specific volume, crust colour, and crumb texture. Similar effects on specific bread volume have been reported with additions of xanthan gum, and α-carrageenan to wheat bread (Rossell et al. 2001), hydroxy-propyl-methyl-cellulose to gluten-free maize-teff bread (Hager et al. 2013), and pectin to gluten-free formulations (Lazaridou et al. 2007). These findings might be a result of the formation of a gel network during oven heating that strengthens the expanding cells of the dough and, as a result, improves gas retention and bread volume (Bell 1990).

Guar gum (GG), which is a galactomannan (Slavin and Greenberg 2003) derived from the seed of a leguminous plant (*Cyamopsis tetragonolobus*), has been widely used as a food additive due to the high viscosity of its aqueous solutions, even at low concentrations. In baked goods, GG is used to improve the mouth feel and changing the rheological properties (Turabi et al. 2008). Results obtained by Mastromatteo et al. (2012) showed that proper amounts of GG improved durum wheat bread enriched with yellow pepper flour.

Xanthan gum (XG) is a branched, anionic polysaccharide produced from the aerobic fermentation of the bacterium *Xanthomonas campestris*. It has been reported to improve dough handling properties, loaf specific volume and crumb softness when incorporated into composite cassava-wheat bread formulations (Shittu et al. 2009). Again, the breads prepared with chestnut/rice flour containing XG blend had higher quality in terms of hardness, specific volume, color and sensory values, as compared to sample without hydrocolloids. Loaf volume increased up to a certain XG supplementation level, but further increase in polymer concentration resulted in volume decrease (Lazaridou et al. 2007). It seems that with XG addition at high concentrations, batter exhibits too high resistance and consistency, which cause a limited gas cell expansion during proofing (Lazaridou et al. 2007). Similarly, Haque and Morris (1994) observed no influence of xanthan incorporation into a rice flour bread and Schober et al. (2005) found a decrease in loaf volume of gluten-free breads from sorghum with increasing xanthan gum levels. Moreover, the introduction of xanthan gum into the mixture of hydrocolloids used as a gluten replacement significantly influenced hardness of gluten-free bread and to the lesser extent their cohesiveness, due to the different interactions between starch and hydrocolloids (Schober et al. 2005). It is of interest to observe that the addition of XG increased batter consistency and improved gluten-free bread quality (100% rice flour), which led to bread with high volume, increased cell average size and lower crumb firmness and staling rate over storage; similarly, it improved bread's overall appearance with respect to the other hydrocolloids (Sciarini et al. 2010).

Locust bean gum (LBG) is a natural hydrocolloid extracted from the seeds of the carob tree (*Ceratonia siliqua* L.) after the removal of testa (seed coat) (Gonçalves and Romano 2005, Bonaduce et al. 2007). Sharadanant and Khan (2003) investigated the effects of various levels of hydrophilic gums such as CMC, arabic gum and LBG on the quality of bread. The best results were reported for LBG producing the highest loaf volume followed by arabic gum and CMC. External appearance of bread and its internal characteristics such as texture, grain, cell wall structure, color and softness were also improved (Selomulyo and Zhou 2007). Moreover, the addition of gums determines increasing of dietary fiber content and decreasing of caloric value by diluting the content especially through the water content increase (Ognean et al. 2007). From the caloric point of view, bread is responsible for a great proportion of energetic daily intake. To increase the dietary fiber levels in gluten-free formulations one of the mainly used hydrocolloids is the oat β-glucan isolate, which has been associated with the reduction of plasma cholesterol and better control of post-prandial serum glucose levels in humans and animals (Wood 2002). The inclusion of β-glucan in gluten-free bread is considered here for the first time, although some evidence has been reported for the behaviour of this hydrocolloid as improver in wheat flour by providing high viscosity to the system (Delcour et al. 1991). Wang et al. (1998) found that incorporation of β-glucan into wheat bread improved the crumb by stabilizing air cells in bread dough and preventing coalescence of the cells.

Keywords: Bread, pseudocereals and legumes, by products and vegetable flours, hydrocolloids

References

Almana, H. and Mahmoud, R. (1994). Palm date seeds as an alternative source of dietary fiber in Saudi bread. Ecology of Food and Nutrition 32: 261–270.

Alvarez-Jubete, L., Auty, M., Arendt, E. and Gallagher, E. (2010a). Baking properties and microstructure of pseudocereal flours in gluten-free formulations. European Food Research and Technology 21: 106–113.

Alvarez-Jubete, L., Wijngaard, H.H., Arendt, E.K. and Gallagher, E. (2010b). Polyphenol composition and *in-vitro* antioxidant activity of amaranth, quinoa and buckwheat as affected by sprouting and bread baking. Food Chemistry 119: 770–778.

Alvarez-Jubete, L., Holse, M., Hansen, A., Arendt, E.K. and Gallagher, E. (2009). Impact of baking on the vitamin E content of the pseudocereals amaranth, quinoa and buckwheat. Cereal Chemistry 86: 511–515.

Alvarez-Jubete, L., Arendt, E.K. and Gallagher, E. (2009). Nutritive value and chemical composition of pseudocereals as gluten-free ingredients. International Journal of Food Sciences and Nutrition 60: 240–257.

Angioloni, A. and Collar, C. (2011a). Polyphenol composition and *"in vitro"* antiradical activity of single and multigrain breads. Journal of Cereal Science 53: 90.96.

Angioloni, A. and Collar, C. (2011b). Nutritional and functional added value of oat, Kamut, spelt, rye and buckwheat versus common wheat in breadmaking. Journal of the Science of Food and Agriculture 91: 1283–1292.

Angioloni, A. and Collar, C. (2012). High legume-wheat matrices: an alternative to promote bread nutritional value meeting dough viscoelastic restrictions. European Food Research and Technology 234: 273–284.

Angioloni, A. and Collar, C. (2009). Small and large deformation viscoelastic behaviour of selected fibre blends with gelling properties. Food Hydrocolloids 23: 742–748.

Arendt, E.K., O'Brien, C.M., Schober, T., Gormley, T.R. and Gallagher, E. (2002). Development of gluten-free cereal products. Farm and Food 12: 21–27.

Armero, E. and Collar, C. (1996). Antistaling additive effect on fresh wheat bread quality. Food Science and Technology International 2: 323–333.

Arvanitoyannis, I.S. and Van Houwelingen-Koukaliaroglou, M. (2005). Functional foods: a survey of health claims, pros and cons and current legislation. Critical Reviews in Food Science and Nutrition 445: 385–404.

Aubrecht, E. and Biacs, P.A. (2001). Characterization of buckwheat grain proteins and its products. Acta Alimentaria 28: 261–268.

Bárcenas, M.E. and Rosell, C.M. (2007). Different approaches for increasing the shelf life of partially baked bread: Low temperatures and hydrocolloid addition. Food Chemistry 100: 1594–1601.

Bárcenas, M.E. and Rosell, C.M. (2005). Effect of HPMC addition on the microstructure, quality and aging of wheat bread. Food Hydrocolloids 19: 1037–1043.

Bárcenas, M.E., Benedito, C. and Rosell, M.C. (2004). Use of hydrocolloids as bread improvers in interrupted baking process with frozen storage. Food Hydrocolloids 18: 769–774.

Belitz, H.D., Grosch, W. and Schieberle, P. (2004). Food Chemistry. Springer, Berlin p. 627.

Bell, D.A. (1990). Methylcellulose as a structure enhancer in bread baking. Cereal Foods World 35: 1001–1006.

Benítez, V., Mollá, E., Marín-Cabrejas, M.A., Aguilera, Y., Lopéz-Andreéu, F.J. and Esteban, R.M. (2011). Effect of sterilisation on dietary fibre and physicochemical properties of onion by-products. Food Chemistry 127: 501–507.

Berghofer, E. and Schoenlechner, R. (2002). Grain amaranth. pp. 219–260. *In*: P.S. Belton and J.R.N. Taylor (eds.). Pseudocereals and Less Common Cereals: Grain Properties and Utilization Potential. Springer, Berlin.

Besbes, S., Drira, L., Blecker, C., Deroanne, C. and Attia, H. (2009). Adding value to hard date (*Phoenix dactylifera* L.): Compositional, functional and sensory characteristics of date jam. Food Chemistry 112: 406–411.

Bonaduce, I., Brecoulaki, Perla Colimbini H., Lluveras, A., Restivo, V. and Ribechini, E. (2007). Gas Chromatographic-Mass Spectrometric characterization of plant gums in samples from painted works of art. Journal of Chromatography 1175: 275–282.

Bouaziz, M., Amara, W., Attia, H., Blecker, C. and Besbes, S. (2010). Effect of the addition of defatted date seeds on wheat dough performance and bread quality. Journal of Texture Study 41: 511–531.

Brodowski, D. and Geisman, J.R. (1980). Protein content and amino acid composition of protein of seeds from tomatoes at various stages of ripeness. Journal of Food Science 45: 228–229, 235.

Canella, M., Cardinali, F., Castriotta, G. and Nappucci, R. (1979). Chemical properties of seeds of different tomato varieties. Riv. Ital. Sostanze Grasse 56: 8–11.

Carlson, B.L., Knorr, D. and Watkins, T.R. (1981). Influence of tomato seed addition on the quality of wheat flour breads. Journal of Food Science 46: 1029–1031, 1042.

Chantaro, P., Devahastin, S. and Chiewchan, N. (2008). Production of antioxidant high dietary fibre powder from carrot peels. LWT—Food Science and Technology 41: 1987–1994.

Chen, C.P. (2009). Quality evaluation of Antrodia mycelium bread and Phellinus mycelium bread. Master's Thesis. National Chung Hsing University, Taichung, Taiwan.

Chenlo, F., Moreira, R., Pereira, G. and Silva, C.C. (2007). Evaluation of the rheological behaviour of chestnut (castanea sativa mill) flour pastes as function of water content and temperature. Electronic Journal of Environmental, Agricultural and Food Chemistry 6: 1794–1802.

Chinachoti, P. (1995). Carbohydrates: functionality in foods. American Journal of Clinical Nutrition 61: 922–929.

Christa, K. and Sorel-Smietana, M. (2008). Buckwheat grains and buckwheat products-nutritional and prophylactic value of their components a review. Czech Journal of Food Science 26: 153–162.

Collar, C. (2016). Role of bread on nutrition and health worldwide. pp. 26–52. *In*: C.M. Rosell, J. Bajerska and A.F. El Sheikha (eds.). Bread and its Fortification, CRC Press, Boca Raton.

Collar, C., Martínez, J.C. and Rosell, C.M. (2001). Lipid binding of fresh and stored formulated wheat breads. Relationships with dough and bread technological performance. Food Science and Technology International 7: 501–510.

Collar, C., Andreu, P., Mart´ınez, J.C. and Armero, E. (1999). Optimization of hydrocolloid addition to improve wheat bread dough functionality: a response surface methodology study. Food Hydrocolloids 13: 467–475.

Collar, C., Armero, E. and Martínez, J. (1998). Lipid binding of formula bread doughs: relationships with dough and bread technological performance. Zeitschrift für Lebensmitteluntersuchung und-Forschung A 207: 110–121.

Davidou, S., Le Meste, M., Debever, E. and Bekaert, D. (1996). A contribution to the study of staling of white bread: effect of water and hydrocolloid. Food Hydrocolloids 10: 375–383.

Delcour, J.A., Vanhamel, S. and Hoseney, R.C. (1991). Physicochemical and functional properties of rye nonstarch polysaccharides. II. Impact of a fraction containing water-soluble pentosans and proteins on gluten–starch loaf volumes. Cereal Chemistry 68: 72–76.

Demin, M.A., Vuceli Radovi, V., Banjac, N.R., Tipsina, N.N. and Milovanovi, M.M. (2013). Buckwheat and Quinoa seeds as supplements in wheat bread Production. Hemijska Industrija 67: 115–121.

Demirkesen, I., Mert, B., Sumnu, G. and Sahin, S. (2010). Rheological properties of glutenfree bread formulations. Journal of Food Engineering 96: 295–303.

Dewettinck, K., Van Bockstaele, F., Kuhne, B., Van de Walle, Courtens, T. and Gellynck, X. (2008). Nutritional value of bread: Influence of processing, food interaction and consumer perception. Review Journal of Cereal Science 48: 243–257.

Dickinson, E. (2003). Hydrocolloids at interfaces and the influence on the properties of dispersed systems. Food Hydrocolloids 17: 25–39.

Drzewiecki, J., Delgado-Licon, E., Haruenkit, R., Pawelzik, E., Martin-Belloso, O. and Park, Y.S. (2003). Identification and differences of total proteins and their soluble fractions in some pseudocereals based on electrophoretic patterns. Journal of Agricultural and Food Chemistry 51: 7798–7804.

Dziezak, J.D. (1991). A focus on gums. Food Technology 45: 115.

Eduardo, M., Svanberg, U.L.F. and Aherné. (2014). Effect of hydrocolloids and emulsifiers on baking quality of composite cassava-maize-wheat breads. International Journal of Food Science 2014: 1–9.

Eidam, D., Kulicke, W., Kuhn, K. and Stute, R. (1995). Formation of starch gels selectively regulated by the addition of hydrocolloids. Starch 47: 378–384.

El Soukkary, F.A.H. (2001). Evaluation of pumpkin seed products for bread fortification. Plant Foods for Human and Nutrition 56: 365–384.

El-Porai, E., Salama, A., Sharaf, A., Hegazy, A. and Gadallah, M. (2013). Effect of different milling processes on Egyptian wheat flour properties and pan bread quality. Annals of Agricultural Science 58: 51–59.

El-Soukkary, F.A.H. (2001). Evaluation of pumpkin seed products for bread Fortification. Plant Foods for Human Nutrition 56: 365–384.

Elleuch, M., Bedigian, D., Roiseux, O., Besbes, S., Blecker, C. and Attia, H. (2011). Dietary fibre and fibre-rich by-products of food processing: Characterisation, technological functionality and commercial applications. Review Food Chemistry 124: 411–421.

Food Standards Agency and Institute of Food Research (2002). McCance and Widdowson's. The Composition of Foods. (6th. Summary Edition). Royal Society of Chemistry, Cambridge.

Fuentes-Alventosa, J.M., Rodriguez-Gutierrez, G., Jaramillo Carmona, S., Espejo Calvo, J.A., Rodriguez-Arcos, R., Fernandez-Bolanos, J., Guillen-Bejarano, R. and Jimenez-Araujo, A. (2009). Effect of extraction method on chemical composition and functional characteristics of high dietary fibre powders obtained from asparagus by-products. Journal of Food Chemistry 113: 665–692.

Funami, T., Kataoka, Y., Omoto, T., Goto, Y., Asai, I. and Nishinari, K. 2005. Effect of non-ionic polysaccharides on the gelatinization and retrogradation behavior of wheat starch. Food Hydrocolloids 19: 1–13.

Gallagher, E., Gormley, T.R. and Arendt, E.K. (2004). Recent advances in the formulation of gluten-free cereal-based products. Trends Food Science and Technology 15: 143–152.

Gallagher, E., Kunkel, A., Gormley, T.R. and Arendt, E.K. (2003a). The effect of dairy and rice powder addition on loaf and crumb characteristics, and on shelf life (intermediate and long-term) of gluten-free breads stored in a modified atmosphere. European Food Research and Technology 218: 44–48.

Gambus, H., Gambus, F. and Sabat, R. (2002). Quality improvement of gluten-free bread by Amaranthus flour. Zywnosc 9: 99–112.

Gambuś, H., Nowotna, A., Ziobro, R., Gumul, D. and Sikora, M. (2001). The effect of use of guar gum with pectin mixture in gluten-free bread. EJPAU 4, http://www.ejpau.media.pl/series/volume4/issue2/food/art-09.html.

Gimeno, E., Morau, C.I. and Kokini, J.L. (2004). Effect of xanthan gum and CMC on the structure and texture of corn flour pellets expanded by microwave heating. Cereal Chemistry 8: 100–107.

Goesaert, H., Brijs, K., Veraverbeke, W.S., Courtin, C.M., Gebruers, K. and Delcour, J.A. (2005). Wheat flour constituents: how they impact bread quality, and how to impact their functionality. Trends Food Science and Technology 16: 12–30.

Gómez, M., Oliete, B., Rosell, C.M., Pando, V. and Fernández, E. (2008). Studies on cake quality made of wheat-chickpea flour blends. LWT Food Science and Technology 41: 1701–1709.

Gómez, M., Ronda, F., Blanco, C.A., Caballero, P.A. and Aspeteguía, A. (2003). Effect of dietary fibre on dough rheology and bread quality. European Food Research and Technology 216: 51–56.

Gonçalve, S. and Romano, A. (2005). Locust bean gum (LBG) as a gelling agent for tissue culture media. Scientia Horticulturae 106: 129–134.

González-Centeno, M.R., Rosselló, C., Simal, S., Garau, M.C., López, F. and Femenia, A. (2010). Physico-chemical properties of cell wall materials obtained from ten grape varieties and their byproducts: grape pomaces and stems. Food Science and Technology LEB 43: 1580–1586.

Górecka, A., Bakuniak, M., Chruszczewski, M.H. and Jezierski, T.A. (2007). A note on the habituation to novelty in horses: handler effect. Animal Science Papers and Reports 25: 143–152.

Górecka, D., Hęś, M., Szymandera-Buszka, K. and Dziedzic, K. (2009). Contents of selected bioactive components in buckwheat groats. ACTA Scientiarum Polonorum—Food Science and Human Nutrition 8: 75–83.

Gorinstein, S., Lojek, A., Ciz, M., Pawelzik, E., Delgado-Licon, E. and Medina, O.J. (2008). Comparison of composition and antioxidant capacity of some cereals and pseudocereals. International Journal of Food Science and Technology 43: 629–637.

Gorinstein, S., Pawelzik, E., Delgado-Licon, E., Haruenkit, R., Weisz, M. and Trakhtenberg, S. (2002). Characterisation of pseudocereal and cereal proteins by protein and amino acid analyses. Journal of the Science of Food and Agriculture, vol. 82, pp. 886–891.

Gray, J.A. and BeMiller, J.N. (2003). Bread staling: molecular basis and control. Comprehensive Reviews in Food Science and Food Safety 2: 1–21.

Greene, J.L. and Bovell-Benjamin, A.C. 2004. Macroscopic and sensory evaluation of bread supplemented with sweet-potato flour. Journal of Food Science 69: 167–173.

Guarda, A., Rosell, C.M., Benedito, C. and Galottoc, M.J. (2004). Different hydrocolloids as bread improvers and antistaling agents. Food Hydrocolloids 18: 241–247.

Gujral, H.S. and Rosell, C.M. (2004). Improvement of the breadmaking quality of rice flour by glucose oxidase. Food Research International 37: 75–81.

Hager, A.S. and Arendt, E.K. (2013). Influence of hydroxypropylmethylcellulose (HPMC), xanthan gum and their combination on loaf specific volume, crumb hardness and crumb grain characteristics of gluten-free breads based on rice, maize, teff and buckwheat. Food Hydrocolloids 32: 195–203.

Hager, A.S., Wolter, A., Jacob, F., Zannini, E. and Arendt, E.K. (2012). Nutritional properties and ultra-structure of commercial gluten free flours from different botanical sources compared to wheat flours. Journal of Cereal Science 56: 239–247.

Haque, A. and Morris, E.R. (1994). Combined use of ispaghula and HPMC to replace or augment gluten in breadmaking. Food Research International 27: 379–393.

Haque, A., Richardson, R.K., Morris, E.R., Gidley, M.J. and Caswell, D.C. (1993). Thermogelation of methylcellulose. Part II: Effect of hydroxypropyl substituents. Carbohydrate Polymers 22: 175–186.

Hasler, C.M. (2002). The cardiovascular effects of soy products. Journal of Cardiovascular Nursing: 16: 50–63.

He, H. and Hoseney, R.C. (1990). Changes in bread firmness and moisture during long-term storage. Cereal Chemistry 67: 603–607.

Heimendinger, J. and Chapelsky, D. (1996). The national 5 a day for better health program. Advances in Experimental Medicine and Biology 401: 199–206.

Hoseney, R.C. (1998). Principles of Cereal Science and Technology. 2nd Ed. AACC International: St. Paul, MN.

Hug-Iten, S., Handschin, S., Conde-Petit, B. and Escher, F. (1999). Changes in starch microstructure on baking and staling of wheat bread. LWT - Food Science and Technology 32: 255–260.

Isserliyska, D., Karadjov, G. and Angelov, A. (2001). Mineral composition of Bulgarian wheat bread. European Food Research and Technology 213: 244–245.

Jayasena, V., Leung, P. and Nasar-Abbas, S.M. (2008). Lupins for health and wealth. pp. 473–477. *In:* J.A. Palta and J.B. Berger (eds.). Proceedings of the 12th International Lupin Conference.

Jones, M.G. (2009). Formation of vegetable flavour. *In:* Fruit and Vegetable Flavour – Recent Advances and Future Prospects. Ed. B. Brücker, W.S. Grant. Woodhead Publication, Sivam, A.S., Sun-Waterhouse, D., Quek, S.Y. and Perera, C.O. (2010). Properties of bread dough with added fibre polysaccharides and phenolic antioxidants: A review. Journal of Food Science 75: 163–174.

Kalinova, J. and Dadakova, E. (2009). Rutin and total quercetin content in amaranth (*Amaranthus* spp.). Plant Foods for Human Nutrition 64: 68–74.

Kawka, A., Gorecka, D. and Gasiorowski, H. (1999). The effects of commercial barley flakes on dough characteristic and bread composition. Journal Polish of Agriculture Vol. 2.

Kiskini, A., Argiri, K., Kalogeropoulos, M., Komaitis, M., Kostaropoulos, A. and Mandala, I. (2007). Sensory characteristics and iron dialyzability of gluten-free bread fortified with iron. Food Chemistry 102: 309–316.

Kohajdová, Z., Karovičová, J., Jurasová, M. and Kukurová, K. (2011). Application of citrus dietary fibre preparations in biscuit preparation. Journal of Food and Nutrition Research 50: 182–190.

Kohajdova, Z., Karovicova, J. and Jurasova, M. (2012). Influence of carrot pomace powder on the rheological characteristics of wheat flour dough and on wheat rolls quality. Acta Scientiarum Polonorum Technology Alimentaria 11: 381–387.

Kohajdova, Z. and Katovicòvà, J. (2009). Review-Application of hydrocolloids as baking improvers. Chemical Papers 63: 26–28.

Kupper, C. (2005). Dietary guidelines and implementation for celiac disease. Gastroenterol 128: 121–127.

Lazaridou, A., Duta, D., Papageorgiou, M., Belc, N. and Biliaderis, C.G. (2007). Effects of hydrocolloids on dough rheology and bread quality parameters in gluten-free formulations. Journal of Food Engineering 79: 1033–1047.

Li, S.Q. and Zhang, Q.H. (2001). Advances in the development of functional foods from buckwheat. Critical Review in Food Science and Nutrition 41: 451–464.

Liang, C.P., Wang, M., Simon, J.E. and Ho, C.T. (2004). Antioxidant activity of plant extracts on the inhibition of citral off-odor formation. Molecular Nutrition and Food Research 48: 308–317.

Lin, L., Liu, H., Yu, Y., Lin, S. and Mau, J. (2009). Quality and antioxidant property of buckwheat enhanced wheat bread. Food Chemistry 112: 987–991.

Lin, C.H., Chiu, Y.C., Cheng, C.M. and Hsieh, J.C. (2008). Brain maps of Iowa gambling task. BMC Neuroscience. doi:10.1186/1471-2202-9-72.

Makinde, F.M. and Akinoso, R. (2014). Physical, nutritional and sensory qualities of bread samples made with wheat and black sesame (*Sesamum indicum* Linn) flours. International Food Research Journal 21: 1635–1640.

Makinde, F.M. and Akinoso, R. (2013). Nutrient composition and effect of processing treatments on antinutritional factors of Nigerian sesame (*Sesamum indicum* Linn) cultivars. International Food Research Journal 20: 2293–2300.

Malcolmson, L.J., Matsuo, R.R. and Balshaw, R. (1993). Textural optimization of spaghetti using response surface methodology. Cereal Chemistry 70: 417–423.

Mandala, I., Karabela, D. and Kostaropoulos, A. (2007). Physical properties of breads containing hydrocolloids stored at low temperature. I. Effect of chilling. Food Hydrocolloids 21: 1397–1406.

Martirosyan, D.M., Miroshnichenko, L.A., Kulakova, S.N., Pogojeva, A.V. and Zoloedov, V.I. (2007). Amaranth oil application for coronary heart disease and hypertension. Lipids in Health and Disease 6: 1–12.

Mase, T., Sato, Y., Miyazaki, S., Kato, Y. and Isshiki, S. (2013). Quality evaluation of bread Containing unripe apple or banana flour. Journal of Sugiyama Jogakuen University 43: 47–52.

Mastromatteo, M., Danza, A., Guida, M. and Del Nobile, M.A. (2012). Formulation optimization of vegetable flour-loaded functional bread. Part II: effect of the flour hydration on the bread quality. International Journal Food Science and Technology 47: 2109–2116.

McCarthy, D.F., Gallagher, E., Gormley, T.R., Schober, T.J. and Arendt, E.K. (2005). Application of response surface methodology in the development of gluten-free bread. Cereal Chemistry 82: 609–615.

Mildner-Szkudlarz, S. and Bajerska, J. (2016). Phytochemicals as functional bread compounds: Physiological effects. pp. 79–101. *In*: C.M. Rosell, J. Bajerska and A.F. El Sheikha (eds.). Bread and its Fortification, CRC Press, Boca Raton.

Miller-Jones, J.M. (2009). Nutrition: more on the GI debate. Cereal Foods World 54: 138–140.

Misra, A. and Kulshrestha, K. (2003). Potato flour incorporation in biscuit manufacture. Plant Foods for Human Nutrition 58: 1–9.

Mokhnacheva, A.I., Ostrovskaya, L. and Rakhmankulov, R.G. (1975). Use of tomato seeds in breadmaking. Zerno Perera Batyrayushchayai Pishchevaya Promyshlnost 5: 44–49.

Moore, M.M., Schober, T.J., Dockery, P. and Arendt, E.K. (2004). Textural comparisons of gluten-free and wheat-based doughs, batters, and breads. Cereal Chemistry 81: 567–575.

Mlakar, S.G., Turinek, M., Tasner, L., Jakop, M., Bavec, M. and Bavec, F. (2008). Organically produced grain amaranth-wheat composite flours: II. Bread quality. pp. 172–176. *In*: Z.U. Hardi (ed.). Proceedings of the 4th International Congress on Flour-Bread '07. J J Strossmayer Univ Osijek-Croatia, Fac Food.

Murphy, P.A., Song, T., Buseman, G., Barua, K., Beecher, G.R. and Trainer, D. (1999). Isoflavones in retail and institutional soy foods. Journal of Agricultural and Food Chemistry 47: 697–704.

Nadeem, M., Anjum, F.M., Arshad, M.U. and Hussain, S. (2010). Chemical characteristics and antioxidant activity of different sunflower hybrids and their utilization in bread. African Journal of Food Science 4: 618–626.

Najafi, M. (2011). Date Seeds: A Novel and Inexpensive Source of Dietary Fiber. International Conference on Food Engineering and Biotechnology, Singapore, pp. 323–326.

Navarro-González, I., García-Valverde, V., García-Alonso, J. and Periago, M.J. (2011). Chemical profile, functional and antioxidant properties of tomato peel fiber. Food Research International 44: 1528–1535.

Ng, S.C., Anderson, A., Coker, J. and Ondrus, M. (2007). Characterization of lipid oxidation products in quinoa (*Chenopodium quinoa*). Food Chemistry 101: 185–192.

Obiegbuna, J.E., Akubor, P.I., Ishiwu, C.N. and Ndife, J. (2013). Effect of substituting sugar with date palm pulp meal on the physicochemical, organoleptic and storage properties of bread. African Journal of Food Science 7: 113–119.

Oboh, H.A. and Erema, V.G. 2010. Glycemic indices of processed unripe plantain (*Musa paradisiaca*) meals. African Journal of Food Science 4: 514–521.

Ognean, M., Ognean, C.F. and Darie, N. (2007). Technological aspects of addition of several types of hydrocolloids in bread. Acta Universitatis Cibiniensis Series E: Food Technology 11: 47–54.

Olaoye, O.A., Onilude, A.A. and Idowu, O.A. (2006). Quality characteristics of bread produced from composite flours of wheat, plantain and soybeans. African Journal of Biotechnology 5: 1102–1106.

Orr, P.H., Toma, R.B., Munson, S.T. and D'appolonia, B. (1982). Sensory evaluation of breads containing various levels of potato peel. American Journal of Potato Research 59: 605–611.

Park, S.H. and Morita, N. (2005). Dough and breadmaking properties of wheat flour substituted by 10% with germinated quinoa flour. Food Science and Technology International 11: 471–476.

Pasko, P., Sajewicz, M., Gorinstein, S. and Zachwieja, Z. (2008). Analysis of selected phenolic acids and flavonoids in *Amaranthus cruentus* and *Chenopodium quinoa* seeds and sprouts by HPLC. Acta Chromomatographica 20: 661–672.

Patterson, C.A., Maskus, H. and Bassett, C.M.C. (2010). Fortifying foods with pulses. Cereal Foods World 55: 56–62.

Peiretti, P.G., Gaia, F. and Tassone, S. (2013). Fatty acid profile and nutritive value of quinoa Chenopodium quinoa Willd. seeds and plants at different growth stages. Animal Feed Science and Technology 183: 56–61.

Peng, X., Maa, J., Cheng, K.W., Jiang, Y., Chen, F. and Wang, M. (2010). The effects of grape seed extract fortification on the antioxidant activity and quality attributes of bread. Food Chemistry 119: 49–53.

Raffo, A., Pasqualone, A., Sinesio, F., Paoletti, F., Quaglia, G. and Simeone, R. (2003). Influence of durum wheat cultivar on the sensory profile and staling rate of Altamura bread. European Food Research and Technology 218: 4955.

Reda, M.M. (2006). The effect of barley and barley fortified bread on diabetic rats. M.Sc. Thesis, of Home Economics. Helwan Univ. Egypt.

Renzetti, S., Dal Bello, F. and Arendt, E.K. (2008). Microstructure, fundamental rheology and baking characteristics of batters and breads from different gluten-free flours treated with a microbial transglutaminase. Journal of Cereral Science 48: 33–45.

Rodge, A.B., Ghatge, P.U., Wankhede, D.B. and Kokate, R.K. (2006). Isolation, purification & rheological study of guar genotypes RGC-1031 and RGC-1038. Journal of Arid Legumes 3: 41–43.

Rojas, J.A., Rosell, C.M. and Benedito de Barber, C. (1999). Pasting properties of different wheat flour-hydrocolloid systems. Food Hydrocolloids 13: 27–33.

Rosell, C.M., Rojas, J.A. and Benedito de Barber, C. (2001). Influence of hydrocolloids on dough rheology and bread quality. Food Hydrocolloids 15: 75–81.

Rosell, C.M., Bajerska, J. and El Sheikha, A.F. (eds.). (2015). Bread and its Fortification, CRC Press, Boca Raton, pp. 407.

Sabanis, D. and Tzia, C. (2011). Effect of hydrocolloids on selected properties of gluten-free dough and bread. Food Science and Technology International 17: 279–291.

Sacchetti, G., Pinnavaia, G.G., Guidolin, E. and Dalla-Rosa, M. (2004). Effects of extrusion temperature and feed composition on the functional, physical and sensory properties of chestnut and rice flour-based snack-like products. Food Research International 37: 527–534.

Saito, M., Hosoyama, H., Ariga, T., Kataoka, S. and Yamaji, N. (1998). Antiulcer activity of grape seed extract and procyanidins. Journal of Agricultural and Food Chemistry 46: 1460–1464.

Sanchez, H., Osella, C. and Torre, M.D.L. (2002). Optimization of gluten-free bread prepared from corn starch, rice flour, and cassava starch. Journal of Food Science 67: 416–419.

Sanful, R.E., Adiza, S. and Sophia, D. (2010). Nutritional and sensory analysis of soyabean and wheat flour composite cake. Pakistan Journal of Nutrition 9: 794–796.

Sarkar, N. and Walker, L.C. (1995). Hydration-dehydration properties of methylcellulose and hydroxyprophylmethylcellulose. Carbohydrate Polymers 27: 177–185.

Schieber, A., Keller, P. and Carle, R. (2001a). Determination of phenolic acids and flavonoids of apple and pear by high performance liquid chromatography. Journal of Chromatography 910: 265–273.

Schieber, A., Stintzing, F.C. and Carle, R. (2001b). By-products of plant food processing as a source of functional compounds—recent developments. Trends in Food Science and Technology 12: 401–413.

Schiraldi, A., Piazza, L. and Riva, M. (1996). Bread staling: a calorimetric approach. Cereal Chemistry 73: 32–39.

Schober, T.J. (2009). Manufacture of gluten-free speciality breads and confectionery products. pp. 130–180. In: E. Gallagher (ed.). Gluten-Free Food Science and Technology. Wiley-Blackwell, Oxford.

Schober, T.J., Messerschmidt, M., Bean, S.R., Park, S.H. and Arendt, E.K. (2005). Gluten-free bread from sorghum: quality differences among hybrids. Cereal Chemistry 82: 394–404.

Schoenlechner, R., Siebenhandl, S. and Berghofer, E. (2008). In: E.K. Arendt and F. Dal Bello (eds.). Gluten-Free Cereal Products and Beverages. Academic Press, London.

Sciarini, S.L., Ribotta, D.P., León, E.A. and Pérez, T.G. (2010). Influence of gluten-free flours and their mixtures on batter properties and bread quality. Food and Bioprocess Technology 3: 773–780.

Selomulyo, V.O and Zhou, W. (2007). Frozen bread dough-Effects of freezing storage and dough improvers. Journal of Cereal Science 45: 1–17.

Serrem, C.A., Kock, H.L. and Taylor, J.R.N. (2011). Nutritional quality, sensory quality and consumer acceptability of sorghum and bread wheat biscuits fortified with defatted soy flour. International Journal of Food Science and Technology 46: 74–83.

Sharadanant, R. and Khan, K. (2003). Effect of hydrophilic gums on the quality of frozen dough: II. Bread characteristics. Cereal Chemistry 80: 773–780.

Shittu, T.A., Aminu, R.A. and Abulude, E.O. (2009). Functional effects of xanthan gum on composite cassava-wheat dough and bread. Food Hydrocolloids 23: 2254–2260.

Shyamala, B.N. and Jamuna, P. (2010). Nutritional content and antioxidant properties of pulp waste from Daucus carota and Beta vulgaris. Malaysian Journal of Nutrition 16: 397–408.

Sivam, A.S., Sun-Waterhouse, D., Quek, S.Y. and Perera, C.O. (2010). Properties of bread dough with added fibre polysaccharides and phenolic antioxidants: A review. Journal of Food Science 75: 163–174.

Škrbić, B. and Filipčev, B. (2008). Nutritional and sensory evaluation of wheat breads supplemented with oleic rich sunflower seed. Food Chemistry 108: 119–129.

Slavin, J.L. and Greenberg, N.A. (2003). Partially hydrolyzed guar gum: Clinical nutrition uses. Nutrition 19: 549–552.

Strinmetz, K.A. and Potter, J.D. (1996). Vegetables fruit and cancer prevention. J Am Diet Assoc 96: 1027–1039.

Sumnu, G.S., Ndife, M.K. and Bayındırlı, L. (2000). Effects of sugar, protein and water content on wheat starch gelatinization due to microwave heating. European Food Research and Technology 211: 169–174.

Taylor, J.R.N. and Parker, M.L. (2002). Quinoa. pp. 93–122. In: P.S. Belton and J.R.N. Taylor (eds.). Pseudocereals and Less Common Cereals: Grain Properties and Utilization. Berlin: Springer Verlag.

Thompson, T., Dennis, M., Higgins, L.A., Lee, A.R. and Sharrett, M.K. (2005). Gluten-free diet survey: are Americans with coeliac disease consuming recommended amounts of Wbre, iron calcium and grain foods? Journal of Human Nutrition Dietetics 18: 163–169.

Thompson, T. (2001). Case problem: questions regarding the acceptability of buckwheat, amaranth, quinoa, and oats from a patient with celiac disease. Journal of the American Dietetic Association 101: 586–587.

Thompson, T. (2000). Folate, Iron, and dietary Wber contents of the gluten-free diet. Journal of the American Dietetic Association 100: 1389–1395.

Thompson, T. (1999). Thiamin, riboXavin, and niacin contents of the gluten free diet: is there cause for concern? Journal of the American Dietetic Association 99: 858–862.

Toma, R.B., Orr, P.H., D'Appolonia, B., Dintzis, F.R. and Tabekhia, M.M. (1979). Physical and chimica properties of potato peel as a source of dietary fiber in bread. Journal of Food Science 44: 805–806.

Tömösközi, S., Gyenge, L., Pelceder, A., Abonyi, T., Schönlechner, R. and Läsztity, A. (2011). Effects of flour and protein preparations from amaranth and quinoa seeds on the rheological properties of wheat-flour dough and bread crumb. Czech Journal of Food Science 29: 109–16.

Tomotake, H., Yamamoto, N., Kitabayashi, H., Kawakami, A., Kayashita, J. and Ohinata, H. (2007). Preparation of tartary buckwheat protein product and its improving effect on cholesterol metabolism in rats and mice fed cholesterol enriched diet. Journal of Food Science 72: 528–533.

Tosi, E.A., Ré, E.D., Masciarelli, R., Sanchez, H., Osella, C. and de la Torre, M.A. (2002). Whole and defatted hyperproteic amaranth flours tested as wheat flour supplementation in mold breads. Lebensmittel-Wissenschaft & Technologie 35: 472–475.

Trainer, D. and Holden, J. (1999). Isoflavones in retail and institutional soy foods. Journal of Agricultural Food Chemistry 47: 697–704.

Tseng, Y.H., Yang, J.H. and Mau, J.L. (2008). Antioxidant properties of polysaccharides from Ganoderma tsugae. Food Chemistry 107: 732–738.

Turabi, E., Summu, G. and Sahin, S. (2008). Rheological properties and quality of rice cakes formulated with defferent gums and an emulsifier blend. Food Hydrocolloids 22: 305–312.

Ulziijargal, E. (2009). Quality evaluation of Agaricus mycelium bread and Hericium mycelium bread. Master's Thesis. National Chung Hsing University, Taichung, Taiwan.

Valcárcel-Yamani, B. and da Silva Lannes, S.C. (2012). Applications of Quinoa (*Chenopodium Quinoa* Willd.) and Amaranth (*Amaranthus* spp.) and their influence in the nutritional value of cereal based foods. Food and Public Health 2: 265–275.

Van poppel, G. and Van den Berg, H. (1997). Vitamins and cancer. Cancer Letters 114: 195–202.

Vitseva, O., Varghese, S., Chakrabarti, S., Folts, J.D. and Freedman, J.E. (2005). Grape seed and skin extracts inhibit platelet function and release of reactive oxygen intermediates. Journal of Cardiovascular Pharmacology 46: 445–451.

Wang, L., Miller, R.A. and Hoseney, R.C. (1998). Effects of (1-3)(1-4)-β-D-glucans of wheat flour on breadmaking. Cereal Chemistry 75: 629–633.

Wasser, S.P. and Weis, A.L. (1999). Medicinal properties of substances occurring in higher Basidiomycetes mushrooms: Current perspective (review). International Journal of Medicinal Mushrooms 1: 31–62.

Wijngaard, H.H. and Arendt, E.K. (2006). Buckwheat. Cereal Chemistry 83: 391–401.

Wood, P.J. (2002). Relationships between solution properties of cereal β-glucans and physiological effects—a review. Trends in Food Science and Technology 13: 313–320.

Wronkowska, M., Troszynska, A., Soral-Smietana, M. and Wo1ejszo, A. (2008). Effects of buckwheat flour (Fagopyrum esculentum Moench) on the quality of gluten free bread. Polish Journal of Food and Nutrition Sciences 58: 211–216.

Wu, Q., Wang, M. and Simon, J.E. (2005). Determination of proanthocyanidins in fresh grapes and grape products using liquid chromatography with mass spectrometric detection. Rapid Communications in Mass Spectrometry 19: 2062–2068.

Yaseen, A.A.E., El-Din, M.H.A. and El-Latif, A.R.A. (1991). Fortification of Balady bread with tomato seed meal. Cereal Chemistry 68: 159–161.

Ylimaki, G., Hawrysh, Z.J., Hardin, R.T. and Thompson, A.B.R. (1991). Response surface methodology in the development of rice flour yeast breads: sensory evaluation. Journal of Food Science 56: 751–755.

Zagibalov, A.F., D'Yakonova, A.K., Gubanov, S.N. and Savchenko, S.N. (1985). Causes of bitterness in protein isolate from tomatoes. Izvestiya Vysshikh Uchebnykh Zavedeni 4: 42–43.

11

Fortification as a Tool in Improving Nutritional Properties of Bread and Controlling Life Style Diseases

Mostafa Aghamirzaei,[1,*] Milad Fathi,[2] Mohammad Sarbazi,[1] Majid Aghajafari[3] and Roya Fathi Til[4]

1. Introduction

Fortification refers to deliberately increasing the content of essential nutrients in a food, irrespective of whether the nutrients were originally in the food before processing or not, to improve the nutritional quality of the food supply and to provide a public health benefit with minimal risk.

[1] Department of Food Science and Technology, College of Agriculture, University of Tabriz, Tabriz 5166616471, Islamic Republic of Iran.
E-mail: M_sarbazi010@yahoo.com
[2] Department of Food Science and Technology, College of Agriculture, Isfahan University of Technology, Isfahan 84156-83111, Islamic Republic of Iran.
E-mail: mfathi@cc.iut.ac.ir
[3] Department of Food Science and Technology, Gorgan University of Agricultural, Science and Natural Resources, Gorgan, Iran.
E-mail: m.aghajafari@yahoo.com
[4] Department of Food Science and Technology, Faculty of Agriculture, University of Urmia, Urmia, Iran.
E-mail: r.fathitil88@tabrizu.ac.ir
* Corresponding author: aghamirzaei.ma88@gmail.com

Enrichment is defined as addition of nutrients to a food which are lost during processing (WHO and FAO 2006). The four main methods of food fortification are: (i) Biofortification (e.g., breeding crops to increase their nutritional value, which can include both conventional selective breeding, and modern genetic modification) (ii) Synthetic biology (e.g., addition of probiotic bacteria to foods) (iii) Commercial and industrial fortification (e.g., flour, rice, oils) (iv) Home fortification (e.g., adding soya and green bean in cooked rice) (Darnton-Hill 1998).

The important factors affecting food fortification are the kind and quantity of micronutrient shortages in society, people's nutrition habits, diseases problems, technological conditions, and economic considerations. Due to changes in society's dietary habits from natural and nourishing foods toward fast foods, especially in some industrialized countries, and increasing variation of diseases, the need for food fortification has become more noticeable.

As outlined by the FAO, the most common fortified foods are cereals and cereal based products, milk and milk products, fats and oils, and tea and infant formulas (Liyanage and Hettiarachchi 2011). Cereals are the main source of food for mankind, particularly in developing countries, where one-half of the calorie intakes are derived from cereal grains. Bread is an excellent source of numerous vitamins and minerals especially phosphorus and copper (Collar 2016). While, cereal proteins are deficient in some essential amino acids such as lysine, tryptophan, and threonine. Bread is good choice for fortification, because it forms the staple of common man's diet in most countries and also has a simple and inexpensive processing technology (Rosell et al. 2016). The aim of this chapter is to determine the nutritional benefits of bread and the effects of adding different nutrients on the health. In the following sub-sections different bioactive components used in fortification of bread are introduced and their effects on physic-chemical properties are discussed. The fortification health risks are also discussed towards the end of this chapter.

2. Amino Acids and Proteins

Wheat proteins are not well balanced and therefore it is necessary to improve essential amino acids in cereal-based diets. Lysine and tryptophan are limiting amino acids of the bread. Addition of cysteine and glycine to the dough resulted in decrease of acrylamide content in an asparagines-glucose model system because they compete with asparagine in the reactions with reducing sugars (Mustafa et al. 2011). Adding amino acids with wheat flours cause increase browning of the baked bread crust. Lysine and glycine fortification of dough lead to lower gas production.

Proteins are added to gluten-free applications to increase elastic modulus by cross linking to enhance Maillard browning and flavor, gelation and foaming. The effect of soy proteins isolate (SPI) addition on gluten-free products showed an increase of elastic modulus (G′), resulting in enhanced gas retention and loaf volume, and improve water binding in the bread loaves (Arendt et al. 2008). Egg proteins show high biological value as it contains all the essential amino acids, and is used to improve color and enhance flavor of bakery products. Easily digested proteins found in eggs are ideal for recovering celiac disease patients. Also, whey protein, milk powder and fish protein concentrates (e.g., Tilapia fish protein concentrate) may be used as a source of proteins, and for increasing and balancing the amino acids such as isoleucine, leucine, lysine and valine (Adeleke and Odedeji 2010, Al-Dmoor 2012).

3. Fat and Fatty Acids

Docosahexaenoic acid (DHA) is considered as the most important omega-3 fatty acids. Breads fortified with algae oils required the least weight addition of omega-3 oils because they contained relatively high levels of DHA/omega-3 fatty acid. The algae oil breads contained approximately 25 or 50 mg DHA/serving, whereas the fish oil supplemented bread contained approximately half of the amount but still maintained the 25–50 mg of total long-chain omega-3 fatty acid, which is contributed by eicosapentaenoic acid (EPA) and DHA. It is necessary to mention that DHA and EPA diminish the cholesterol and serum triglycerides because they bind to peroxisome proliferator-activated receptors (PPARs), which act as transcription factors of specific genes (Serna-Saldivar and Abril 2011).

4. Bran and Fibers

Dietary fibers with larger particle size resulted in highly sensory acceptable breads with higher amounts of resistant starch and slightly lower protein digestibility (Salehifar 2011, Rosell 2016). Fiber sources exhibiting high viscoelasticity-G′, G″- and complex viscosity -ή*- in concentrated solutions yielded breads with better sensory perception, lower digestible starch and higher resistant starch contents bringing to lower *in vitro* expected Glycaemic Index. All dietary fibers are presumed to have physiological and health effects (Table 1).

5. Cereal Grains

Fortification of wheat flour with non-wheat flours (cereal grains) has been advocated for nutritional enrichment of these products. Cereal grains

Table 1. Type of fiber and their physiological and health effects.

Type of fiber	Fiber component	Main food sources	Physiological and health effects
Water insoluble	Cellulose	Plants (vegetables, sugar beet, various brans)	Increase insulin sensitivity, decrease risk of type 2 diabetes, increase gut transit time, lower production of short-chain fatty acids, promote normal laxation, low energy density, increase bulking effects, decrease energy intake, decrease weight gain, promote satiation and satiety, increase/decrease gut hormones release, diminish inflammation markers (CRP), increase insulin sensitivity, decrease risk of cardiovascular disease (reduce levels of PAI-1), improve blood pressure, reduce risk of certain types of cancer (colon, rectum, and breast)
	Hemicellulose	Cereal grains	
	Lignin	Woody plants	
Water soluble	Pectin	Fruits, vegetables, legumes, sugar beet, potato	Decrease postprandial glucose response, decrease postprandial insulin response, reduce total and LDL cholesterol, delay gastric emptying, delay intestinal absorption, higher production of short-chain fatty acids, low energy density, increase bulking effects, decrease energy intake, decrease weight gain, promote satiation and satiety, increase/decrease gut hormones release, diminish inflammation markers (CRP), increase insulin sensitivity, decrease risk of type 2 diabetes, decrease risk of cardiovascular disease (reduce levels of PAI-1), improve blood pressure, reduce risk of certain types of cancer (colon, rectum, and breast)
	Gums	Leguminous seed plants (guar, locust bean), seaweed extracts (carrageenan, alginates), microbial gums (xanthan, gellan)	
	Mucilages	Plant extracts (gum acacia, gum karaya, gum tragacanth)	

(barley, corn, maize, millet, oat, rice, rye, and sorghum) have always been the most important plant group for the human diet because provide starch, dietary fiber, protein, antioxidants (such as tocopherols, tocoterienols and phytochemicals), minerals, vitamins especially B vitamins, and essential fatty acids (Aghamirzaei et al. 2013). The main parameters that have to be considered are the composition and processing of the cereal grains in bakery products, substrate formulation, the growth capability and productivity of the starter culture, the organoleptic properties, the stability of the probiotic strain during storage, and the nutritional value of the bakery products. Furthermore, cereal grains can be used as sources of non-digestible carbohydrates that besides promoting various beneficial effects can also selectively stimulate the growth of bifidobacteria and lactobacilli in colon.

Barley and oat contain water-soluble fiber such as β-glucan. Thondre and Henry (2009) mentioned that supplementation of whole wheat flour with a commercial barley β-glucan fiber preparation created palatable chapatis for diabetic patients, and the glycemic index (GI) of chapatis with 4 g of β-glucan per serving was significantly reduced. Also, β-Glucans are known to reduce the rate of lipid absorption, increase bile acid transport toward the lower parts of the intestinal tract and reduced the total and LDL cholesterol (Queenan et al. 2007, Alminger and Eklund-Jonsson 2008). The effect of β-glucans on the postprandial glucose metabolism, lowering cholesterol and serum lipids is related to their ability to increase viscosity in the gut causing a delayed or decreased absorption of glucose into the blood stream.

Sorghum may be a particularly functional food for diabetes and obesity. Celiac disease, a syndrome characterized by damage to the mucosa of the small intestine, is caused by ingestion of gluten protein. The only treatment is lifelong avoidance of foods containing wheat and similar cereals such as rye, corn, millet, oat, rice and barley. Because, sorghum is a gluten-free cereal, and it contains various phenolic compounds that appear to have health benefits, which makes the grain suitable for developing functional foods for celiac patients.

Rye bread in comparison to wheat bread improves bowel function and increases the concentration of plasma enterolactone in postmenopausal women. Also, whole grain rye breads and endosperm rye products induced significantly lower insulinemic indices compared to white wheat bread (Rosén et al. 2009). These beneficial effects may be mediated through glucose-dependent insulinotropic polypeptide and glucagon-like peptide 1 (GLP-1), which is the most important insulin tropicincretins. Thus, there were a good potential in increasing nutritional and health value of bread by incorporating cereal grains (Singh and Singh 2011).

6. Legumes

The more important legumes are fenugreek, amaranth, quinoa, pea, chickpea, cow pea, pigeon pea, bean, faba bean, broad bean, lentil, spelt and lupin. It is necessary to mention that non-protein tryptophan in legumes is important because this fraction is easily absorbable in the gastrointestinal media, increasing its availability for brain serotonin synthesis (Comai et al. 2011). Additionally, legumes are rich in lysine and deficient in sulphur-containing amino acids, whereas cereal proteins are deficient in lysine, but have adequate amounts of sulphur amino acids. In addition, these contain many health-promoting components, such as dietary fiber, abundance of carbohydrates, low fat, high concentration of polyunsaturated fatty acids, B complex vitamins, resistant starch, minerals, and numerous phytochemicals endowed with useful biological activities. Legumes such as lupine have a positive impact on the risk factors of colon cancer. Because, it reduces carbohydrate content and caloric density and also, decreases plasma cholesterol, triacylglyceroles, and C-reactive protein in moderate hypercholesterolemic subjects could be used as a substitute for wheat flour.

Fenugreek seed also possesses hypocholesterolemic and hypoglycemic properties (Neeraja and Rajyalakshmi 1996). Hence, development and consumption of such therapeutic bakery products such as bread, pasta and cookie would help to improve the nutritional status of population. Also, it is necessary to mention that fenugreek seed (raw, soaked and germinated) significantly reduced serum total cholesterol, total lipids, LDL-cholesterol while serum HDL-cholesterol and triglycerides didn't showed significant changes. Ibrahim and Hegazy (2009) reported that maximum decrease (44%) in phytic acid content was found in biscuit containing 10% germinated fenugreek seed flour, while minimum decrease (20%) was found in biscuit containing 5% soaked fenugreek seed flour. This might be due to phytate leaching or phytate hydrolysis during soaking and germination processes by phytase and phosphatase enzymes.

Amaranth is a grain with high glycemic index, attributed to its small starch granule size, low resistant starch content and amylase, along with a tendency to completely lose its crystalline and granular starch structure. The glycemic indexes of various amaranths are compared in Table 2. Raw amaranth seeds have a glycemic index of 87.2 and a rapidly digestible starch content of 30.7% (dry weight basis) (Capriles et al. 2008). However, lipids of amaranth are rich in squalene and tocotrienols, which are natural materials positively involved in lowering low-density lipoprotein blood cholesterol.

Anton et al. (2008) showed increases in total phenolics and antioxidant activity of bean/wheat composite tortillas with increased substitution with bean flour. Regarding anti-nutritional factors, they mentioned that there was an increase in levels of phytic acid and trypsin inhibitor activity in bean/wheat composite flours as the rate of substitution with bean flour increased. They reported that processing flours into tortillas significantly reduced phytic acid and trypsin inhibitor levels (Anton et al. 2008).

Table 2. Glycemic Index of Various Amaranths.

Type of Amaranth	Glycemic Index (GI)
Raw	87
Popped	94
Roasted	101
Flaked	106
Extruded	106

Source: Capriles et al. (2008)

Table 3. Summary of research studies on the use of composite flours in baked foods.

Type of Product	Composite flours used	Properties investigated	Reference
Bread	Wheat-cracked broad Bean	Proximate analyses on individual flours, rheological properties of flour blends and sensory evaluation of breads	Abdel-Kader (2000)
Bread	Fermented/ germinated Cowpea-wheat	Composite flours analyzed for ash, protein, gluten content, α-amylase activity; color and rheological characteristics Breads analyzed for loaf volume and weight, texture, crumb structure and color	Hallén et al. (2004)
Bread	Wheat-cassava	α-amylase activity and rheological properties Physical characteristics and sensory properties of the breads determined	Khalil et al. (2000)
Biscuit	Millet-pigeon pea	Proximate analyses of flours and biscuits	Eneche (1999)
Biscuit	Wheat-moth bean	Biscuits analyzed for proximate composition, physical characteristics, sensory quality and biological parameters using weanling albino rats	Awan et al. (1995)
Cake	Wheat-chickpea	Physical properties of the batters and cakes	Gómez et al. (2008)
Tortillas	Wheat-bean	Rheological properties of the dough and Physical properties. Protein content, total phenolics and antioxidant activity of the tortillas, phytic acid content and trypsin inhibitor activity of the flours and tortillas	Anton et al. (2008)

Table 3 provides a summary of studies on the use of composite flours in baked foods and details of these studies are discussed here.

7. Oil Seeds

Different oil seeds such as soybean, flaxseed, sesame seed, grape seed, coconut flour and cottonseed can be used for fortification of bread. Oilseeds have high content of polyunsaturated fatty acids, vitamin E, protein, niacin, iron, magnesium, and phosphorus (Chen et al. 1992). Soybean flour diets decrease relative weight of fat deposits and even reduce the energetic expenditure without alterations in food intake in rat (Jansen and Monte 1997). Also, soybean flour activates the β-cell cAMP/PKA pathway, increasing insulin secretion in response to glucose. In addition, muscle insulin-reduced phosphorylation of GSK-3 is improved by soy proteins and isoflavones, partially preventing deleterious effects of fat feeding (Latorraca et al. 2011). Therefore, the enhancement of the nutritional value of breads with the addition of soybean flour could help to alleviate the problem of protein – energy malnutrition prevalent in developing countries.

Results showed that bread with the addition of grape seed extract (GSE) had stronger antioxidant activity than control bread. Other benefit for grape seed extract (GSE) was reduction of carboxymethyl lysine (CML) in bread. Incorporating grape seed powder (GSP) in wheat flour decreased total protein, whereas an increase in total dietary fiber, fat and total phenol was observed (Peng et al. 2010, Peighambardoust and Aghamirzaei 2014, Aghamirzaei et al. 2015).

8. Edible Nut

Fortification of wheat flour with 10% pumpkin seed flour led to an increase of 80.8% in crude protein, 43.9% in calcium, 71.9% in potassium and 63.0% in phosphorus contents of composite control bread (Giami et al. 2003).

The antioxidant potential of oil sunflower seed is of great concern with ever increasing use of this oils seed in various food products. The addition of sunflower seed resulted in increasing chemical composition (copper, zinc, essential fatty acids, crude fat, crude fiber and crude protein) of breads.

Apricot kernel is mainly used in the production of oils, and the kernels are also added to bakery products either whole or grounded and also consumed as an appetizer (Hyta and Alpaslan 2011). However, the use of peanut (Rao and Vakil 1980), groundnuts (Akubor 2003) and breadnut flour (Malomo et al. 2011) as functional ingredients in bakery products has been successfully demonstrated to give extra protein and changes in texture of the product.

9. Antioxidants and Vitamins

Free radicals derived from a wide range of biological reactions in the body can damage essential biomolecules. Excess of un-scavenged free radicals cause unhealthy conditions as well as diseases, for example, reactive oxygen species (ROS) including superoxide (O_2^-), hydroxyl radical (OH), hydrogen peroxide (H_2O_2), and lipid peroxide radicals have been associated with chronic degenerative diseases such as inflammatory, cancer, cardiovascular and aging disease. Seidel et al. (2007) studied the influence of antioxidants in bread on the immune system, in male smokers and non-smokers. They reported that total radical-trapping antioxidant parameter (TRAP) and photosensitive chemoluminescence (PCL) measured in urine increased significantly whereas ferric reducing ability of plasma (FRAP) and gallic acid equivalents (GAE) in urine remained unchanged after consumption of breads fortified with antioxidant.

Green tea (GT) contains antioxidants that may be used as health promoting ingredients. The results revealed that, feeding chronic renal failure rats on arginine diet containing GT-fortified bread resulted in significant decrease in serum total cholesterol (TC), triglycerides (TG), low-density lipoprotein cholesterol (LDL-C), Aspartate Amino Transferase (AST), Alanine Amino Transferase (ALT), uric acid, urea nitrogen and creatinine and significant increase in serum high-density lipoprotein cholesterol (HDL-C). These beneficial effects are attributed to a reduction in cholesterol absorption and to an increased excretion of biliary acids and of cholesterol synthesis in the liver (Abd El-Megeid et al. 2009, Somboonvechakarn 2008). Also, antioxidants could significantly reduce the acrylamide content generated in bread sticks and keep original flavor and crispness of fried bread sticks. The extensive fermentation with yeast may be one of possible ways to reduce acrylamide content in fortified breads.

Major bioactive in turmeric (*Curcuma longa* L.) are polyphenols, including curcumin, which is well known, besides other polyphenols, for it strong antioxidant activity. Thus, substitution of wheat flour in the bread with turmeric powder can result in development of bread with additional health benefits.

Vitamin recoveries in bread made from fortified flour range from 75% to 90% for pyridoxine and thiamine and from about 70% to 95% for niacin. On this basis, and assuming that any added B vitamins are 100% absorbed, in flour an average of approximately 20–30% is thus usually sufficient to provide the desired amount in bakery products (Allen et al. 2006). Cereals such as wheat contain no vitamin B12 and this vitamin is present only in animal products. Vitamin B_{12} deficiency is a common clinical condition, sometimes leading to severe neurologic disease. Fortification of bread with this vitamin is expected to reduce the vitamin deficiency-related diseases and prevent the folic acid masking effect. It may also offer an opportunity to increase the folic acid dose in different fortified food. Previous studies demonstrated that co-fortification of bakery products was associated with increased folic acid and vitamin B_{12} levels and with decreased homocysteine concentrations in a significant portion of the population (Winkels et al. 2008). Therefore, two conclusions can be made. First, a fortification program with 140 mg of folic acid per every 100 g of flour, probably achieved some results in terms of reduction of neural tube defect, but a greater reduction could probably be achieved with higher levels of folic acid fortification. Second, the imbalance between vitamin B_{12} and folic acid levels can produce adverse events, at least in a specific segment of the population. Figure 1 shows the health benefits of vitamin B in bread and flour.

Vitamin E can be added to bread without adversely affecting sensory or other characteristics of bread. About 1/3 of the added vitamin E is lost during baking period (Ranhotra et al. 2000). Vitamin C is routinely added to bread flour at levels from 15 to 100 ppm to improve the flour protein

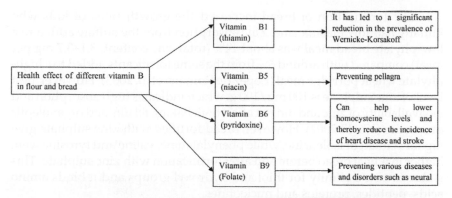

Figure 1. Health affects other vitamin B in bread and flour.

functionality during bread baking (Al-Dmoor 2012). Vitamin C provides a number of important nutritional benefits such as enhancement the several fold absorption of both native and added iron. Vitamin A has been added to wheat flour buns and an increase in the initial serum retinol concentrations was observed in school-age children after a daily consumption of these buns (Solon et al. 2000).

10. Minerals

Adequate intake and availability of minerals are closely related to survival, cognitive and reproductive functions and immunity. Fortification of bread with metal ions to control the formation of acrylamide was proposed by Gökmen and Senyuva (2007). The reducing effect was related to the ability of these metal ions (especially Ca^{+2}, Mg^{+2}, Na^+ and K^+) to stop production of Schiff base, which is an important intermediate in the Maillard reaction leading to formation of acrylamide. A low dietary intake of iron and its poor bioavailability are majorly responsible for the wide prevalence of anaemia (Demirozut et al. 2003). Certain amino acids such as cycteine, histidine and lysine have an important role in absorption, improvement and utilization of iron. Previous studies have shown that intake of fortified whole meal rye bread resulted in stabilization of the iron status in young women (Hansen et al. 2005).

Calcium is essential for formation of bones and teeth, for clotting of blood, where in calcium acts as a catalyst in conversion of prothrombin to thrombin, for normal functioning of nerves and muscles, and plays a vital role in prevention and management of osteoporosis. Bioavailability of calcium in breads fortified with different calcium sources was reported by Ranhotra et al. (1997).

Zinc fortification of bread increased the growth rates of kids who initially had low plasma zinc. Zinc absorption from the sulfate added to a low-phytate bread meal was about 14% (total zinc content, 3.1–3.7 mg per meal) compared with around 6% from the same fortificants added to a high-phytate wheat porridge meal (Lopez de Romana et al. 2003). However, best level of zinc sulphate is 100 mg/100 g, which indicates high absorption and bioavailability of zinc and did not affect both baking results and organoleptic properties (Khalil 2002). However, bread fortified with zinc sulphate give higher content of isoleucine, while phenylalanine, valine and tyrosine were higher in wheat bread before and after fortification with zinc sulphate. This is due to zinc's affinity for thiol and hydroxyl groups and it binds amino acids, peptides, proteins and nucleotides.

The bioaccessibility of five essential micronutrients (iron, zinc, copper, manganese and molybdenum) from the Lebanese food basket including bread, different varieties of white cheese, fruit and vegetables was evaluated by Khouzam et al. (2011). Only very small fraction of Fe and Zn (*ca.* 10%) was found bioaccessible from bread. High bioaccessibility (>50%) was also observed for manganese in fruit and vegetables whereas that from bread was fair (25–30%). The fortification of bread with iodine is being achieved by the addition of iodized salt which has resulted in a desirable increase in iodine intake, and prevents Gutter disease (Miki et al. 1993).

11. Prebiotics, Probiotics and Synbiotics

Probiotics are microorganisms being used for different purposes such as enhancing growth, facilitating digestion and absorption and quelling infectious diseases. Prebiotics are defined as non-digestible or low-digestible food ingredients that benefit the host organism by selectively stimulating the growth or activity of one or a limited number of probiotic bacteria in the colon. Prebiotics showed a beneficial influence on mineral absorption, risk of colon cancer and immune system.

Combined use of wheat flour, dextran-producing lactic acid bacteria strains, and the sourdough process stimulated lactic acid bacteria and yeast metabolism and led to shorter leavening times and higher exopolysaccharide production in dough. In addition, exopolysaccharides are reported to be able to replace hydrocolloids currently used for texturizing, anti-staling, cholesterol-lowering, immunomodulating, antitumoral, and prebiotic activities (Tieking et al. 2003). As a consequence, it was possible to develop prebiotic bread able to satisfy about 30% of the daily requirement for fructo-oligosaccharides and supplement the intake of prebiotics in the daily diet (Pepe et al. 2013).

It was indicated that the inclusion of lactic acid in bread reduces the rate of starch digestion by creating interactions between the gluten and starch. Also, presence of lactic acid during starch gelatinization appeared to be a prerequisite for a reduced starch bioavailability. Shakeri et al. (2014) evaluated the effects of the daily consumption of synbiotic bread (*Lactobacillus sporogenes*/inulin) on blood lipid profiles of patients with type 2-diabetic. Their results showed an increase in serum HDL-C levels and decrease in serum TG, VLDL-C; TC/HDL-C ratio.

12. Health and Technological Risk of Fortification

Most of the legume anti-nutritional factors are heat-labile. These anti-nutritional factors include: protease inhibitors, lectins, goitrogens, antivitamins, phytates, saponins, estrogens, flatulence factors, allergens and lysinoalanine. Heat-stable anti-nutritional factors (e.g., phytate and polyphenols) are not eliminated by simple soaking and heating, but through germination or fermentation. Nowadays, some of the anti-nutritional factors (e.g., tannins) are of much interest due to antioxidant activity as a potential health benefit. However, legumes are generally cooked, which improves their nutritional value and reduces the risk of toxicity. Also, fenugreek can cause increase milk supply in breast feeding women, it may stimulate the uterus and should not be taken during pregnancy. A negative correlation among phytate and Fe availability was found in high concentration of amaranth.

Didar and Haddad-Khodaparast (2011) showed that bread making with *Lactobacillus plantarum* sourdough had lower phytic acid. This might be related to microbial phytase enzyme and dough acidification that provided suitable condition for endogenic and microbial phytase activity and solubility increase of phytate complexes.

Maximum tolerable level is the maximum micro/macro nutrients content that a fortified food can present as it is established in food law, in order to minimize the risk of overdose. Fortified food for example bread as a resource of vitamins, minerals herbal supplements that are potentially harmful when taken in large doses by infants, children and pregnant or breastfeeding women. When antioxidants are taken in greater than certain concentrations, they may cause deadly physiological effects. For example, while an overdose of vitamin C is fairly difficult to achieve, as the concentration ingested in order to produce overdose-like effects is approximately 3,000 mg/day, exceeding this dose can lead to some negative physiological effects such as kidney stones and an increased need for oxygen (Maldonado 2000). Moreover, some polyphenols have been shown to interfere with the metabolism of certain medicines when taken in excess (Shahidi 2008). In doses of 1,600–3,200 mg/day for extended periods of time, vitamin A overdoses can lead to symptoms such as breast soreness, gastrointestinal stress, vascular inflammation, and fatigue and thyroid problems (Goodman 1980). Table 4 shown toxicity symptoms of different micro/macro nutrients.

Table 4. Toxicity symptoms of different micro/macro nutrients (Kleinman 2004, Maher and Escott-Stump 2004, McKinley Health Center 2009).

Nutrient	Toxicity symptoms	Nutrient	Toxicity symptoms	Nutrient	Toxicity symptoms
Protein	Azotemia; acidosis; hyperammonemia	Vitamin B6	Sensory neuropathy with progressive ataxia; photosensitivity	Copper	Wilson's disease—copper deposits in the cornea; cirrhosis of liver; deterioration of neurological processes
Carbohydrate		Vitamin B1		Iodine	Possible thyroid enlargement
Fat		Vitamin B2		Magnesium	Diarrhea; transient hypocalcemia
Vitamin D	Abnormally high blood calcium (hypercalcemia), retarded growth, vomiting, nephrocalcinosis	Niacin	Transient due to the vasodilating effects of niacin (does not occur with niacinamide)- flushing, tingling, dizziness, nausea; liver abnormalities; hyperuricemia; decreased LDL and increased HDL cholesterol	Manganese	In extremely high exposure from contamination: severe psychiatric and neurologic disorders
Vitamin A	Fatigue; night sweats; vertigo; headache; dry and fissured skin; lips; hyperpigmentation; retarded growth; bone pain; abdominal pain; vomiting; jaundice; hypercalcemia	Calcium	Excessive calcification of bone; calcification of soft tissue; hypercalcemia; vomiting; lethargy	Molybdenum	Gout like syndrome

Vitamin E	May interfere with vitamin K activity leading to prolonged clotting and bleeding time; in anemia, suppresses the normal hematologic response to iron	Iron	Hemochromatosis; hemosiderosis	Phosphorus	Hypocalcemia (when parathyroid gland not fully functioning)
Vitamin K	Possible hemolytic anemia; hyperbilirubinemia (jaundice)	Zinc	Acute gastrointestinal upset; vomiting; sweating; dizziness; copper deficiency	Potassium	
Vitamin C	Nausea, abdominal cramps, diarrhea, possible formation of kidney stones	Fluoride	Mottled, discolored teeth; possible increase in bone density; calcified muscle insertions and exotosis	Selenium	
Vitamin B12		Chloride		Sodium	
Folate	Masking of B12 deficiency symptoms in those with pernicious anemia not receiving cyanocobalamin	Chromium		Pantothenic Acid	Diarrhea; water retention

It is necessary to mention that excess omega 3 will easily suppress the omega 6, whereas the reverse does not seem to occur, 6 does not suppress the 3.

Celiac disease is the immune system's reaction to gluten which causes damage to the lining of the gut. People with celiac disease must follow a strict gluten free diet for life (substitute's wheat with corn, potato, sorghum, lentil, and arrowroot starch).

Autism is the consequence of the action of peptides of exogenous origin affecting neurotransmission within the central nervous system (CNS). Peptides, formed through the incomplete breakdown of foods containing gluten, exhibit direct opioid activity or form ligands for the peptidase enzymes which break down endogenous endorphins and enkephalins. Results indicated that autism patients on a gluten free diet showed an improvement on a number of behavioral markers/patterns (Whiteley et al. 1999).

13. Fortification Using Micro- and Nano-Delivery Systems

Micro- or Nano-encapsulations are an important tool for food processing engineers. They can enhance the delivery of bioactive compounds, such as vitamins, antioxidants and viable microorganisms in foods and to the gastro-intestinal tract. Encapsulation can protect components during storage or processing.

The encapsulation of micronutrients such as iron, zinc, iodine, and vitamin A for fortification may be advantageous in preventing the unwanted sensory changes and the interactions of nutrients with the components of flour and bread. Souto et al. (2008) showed that children's acceptance of bread rolls fortified with microcapsulated iron was significantly lower than that of rolls without iron fortification.

Long chain omega-3 polyunsaturated fatty acids (PUFAs) such as fish oil (e.g., tuna oil) and flax seed oil, are very susceptible to oxidation during processing and storage resulting in decreased nutritional value and sensory quality. Nanoencapsulation was possible to make shelf-stable cream-filled sandwich cookies with high levels of long-chain omega-3 fatty acids without any adverse effect on sensory properties. Nanoencapsulated flax seed oil increased final product quality and safety by lowering lipid oxidation and formation of harmful compounds such as acrylamide and hydroxymethyl furfural in breads during baking. This probably is considering that carbonyls arising from the thermoxidation of free flax seed oil during baking can promote the conversion of asparagine into acrylamide (Gökmen et al. 2011).

Folate (vitamin B9) compounds, 5-methyltetrahydrofolic acid in particular, are susceptible to thermal degradation, high hydrostatic pressure, presence of various levels of oxygen, extreme pH, presence of various chemicals, etc. Microencapsulation of 5-methyltetrahydrofolic acid slightly improved the retention of 5-methyltetrahydrofolic acid over unencapsulated biscuits at 180°C for 5 min (Shrestha et al. 2012).

Noort et al. (2012) investigated a technological approach to reduce the sodium content of bread whilst retaining its sensory profile by creating taste contrast using encapsulated salt. They demonstrated that sensory contrast in bread induced by encapsulated salt can enhance saltiness and allows for a salt reduction of up to 50% while maintaining saltiness intensity. Also, large encapsulates lead to large concentration gradients which enhance saltiness significantly and reduce consumer acceptance.

14. Effect of Fortification on Product Quality

Fortification of bread should not have adverse effects on consumer acceptability. Iron may cause a slight darkening of flour, while high levels of riboflavin and folic acid can cause a slight yellowing, but these changes are accepted. The use of fortified flour with up to 0.3% lysine in bread making did not alter the organoleptic values for appearance, texture, flavor and taste, and the overall acceptability of the resulting breads (Yasoda-Devi and Geervani 1979). Pseudocereal-fortified breads were characterized by a significantly softer crumb texture effect that was attributed to the presence of natural emulsifiers in the pseudocereal flours and confirmed by the confocal images (Alvarez-Jubete et al. 2010). However, addition of such compounds promotes detrimental effect on bread quality in terms of loaf volume, texture, color and sometimes taste. Baking properties, color and sensory evaluation tests showed that 15% of wheat flour could be replaced with germinated legumes and still providing good quality of bread and biscuits. Also, mixing tolerance index values as an indicator for staling test revealed that wheat bread was better than wheat-germinated legumes bread regarding freshness (Eissa et al. 2007). Breads containing sourdough *Lb. plantarum* had better aroma compared to soy flour breads due to presence of proteolysis enzymes, decomposition of some of dough proteins and formation of different aromatic compounds. Thermal processing exerts a significant effect on the antioxidant activity of cell-free systems (Peng et al. 2010). Use of high level of added gums (> 10%) and phenolic compounds in food formulations may lead to negative effects on the sensory attributes of finished foods such as increased bitterness and astringency (Ognean et al. 2007, Jaeger et al. 2009).

Vitamin B5 is very stable and has no problem with cooking or baking losses. Also, carotene in whole wheat bread and crackers is highly stable during the prebaking steps and also during the typical market shelf life of

Table 5. Effect of different fiber sources in quality properties.

Different fiber source	Results	References
Date	Dough rheological characteristics showed that water absorption, stability, index quality, resistance to deformation increased with the amount of added date fiber, whereas degree of softening and extensibility decreased in all levels. Bread evaluation revealed that date fiber addition caused an increase of bread yield especially for the highest levels, a change of crumb color and an insignificant decrease of bread volume.	Borchani et al. 2011
Bran	Wheat bran had no effect on total, Low-density lipoprotein or high-density lipoprotein cholesterol irrespective of particle size or level of gluten in the diet. Rice bran hemicellulose exhibited high fat binding capacity. However, rice bran hemicellulose was found to be low viscous. Addition rice bran hemicellulose preparation reduced loaf volume significantly and increased the firmness of the breads. Sensory evaluations revealed that breads with rice bran hemicellulose were overall acceptable.	Jenkins et al. (1999), Hu et al. (2009)
Jackfruit	Increasing the level of Jackfruit rind flour incorporated into wheat flour caused an increase in hardness and darkness of bread samples, and decrease in their volume compared to the control. Bread samples substituted with 5% Jackfruit rind flour had the highest mean scores of overall acceptance.	Feili et al. (2013)
Apple	Water absorption increased significantly with increase in pomace. Also, dough stability decreased and mixing tolerance index increased, indicating weakening of the dough.	Sudha et al. (2007)
Coconut	Water absorption decreased with the increase in substitution, whereas dough development time and stability were increased up to 20% substitution level. Coconut flour addition up to 10% was ranked 'good', whereas 30% substitution negatively affected appearance, texture and overall acceptability of the product, ranked 'poor' in sensory evaluation.	Gunathilake et al. (2009)
Hulls and Cotyledon	Breads containing hull fibers exhibited the lowest starch transition enthalpies as determined by differential scanning calorimeter after 7 days of storage, while the starch transition enthalpies of breads containing added soluble or insoluble fiber were not significantly different from the control bread.	Dalgetty and Baik (2006)
Xanthan gum	Xanthan gum had significant effects on the dough tenacity and extensibility and sensory acceptability of fresh composite bread. The oven spring, specific volumes of bread loaf and crumb softness were higher at 1% xanthan gum content. Also, addition of xanthan gum made the composite bread samples had more open crumb structure and better sensory acceptability. However, moisture loss and crumb firming during bread storage were best reduced when 1% xanthan gum was added to bread formulation.	Shittu et al. (2009)

the products. However, the usefulness of adding vitamin A to wheat flour products is limited because of its poor stability in the presence of oxygen or air. The loss of vitamin A during 70 mins of baking at 200°C could be greater than 50% (Cakirer and Lachance 1975).

Fat and oil seems to interact with dough components (starch and gluten) and delay the reactions that end loaf expansion during baking. The measurement of texture of the cookie dough in the texture analyzer revealed that dough containing the hydrogenated fat, needed more force to compress it than those containing either the sunflower oil or the other types of fats (Leelavathi 2007).

Oil seed flours decrease mixing tolerance whereas increased absorption of dough's concomitantly with increases in replacement levels of wheat flour. Therefore, less severe and shorter mixing times must be adopted to minimize damage to the gluten structure. These steps can prevent excessive stretching and tearing of the gluten.

Dietary fibers, in general, had pronounced effects on dough properties yielding higher water absorption, mixing tolerance and tenacity, and shelf life. Effect of different fiber sources in quality properties are shown in Table 5.

15. Conclusion

Bread is an important part of the diet for millions of people worldwide. Wheat bread represents the main source of carbohydrate, minerals and vitamins for most of the people. However, white breads are considered to be nutritionally poor, as the wheat proteins are deficient in essential amino acids such as lysine, tryptophan, and threonine. Fortification of bread with different micro/macro nutrients increases the content of essential nutraceuticals of the bread. Composite breads improve the nutritional quality of bread and satisfy the increasing interest among vegetarians to consume protein-enriched food from plant sources, which are rich in lysine and have great potential in overcoming protein–calorie malnutrition. However, Fortification might cause different effects in physicochemical, shelf life, sensory, quality and rheological properties of bread.

Acknowledgments

The authors thank Younes Sobhi-Sarabi, Alireza Neissi, Saeed Moodi, Elyas Mohammadi-Gouraji for suggestions on the manuscript and encouragement offered.

Keywords: Food, bread, fortification, nutritional quality, health benefits, controlling diseases

References

Abd El-Megeid, A.A., AbdAllah, I.Z.A., Elsadek, M.F. and Abd El-Moneim, Y.F. (2009). The Protective effect of the fortified bread with green tea against chronic renal failure induced by excessive dietary arginine in male albino rats. World Journal of Dairy and Food Sciences 4: 107–117.

Abdel-Kader, Z.M. (2000). Enrichment of Egyptian "Balady" bread: Part 1. Baking studies, physical and sensory evaluation of enrichment with decorticated cracked broad bean flour (*Vicia faba* L.). Nahrung 44: 418–421.

Adeleke, R.O. and Odedeji, J.O. (2010). Acceptability studies on bread fortified with tilapia fish flour. Pakistan Journal of Nutrition 9: 531–534.

Aghamirzaei, M., Peighambardoust, S.H. and Azadmard-Damirchi, S. (2015). Effects of grape seed powder as a functional ingredient on flour physicochemical characteristics and dough rheological properties. Journal of Agricultural Science and Technology 17: 365–373.

Aghamirzaei, M., Heydari-Dalfard, A., Karami, F. and Fathi, M. (2013). Pseudo-cereals as a functional ingredient: effects on bread nutritional and physiological properties-Review. International Journal of Agriculture and Crop Sciences IJACS/2013/5–0/00–00.

Akubor, P.I. (2003). Functional properties and performance of cowpea/plantain/wheat flour blends in biscuits. Plant Foods for Human Nutrition 58: 1–8.

Al-Dmoor, H.M. (2012). Flat bread: ingredients and fortification. Quality Assurance and Safety of Crops and Foods 4: 2–8.

Allen, L., de Benoist, B., Dary, O. and Hurrell, R. (2006). Guidelines on food fortification with micronutrients. World Health Organization and Food and Agriculture Organization of the United Nations pp. 1–376.

Alminger, M. and Eklund-Jonsson, C. (2008). Whole-grain cereal products based on a high-fibre barley or oat genotype lower post-prandial glucose and insulin responses in healthy humans. European Journal of Nutrition 47: 294–300.

Alvarez-Jubete, L., Auty, M., Arendt, E.K. and Gallagher, E. (2010). Baking properties and microstructure of pseudocereal flours in gluten-free bread formulations. European Food Research and Technology 230: 437–445.

Anton, A.A., Ross, K.A., Lukow, O.M., Fulcher, R.G. and Arntfield, S.D. (2008). Influence of added bean flour (*Phaseolus vulgaris* L.) on some physical and nutritional properties of wheat flour tortillas. Food Chemistry 109: 33–41.

Arendt, E.K., Morrissey, A., Moore, M.M. and Bello, F.D. (2008). Gluten-free breads. pp. 289–319. *In*: B.R. Hamaker (ed.). Technology of Functional Cereal Products, CRC Press, New York.

Awan, J.A., Ateeq-ur-Rehman, Saleem-ur-Rehman, Siddique, M.I. and Hashmi, A.S. (1995). Evaluation of biscuits prepared from composite flour containing moth bean flour. Pakistan Journal of Agricultural Science 32: 211–217.

Borchani, C., Masoudi, M., Besbes, S., Attia, H., Deroanne, C. and Blecker, C. (2011). Effect of date flesh fiber concentrate addition on dough performance and bread quality. Journal of Texture Studies 42: 300–308.

Boschin, G. and Arnoldi, A. (2011). Legumes are valuable sources of tocopherols. Food Chemistry 127: 1199–1203.

Cakirer, O.M. and Lachance, P.A. (1975). Added micronutrients: their stability in wheat flour during storage and the baking process. Bakers' Digest 49: 53–57.

Capriles, V.D., Coelho, K.D., Guerra-Matias, A.C. and Areas, J.A. (2008). Effects of processing methods on amaranth starch digestibility and predicted glycemic index. Journal of Food Science 73: 160–164.

Chen, Z.Y., Ratnayake, W.M.N. and Cunnane, S.C. (1992). Stability of Flaxseed during baking. Journal of the American Oil Chemists' Society 71: 629–632.

Collar, C. (2016). Role of bread on nutrition and health worldwide. pp. 26–52. *In*: C.M. Rosell, J. Bajerska and A.F. El Sheikha (eds.). Bread and its Fortification, CRC Press, Boca Raton.

Comai, S., Bertazzo, A., Costa, C.V.L. and Allegri, G. (2011). Quinoa: Protein and nonprotein tryptophan in comparison with other cereal and legume flours and bread. pp. 115–126.

In: V.R. Preedy, R.R. Watson and B.P. Patel (eds.). Flour and Breads and Their Fortification in Health and Disease Prevention. Academic Press.

Dalgetty, D.D. and Baik, B.K. (2006). Fortification of Bread with Hulls and Cotyledon Fibers Isolated from Peas, Lentils, and Chickpeas. Cereal Chemistry 83: 269–274.

Darnton-Hill, E. (1998). Overview: Rationale and elements of a successful food-fortification programme. Food and Nutrition Bulletin 19: 92–100.

Demirozut, B., Saldamli, I., Gurselt, B., Uçak, A., Cetinyoku, F. and Yuzbasit, N. (2003). Determination of some metals which are important for food quality control in bread. Journal of Cereal Science 37: 171–177.

Didar, Z. and Haddad-Khodaparast, M.H. (2011). Effect of different lactic acid bacteria on phytic acid content and quality of whole wheat toast bread. Journal of Food Biosciences and Technology 1: 1–10.

Eissa, H.A., Hussein, A.S. and Mostafa, B.E. (2007). Rheological properties and quality evaluation of egyption bread and biscuits supplemented with flours of ungerminated and germinated legume seeds or mushroom. Polish Journal of Food and Nutrition Sciences 57: 487–496.

Eneche, E.H. (1999). Biscuit-making potential of millet/pigeon pea flour blends. Plant Foods for Human Nutrition 54: 21–27.

Feili, R., Zzaman, W., Wan Abdullah, W.N. and Yang, T.A. (2013). Physical and sensory analysis of high fiber bread incorporated with jackfruit rind flour. Food Science and Technology 1: 30–36.

Giami, S.Y., Mepba, H.D., Kin-Kabari, D.B. and Achinewhu, S.C. (2003). Evaluation of the nutritional quality of breads prepared from wheat-fluted pumpkin seed flour blends (*Telfairia occidentalis* Hook). Plant Foods for Human Nutrition 58: 1–8.

Gökmen, V. and Senyuva, H.Z. (2007). Acrylamide formation is prevented by divalent cations during the Maillard reaction. Food Chemistry 103: 196–203.

Gökmen, V., Mogol, B.A., Lumaga, R.B., Fogliano, V., Kaplun, Z. and Shimoni, E. (2011). Development of functional bread containing nanoencapsulated omega-3 fatty acids. Journal of Food Engineering 105: 585–591.

Gómez, M., Oliete, B., Rosell, C. M., Pando, V. and Fernández, E. (2008). Studies on cake quality made of wheat-chickpea flour blends. LWT Food Science and Technology 41: 1701–1709.

Goodman, D.W.S. (1980). Vitamin A metabolism. Federation Proceedings-Fed of American Societies for Experimental Biology 39: 2716–2722.

Gunathilake, K.D.P.P., Yalegama, C. and Kumara, A.A.N. (2009). Use of coconut flour as a source of protein and dietary fibre in wheat bread. Asian Journal of Food and Agro-Industry 2: 382–391.

Hallén, E., Ibanoğu, S. and Ainsworth, P. (2004). Effect of fermented/germinated cowpea flour addition on the rheological and baking properties of wheat flour. Journal of Food Engineering 63: 177–184.

Hansen, M., Bæch, S.B., Thomsen, A.D., Tetens, I. and Sandström, B. (2005). Long-term intake of iron fortified whole meal rye bread appears to benefit iron status of young women. Journal of Cereal Science 42: 165–171.

Hu, G., Huang, S., Cao, S. and Ma, Z. (2009). Effect of enrichment with hemicellulose from rice bran on chemical and functional properties of bread. Food Chemistry 115: 839–842.

Hyta, M. and Alpaslan, M. (2011). Apricot kernel flour and its use in maintaining health. pp. 213–221. *In*: V.R. Preedy, R.R. Watson and B.P. Patel (eds.). Flour and Breads and their Fortification in Health and Disease Prevention. Academic Press.

Ibrahim, M.I. and Hegazy, A.I. (2009). Iron bioavailability of wheat biscuit supplemented by fenugreek seed flour. World Journal of Agricultural Sciences 5: 769–776.

Jaeger, S.R., Axten, L.G., Wohlers, M.W. and Sun-Waterhouse, D. (2009). Polyphenol-rich beverages: Insights from sensory and consumer science. Journal of the Science of Food and Agriculture 89: 2356–2363.

Jansen, G.R. and Monte, W.C. (1977). Amino acid fortification of bread fed at varying levels during gestation and lactation in rats. Journal of Nutrition 107: 300–309.

Jenkins, D.J.A., Kendall, C.W.C., Vuksan, V., Augustin, L.S.A., Mehling, C., Parker, T., Vidgen, E., Lee, B., Faulkner, D., Seyler, H., Josse, R, Leiter, L.A., Connelly, P.W. and Fulgoni, V. (1999). Effect of wheat bran on serum lipids: influence of particle size and wheat protein. Journal of the American College of Nutrition 18: 159–165.

Khalil, M.M. (2002). Bioavailability of zinc in fiber-enriched bread fortified with zinc sulphate. Nahrung/Food 46: 389–393.

Khalil, A.H., Mansour, E.H. and Dawoud, F.M. (2000). Influence of malt on rheological and baking properties of wheat-cassava composite flours. LWT-Food Science and Technology 41: 1701–1709.

Khouzam, R.B., Pohl, P. and Lobinski, R. (2011). Bioaccessibility of essential elements from white cheese, bread, fruit and vegetables. Talanta 86: 425–428.

Kleinman, R.E. (2004). Vitamins. *In*: I.L. Elk Grove Village (ed.). Pediatric Nutrition Handbook. American Academy of Pediatrics.

Latorraca, M.Q., Stoppiglia, L.F., Gomes-da-Silva, M.H.G., Martins, M.S.F., de Barros Reis, M.A., Veloso, R.V. and Arantes, V.C. (2011). Effects of the soybean flour diet on insulin secretion and action. pp. 495–506. *In*: V.R. Preedy, R.R. Watson and B.P. Patel (eds.). Flour and Breads and their Fortification in Health and Disease Prevention. Academic Press.

Leelavathi, J.J. (2007). Effect of fat-type on cookie dough and cookie quality. Journal of Food Engineering 79: 299–305.

Liyanage, C. and Hettiarachchi, M. (2011). Food fortification. Ceylon Medical Journal 56: 124–127.

Lopez de Romana, D., Lonnerdal, B. and Brown, K.H. (2003). Absorption of zinc from wheat products fortified with iron and either zinc sulfate or zinc oxide. American Journal of Clinical Nutrition 78: 279–283.

Maher, L.K. and Escott-Stump, S. (2004). Krause's Food, Nutrition, and Diet Therapy. 11th ed. USA, Elsevier.

Maldonado, A. (2000). Mineral enhanced bakery products. United States Patent.

Malomo, S.A., Eleyinmi, A.F. and Fashakin, J.B. (2011). Chemical composition, rheological properties and bread making potentials of composite flours from breadfruit, breadnut and wheat. African Journal of Food Science 5: 400–410.

McKinley Health Center. Vitamins & Minerals. (2009). Web site at: http://www.mckinley.uiuc.edu.

Miki, T., Fukatsu, M. and Harada, S. (1993). Processed food made of iodine-enriched wheat flour. European Patent Application EP 0 394 904 A2, JP89-108970 (19890427).

Mustafa, A., Andersson, R., Kamal-Eldin, A. and Åman, P. (2011). Fortification with free amino acids affects acrylamide content in yeast leavened bread. pp. 325–335. *In*: V.R. Preedy, R.R. Watson and B.P. Patel (eds.). Flour and Breads and their Fortification in Health and Disease Prevention. Academic Press.

Neeraja, A. and Rajyalakshmi, P. (1996). Hypoglycemic effect of processed fenugreek seeds in humans. Journal of Food Science and Technology 33: 427–430.

Noort, M.W.J., Bult, J.H.F. and Stieger, M. (2012). Saltiness enhancement by taste contrast in bread prepared with encapsulated salt. Journal of Cereal Science 55: 218–225.

Ognean, M., Jâșcanu, V., Darie, N., Popa, L.M., Kurti, A. and Ognean, C.F. (2007). Technological and nutritional and sensorial influences on using different types of hydrocolloids on bread. Journal of Agroalimentary Processes and Technologies 8: 149–156.

Peighambardoust, S.H. and Aghamirzaei, M. (2014). Physicochemical, nutritional, shelf life and sensory properties of Iranian Sangak bread fortified with grape seed powder. Food Processing and Technology 5: 381. doi:10.4172/2157–7110.1000381.

Peng, X., Ma, J., Cheng, K.W., Jiang, Y., Chen, F. and Wang, M. (2010). The effects of grape seed extract fortification on the antioxidant activity and quality attributes of bread. Food Chemistry 119: 49–53.

Pepe, O., Ventorino, V., Cavella, S., Fagnano, M. and Brugno, R. (2013). Prebiotic content of bread prepared with flour from immature wheat grain and selected dextran-producing lactic acid bacteria. Applied and Environmental Microbiology 79: 3779. doi: 10.1128/ AEM.00502–13.

Queenan, K.M., Stewart, M.L., Smith, K.N., Thomas, W., Fulcher, R.G. and Slavin, J.L. (2007). Concentrated oat β-glucan, a fermentable fiber, lowers serum cholesterol in hypercholesterolemic adults in a randomized controlled trial. Journal of Nutrition 6, 6.

Ranhotra, G.S., Gelroth, J.A. and Okot-Kotber, B.M. (2000). Stability and Dietary Contribution of Vitamin E Added to Bread. Cereal Chemistry 77: 159–162.

Ranhotra, G.S., Gelroth, J.A., Leinen, S.D. and Schneller, F.E. (1997). Bioavailability of calcium in breads fortified with different calcium sources. Cereal Chemistry 74: 361–363.

Ranhotra, G.S., Gelroth, J.A., Langemeier, J. and Rogers, D.E. (1995). Stability and contribution of beta carotene added to whole wheat bread and crackers. Cereal Chemistry 72: 139–141.

Rao, V.S. and Vakil, U.K. 1980. Improvement of baking quality of oilseed enriched wheat flour by addition of gluten and soy lecithin. J Food Science and Technology 17: 259–262.

Rosell, C.M. (2016). Bread fortification. pp. 163–186. *In*: C.M. Rosell, J. Bajerska and A.F. El Sheikha (eds.). Bread and its Fortification, CRC Press, Boca Raton.

Rosell, C.M., Bajerska, J. and El Sheikha, A.F. (eds.). (2015). Bread and its Fortification, CRC Press, Boca Raton, pp. 407.

Rosén, L.A.H., Silva, L.O.B., Andersson, U.K., Holm, C., Östman, E.M. and Björck, I.M.E. (2009). Endosperm and whole grain rye breads are characterized by low post-prandial insulin response and a beneficial blood glucose profile. Nutrition Journal 8: 42.

Salehifar, M. (2011). Effects of protein variation on starch cystallinity and bread staling. International Conference on Food Engineering and Biotechnology 9: 300–304.

Seidel, C., Boehm, V., Vogelsang, H., Wagner, A., Persin, C., Glei, M., Pool-Zobel, B.L. and Jahreis, G. (2007). Influence of prebiotics and antioxidants in bread on the immune system, antioxidative status and antioxidative capacity in male smokers and non-smokers. British Journal of Nutrition 97: 349–356.

Serna-Saldivar, S.O. and Abril, R. (2011). Production and nutraceutical properties of breads fortified with DHA- and Omega-3-containing oils. pp. 313–323. *In*: V.R. Preedy, R.R. Watson and B.P. Patel (eds.). Flour and Breads and their Fortification in Health and Disease Prevention. Academic Press.

Shahidi, F. (2008). Antioxidants: Extraction, identification, application and efficacy measurement. Electronic Journal of Environmental, Agricultural and Food Chemistry 8: 3325–3330.

Shakeri, H., Hadaegh, H., Abedi, F., Tajabadi-Ebrahimi, M., Mazroii, N., Ghandi, Y. et al. (2014). Consumption of synbiotic bread decreases triacylglycerol and VLDL levels while increasing HDL levels in serum from patients with type-2 diabetes. Lipids 49: 695–701.

Shittu, T.A., Aminu, R.A. and Abulude, E.O. (2009). Functional effects of xanthan gum on composite cassava-wheat dough and bread. Food Hydrocolloids 23: 2254–2260.

Shrestha, A.K., Arcot, J., Dhital, S. and Crennan, S. (2012). Effect of biscuit baking conditions on the stability of microencapsulated 5-Methyltetrahydrofolic acid and their physical properties. Food and Nutrition Sciences 3: 1445–1452.

Singh, N. and Singh, P. (2011). Amaranth: Potential source for flour enrichment. pp. 101–112. *In*: V.R. Preedy, R.R. Watson and B.P. Patel (eds.). Flour and Breads and their Fortification in Health and Disease Prevention. Academic Press.

Solon, F.S., Klemmr, D., Sanchez, L., Darnton-Hill, I., Craft, N.E., Christan, P. and Westk, P. (2000). Efficacy of a vitamin A fortified wheat flour bun on the vitamin A status of Filipino school children. The American Journal of Clinical Nutrition 72: 738–744.

Somboonvechakarn, C. (2008). The Effects of Green Tea Extract on Soy Bread Physical Properties and Total Phenolic Content. Thesis. The Ohio State University, Ohio, pp. 1–18.

Souto, T.S., Brasil, A.L.D. and Taddei, J.A. deA. C. (2008). Acceptability of bread fortified with microencapsulated iron by children of daycare centers in the south and east regions of São Paulo city, Brazil. Revista de Nutrição 21: 647–657.

Sudha, M.L., Baskaran, V. and Leelavathi, K. (2007). Apple pomace as a source of dietary fiber and polyphenols and its applications on the rheological characteristics and cake making. Food Chemistry 104: 686–692.

Thondre, P.S. and Henry, C.J.K. (2009). High-molecular-weight barley b-glucan in chapatis (unleavened Indian flatbread) lowers glycemic index. Nutrition Research 29: 480–486.

Tieking, M., Korakli, M., Ehrmann, M.A., Gänzle, M.G. and Vogel, R.F. (2003). *In situ* production of exopolysaccharides during sourdough fermentation by cereal and intestinal isolates of lactic acid bacteria. Applied and Environmental Microbiology 69: 945–952.

Whiteley, P., Rodgers, J., Savery, D. and Shattock, P. (1999). A gluten-free diet as an intervention for autism and associated spectrum disorders: preliminary findings. Autism 3: 45–65.

Winkels, R.M., Brouwer, I.A., Carke, R., Katan, M.B. and Verhoef, P. (2008). Bread co-fortified with folic acid and vitamin B12 improves the folate and vitamin B12 status of healthy older people: A randomized controlled trial. American Journal of Clinical Nutrition 88: 348–355.

World Health Organization and Food and Agriculture Organization of the United Nations Guidelines on food fortification with micronutrients. (2006). Micronutrient Fortification of Food: Technology and Quality Control. [Cited on 2011 Oct 30].

Yasoda-Devi, M. and Geervani, P. (1979). Acceptability of fortified wheat products. The Indian Journal of Nutrition and Dietetics 16: 49–51.

12

The Sourdough Micro-ecosystem
An Update
Spiros Paramithiotis* and Eleftherios H. Drosinos

1. Introduction

Cereal grains constitute a major source of nutrients throughout the world. Different cereals flourish in different geographical areas, e.g., wheat and rye in Europe, and maize, millet, sorghum, tef and occasionally rice and wheat in Africa. Due to this large variety of raw materials, a great diversity of cereal-based lactic acid fermented products exists.

Many fermented cereal-based foods have been studied, particularly sourdoughs intended for bread making. Sourdough production can be as simple as mixing flour and water and placing in a warm place; sourdough will be formed after several replenishments. Depending on the desired characteristics, such as acidity, fermentation temperature or microbial activity, several variations of this process may exist (Spicher and Bruemmer 1995, Behera and Ray 2016).

Department of Food Science and Human Nutrition, Agricultural University of Athens, Athens, Greece.
* Corresponding author: sdp@aua.gr

Sourdoughs can be classified into three types according to the technology applied for their production (De Vuyst and Neysens 2005, Behera and Ray 2016):

Type I   sourdoughs are produced with traditional techniques; the microorganisms are kept in an active state with continuous, daily refreshments. This type can be further subdivided into Type Ia that include utilisation of pure starter culture from different origin, Type Ib that include all spontaneously fermented sourdoughs and Type Ic that include sourdoughs made in tropical regions and are fermented at high temperatures.

Type II   sourdoughs are semi-fluid preparations that have been created in order to assist the industrialised bread making process and serve mainly as dough acidifiers.

Type III sourdoughs are dried doughs in powder form. They are initiated by defined starter cultures and used as acidifier supplements and aroma carriers during bread making.

The increased interest in sourdough ecosystem results from the positive effect on the quality of the final product as well as the potential health benefits that have been claimed; most of these being directly related to the sourdough microbiota. Scientific research and industrial application have shared their interest regarding the latter and therefore it was brought to the epicenter of intensive research with the ultimate aim of designing suitable functional starter cultures. The requirements that a starter culture should fulfill have been adequately reviewed (Leroy and de Vuyst 2004, Gaenzle 2009, Coda et al. 2014). Thus, the lactic acid bacteria (LAB) and yeasts consisting the sourdough microbiota, the metabolic properties that justify adaptation to this micro-environment, the trophic relationships between the members of this micro-community as well as the potential of these microorganisms, from a technological point of view, have been extensively studied and critically reviewed in the next pages.

2. Sourdough Micro-ecosystem

The micro-ecosystem of spontaneously fermented sourdoughs has been extensively studied. Wheat and rye flours are most widely used; the exploitation of other cereals, pseudocereals and legumes such as sorghum, rice, barley, oat, spelt, buckwheat, millet, quinoa, amaranth, chickpea, lentil, bean and mixtures with wheat and rye has also been reported (Sterr et al. 2009, Vogelmann et al. 2009, Weckx et al. 2010, Flander et al. 2011, Moroni et al. 2012, Rieder et al. 2012, Mariotti et al. 2014, Rizzello et al. 2014, Farahmand et al. 2015, Ogunsakin et al. 2015, Kuligowski et al. 2016, Slukova et al. 2016).

LAB and yeasts dominate this micro-ecosystem (Behera and Ray 2016). The population of the former usually ranges from 10^7 to 10^9 and the latter from 10^6 to 10^7 CFUg^{-1}. Sourdoughs prepared in bakeries are often characterized by elevated yeast population that may even dominate over the population of LAB due to the ubiquitous presence of baker's yeast.

Lactobacillus sanfranciscensis, Lb. brevis, Lb. plantarum and *Lb. fermentum* dominate the LAB microbiota. However, a wide range of other LAB species belonging to the *Lactobacillus, Leuconostoc, Pediococcus, Weissella, Enterococcus* and *Lactococcus* genera is very often present at lower populations and therefore characterized as secondary microbiota. In Table 1, the diversity of LAB isolated from wheat and rye sourdoughs is exhibited; as many as 64 different species have been recovered from sourdoughs around the world.

Yeast biodiversity, on the other hand, has not been as exhaustively studied. Despite that, a remarkable biodiversity has been exposed since 24 different species belonging to 12 genera have been isolated from wheat and rye sourdoughs (Table 2). *Saccharomyces cerevisiae* and *Candida humilis* are the species most frequently isolated from wheat and rye sourdoughs.

The interactions between LAB and yeasts in sourdough have been extensively studied. In several cases a stable association between *Lb. sanfranciscensis* and *Kazachstania exigua* was observed. More accurately, this association is characteristic for the San Francisco French bread and Panettone (Corsetti and Settanni 2007). The stability of this association has been assigned to the lack of antagonism for the main carbon source, namely maltose, because *K. exigua* is a maltose-negative yeast. However, in the majority of the cases, sourdough micro-ecosystems are much more complex as is the metabolic interactions between their members. Yeast growth is generally negatively affected by LAB due to the rapid pH drop caused by the metabolites produced by the LAB, i.e., lactic and acetic acid. The latter at the usual sourdough pH (<4.0) is mainly at the un-dissociated lipophilic and membrane-diffusible form and when combined with ethanol may have detrimental effect to yeast growth (Casal et al. 1996, 1998). On the other hand, growth of LAB is affected by a species-specific manner (Paramithiotis et al. 2006).

Regarding the metabolites, production of lactic acid was not significantly affected. On the contrary, production of acetic acid and mannitol was significantly improved when *Lb. sanfranciscensis* or *Lb. brevis* were co-cultured with *S. cerevisiae*, most probably due to the antagonism for the main carbon source. Yeast metabolite, more accurately ethanol and glycerol production was also significantly affected but only according to the accompanying lactic acid bacterium (Meignen et al. 2001, Paramithiotis et al. 2006); this effect has been related to a change in the energetic potential of the yeast cells through the NADH/NAD balance.

Table 1. Lactic acid bacteria species isolated from wheat and rye sourdoughs.

Origin	Lactic acid bacteria	References
Wheat sourdoughs		
Albania	*Lb. plantarum, Lc. lactis, Ln. citreum, Ln. mesenteroides, Pd. pentosaceus*	Nionelli et al. (2014)
Belgium	*Lb. acidifarinae, Lb. brevis, Lb. buchneri, Lb. crustorum, Lb. hammesii, Lb. helveticus, Lb. nantensis, Lb. parabuchneri, Lb. paracasei, Lb. paralimentarius, Lb. plantarum, Lb. pontis, Lb. rossiae, Lb. sakei, Lb. sanfranciscensis, Lb. spicheri, Ln. mesenteroides, Pd. pentosaceus, W. cibaria, W. confusa*	Scheirlinck et al. (2007a,b, 2008)
China	*Ec. durans, Lb. brevis, Lb. crustorum, Lb. curvatus, Lb. fermentum, Lb. guizhouensis, Lb. helveticus, Lb. mindensis, Lb. paralimentarius, Lb. plantarum, Lb. rossiae, Lb. sanfranciscensis, Lb. zeae, Lc. lactis, Ln. citreum, Ln. mesenteroides, W. cibaria, W. confusa*	Zhang et al. (2011)
France	*Ec. hirae, Lb. acidophilus, Lb. brevis, Lb. casei, Lb. curvatus, Lb. delbrueckii, Lb. diolivorans, Lb. farraginis, Lb. frumenti, Lb. hammesii, Lb. hilgardii, Lb. kimchii, Lb. koreensis, Lb. nantensis, Lb. panis, Lb. paracasei, Lb. paralimentarius, Lb. paraplantarum, Lb. pentosus, Lb. plantarum, Lb. pontis, Lb. sakei, Lb. sanfranciscensis, Lb. spicheri, Lb. xiangfangensis, Lc. lactis, Ln. citreum, Ln. mesenteroides, Pd. pentosaceus, W. cibaria, W. confusa*	Infantes and Tourneur (1991), Valcheva et al. (2005), Ferchichi et al. (2008), Robert et al. (2009), Lhomme et al. (2015a,b, 2016)
Germany	*Lb. brevis, Lb. buchneri, Lb. casei, Lb. delbrueckii, Lb. fermentum, Lb. plantarum*	Spicher (1959)
Greece	*Ec. faecium, Lb. brevis, Lb. paralimentarius, Lb. plantarum, Lb. sanfranciscensis, Lb. zymae, Pd. pentosaceus, W. cibaria*	De Vuyst et al. (2002), Vancanneyt et al. (2005), Paramithiotis et al. (2010)
Italy	*Ec. faecalis, Lb. acidophilus, Lb. alimentarius, Lb. belbrueckii, Lb. brevis, Lb. casei, Lb. cellobiosus, Lb. curvatus, Lb. farciminis, Lb. fermentum, Lb. fructivorans, Lb. gallinarum, Lb. graminis, Lb. helveticus, Lb. paracasei, Lb. paralimentarius, Lb. paraplantarum, Lb. pentosus, Lb. plantarum, Lb. rhamnosus, Lb. rossiae, Lb. rossii, Lb. sakei, Lb. salivarius, Lb. sanfranciscensis, Lc. lactis, Ln. citreum, Ln. durionis, Ln. fructosus, Ln. mesenteroides, Ln. pseudomesenteroides, Pd. pentosaceus, Pd. argentinicus, Pd. inopinatus, Pd. parvulus, W. cibaria, W. confusa, W. paramesenteroides*	Gobbetti et al. (1994), Ottogalli et al. (1996), Corsetti et al. (2001, 2005), Randazzo et al. (2005), Garofalo et al. (2008), Zotta et al. (2008), Iacumin et al. (2009), Osimani et al. (2009), Minervini et al. (2012, 2015), Ventimiglia et al. (2015)
Spain	*Lb. brevis, Lb. plantarum*	Barber et al. (1983)
Rye sourdoughs		
Belgium	*Lb. brevis, Lb. hammesii, Lb. nantensis, Lb. paralimentarius, Lb. plantarum*	Scheirlinck et al. (2007b, 2008)
Bulgaria	*Lb. kimchii, Lb. paralimentarius, Lb. sanfranciscensis, Lb. spicheri*	Ganchev et al. (2014)
Denmark	*Lb. amylovorus, Lb. panis, Lb. reuteri*	Rosenquist and Hansen (2000)
Finland	*Lb. acidophilus, Lb. casei, Lb. plantarum*	Salovaara and Katunpaa (1984)
Germany	*Lb. amylovorus, Lb. frumenti, Lb. pontis, Lb. reuteri*	Mueller et al. (2001)

Ec.: Enterococcus; Lb.: Lactobacillus; Lc.: Lactococcus; Ln.: Leuconostoc; Pd.: Pediococcus; W.: Weissella

Table 2. Yeast species isolated from wheat and rye sourdoughs.

Origin	Yeasts	References
Wheat sourdoughs		
Belgium	*K. barnetti, K. unispora, S. cerevisiae, T. delbrueckii, W. anomalus*	Vrancken et al. (2010)
China	*C. humilis, C. parapsilosis, K. exigua, Mz. guilliermondii, P. kudriavzevii, S. cerevisiae, T. delbrueckii, W. anomalus*	Zhang et al. (2011)
France	*C. carpophila, C. humilis, H. pseudoburtonii, K. bulderi, K. exigua, K. servazzii, K. unispora, R. mucilaginosa, S. cerevisiae, T. delbrueckii*	Lhomme et al. (2015a, 2016)
Greece	*D. hansenii, P. membranifaciens, S. cerevisiae, T. delbrueckii, Y. lipolytica*	Paramithiotis et al. (2000, 2010)
Italy	*C. glabrata, C. humilis, C. milleri, C. stellata, K. exigua, Me. pulcherrima, P. kudriavzevii, P. membranifaciens, S. cerevisiae, S. pastorianus, T. delbrueckii, W. anomalus*	Gobbetti et al. (1994), Galli et al. (1988), Corsetti et al. (2001), Gullo et al. (2003), Succi et al. (2003), Foschino et al. (2004), Vernocchi et al. (2004a,b), Garofalo et al. (2008), Iacumin et al. (2009), Minervini et al. (2015)
Spain	*S. cerevisiae*	Barber et al. (1983)
Rye sourdoughs		
Belgium	*S. cerevisiae, W. anomalus*	Vrancken et al. (2010)
Denmark	*S. cerevisiae*	Rosenquist and Hansen (2000)
Finland	*C. humilis, C. stellata, K. exigua, K. unispora, S. cerevisiae, W. anomalus*	Salovaara and Katunpaa (1984), Mantynen et al. (1999)

C.: *Candida*, D.: *Debaryomyces*, H.: *Hyphopichia*, K.: *Kazachstania*, Me.: *Metschnikowia*, Mz.: *Meyerozyma*, P.: *Pichia*, R.: *Rhodotorula*, S.: *Saccharomyces*, T.: *Torulaspora*, W.: *Wickerhamomyces*, Y.: *Yarrowia*

2. Metabolic Activities of Yeasts and LAB

The metabolic activities of yeasts and LAB play pivotal roles in sourdough properties.

2.1 Carbohydrate Metabolism

Glucose, fructose, sucrose and maltose are the main flour carbohydrates. The LAB may catabolize them homo-fermentatively or hetero-fermentatively. The former occurs *via* the Embden-Meyerhoff pathway and the principal end product is lactate. Homo-fermentative LAB do not normally dominate spontaneous sourdough fermentations for reasons that will be explained in the following paragraphs and are only present as secondary microbiota, therefore they have not been thoroughly studied. Paramithiotis et al. (2007) assessed the flour carbohydrate catabolism of *Lb. paralimentarius* and *Pediococcus pentosaceus*, two homo-fermentative species that were reported to participate in the secondary LAB microbiota of spontaneous Greek sourdoughs (De Vuyst et al. 2002). In the case of *Lb. paralimentarius*, preference on glucose over fructose was observed when these sugars were present as carbon sources and their simultaneous catabolism took place only when glucose reached levels lower than 97.1 mM. On the other hand, this was not observed when glucose was combined with maltose or maltose and fructose; in both cases simultaneous fermentation of all carbohydrates was evident. In the case of *Pd. pentosaceus*, a definite preference to glucose over fructose and/or maltose was reported; simultaneous fermentation initiated when glucose concentration was lower than 82.8 and 90.0 mM, respectively.

Hetero-fermentative catabolism takes place *via* the phosphoketolase pathway with lactate, acetate, ethanol and CO_2 as main end products. The dominant sourdough lactic acid microbiota is either obligate (*Lb. sanfranciscensis, Lb. brevis, Lb. fermentum*) or facultative (*Lb. plantarum*) hetero-fermentative, mainly due to the ability to obtain extra energy in the form of ATP through alternative pathways; therefore their metabolic traits have been extensively studied.

Maltose is often referred to as the preferred carbon and energy source of *Lb. sanfranciscensis*. It enters the cell through a maltose/H^+ symporter and cleaves to glucose and glucose-1-phosphate by maltose phosphorylase (Stolz et al. 1996). The latter has also been reported for other hetero-fermentative sourdough LAB such as *Lb. brevis, Lb. pontis, Lb. reuteri* and *Lb. fermentum* (Stolz et al. 1996). Glucose-1-phosphate is further catabolized while glucose may be excreted or immediately phosphorylated by hexokinase activity, which is induced by glucose presence, and further catabolized (Stolz et al. 1996, Paramithiotis et al. 2007). The fate of glucose may play an important role regarding the metabolic interactions within this micro-ecosystem; if glucose is not replenished by maltose catabolism, it will soon be depleted, leading to an increased antagonism over maltose. In yeasts, maltose enters the cell through an induced proton symport mechanism and hydrolyzed by α-glucosidase, which is also inducible, to glucose that is further catabolized through glucolysis. This occurs at least in *S. cerevisiae* and the closely related species and is subjected to some extent to glucose repression (Barnett 1997, Entian and Schueller 1997).

The fate of sucrose during sourdough fermentation is more complex than of maltose. LAB may either catabolize it or convert it to homopolysaccharides. For the former, sucrose may enter the cell by the sucrose phosphoenolpyruvate-dependent phosphotransferase system (PTS) that converts sucrose to sucrose-6-phosphate, which in turns may be further hydrolyzed by sucrose-6-phosphate hydrolase to fructose and glucose-6-phosphate (Hiratsuka et al. 1998). Alternatively, extracellular glycosyltransferases may hydrolyze sucrose and synthesize fructo- or gluco-oligosaccharides from the respective sucrose moieties (Monsan et al. 2001). Regarding yeasts, sucrose may either be hydrolyzed by invertase that is held in the cell wall or sucrose inside the cell by a constitutive proton symport mechanism and hydrolyzed by an inducible cytosolic α-glucosidase (Barnett 1997).

The mode of fructose utilization is among the factors that account for dominance in this micro-ecosystem. Apart from carbon source, fructose may be utilized by sourdough LAB such as *Lb. sanfranciscensis*, *Lb. brevis* and *Lb. fermentum* as an electron acceptor. At the same time NAD+ is regenerated and therefore conversion of acetyl-CoA to acetaldehyde (and consecutively to ethanol) is not necessary. On the contrary, extra ATP is gained from the production of acetate from acetyl phosphate. In that way, catabolism is shifted towards acetate production with concomitant arrest of ethanol production for as long as fructose is available. Upon depletion of fructose, ethanol production re-commences. According to Paramithiotis et al. (2007), when fructose is the sole carbon and energy source, it is mainly utilized as an electron acceptor and only partially as a carbon source, leading to a nearly equimolar production of lactic and acetic acid. On the contrary, in the presence of another carbon source, such as glucose or maltose, fructose is almost exclusively used as an electron acceptor and the resulting fermentation quotient depends upon the carbohydrate concentration. Alternatively, oxygen, citrate, malate, fumarate, short-chain aldehydes and oxidized glutathione may serve as electron acceptors as well, offering competitive advantage to the utilizing strains (Stolz et al. 1995a,b, Gobbetti and Corsetti 1996, Gaenzle et al. 2007).

2.2 Amino Acid Metabolism

Proteolysis during sourdough fermentation is very important since it may result in dough softening, enhancement of microbial growth and flavor development. Proteolytic activities are not common among sourdough LAB. However, presence of peptide transport systems and intracellular peptidases seem to be important for their amino acid metabolism. Indeed, Vermeulen et al. (2005) reported detection of *opp* and *dtpT*, the genes encoding for the respective peptide transport systems as well as *pepT, pepR, pepC, pepN* and *pepX*, the genes encoding for the respective intracellular peptidases. Moreover, downregulation of *opp-pepN* and *dtpT* in exponentially grown cells upon supplementation of the dough with peptides was attributed to the limited supply rather than increased nitrogen requirements. Absence of extracellular protease and presence of the *dtpT,* the genes encoding for the complete Opp transport system as well as amino acid transporters with unknown specificity, ABC transporters for glutamine, methionine and cysteine, a lysine-specific permease, a serine-threonine and an arginine-ornithine antiporter, a choline-glycine betaine transporter and a γ-aminobutyrate permease, along with 20 genes encoding for cytoplasmic peptidases were reported to be predicted in the genome of *Lb. sanfranciscensis* TMW 1.1304 (Vogel et al. 2011).

Several proteolytic strains have been isolated and thoroughly studied. Gobbetti et al. (1996) characterized the proteolytic system of *Lb. sanfranciscensis* CB1 consisting of a cell envelope-associated serine protease, a metal-dependent dipeptidase and a general aminopeptidase. Additionally, an X-propyl dipeptidyl aminopeptidase was biochemically characterized (Gallo et al. 2005). Di Cagno et al. (2002) reported that two-dimensional gel electrophoresis of sourdoughs started with *Lb. alimentarius* 15M, *Lb. brevis* 14G, *Lb. sanfranciscensis* 7A or *Lb. hilgardii* 51B revealed that 37 to 42 polypeptides, with pI values ranging from 3.5 to 9.0 and molecular masses from 10 to 200 kDa were hydrolyzed. Interestingly, albumin, globulin and gliadin but not glutenin fractions were degraded. Moreover, the 31–43 fragment of A-gliadin, a toxic peptide for patients suffering from celiac disease, was hydrolyzed by enzyme preparations of these lactobacilli. These proteolytic activities were further exploited; a mixture consisting of wheat (30%), oat, millet and buckwheat flours was fermented by these LAB and baked using common bread making procedure. These products were tolerated by celiac sprue patients on the basis of determination of intestinal permeability during an acute *in vivo* challenge (Di Cagno et al. 2004).

Amino acid conversions are very important since they contribute to pH homeostasis as well as the organoleptic profile of the final product. The arginine deiminase (ADI) pathway is of manifold importance: it allows proton removal and energy production with concomitant formation of important flavor precursor. More accurately, arginine is first degraded to citrulline *via* arginine deiminase activity. Citrulline is then converted to ornithine and carbamoyl-phosphate by ornithine transcarbamoylase. Finally, carbamoyl-phosphate is further catabolized by carbamate kinase to ammonia and carbon dioxide with simultaneous ATP production. Ornithine is an important flavor precursor; it is converted during baking to 2-acetyl-1-pyrrolin that is responsible for the roasty note of bread crust (Gobbetti et al. 2005). Presence of ADI pathway has been reported in several strains belonging to species that either dominate or just participate in the secondary sourdough microbiota, such as *Lb. sanfanciscensis, Lb. brevis, Lb. fermentum, Lb. hilgardii, Lb. pontis, Lb. reuteri* and *Ec. faecium* (De Angelis et al. 2002, Thiele et al. 2002, Rollan et al. 2003, Hiraga et al. 2008, Vracken et al. 2009, Lamberti et al. 2011, Su et al. 2011, Kaur and Kaur 2015).

Glutamine is the most abundant amino acid in wheat proteins. Glutamine metabolism is of equal importance with respect to arginine: it allows extracellular pH regulation, flavor compound formation and is a source of α-ketoglutarate, a compound of central importance for amino acid catabolism. More accurately, glutamine is taken up by the cell and deaminated to glutamate by glutaminase. The latter may be excreted to the growth medium, decarboxylated to γ-aminobutyrate or converted to α-ketoglutarate by glutamate dehydrogenase. In the first case, extracellular pH is increased through proton consumption. Moreover, an effect on bread flavor is possible since glutamate accumulation has been linked to the umami taste of foods (Vermeulen et al. 2007). In the latter case, α-ketoglutarate, apart from NADH+H$^+$ regeneration, serves as an amino acceptor in the transamination of several amino acids (Gaenzle et al. 2007, Teixeira et al. 2014). Furthermore, use of α-ketoglutarate as an electron acceptor has also been reported (Zhang and Gaenzle 2010).

2.3 Exopolysaccharide Production by LAB

Biosynthesis of exopolysaccharides (EPS) is common among bacteria in general and LAB in particular. They are divided into heteropolysaccharides and homopolysaccharides according to their composition, i.e., the number of monosaccharide types that constitute the repeating units, the glycosidic linkages between them and the mode of branching (van Hijum et al. 2006).

Several reports regarding homopolysaccharide production from cereal-associated lactobacilli such as *Lb. sanfranciscensis*, *Lb. reuteri*, *Lb. pontis*, *Lb. frumenti*, *Lb. plantarum*, *Lb. panis*, *Weissella confusa* and *W. cibaria* are available in the literature (Korakli et al. 2001, Tieking et al. 2003, Di Cagno et al. 2006, Kaditzky et al. 2008, Waldherr et al. 2008, Katina et al. 2009, Galle et al. 2010a). From a physiological perspective, EPS production enhances survival of the producer strain against adverse environmental conditions (Looijesteijn et al. 2001, Boke et al. 2010, Sims et al. 2011). The essential details regarding their biosynthesis and structure have been reviewed by De Vuyst and Degeest (1999), Welman and Maddox (2003) and van Hijum et al. (2006).

The benefits of using hydrocolloids as bread improvers have been well documented. EPS produced by sourdough LAB exhibit similar positive effects and furthermore, for some of them, prebiotic properties have been claimed. Comparison between *in situ* and *ex situ* production and addition revealed that the former was more effective both in terms of compliance with consumer demand for additive reduction and quality of final product (Brandt et al. 2003). *In situ* production of EPS has been extensively studied; factors such as LAB strain, pH, sucrose concentration, temperature, dough yield, fermentation substrate composition and level of sourdough addition to bread making have been reported to exert profound effect and therefore the need for optimization has been highlighted (Korakli et al. 2001, 2003, Tieking et al. 2003, 2005, Di Cagno et al. 2006, Schwab et al. 2007, 2008, Kaditzky and Vogel 2008, Katina et al. 2009, Galle et al. 2010a,b, Wolter et al. 2014, Di Monaco et al. 2015). The most effective EPS producing strains belong to the *W. cibaria*, *W. confusa* and *Ln. mesenteroides* species. Elevated sucrose levels that are necessary for EPS production may lead to increased fructose concentration. The latter will result in extensive acidification that may compromise the positive effects of *in situ* EPS production, if strains capable to utilize fructose as an electron acceptor are used (Kaditzky et al. 2008, Galle et al. 2012).

A new insight was recently offered by Hermann et al. (2015). In that study the ability of *Kozakia baliensis* DSM 14400 and *Neoasaia chiangmaiensis* NBRC 101099 to grow in wheat, wholewheat, spelt and rye sourdoughs and produce EPS upon various sucrose concentrations was examined. A positive correlation between sucrose concentration and EPS formation was identified. Regarding acetic acid production, it was negatively correlated with sucrose concentration and positively correlated with fermentation time. The maximum EPS yield was obtained by *K. baliensis* in spelt dough and reached 49 g Kg^{-1}, which is significantly higher than the one reported for lactobacilli and may even reach approx. 15 g Kg^{-1} (Ruhmkorf et al. 2012). Additionally, the ability of *K. baliensis* to become dominant in aerobic type I sourdoughs was exhibited.

The prebiotic potential of the produced EPS has offered another interesting insight. Indeed, the *in vitro* prebiotic properties of EPS from two *Lb. sanfranciscensis* strains have been exhibited (Dal Bello et al. 2001, Korakli et al. 2002). However, further research is still necessary in order to verify *in situ* the improvement of the nutritional value.

3. Antifungal Activity of Sourdoughs

Increase of mold-free shelf-life of bakery goods has attracted significant amount of research over the last decades. Typically, it may be achieved by the addition of chemical preservatives such as acetic, propionic and sorbic acids as well as their salts or exposure to UV light, IR or microwave irradiation. However, current consumer trend towards minimal, if any, use of chemical preservatives or processes, along with several drawbacks of technological nature and the adverse health effects that have been linked to their use, resulted in an increased interest for the use of natural antifungal compounds. For that purpose, application of several essential oils, plant extracts as well as sourdough has been extensively studied.

Essential oils are volatile oily liquids obtained from different parts of plants and fruits. Their antimicrobial properties have been known for centuries. In the food industry, they are widely used as flavoring agents while their antimicrobial capacity has been displayed in numerous studies. However, the latter is usually achieved in concentrations that inflict changes in the natural taste and odor of the product. In order to reduce this effect, the use of vapors instead of the direct addition of the essential oil has been proposed (Velazquez-Nunez et al. 2013). In that case, microbial growth inhibition is achieved at relatively lower concentrations compared to liquid phase application and therefore the effect on sensory characteristics is reduced (Tyagi and Malik 2011, Avila-Sosa et al. 2012, Hadjilouka et al. 2015). This was also the case in bakery products. Only a few studies currently exist assessing *in situ* the antifungal effect of essential oils. In all cases, the effect on the organoleptic properties of the products was noticed. Rehman et al. (2007) applied citrus peel essential oils in the dough or directly on the surface of the bread by spraying. In all cases the antifungal effect was evident but in the first case the effect of the essential oil on the color and the texture of the bread was also highlighted. On the other hand, Nielsen and Rios (2000), Suhr and Nielsen (2005) and Krisch et al. (2013) assessed the antifungal capacity of essential oils using the second approach. In the first two studies, a combination strategy with modified atmosphere packaging and mustard essential oil was reported as optimal for growth inhibition of *Penicillium commune, P. roqueforti, Aspergillus flavus* and *Endomyces fibuliger* artificially inoculated on wheat and rye bread. In the latter study, a significant reduction of *A. niger, P. chrysogenum* and *Rhizopus* sp. growth was observed on wheat, rye and mixed bread slices that were treated with marjoram or clary sage essential oils.

An extended variety of plant extracts have exhibited antimould activity (Ng 2004). The efficacy of *Phaseolus vulgaris* cv. Pinto water-soluble extract and *Pisum sativum* flour hydrolyzate was assessed by Coda et al. (2008) and Rizzello et al. (2015). In both cases, slices of wheat bread artificially inoculated with *P. roqueforti* and packed in polyethylene bags did not show any visible fungi growth for at least 21 days at room temperature. This antifungal activity was assigned in the first study to phaseolin alpha-type precursor, phaseolin and erythroagglutinating phytohemogglutinin precursor and in the latter study to pea defensins 1 and 2, a nonspecific lipid transfer protein and a mixture of peptides released during hydrolysis. On the contrary, the water-soluble extract from *Amaranthus* spp. seeds failed to increase the shelf-life for more than 7 days. However, the level of contamination was reduced, compared to the control. This was assigned to four novel antifungal peptides that were sequenced (Rizzello et al. 2009).

Sourdough contains a wide range of compounds with antifungal activity. These compounds result from the metabolic activities of both LAB and yeasts that constitute the sourdough ecosystem. Lactic, acetic and propionic acids are metabolic end products of LAB whose antifungal activity has been well documented. Furthermore, several other metabolites, such as phenyllactic acid, 4-hydroxy-phenyllactic acid, hydroxyl fatty acids and peptides have exhibited significant antifungal activity and therefore have been studied to some extent.

Production of phenyllactic and 4-hydroxy-phenyllactic acids was first reported by Lavermicocca et al. (2000). Both compounds were produced by *Lb. plantarum* strains isolated from sourdough and resulted in the extension of mold-free shelf-life of wheat bread prepared with the use of a producer strain. Since then, strains belonging to several species such as *Lb. coryniformis, Lb. reuteri, Lb. brevis, Lb. amylovorus, Pd. acidilactici, W. cibaria* and *W. paramesenteroides* have been reported to produce one or both of these acids (Magnusson et al. 2003, Gerez et al. 2009, Ndagano et al. 2011, Ryan et al. 2011, Mu et al. 2012).

Another class of compounds, namely hydroxy fatty acids, has been recognized with potent antifungal functions (Hou 2008). Black et al. (2013) reported that *Lb. hammessi* strain DSM16381 was able to convert linoleic acid to a monohydroxy octadecenoic acid with antifungal activity. This was also observed during sourdough fermentation upon supplementation with linoleic acid resulting in mold-free shelf-life of the final product comparable to the respective obtained with the addition of 0.15% coriolic acid or 0.4% Ca-propionate. The latter was evident only regarding environmental contaminants; when artificial inoculation with *P. roqueforti* or *A. niger* was the case, sourdough addition was not as effective.

Cyclic dipeptides is another class of compounds whose antifungal activity has been studied. The production of cyclo (L-Phe-L-Pro) and cyclo (L-Phe-trans-4-OH-L-Pro) along with phenyllactic acid by *Lb. plantarum* strain MiLAB 393 was reported by Strom et al. (2002). Similarly, Magnusson et al. (2003) reported the production of, besides phenyllactic acid, cyclo (Phe-Pro) and cyclo (Phe-4-OH-Pro) by *Lb. coryniformis* strain Si3. Both cyclic dipeptides appeared to be very similar to the ones reported by Strom et al. (2002) however, the complete stereochemistry was not completely elucidated. Dal Bello et al. (2007) used *Lb. plantarum* strain FST 1-7, producer of phenyllactic acid, cyclo (L-Leu-L-Pro) and cyclo (L-Phe-L-Pro), in bread making experiments and concluded that the quality of the final product and the antifungal effect against *Fusarium culmorum* and *F. graminearum* were comparable to the ones obtained by *Lb. sanfranciscensis*.

An important aspect that should be taken into consideration when the use of the producer strains is considered is that the amounts of these anti-fungal compounds that are produced *in situ* are much less than the MIC calculated *in vitro*. Therefore, it is suggested that the antifungal activity observed in bread making experiments results from the interaction of these compounds and possibly others yet to be identified.

Sourdough yeasts may also produce compounds with antifungal activity. Coda et al. (2013) studied 146 yeast strains for antifungal activity and reported that the *Meyerozyma guilliermondii* strain LCF1353 produced the extracellular cell wall degrading enzyme β-1,3-glucanase that combined with ethyl acetate and ethanol, two major yeast metabolic end products, exhibited fungistatic activity against *P. roqueforti*. Indeed, bread making experiments with the use of the above strain as well as strains of *Wickeramomyces anomalus* and *Lb. plantarum*, previously selected for their antifungal potential, prolonged the shelf-life of the resulting bread for at least 14 days upon artificial inoculation.

4. Conclusions and Future Perspectives

Sourdough micro-ecosystem is very well studied. A wealth of knowledge regarding the micro-community composition, the trophic relationships between its members as well as their capabilities is currently available. However, from a geographical point of view, sourdoughs from many regions have still not been adequately assessed. Advances in the field of molecular biology and the widespread use of modern molecular techniques along with scientific curiosity are very likely to trigger such studies and sourdough micro-ecosystems still unexplored may be characterized. Moreover, the potential of sourdough bread making to promote health has already been exhibited. This is an issue of increased interest, and thus it is very likely that our knowledge on that subject will be expanded in the next few years.

Keywords: Sourdough, microbiota, yeast, lactic acid bacteria, antifungal activity

References

Avila-Sosa, R., Palou, E., Jimenez-Munguia, M.T., Nevarez-Moorillon, G.V., Navarro Cruz, A.R. and Lopez-Malo, A. (2012). Antifungal activity by vapor contact of essential oils added to amaranth, chitosan, or starch edible films. International Journal of Food Microbiology 153: 66–72.

Barber, S., R. Baguena, M.A. Martinez-Anaya and M.J. Torner. (1983). Microflora de la masa madre panaria. I. Identificacion y propiedades funcionales de microorganismos de masas madre industriales, elaboradas con harina de trigo. Revista Agrociencia e Tecnologia de Alimentos 23: 552–562.

Barnett, J.A. (1997). Sugar utilization by Saccharomyces cerevisiae. pp. 35–44. In: F.K. Zimmermann and K.-D. Entian (eds.). Yeast Sugar Metabolism. Technomic Publishing AG, Basel, Switzerland.

Behera, S.S. and Ray, R.C. (2016). Sourdough bread. pp. 53–67. In: C.M. Rosell, J. Bajerska and A.F. El Sheikha (eds.). Bread and its Fortification, CRC Press, Boca Raton.

Black, B.A., Zannini, E., Curtis, J.M. and Gaenzle, M.G. (2013). Antifungal hydroxy fatty acids produced during sourdough fermentation: microbial and enzymatic pathways, and antifungal activity in bread. Applied and Environmental Microbiology 79: 1866–1873.

Boke, H., Aslim, B. and Alp, G. (2010). The role of resistance to bile salts and acid tolerance of exopolysaccharides (EPSs) produced by yogurt starter bacteria. Archives of Biological Sciences 62: 323–328.

Brandt, M.J., Roth, K. and Hammes, W.P. (2003). Effect of an exopolysaccharide produced by *Lactoabacillus sanfranciscensis* LTH1729 on dough and bread quality. p. 80. *In*: L. De Vuyst (ed.). Sourdough: From Fundamentals to Application. Vrije Universiteit Brussels (VUB), Brussels.

Casal, M., Cardoso, H. and Leao, C. (1996). Mechanisms regulating the transport of acetic acid in *Saccharomyces cerevisiae*. Microbiology 142: 1385–1390.

Casal, M., Cardoso, H. and Leao, C. (1998). Effects of ethanol and other alkanols on transport of acetic acid in *Saccharomyces cerevisiae*. Applied and Environmental Microbiology 64: 665–668.

Coda, R., Rizzello, C.G., Di Cagno, R., Trani, A., Cardinali, G. and Gobbetti, M. (2013). Antifungal activity of *Meyerozyma guilliermondii*: Identification of active compounds synthesized during dough fermentation and their effect on long-term storage of wheat bread. Food Microbiology 33: 243–251.

Coda, R., Di Cagno, R., Gobbetti, M. and Rizzello, C.G. (2014). Sourdough lactic acid bacteria: Exploration of non-wheat cereal-based fermentation. Food Microbiology 37: 51–58.

Coda, R., Rizzello, C.G., Nigro, F., De Angelis, M., Arnault, P. and Gobbetti, M. (2008). Long-term fungal inhibitory activity of water-soluble extracts of Phaseolus vulgaris cv. Pinto and sourdough lactic acid bacteria during bread storage. Applied and Environmental Microbiology 74: 7391–7398.

Corsetti, A. and Settanni, L. (2007). Lactobacilli in sourdough fermentation. Food Research International 40: 539–558.

Corsetti, A., L. Settanni, D. van Sinderen, G.E. Felis, F. Dellaglio and M. Gobbetti. (2005). *Lactobacillus rossii* sp. nov., isolated from wheat sourdough. International Journal of Systematic and Evolutionary Microbiology 55: 35–40.

Corsetti, A., Lavermicocca, P., Morea, M., Baruzzi, F., Tosti, N. and Gobbetti, M. (2001). Phenotypic and molecular identification and clustering of lactic acid bacteria and yeasts from wheat (species *Triticum durum* and *Triticum aestivum*) sourdoughs of southern Italy. International Journal of Food Microbiology 64: 95–104.

Dal Bello, F., Clarke, C.I., Ryan, L.A.M., Ulmer, H., Schober, T.J., Strom, K., Sjogren, J., van Sinderen, D., Schnurer, J. and Arendt, E.K. (2007). Improvement of the quality and shelf life of wheat bread by fermentation with the antifungal strain *Lactobacillus plantarum* FST 1.7 Journal of Cereal Science 45: 309–318.

Dal Bello, F., Walter, J., Hertel, C. and Hammes, W.P. (2001). *In vitro* study of prebiotic properties of levan-type exopolysaccharides from Lactobacilli and non-digestible carbohydrates using denaturing gradient gel electrophoresis. Systematic and Applied Microbiology 24: 232–237.

De Angelis, M., Mariotti, L., Rossi, J., Servili, M., Fox, P.F., Rollan, G. and Gobbetti, M. (2002). Arginine catabolism by sourdough lactic acid bacteria: purification and characterization of the arginine deiminase pathway enzymes from *Lactobacillus sanfranciscensis* CB1. Applied and Environmental Microbiology 68: 6193–6201.

De Vuyst, L. and Degeest, B. (1999). Heteropolysaccharides from lactic acid bacteria. FEMS Microbiology Reviews 23: 153–177.

De Vuyst, L. and Neysens, P. (2005). The sourdough microflora: biodiversity and metabolic interactions. Trends in Food Science and Technology 16: 43–56.

De Vuyst, L., Schrijvers, V., Paramithiotis, S., Hoste, B., Vancanneyt, M., Swings, J., Kalantzopoulos, G., Tsakalidou, E. and Messens, W. (2002). The biodiversity of lactic acid bacteria in Greek traditional wheat sourdough is reflected in both composition and metabolite formation. Applied and Environmental Microbiology 68: 6059–6069.

Di Cagno, R., De Angelis, M., Limitone, A., Minervini, F., Carnevali, Corsetti, P.A., Gaenzle, M., Ciati, R. and Gobbetti, M. (2006). Glucan and fructan production by sourdough *Weissella cibaria* and *Lactobacillus plantarum*. Journal of Agricultural and Food Chemistry 54: 9873–9881.

Di Cagno, R., De Angelis, M., Lavermicocca, P., De Vincenzi, M., Giovannini, C., Faccia, M. and Gobbetti, M. (2002). Proteolysis by sourdough lactic acid bacteria: effects on wheat flour protein fractions and gliadin peptides involved in human cereal intolerance. Applied and Environmental Microbiology 68: 623–633.

Di Cagno, R., De Angelis, M., Auricchio, S., Greco, L., Clarke, C., De Vincenzi, M., Giovannini, C., D'Archivio, M., Landolfo, F., Parrilli, G., Minervini, F., Arendt, E. and Gobbetti, M. (2004). Sourdough bread made from wheat and nontoxic flours and started with selected lactobacilli is tolerated in celiac sprue patients. Applied and Environmental Microbiology 70: 1088–1096.

Di Monaco, R., E. Torrieri, O. Pepe, P. Masi and S. Cavella. (2015). Effect of sourdough with exopolysaccharide (EPS)-producing lactic acid bacteria (LAB) on sensory quality of bread during shelf life. Food and Bioprocess Technology 8: 691–701.

Entian, K.-D. and Schueller, H.-J. (1997). Glucose repression (carbon catabolite repression) in yeast. pp. 409–434. In: F.K. Zimmermann and K.-D. Entian (eds.). Yeast Sugar Metabolism. Technomic Publishing AG, Basel, Switzerland.

Farahmand, E., Razavi, S.H., Yarmand, M.S. and Morovatpour, M. (2015). Development of Iranian rice-bran sourdough breads: physicochemical, microbiological and sensorial characterization during the storage period. Quality Assurance and Safety of Crops and Foods 7: 295–303.

Ferchichi, M., Valcheva, R., Oheix, N., Kabadjova, P., Prevost, H., Onno, B. and Dousset, X. (2008). Rapid investigation of French sourdough microbiota by restriction fragment length polymorphism of the 16S-23S rRNA gene intergenic spacer region. World Journal of Microbiology and Biotechnology 24: 2425–2434.

Flander, L., Suortti, T., Katina, K. and Poutanen, K. (2011). Effects of wheat sourdough process on the quality of mixed oat-wheat bread. LWT—Food Science and Technology 44: 656–664.

Foschino, R., Gallina, S., Andrighetto, C., Rossetti, L. and Galli, A. (2004). Comparison of cultural methods for the identification and molecular investigation of yeasts from sourdoughs for Italian sweet baked products. FEMS Yeast Research 4: 609–618.

Gaenzle, M.G. (2009). From gene to function: metabolic traits of starter cultures for improved quality of cereal foods. International Journal of Food Microbiology 134: 29–36.

Gaenzle, M.G., Vermeulen, N. and Vogel, R.F. (2007). Carbohydrate, peptide and lipid metabolism of lactic acid bacteria in sourdough. Food Microbiology 24: 128–138.

Galle, S., Schwab, C., Arendt, E. and Gaenzle, M.G. (2010a). Exopolysaccharide forming *Weissella* strains as starter cultures for sorghum and wheat sourdoughs. Journal of Agricultural and Food Chemistry 58: 5834–5841.

Galle, S., Schwab, C., Arendt, E.K. and Gaenzle, M.G. (2010b). Structural and rheological characterisation of heteropolysaccharides produced by lactic acid bacteria in wheat and sorghum sourdough. Food Microbiology 28: 547–553.

Galle, S., Schwab, C., Dal Bello, F., Coffey, A., Gaenzle, M. and Arendt, E. (2012). Comparison of the impact of dextran and reuteran on the quality of wheat sourdough bread. Journal of Cereal Science 56: 531–537.

Galli, A., Franzetti, L. and Fortina, M.G. (1988). Isolation and identification of sour dough microflora. Microbiologie, Aliments, Nutrition 6: 345–351.

Gallo, G., De Angelis, M., McSweeney, P.L.H., Corbo, M.R. and Gobbetti, M. 2005. Partial purification and characterization of an X-prolyl dipeptidyl aminopeptidase from *Lactobacillus sanfranciscensis* CB1. Food Chemistry 91: 535–544.

Ganchev, I., Koleva, Z., Kizheva, Y., Moncheva, P. and Hristova, P. (2014). Lactic acid bacteria from spontaneously fermented rye sourdough. Bulgarian Journal of Agricultural Science 20: 69–73.

Garofalo, C., Silvestri, G., Aquilanti, L. and Clementi, F. (2008). PCR-DGGE analysis of lactic acid bacteria and yeast dynamics during the production processes of three varieties of Panettone. Journal of Applied Microbiology 105: 243–254.

Gerez, C.L., Torino, M.I., Rollan, G. and Font de Valdez, G. (2009).Prevention of bread mould spoilage by using lactic acid bacteria with antifungal properties. Food Control 20: 144–148.

Gobbetti, M., De Angelis, M., Corsetti, A. and Di Cagno, R. (2005). Biochemistry and physiology of sourdough lactic acid bacteria. Trends in Food Science and Technology 16: 57–69.

Gobbetti, M. and Corsetti, A. (1996). Co-metabolism of citrate and maltose by *Lactobacillus brevis* subsp. *lindneri* CB1 citrate-negative strain: effect on growth, end-products and sourdough fermentation. Zeitschrift fuer Lebensmittel Untersuchung und Forschung 203: 82–87.

Gobbetti, M., Smacchi, E. and Corsetti, A. (1996). The proteolytic system of *Lactobacillus sanfrancisco* CB1: purification and characterization of a proteinase, a dipeptidase, and an aminopeptidase. Applied and Environmental Microbiology 62: 3220–3226.

Gobbetti, M., Corsetti, A., Rossi, J., La Rosa, F. and De Vincenzi, S. (1994). Identification and clustering of lactic acid bacteria and yeasts from wheat sourdoughs of central Italy. Italian Journal of Food Science 6: 85–94.

Gullo, M., Romano, A.D., Pulvirenti, A. and Giudici, P. (2003). *Candida humilis*—dominant species in sourdoughs for the production of durum wheat bran flour bread. International Journalof Food Microbiology 80: 55–59.

Hadjilouka, A., Polychronopoulou, M., Paramithiotis, S., Tzamalis, P. and Drosinos, E.H. (2015). Effect of temperature, atmosphere and lemongrass vapors on microbial dynamics and *L. monocytogenes* survival during storage of fresh-cut rocket and melon. Microorganisms 3: 535–550.

Hermann, M., Petermeier, H. and Vogel, R.F. (2015). Development of novel sourdoughs with *in situ* formed exopolysaccharides from acetic acid bacteria. European Food Research and Technology 241: 185–197.

Hiraga, K., Ueno, Y. and Oda, K. (2008). Glutamate decarboxylase from *Lactobacillus brevis*: activation by ammonium sulfate. Bioscience Biotechnologyand Biochemistry 72: 1299–1306.

Hiratsuka, K., Wang, B., Sato, Y. and Kuramitsu, H. (1998). Regulation of sucrose-6-phosphate hydrolase activity in *Streptococcus mutans*: characterization of the scrR. Gene Infection and Immunity 66: 3736–3743.

Hou, C.T. (2008). New bioactive fatty acids. Asia Pacific Journal of Clinical Nutrition 17: 192–195.

Iacumin, L., Cecchini, F., Manzano, M., Osualdini, M., Boscolo, D., Orlic, S. and Comi, G. (2009). Description of the microflora of sourdoughs by culture-dependent and culture-independent methods. Food Microbiology 26: 128–135.

Infantes, M. and C. Tourneur. 1991. Etude de la flore lactique de levains naturels de panification provenant de differentes regions francaises. Sciences des Aliments 11: 527–545.

Kaditzky, S. and Vogel, R.F. (2008). Optimization of exopolysaccharide yields in sourdoughs fermented by lactobacilli. European Food Research and Technology 228: 291–299.

Kaditzky, S., Seitter, M., Hertel, C. and Vogel, R.F. (2008). Performance of *Lactobacillus sanfranciscensis* TMW 1.392 and its levansucrase deletion mutant in wheat dough and comparison of their impact on bread quality. European Food Research and Technology 227: 433–442.

Katina, K., Maina, N.H., Juvonen, R., Flander, L., Johansson, L., Virkki, L., Tenkanen, M. and Laitila, A. (2009). *In situ* production and analysis of *Weissella confusa* dextran in wheat sourdough. Food Microbiology 26: 734–743.

Kaur, B. and Kaur, R. (2015). Isolation, identification and genetic organization of the ADI operon in *Enterococcus faecium* GR7. Annals of Microbiology 65: 1427–1437.

Korakli, M., Rossmann, A., Gaenzle, M.G. and Vogel, R.F. (2001). Sucrose metabolism and exopolysaccharide production in wheat and rye sourdoughs by *Lactobacillus sanfranciscensis*. Journal of Agricultural and Food Chemistry 49: 5194–5200.

Korakli, M., Pavlovic, M., Gaenzle, M.G. and Vogel, R.F. (2003). Exopolysaccharide and kestose production by *Lactobacillus sanfranciscensis* LTH2590. Applied and Environmental Microbiology 69: 2073–2079.

Korakli, M., Gaenzle, M.G. and Vogel, R.F. (2002). Metabolism by bifidobacteria and lactic acid bacteria of polysaccharides from wheat and rye, and exopolysaccharides produced by *Lactobacillus sanfranciscensis*. Journal of Applied Microbiology 92: 958–965.

Krisch, J., Rentskenhand, T., Horvath, G. and Vagvolgyi, C. (2013). Activity of essential oils in vapor phase against bread spoilage fungi. Acta Biologica Szegediensis 57: 9–12.

Kuligowski, M., Nowak, J. and Jasinska-Kuligowska, I. (2016). Fermentation process and bioavailability of phytochemicals from sourdough bread. pp. 68–78. *In*: C.M. Rosell, J. Bajerska and A.F. El Sheikha (eds.). Bread and its Fortification, CRC Press, Boca Raton.

Lamberti, C., Purrotti, M., Mazzoli, R., Fattori, P., Barello, C., Coisson, J.D., Giunta, C. and Pessione, E. (2011). ADI pathway and histidine decarboxylation are reciprocally regulated in *Lactobacillushilgardii* ISE 5211: proteomic evidence. Amino Acids 41: 517–527.

Lavermicocca, P., Valerio, F., Evidente, A., Lazzaroni, S., Corsetti, A. and Gobbetti, M. (2000). Purification and characterization of novel antifungal compounds from the sourdough *Lactobacillus plantarum* strain 21B. Applied and Environmental Microbiology 66: 4084–4090.

Leroy, F. and De Vuyst, L. (2004). Lactic acid bacteria as functional starter cultures for the food fermentation industry. Trends in Food Science and Technology 15: 67–78.

Lhomme, E., Lattanzi, A., Dousset, X., Minervini, F., De Angelis, M., Lacaze, G., Onno, B. and Gobbetti, M. (2015a). Lactic acid bacterium and yeast microbiotas of sixteen French traditional sourdoughs. International Journal of Food Microbiology 215: 161–170.

Lhomme, E., Orain, S., Courcoux, P., Onno, B. and Dousset, X. (2015b). The predominance of *Lactobacillus sanfranciscensis* in French organic sourdoughs and its impact on related bread characteristics. International Journal of Food Microbiology 213: 40–48.

Lhomme, E., Urien, C., Legrand, J., Dousset, X., Onno, B. and Sicard, D. (2016). Sourdough microbial community dynamics: An analysis during French organic bread-making processes. Food Microbiology 53: 41–50.

Looijesteijn, P.J., Trapet, L., de Vries, E., Abee, T. and Hugenholtz, J. (2001). Physiological function of exopolysaccharides produced by *Lactococcus lactis*. International Journal of Food Microbiology 64: 71–80.

Magnusson, J., Strom, K., Roos, S., Sjogren, J. and Schnurer, J. (2003). Broad and complex antifungal activity among environmental isolates of lactic acid bacteria. FEMS Microbiology Letters 219: 129–135.

Mantynen, V.H., Korhola, M., Gudmundsson, H., Turakainen, H., Alfredsson, G.A., Salovaara, H. and Lindstrom, K. (1999). A polyphasic study on the taxonomic position of industrial sour dough yeasts. Systematic and Applied Microbiology 22: 87–96.

Mariotti, M., Garofalo, C., Aquilanti, L., Osimani, A., Fongaro, L., Tavoletti, S., Hager, A.-S. and Clementi, F. (2014). Barley flour exploitation in sourdough bread-making: A technological, nutritional and sensory evaluation. LWT—Food Science and Technology 59: 973–980.

Meignen, B., Onno, B., Gelinas, P., Infantes, M., Guilois, S. and Cahagnier, B. (2001). Optimization of sourdough fermentation with *Lactobacillus brevis* and baker's yeast. Food Microbiology 18: 239–245.

Minervini, F., Lattanzi, A., De Angelis, M., Celano, G. and Gobbetti, M. (2015). House microbiotas as sources of lactic acid bacteria and yeasts in traditional Italian sourdoughs. Food Microbiology 52: 66–76.

Minervini, F., Di Cagno, R., Lattanzi, A., De Angelis, M., Antonielli, L., Cardinali, G., Cappelle, S. and Gobbetti, M. (2012). Lactic acid bacterium and yeast microbiotas of 19 sourdoughs used for traditional/typical Italian breads: Interactions between ingredients and microbial species diversity. Applied and Environmental Microbiology 78: 1251–1264.

Monsan, P., Bozonnet, S., Albenne, C., Joucla, G., Willemot, R.-M. and Remaud-Simeon, M. (2001). Homopolysaccharides from lactic acid bacteria. International Dairy Journal 11: 675–685.

Moroni, A.V., Zannini, E., Sensidoni, G. and Arendt, E.K. (2012). Exploitation of buckwheat sourdough for the production of wheat bread. European Food Research and Technology 235: 659–668.

Mu, W., Yu, S., Zhu, L., Jiang, B. and Zhang, T. (2012). Production of 3-phenyllactic acid and 4-hydroxyphenyllactic acid by *Pediococcus acidilactici* DSM 20284 fermentation. European Food Research and Technology 235: 581–585.

Mueller, M.R.A., Wolfrum, G., Stolz, P., Ehrmann, M.A. and Vogel, R.F. (2001). Monitoring the growth of *Lactobacillus* species during a rye flour fermentation. Food Microbiology 18: 217–227.

Ndagano, D., Lamoureux, T., Dortu, C., Vandermoten, S. and Thonart, P. (2011). Antifungal activity of 2 lactic acid bacteria of the *Weissella* genus isolated from food. Journal of Food Science 76: M305–M311.

Ng, T.B. (2004). Antifungal proteins and peptides of leguminous and nonleguminous origins. Peptides 25: 1215–1222.

Nielsen, P.V. and Rios, R. (2000). Inhibition of fungal growth on bread by volatile components from spices and herbs, and the possible application in active packaging, with special emphasis on mustard essential oil. International Journal of Food Microbiology 60: 219–229.

Nionelli, L., Curri, N., Curiel, J.A., Di Cagno, R., Pontonio, E., Cavoski, I., Gobbetti, M. and Rizzello, C.G. (2014). Exploitation of Albanian wheat cultivars: Characterization of the flours and lactic acid bacteria microbiota, and selection of starters for sourdough fermentation. Food Microbiology 44: 96–107.

Ogunsakin, O.A., Banwo, K., Ogunremi, O.R. and Sanni, A.I. (2015). Microbiological and physicochemical properties of sourdough bread from sorghum flour. International Food Research Journal 22: 2610–2618.

Osimani, A., Zannini, E., Aquilanti, L., Mannazzu, I., Comitini, F. and Clementi, F. (2009). Lactic acid bacteria and yeasts from wheat sourdoughs of the Marche Region. Italian Journal of Food Science 21: 269–286.

Ottogalli, G., Galli, A. and Foschino, R. (1996). Italian bakery products obtained with sour dough: characterization of the typical microflora. Advances in Food Sciences 18: 131–144.

Paramithiotis, S., Tsiasiotou, S. and Drosinos, E.H. (2010). Comparative study of spontaneously fermented sourdoughs originating from two regions of Greece: Peloponnesus and Thessaly. European Food Research and Technology 231: 883–890.

Paramithiotis, S., Sofou, A., Tsakalidou, E. and Kalantzopoulos, G. (2007). Flour carbohydrate catabolism and metabolite production by sourdough lactic acid bacteria. World Journal of Microbiology and Biotechnology 23: 1417–1423.

Paramithiotis, S., Gioulatos, S., Tsakalidou, E. and Kalantzopoulos, G. (2006). Interactions between *Saccharomyces cerevisiae* and lactic acid bacteria in sourdough. Process Biochemistry 41: 2429–2433.

Paramithiotis, S., Mueller, M.R.A., Ehrmann, M.A., Tsakalidou, E., Seiler, H., Vogel, R.F. and Kalantzopoulos, G. (2000). Polyphasic identification of wild yeast strains isolated from Greek sourdoughs. Systematic and Applied Microbiology 23: 156–164.

Randazzo, C.L., Heilig, H., Restuccia, C., Giudici, P. and Caggia, C. (2005). Bacterial population in traditional sourdough evaluated by molecular methods. Journal of Applied Microbiology 99: 251–258.

Rehman, S., Hussain, S., Nawaz, H., Ahmad, M.M., Murtaza, M.A. and Rizvi, A.J. (2007). Inhibitory effect of citrus peel essential oils on the microbial growth of bread. Pakistan Journal of Nutrition 6: 558–561.

Rieder, A., Holtekjolen, A.K., Sahlstrom, S. and Moldestad, A. (2012). Effect of barley and oat flour types and sourdoughs on dough rheology and bread quality of composite wheat bread. Journal of Cereal Science 55: 44–52.

Rizzello, C.G., Lavecchia, A., Gramaglia, V. and Gobbetti, M. (2015). Long-term fungal inhibition by *Pisum sativum* flour hydrolysate during storage of wheat flour bread. Applied and Environmental Microbiology 81: 4195–4206.

Rizzello, C.G., Calasso, M., Campanella, D., De Angelis, M. and Gobbetti, M. (2014). Use of sourdough fermentation and mixture of wheat, chickpea, lentil and bean flours for enhancing the nutritional, texture and sensory characteristics of white bread. International Journal of Food Microbiology 180: 78–87.

Rizzello, C.G., Coda, R., De Angelis, M., Di Cagno, R., Carnevali, P. and Gobbetti, M. (2009). Long-term fungal inhibitory activity of water-soluble extract from *Amaranthus* spp. seeds during storage of gluten-free and wheat flour breads. International Journal of Food Microbiology 131: 189–196.

Robert, H., Gabriel, V. and Fontagné-Faucher, C. (2009). Biodiversity of lactic acid bacteria in French wheat sourdough as determined by molecular characterization using species-specific PCR. International Journal of Food Microbiology 135: 53–59.

Rollan, G., Lorca, G.L. and Font de Valdez, G. (2003). Arginine catabolism and acid tolerance response in *Lactobacillus reuteri* isolated from sourdough. Food Microbiology 20: 313–319.

Rosenquist, H. and Hansen, A. (2000). The microbial stability of two bakery sourdoughs made from conventionally and organically grown rye. Food Microbiology 17: 241–250.

Ruehmkorf, C., Jungkunz, S., Wagner, M. and Vogel, R.F. (2012). Optimization of homoexopolysaccharide formation by lactobacilli in gluten-free sourdoughs. Food Microbiology 32: 286–294.

Ryan, L.A., Zannini, E., Dal Bello, F., Pawlowska, A., Koehler, P. and Arendt, E.K. (2011). *Lactobacillusamylovorus* DSM 19280 as a novel food-grade antifungal agent for bakery products. International Journal of Food Microbiology 29: 276–83.

Salovaara, H. and Katunpaa, H. (1984). An approach to the classification of Lactobacilli isolated from Finnish sour rye dough ferments. Acta Alimentaria Polonorum 10: 231–239.

Scheirlinck, I., Van der Meulen, R., Van Schoor, A., Vancanneyt, M., De Vuyst, L., Vandamme, P. and Huys, G. (2008). Taxonomic structure and stability of the bacterial community in Belgian sourdough ecosystems as assessed by culture and population fingerprinting. Applied and Environmental Microbiology 74: 2414–2423.

Scheirlinck, I., Van der Meulen, R., Van Schoor, A., Vancanneyt, M., De Vuyst, L., Vandamme, P. and Huys, G. (2007b). Influence of geographical origin and flour type on diversity of lactic acid bacteria in traditional Belgian sourdoughs. Applied and Environmental Microbiology 73: 6262–6269.

Scheirlinck, I., Van der Meulen, R., Van Schoor, A., Huys, G., Vandamme, P., De Vuyst, L. and Vancanneyt, M. (2007a). *Lactobacillus crustorum* sp. nov., isolated from two traditional Belgian wheat sourdoughs. International Journal of Systematic and Evolutionary Microbiology 57: 1461–1467.

Schwab, C., Mastrangelo, M., Corsetti, A. and Gaenzle, M. (2008). Formation of oligosaccharides and polysaccharides by *Lactobacillus reuteri* LTH5448 and Weissella cibaria 10 M in sorghum sourdoughs. Cereal Chemistry 85: 679–684.

Schwab, C., Walter, J., Tannock, G.W., Vogel, R.F. and Gaenzle, M.G. (2007). Sucrose utilization and impact of sucrose on glycosyltransferase expression in *Lactobacillus reuteri*. Systematic and Applied Microbiology 30: 433–443.

Sims, I.M., Frese, S.A., Walter, J., Loach, D., Wilson, M., Appleyard, K., Eason, J., Livingston, M., Baird, M., Cook, G. and Tannock, G.W. (2011). Structure and functions of exopolysaccharide produced by gut commensal *Lactobacillus reuteri* 100-23. ISME Journal 5: 1115–1124.

Slukova, M., Hinkova, A., Henke, S., Smrz, F., Lukacikova, M., Pour, V. and Bubnik, Z. (2016). Cheese whey treated by membrane separation as a valuable ingredient for barley sourdough preparation. Journal of Food Engineering 172: 38–47.

Spicher, G. (1959). Die Mikroflora des Sauerteiges. I. Mitteilung: Untersuchungen uber die Art der in Sauerteigen anzutreffenden stabchenformigen Milchsaurebakterien (Genus *Lactobacillus* Beijerinck). Zeitblatt fur Bakteriologie II Abt 113: 80–106.

Spicher, G. and Bruemmer, J.-M. (1995). Baked goods. pp. 243–319. In: G. Reed and T.W. Nagodawithana (eds.). Biotechnology Second, Completely Revised Edition, Vol. 9 Enzymes, Biomass, Food and Feed. VCH, Weinheim.

Sterr, Y., Weiss, A. and Schmidt, H. (2009). Evaluation of lactic acid bacteria for sourdough fermentation of amaranth. International Journal of Food Microbiology 136: 75–82.

Stolz, P., Boecker, G., Hammes, W.P. and Vogel, R.F. (1995a). Utilization of electron acceptors by lactobacilli isolated from sourdough. I. *Lactobacillus sanfrancisco*. Zeitschrift fuer Lebensmittel Untersuchung und Forschung 201: 91–96.

Stolz, P., Vogel, R.F. and Hammes, W.P. (1995b). Utilization of electron acceptors by lactobacilli isolated from sourdough. II. *Lactobacillus pontis, L. reuteri, L. amylovorus*, and *L. fermentum*. Zeitschrift fuer Lebensmittel Untersuchung und Forschung 201: 402–410.

Stolz, P., Hammes, W.P. and Vogel, R.F. (1996). Maltose-phosphorylase and hexokinase activity in lactobacilli from traditionally prepared sourdoughs. Advances in Food Sciences 18: 1–6.

Strom, K., Sjogren, J., Broberg, A. and Schnurer, J. (2002). *Lactobacillus plantarum* MiLAB 393 produces the antifungal cyclic dipeptides cyclo(L-Phe-L-Pro) and cyclo(L-Phe-trans-4-OH-L-Pro) and phenyllactic acid. Applied and Environmental Microbiology 68: 4322–4327.

Su, M.S., Schlicht, S. and Gaenzle, M.G. (2011). Contribution of glutamate decarboxylase in *Lactobacillus reuteri* to acid resistance and persistence in sourdough fermentation. Microbial Cell Factories 10(Suppl 1): S8.

Succi, M., Reale, A., Andrighetto, C., Lombardi, A., Sorrentino, E. and Coppola, R. (2003). Presence of yeasts in southern Italian sourdoughs from *Triticum aestivum* flour. FEMS Microbiology Letters 225: 143–148.

Suhr, K.I. and P.V. Nielsen. (2005). Inhibition of fungal growth on wheat and rye bread by modified atmosphere packaging and active packaging using volatile mustard essential oil. Journal of Food Science 70: M37–M44.

Teixeira, J.S., Seeras, A., Sanchez-Maldonado, A.F., Zhang, C., Su, M.S.-W. and Gaenzle, M.G. (2014). Glutamine, glutamate, and arginine-based acid resistance in *Lactobacillus reuteri*. Food Microbiology 42: 172–180.

Thiele, C., Gaenzle, M.G. and Vogel, R.F. (2002). Contribution of sourdough lactobacilli, yeast, and cereal enzymes to the generation of amino acids in dough relevant for bread flavor. Cereal Chemistry 79: 45–51.

Tieking, M., Korakli, M., Ehrmann, M.A., Gaenzle, M.G. and Vogel, R.F. (2003). *In situ* production of exopolysaccharides during sourdough fermentation by cereal and intestinal isolates of lactic acid bacteria. Applied and Environonmental Microbiology 69: 945–952.

Tieking, M., Kuhnl, W. and Gaenzle, M.G. (2005). Evidence for formation of heterooligosaccharides by *Lactobacillus sanfranciscensis* during growth in wheat sourdough. Journal of Agricultural and Food Chemistry 53: 2456–2461.

Tyagi, A.K. and Malik, A. (2011). Antimicrobial potential and chemical composition of *Mentha piperita* oil in liquid and vapour phase against food spoiling microorganisms. Food Control 22: 1707–1714.

Valcheva, R., Korakli, M., Onno, B., Prevost, H., Ivanova, I., Ehrmann, M.A., Dousset, X., Gaenzle, M.G. and Vogel, R.F. (2005). *Lactobacillus hammesii* sp. nov., isolated from French sourdough. International Journal of Systematic and Evolutionary Microbiology 55: 763–767.

van Hijum, S.A.F.T., Kralj, S., Ozimek, L.K., Dijkhuizen, L. and van Geel-Schutten, I.G.H. (2006). Structure-function relationships of glucansucrase and fructansucrase enzymes from lactic acid bacteria. Microbiology and Molecular Biology Reviews 70: 157–176.

Vancanneyt, M., Neysens, P., De Wachter, M., Engelbeen, K., Snauwaert, C., Cleenwerck, I., Van der Meulen, R., Hoste, B., Tsakalidou, E., De Vuyst, L. and Swings, J. (2005). *Lactobacillus acidifarinae* sp. nov. and *Lactobacillus zymae* sp. nov., from wheat sourdoughs. International Journal of Systematicand Evolutionary Microbiology 55: 615–620.

Velazquez-Nunez, M.J., Avila-Sosa, R., Palou, E. and Lopez-Malo, A. (2013). Antifungal activity of orange (*Citrus sinensis* var. Valencia) peel essential oil applied by direct addition or vapor contact. Food Control 31: 1–4.

Ventimiglia, G., Alfonzo, A., Galluzzo, P., Corona, O., Francesca, N., Caracappa, S., Moschetti, G. and Settanni, L. (2015). Codominance of *Lactobacillus plantarum* and obligate heterofermentative lactic acid bacteria during sourdough fermentation. Food Microbiology 51: 57–68.

Vermeulen, N., Ganzle, M.G. and Vogel, R.F. (2007). Glutamine deamidation by cereal-associated lactic acid bacteria. Journal of Applied Microbiology 103: 1197–1205.

Vermeulen, N., Pavlovic, M., Ehrmann, M.A., Gaenzle, M.G. and Vogel, R.F. (2005). Functional characterization of the proteolytic system of *Lactobacillus sanfranciscensis* DSM 20451T during growth in sourdough. Applied and Environmental Microbiology 71: 6260–6266.

Vernocchi, P., Valmorri, S., Dalai, I., Torriani, S., Gianotti, A., Suzzi, G., Guerzoni, M.E., Mastrocola, D. and Gardini, F. (2004a). Characterization of the yeast population involved in the production of a typical italian bread. Journal of Food Science 69: 182–186.

Vernocchi, P., Valmorri, S., Gatto, V., Torriani, S., Gianotti, A., Suzzi, G., Guerzoni, M.E. and Gardini, F. (2004b). A survey on yeast microbiota associated with an Italian traditional sweet leavened baked good fermentation. Food Research International 37: 469–476.

Vogel, R.F., Pavlovic, M., Ehrmann, M.A., Wiezer, A., Liesegang, H., Offschanka, S., Voget, A. Angelov, S., Boecker, G. and Liebl, W. (2011). Genomic analysis reveals *Lactobacillus sanfranciscensis* as stable element in traditional sourdoughs. Microbial Cell Factories 10(Suppl 1): S6.

Vogelmann, S.A., Seitter, M., Singer, U., Brandt, M.J. and Hertel, C. (2009). Adaptability of lactic acid bacteria and yeasts to sourdoughs prepared from cereals, pseudocereals and cassava and use of competitive strains as starters. International Journal of Food Microbiology 130: 205–212.

Vrancken, G., De Vuyst, L., van der Meulen, R., Huys, G., Vandamme, P. and Daniel, H.M. (2010). Yeast species composition differs between artisan bakery and spontaneous laboratory sourdoughs. FEMS Yeast Research 10: 471–481.

Vrancken, G., Rimaux, T., Wouters, D., Leroy, F. and De Vuyst, L. (2009). The arginine deiminase pathway of *Lactobacillus fermentum* IMDO 130101 responds to growth under stress conditions of both temperature and salt. Food Microbiology 26: 720–727.

Waldherr, F.W., Meissner, D. and Vogel, R.F. (2008). Genetic and functional characterization of *Lactobacillus panis* levansucrase. Archives of Microbiology 190: 497–505.

Weckx, S., Van der Meulen, R., Maes, D., Scheirlinck, I., Huys, G., Vandamme, P. and De Vuyst, L. (2010). Lactic acid bacteria community dynamics and metabolite production of rye sourdough fermentations share characteristics of wheat and spelt sourdough fermentations. Food Microbiology 27: 1000–1008.

Welman, A.D. and Maddox, I.S. (2003). Exopolysaccharides from lactic acid bacteria: perspectives and challenges. Trends in Biotechnology 21: 269–274.

Wolter, A., Hager, A.-S., Zannini, E., Galle, S., Gaenzle, M.G., Waters, D.M. and Arendt, E.K. (2014). Evaluation of exopolysaccharide producing *Weissella cibaria* MG1 strain for the production of sourdough from various flours. Food Microbiology 37: 44–50.

Zhang, C. and Gaenzle, M.G. (2010). Metabolic pathway of a-ketoglutarate in *Lactobacillus sanfranciscensis* and *Lactobacillus reuteri* during sourdough fermentation. Journal of Applied Microbiology 109: 1301–1310.

Zhang, J., Liu, W., Sun, Z., Bao, Q., Wang, F., Ju, Y., Chen, W. and Zhang, H. (2011). Diversity of lactic acid bacteria and yeasts in traditional sourdoughs collected from western region in Inner Mongolia of China. Food Control 22: 767–774.

Zotta, T., Piraino, P., Parente, E., Salzano, G. and Ricciardi, A. (2008). Characterization of lactic acid bacteria isolated from sourdoughs for Cornetto, a traditional bread produced in Basilicata (Southern Italy). World Journal of Microbiology and Biotechnology 24: 1785–1795.

13

Traditional Cereal Beverage Boza
Fermentation Technology, Microbial Content and Healthy Effects

Penka Petrova[1],* and *Kaloyan Petrov*[2]

1. Introduction: The Ancient History and Modern Production of Boza

Boza is a cereal beverage that is obtained from the fermentation of whole grain or flour. It is popular in Turkey, Kazakhstan, Kyrgyzstan, Albania, Bulgaria, Macedonia, Montenegro, Bosnia and Herzegovina, and parts of Romania and Serbia. The Turkish name, Boza, comes from the Persian word, *buze*, meaning millet. The Turks who lived in Middle Asia called this beverage *bassoi* (Arici and Daglioglu 2002); there are also similar beverages produced in East European countries (*braga* or *brascha*), the Balkans (*busa*), and Egypt (*bouza*).

Boza's origin dates back to the ancient populations that lived in Anatolia and Mesopotamia. The Greek historian Xenophon records that Boza was made in eastern Anatolia in 401 B.C. and stored in clay jars that were buried beneath the ground (LeBlanc and Todorov 2011). However, *Boza* name specifically, is known from the 10th century, and the preparation

---
[1] Institute of Microbiology, Bulgarian Academy of Sciences, 26, Acad. G. Bontchev str., 1113 Sofia, Bulgaria.
[2] Institute of Chemical Engineering, Bulgarian Academy of Sciences, 103, Acad. G. Bontchev str., 1113 Sofia, Bulgaria.
  E-mail: kaloian04@yahoo.com
* Corresponding author: pepipetrova@yahoo.com

formula of the contemporary beverage dates from 16th century, where its production spread over the countries conquered by the Ottomans. The famous traveler Evliya Çelebi described the consumption of sweet and sour Boza as two different commodities in 17th century Istanbul (Selçuk 2016). Several varieties of sweet Boza existed, such as the rice *sübya* of Egypt, or the *maksıma* kind of Crimea. Sweet Boza was white, topped with a dense layer (*kaymak*), and drunk by pious, pregnant, and the elders. Sour Boza, on the other hand, was a "questionable" substance with higher alcoholic content and other additives, and was mostly consumed by poorer artisans, soldiers and dockers. That was why in the 17th century, Sultan Mehmed IV prohibited alcoholic drinks and, given that Boza originally had a fair dose of alcohol its sale was forbidden. However, at the edges of the Empire, people continued producing it, but with far less alcohol content.

Today, Boza is commercially produced in hundreds of tons in all countries of Balkan Peninsula. Its industrial manufacturing started in 1876 by Haji Sadiq, who opened the specialised stores, Vefa, in Istanbul, followed by Ömür Bozacısı of Bursa, Karakedi Bozacısı of Eskişehir, and Akman Boza of Ankara. The largest Boza companies in Bulgaria are Bomax of Haskovo, Gribash (Peshtera), Harmonica and Radomirska Boza (Radomir), which are made from rye, millet or einkorn. In honor of Bulgarian Boza, a monument of a Boza street vendor was raised in Radomir town, the only one of its kind worldwide (Fig. 1). In Albania, Boza was produced by the Pacara Company

Figure 1. The monument of Boza street vendor, Radomir, Bulgaria.

(Tirana); in Macedonia by the famous Akman (Skopje), and in Thessaloniki, Greece by the specialized confectionery Hatzis, that was founded in 1908.

2. Fermentation Technology of Boza

The unique taste and odor of Boza, as well as its ability to positively influence the human health depend on the following factors that affect the quality of the final product: the usage of selected grain or flour sources, inoculation with preferably probiotic microbial starters, and adherence to the proper fermentation conditions.

2.1 Boza Appearance and Composition

Boza is a viscous low-alcoholic drink with a pleasant sweet taste and slightly acid savour, varying in color from creamy-white and beige to light brownish. It contains about 0.50–1.61% protein, about 12.3% carbohydrate (8–12% sugar), and 75–85% moisture (Zorba et al. 2003, Yegin and Uren 2008). According to the manufacturers (Bomax, Ömür Bozasi), their final products contain 8–57.5% carbohydrates, 0.5–3.5% proteins, 0.4–0.5% lipids, 29 mg Ca, 1.3 mg Fe, 95 mg phosphorus, 1 mg Zn, and vitamins (tiamin, riboflavin and niacin). Boza, prepared with leblebi (yellow chickpea) flour alone, has higher protein (8.46%) and minerals content, Ca (268.1 ppm), K (525.3 ppm), P (199.2 ppm), Zn (3.13 ppm), Mn (1.14 ppm), and less sucrose (0.05%) (Çelik et al. 2016). Oxalic, lactic, pyruvic, acetic and malic acids were found in different Boza samples. Citric acid was detected in rice and maize Boza whereas orotic acid was found in rice, maize and millet Boza (Akpinar-Bayizit et al. 2010).

In general, the pH of the Boza samples ranges initially from 6.7 to 3.4–3.9 at the end of fermentation, as the content of lactic acid (LA) varies from 0.02 to 0.3 mmol/g. LA amount was found to be the lowest in millet with 0.3% (w/v) and the highest in wheat Boza (0.6%), due to the probable higher fermentable carbohydrate content of wheat compared to other raw materials (Akpinar-Bayizit et al. 2010). However, the lactic acid causes the fresh and nice flavor of Boza.

The alcohol content of Boza varies from 0.02 to 0.79% (Hancioğlu and Karapinar 1997, Köse and Yücel 2003), although the product in Egypt has high alcohol content (up to 7% by volume) and it is consumed as beer (LeBlank and Todorov 2011). The alcohol content is lower in wheat Boza (0.5%) and shows fluctuations during storage. The acidity and alcohol content depends mainly on the fermentation period; for longer fermentation periods the acidity and the alcohol concentration increase (Akpinar-Bayizit et al. 2010).

The final product of Boza may contain also water, sugar or saccharine (15–20% w/w), and cinnamon or cocoa (Kabak and Dobson 2011). According to the Turkish Standard Institute, total dry matter and total sugar (as sucrose) content should be a minimum of 20% and 10%, respectively. Ethyl alcohol content should not exceed 2% by volume and total acidity in Boza as lactic acid should be 0.2–1.0% (Tangüler 2014).

2.2 Grain Sources for Boza Production

Boza is made of various kinds of cereals, but the variety with the best quality and taste is made of millet flour. Albanian Boza differs from Boza produced in other regions, as the main ingredient in its production is corn, while in other countries, such as in Turkey, Bulgaria, and Macedonia, the main component is usually either bulgur, millet, barley, chick pea or rice semolina/flour. The grains of all cereal plants have high nutritive value that depends on the species and growth conditions. The most important from a technological point of view is the content of the wheat proteins glyadines and glutenines (water insoluble, forming the gluten), starch, cellulose and lipids. The most used chemical composition for Boza produced from wheat grains is: proteins–12.6%, starch–67.8%, lipids–1.6%, minerals–1.7%, and water–15.5%.

Commonly, Boza is made using mixed milled grains or flours (maize, barley, rye, wheat, millet or rice). Cracked leblebi flour could successfully substitute the wheat, rice or maize flour; however, full substitution of cereal flours negatively influences the sensory properties of Boza (Çelik et al. 2016). The chemical composition of the grains frequently used in Boza manufacturing is presented in Table 1.

Einkorn, frequently used in Bulgarian Boza, has a higher percentage of protein than modern red wheats and is considered as more nutritious

Table 1. Chemical composition of the grains commonly used for Boza production.

Grain	Composition (%)			
	Carbohydrates	*Proteins*	*Fat*	*Dietary fibre*
Wheat	71.2	12.6	1.5	12.2
Maize	18.7	3.3	1.4	2.0
Millet	72.8	11.0[1]	4.2[2]	8.5
Barley	77.7	9.9	1.2	15.6
Rye	60.7	8.8	1.7	13.2
Rice	80.0	7.1	0.7	1.3
Leblebi[3]	41.1–47.4	21.7–23.4	2.6	13.9–17.6

[1] No gluten
[2] Of them - 2% are the polyunsaturated Ω-6 fat acids
[3] According to Çelik et al. 2016

because it also has higher levels of phosphorus, potassium, pyridoxine, and beta-carotene. In contrast to modern forms of wheat, the gliadin protein of einkorn may not be as toxic to sufferers of celiac disease. That is why it is recommended in any gluten-free diet (Stallknecht et al. 1996, Pizzuti et al. 2006).

2.3 Recipes, Production and Storage of Boza

There are three Bulgarian and one European patents that protect the method for Boza production (Sendov 1996, Petrov 1998, Enchev 2003, Borcakli et al. 2014). The scheme of technological process is presented at Fig. 2, the several production steps can be summarized: (i) preparation of the raw materials, (ii) boiling, (iii) cooling, (iv) sugar addition, and (v) fermentation (Arici

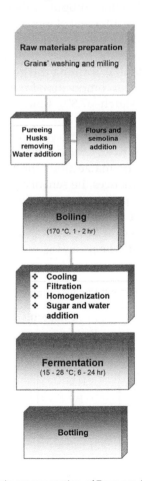

Figure 2. Schematic representation of Boza production technology.

and Daglioglu 2002). Boza (2–3%) from a previous batch is usually used as a starter culture; its ratio depends on the season and temperature at which it was produced. Inoculated mixture is incubated at 15–25°C to ferment for nearly 24 hr before it is ready for use (LeBlank and Todorov 2011). During the course of the fermentation the populations of lactic acid bacteria (LAB) and yeast significantly raise (Hancioğlu and Karapinar 1997). Besides using previously produced Boza, sourdough or yoghurt could be also used as Boza starter culture.

Bulgarian Boza is prepared from gruel of boiled and pressed combination of cereal grains. First, the alien seeds, weeds, dirt and rough admixture are removed from the grains by the use of air-sieve separator and cylindrical grain-cleaner. Then, the grains are washed in a drum washing machine to eliminate any dirt that is stuck to the grains, prior to milling. Next, the processes of boiling (in autoclave, 170°C), cooling, diluting and sieving of the gruel are carried out.

During the heating of the grains a number of physico-chemical processes occur: the softness and swelling of grains, tearing of the husk, swelling of the starch granules, hydrolyses of the starch, proteins denaturizing. The starch is hydrolyzed to dextrins, which leads to important viscosity changes (from 1137 to 500 cP, Gotcheva et al. 2001). Glucose concentration increases from 0.48 g/l to 7.94 g/l. Part of lipids (in the grain) dissociates to glycerin and butyric acids or is emulsified. The boiling of the raw materials continues for 30–35 min, by direct heating, or for 60–120 min, by indirect heating. Samples are taken to gauge when the process ends. When the gruel obtains the color, similar to the color of a well-baked (bread) crust, and the nice bread aroma with a slightly nuance of caramel occurs, it is immediately transferred from the presscooker to the cooler in order to stop the thermal effect that could drive over caramelization and embittering of the product. When the temperature falls under 35°C, the gruel is diluted with water to 8% solid content, and sweetened to 10% sugar content. The mix is inoculated and fermented at the optimal temperature 26–28°C for several hours. During the fermentation a significant increase in glucose content is observed, while the pH, viscosity, free amino nitrogen content and dry matter decreases (Blandino et al. 2003).

When whole-wheat flour was used, the average particle diameter in the flour was found to be 12.3 mm, and the moisture, starch, protein, phosphorus and ash contents measured were 13.6%, 65.1%, 10.1%, 3.1% and 1.6%, respectively (Gotcheva et al. 2001). Higher viscosity values were observed when bran particles were present in the flour.

The sieved Boza could be additionally sweetened, and must be kept at a temperature not higher than 4°C. Boza should not be allowed to freeze.

Usually, in Turkey people consume Boza during the winter season. In Bulgaria, Boza is preferred also in summer. The best drink is the well-cooled

Boza, in which the possibility of additional spontaneous fermentation is prevented; however, it is may be consumed at every stage of the fermentation as well. Generally, Boza is served pure, but it might also be served with cinnamon and/or unsalted roasted chickpeas sprinkled over the fermented cereal (Genç et al. 2002, Yegin and Fernández-Lahore 2012). In Bulgaria it may be supplemented with cocoa.

3. Microbiota of Boza

Two different types of fermentation occur simultaneously during Boza fermentation. The alcohol fermentation produces carbon dioxide bubbles and increases the volume. The lactic acid fermentation produces lactic acid and provides the acidic character of Boza (Cosansu 2009). The microbiota responsible for the fermentation of Boza seems to be particularly heterogeneous, including both *homo*- or *hetero*-fermentative LAB, and yeasts (Gotcheva et al. 2000, Altay et al. 2013). Diversity in fermentation flora and variations in numbers of LAB and yeast from the different samples could be due to different raw materials, production processes and storage conditions (Gotcheva et al. 2001, Botes et al. 2007).

The early studies of microorganisms that were present in Boza were carried out by examination of their morphological and physiological characteristics (Gotcheva et al. 2000, 2001, Todorov and Dicks 2004, 2005). Later, molecular methods for the strains identification by 16S rDNA sequencing were applied (Botes et al. 2007, Caputo et al. 2012, Todorov and Dicks 2006, Todorov 2010, Kivanç et al. 2011). Besides with regard to the traditional methods, Osimani et al. (2015) based the analysis of Boza biodiversity on the Polymerase Chain Reaction-Denaturing Gradient Gel Electrophoresis (PCR-DGGE).

3.1 Representatives of Genus Lactobacillus

The first comprehensive investigation of microbial content of Boza was carried out by Hancioğlu and Karapinar (1997), who isolated 77 different strains of LAB from Boza samples. *Leuconostoc paramesenteroides* (25.6%) was found to be predominant, followed by *Leuc. sanfrancisco* (21.9%) and *Leuc. mesenteroides* subsp. *mesenteroides* (18.6%), *Lactobacillus coryniformis* (9.1%), *Lb. confusis* (7.8%), *Leuc. mesenteroides* subsp. *dextranicum* (7.3%), and *Lb. fermentum* (6.5%). Although the total microbial count of Boza vary from one sample to another, the average value in Bulgarian Boza, obtained by Gotcheva et al. (2000, 2001) was around 10^7 colony-forming units (CFU) per cm^3, of which 70% were LAB. The microbial load of Bulgarian Boza was approximately 10 times smaller than that observed in Boza prepared in Turkey, and the percentage of LAB is also considerably lower (in Turkish

Boza LAB make up 98% of the microflora). The most proliferate lactobacilli in Bulgarian Boza which have been reported are *Lb. plantarum* (24% of LAB), *Lb. acidophilus* (23% of LAB) and *Lb. fermentum* (19% of LAB) (Gotcheva et al. 2000). *Lb. plantarum* was frequently isolated from Turkish Boza samples too (Kivanç et al. 2011).

Lb. paracasei subsp. *paracasei*, *Lb. coryniformis* and *Lb. fermentum* have been isolated from Boza produced in both Turkey and Bulgaria (Hancioğlu and Karapinar 1997, Gotcheva et al. 2000, 2001, Todorov and Dicks 2006, Botes et al. 2007, Kivanç et al. 2011, Petrova and Petrov 2012, Blagoeva and Gotcheva 2013, Osimani et al. 2015). Osimani et al. (2015) also reported about the isolation of *Lb. casei*, *Lb. buchneri* and *Lb. parabuchneri*, along with *Weissella confusa* and *Leuc. citreum*. The representatives of lactobacilli found in Boza samples are listed in Table 2.

3.2 Lactic Acid Cocci Isolated from Boza Samples

Lactic acid cocci are rarely common in Boza. Only 3.7% of the strains, isolated by Hancioğlu and Karapinar (1997) belonged to lactic acid cocci, namely, to *Oenococcus oeni*. Later, multiple LAB species ascribed to *Lactococcus* and *Pediococcus*, which were unrevealed using the culture-dependent approach, were identified by PCR-DGGE analysis (Osimani et al. 2015). Thus, *Pediococcus parvulus*, *Pd. ethanolidurans* and *Lc. lactis* were found in Boza samples (Table 3).

3.3 Yeasts and Molds in Boza

Hancioğlu and Karapinar (1997) reported that 83% of yeast isolates in Turkish Boza were *Saccharomyces uvarum* and 17% - *S. cerevisiae*. According to Gotcheva et al. (2000), the dominant species in Bulgarian Boza was found to be *S. cerevisiae* accounting for 47% of the yeast population. Species of *Candida* and *Geotrichum* were also identified, as *Geotrichum penicillatum* (12%) and *Geotrichum candidum* (8%) were detected. Yeasts of the genus *Candida* represent approximately one third of Bulgarian Boza yeast population (*Candida tropicalis* - 19% and *Candida glabrata* - 14%). Göcmen et al. (2000) isolated 50 yeasts from Boza samples obtained from Turkish local markets and *Candida* spp. was predominant (26 species). However, the major part of the strains consists of unwanted wild yeasts, which negatively affect the quality of Boza: 6 strains were classified as *Trichosporon*, 2—as *Torulaspora* and 1 species as *Rhodotorula*. Only 11 species were classified as *Saccharomyces* genera which are necessary for alcoholic fermentation. Recently, the presence of *Pichia fermentans* and *P. norvegensis* was also reported in Boza (Caputo et al. 2012, Osimani et al. 2015).

Table 2. The representatives of order *Lactobacillales*, isolated from Boza.

Genus	Species	Reference
Lactobacillus	*Lb. plantarum*	Gotcheva et al. 2000, 2001, Todorov and Dicks 2006, Botes et al. 2007, Todorov 2010, Petrova et al. 2010, Kivanç et al. 2011, Blagoeva and Gotcheva 2013
	Lb. paraplantarum	Kivanç et al. 2011
	Lb. pentosus	Botes et al. 2007, Todorov and Dicks 2006, Petrova et al. 2010, Blagoeva and Gotcheva 2013
	Lb. acidophilus	Gotcheva et al. 2000, 2001
	Lb. casei	Blagoeva and Gotcheva 2013, Osimani et al. 2015
	Lb. paracasei	Todorov and Dicks 2006, Botes et al. 2007, Kivanç et al. 2011, Petrova and Petrov 2012, Blagoeva and Gotcheva 2013, Osimani et al. 2015
	Lb. rhamnosus	Todorov and Dicks 2006, Botes et al. 2007
	Lb. brevis	Gotcheva et al. 2000, Botes et al. 2007, Kivanç et al. 2011
	Lb. buchneri	Osimani et al. 2015
	Lb. parabuchneri	Osimani et al. 2015
	Lb. fermentum	Hancioğlu and Karapinar 1997, Gotcheva et al. 2000, 2001, Botes et al. 2007, Osimani et al. 2015
	Lb. coryniformis	Hancioğlu and Karapinar 1997, Osimani et al. 2015
	Lb. graminis	Kivanç et al. 2011
	Lb. sanfrancisco	Hancioğlu and Karapinar 1997
	Lb. coprophilus	Gotcheva et al. 2000
Leuconostoc	*Leuc. citreum*	Kivanç et al. 2011, Osimani et al. 2015
	Leuc. lactis	Todorov 2010
	Leuc. mesenteroides	Gotcheva et al. 2000
	Leuc. mesenteroides subsp. *dextranicum*	Hancioğlu and Karapinar 1997, Todorov and Dicks (2004)
	Leuc. mesenteroides subsp. *mesenteroides*	Hancioğlu and Karapinar 1997
	Leuc. paramesenteroides	Hancioğlu and Karapinar 1997
	Leuc. raffinolactis	Gotcheva et al. 2000
Weissella	*Weissella oryzae*	Osimani et al. 2015
	Weissella confusa	Hancioğlu and Karapinar 1997, Gotcheva et al. 2000, Osimani et al. 2015

Table 3. The representatives of lactic acid cocci isolated from Boza.

Genus	Species	Reference
Lactococcus	*Lc. lactis*	Tuncer and Ozden 2010, Kivanç et al. 2011, Osimani et al. 2015
Oenococcus	*O. oeni*	Hancioğlu and Karapinar 1997
Enterococcus	*Ent. faecium*	Todorov 2010
Pediococcus	*Pediococcus* spp.	Kivanç et al. 2011
	Pd. pentosaceus	Todorov and Dicks 2005
	Pd. parvulus	Osimani et al. 2015
	Pd. ethanolidurans	Osimani et al. 2015

The analysis of the mycobiota isolated from Boza samples show the occurrence of several taxa: *P. fermentans*, *P. guillermondii*, *P. norvegensis* and *Torulaspora* spp. (Botes et al. 2007, Caputo et al. 2012, Osimani et al. 2015). The PCR-DGGE analysis performed on the microbial DNA extracted directly from the three Boza samples permitted the identification of *Galactomyces geotricum*, *G. candidum* and *G. fragrans*, not recovered with the culture-dependent method (Table 4).

Aspergillus fumigatus, *Penicillum chrysogenum*, *Fusarium oxysporum*, *Acremonium* spp., *Geotrichum candidum* and *Geotrichum capitatum* were isolated from Boza. Although the ochratoxigenic molds have not been reported in the Boza, ochratoxin A was determined in two of five samples. The source of ochratoxin A might be the raw materials used (Uysal et al. 2009).

4. Healthy Effects of Boza Consumption

The development of new functional foods and beverages without a significant decrease in their flavor and sensory acceptability has become the modern-day challenge for researchers. Boza is a natural product, which may meet these requirements, as it is made of natural raw material and its processing technology allows the obtaining of preventive and therapeutic healthy effects. Boza has been found to contribute to a number health benefits: it helps to balance blood pressure, improves the colonic health, lowers plasma cholesterol, increases milk production in lactating women, facilitates digestion by enhancing the production of gastric juice and by stimulating the secretion of pancreatic and hepatic cells.

Boza is a valuable nutrient to physically active people, as it contains vitamins A, C, E and four types of vitamin B, and is especially suitable for vegetarians and vegans, as it is entirely plant-based and a good source of vitamins and thus constitutes a good substitute for dairy-based drinks.

The beneficial effects of Boza on the human health are due to two main properties of this drink: the prebiotic features of its production cereal source, combined with direct consumption of probiotic LAB.

Table 4. The representatives of yeasts and molds, isolated from Boza.

Genus	Species	Reference
Candida	*C. diversa*	Botes et al. 2007
	C. glabrata	Gotcheva et al. 2000, 2001
	C. inconspicua	Botes et al. 2007, Caputo et al. 2012
	C. pararugosa	Botes et al. 2007
	C. quercitrusa	Caputo et al. 2012
	C. silvae	Caputo et al. 2012
	C. tropicalis	Gotcheva et al. 2000, 2001
Clavispora	*Cl. lusitaniae*	Caputo et al. 2012
Coniochaeta	*Con. pulveracea*	Caputo et al. 2012
Cryptococcus	*Cr. albidus*	Caputo et al. 2012
Cystofilobasidium	*Cyst. infirmominiatum*	Caputo et al. 2012
Galactomyces	*Gal. geotricum*	Osimani et al. 2015
	Gal. candidum	Osimani et al. 2015
Geotrichum	*Geotrichum* spp.	Caputo et al. 2012
	G. candidum	Gotcheva et al. 2000, Uysal et al. 2009, Caputo et al. 2012
	G. klebahnii	Caputo et al. 2012
	G. penicillatum	Gotcheva et al. 2000
	G. fragrans	Osimani et al. 2015
	G. capitatum	Uysal et al. 2009
Issatchenkia	*I. orientalis*	Botes et al. 2007
Rhodotorula	*Rhodotorula araucariae*	Göcmen et al. 2000
	R. mucilaginosa	Botes et al. 2007, Caputo et al. 2012
Saccharomyces	*S. uvarum*	Hancıoğlu and Karapinar 1997
	S. cerevisiae	Hancıoğlu and Karapinar 1997, Gotcheva et al. 2001, Caputo et al. 2012, Osimani et al. 2015
	S. kluyveri	Göcmen et al. 2000
Pichia	*P. norvegensis*	Botes et al. 2007, Osimani et al. 2015
	P. guilliermondii	Botes et al. 2007, Osimani et al. 2015
	P. fermentans	Botes et al. 2007, Caputo et al. 2012, Osimani et al. 2015
Torulaspora	*Torulaspora* spp.	Osimani et al. 2015
	T. delbrueckii	Göcmen et al. 2000, Botes et al. 2007, Caputo et al. 2012
Trichosporon	*Tr. moniliforme*	Caputo et al. 2012
	Tr. coremiiforme	Caputo et al. 2012
	Tr. cutaneum	Göcmen et al. 2000
Aspergillus	*A. fumigatus*	Uysal et al. 2009
Penicillum	*P. chrysogenum*	Uysal et al. 2009
Fusarium	*F. oxysporum*	Uysal et al. 2009
Acremonium	*Acremonium* spp.	Uysal et al. 2009

4.1 Dietary Fibre in Cereal Sources for Boza Production

The cereal-based products with health benefits can help to overcome chronic diseases as type 2 diabetes, cardiovascular diseases, and certain types of cancers by healthy diet that is based on fat intake decrease and dietary fiber consumption increase (Köksel and Cetiner 2015). Dietary fibre is a general term of different types of carbohydrates derived from plant cell walls that are not hydrolysed by human digestive enzymes (Charalampopoulos et al. 2002). It could affect the weight control, cholesterol level, diabetes, and is factor in the prevention of cardio-vascular diseases, diverticulitis, varicose veins, hiatus of the hernia, and colon cancer (Shahidi and Ambigaipalan 2016).

According to the American association of cereal chemists, the dietary fibre could be divided to water insoluble (cellulose, chitin, hemicellulose, hexoses, pentoses, lignin, xanthan gum and resistant starch), and water-soluble (beta-glucan and arabinoxylan, oilgosaccharides, such as galacto- and fructo-oligosaccharides). Cereals, used for Boza production contain both types of fibres. Cellulose, hemicellulose and lignin are presented in all cereals; the hexoses are abundant in wheat and barley, pentoses–in rye and oat; the resistant starch–in high amylose corn, barley, and high amylose wheat.

Dietary fibres have three main mechanisms of action: bulking, viscosity and fermentation (Gallaher 2006). The different fibres have various effects, suggesting that a variety of dietary fibres contribute to overall health. Most bulking fibres are not fermented, or are minimally fermented throughout the intestinal tract. They absorb water and can significantly increase stool weight and regularity. Viscous fibres thicken the contents of the intestinal tract and may attenuate the absorption of sugar, reduce sugar response after eating, and reduce lipid absorption (notably shown with cholesterol absorption). Their use in food formulations is often limited to low levels, due to their viscosity and thickening effects. Some viscous fibres may also be partially or fully fermented within the intestinal tract (guar gum, ß-glucan, glucomannan and pectins), but some viscous fibres are minimally or not fermented (Gallaher 2006).

Fermentable fibres are consumed by the microbiota within the large intestines. The prebiotics are fermentable fibres that are not hydrolysed by the human digestive enzymes in the upper gastrointestinal tract and beneficially affect the host by selectively stimulating the growth and/or activity of one or a limited number of bacteria in the colon that can improve host health (Gibson and Roberfroid 1995).

β-glucan. This is a major component of water soluble cereal fibre, important part of barley, wheat and oat dietary fibre. β-glucans are non-starch polysaccharides composed of glucose molecules in long linear glucose polymers with mixed β-(1→4) and β-(1→3) links with molecular weight between 50 and 3000 kDa. The soluble β-glucans make viscous, shear thinning solutions even at low concentrations. The mean values of β-d-glucan content in cereals are 41.6 g per kg barley, 34.9 g per kg oat, and 4.8 g per kg wheat (Havrlentová and Kraic 2006). The activity of cereal β-glucan on the human organism has a broad range of effects. It has been shown that β-glucan from barley is hypocholesterolemic, and this property may result of its ability to increase viscosity of intestinal content (Mälkki 2004). β-glucan is a potent inductor of humoral and cell-mediated immunity, and regular daily consumption of β-glucan significantly increases immunological activity (Estrada et al. 1997). Oat β-glucans increase the stability of the drink (Angelov et al. 2006) and support the development of probiotic LAB (Charalampopoulus et al. 2002).

Resistant starch (RS). It provides the benefits of both insoluble and soluble fibers. It is defined as starch that escapes digestion in the small intestine and provides a source of fermentable substrate for caecal and colonic microflora, thus positively modulating the intestinal microbiota. RS can be classified into four subtypes: RS1, physically inaccessible starch granules locked within whole grains (or partially milled grains); RS2, granular starch that is tightly packed, consisting of ungelatinised granules, as in cereals; RS3, highly RS fraction, mainly composed of retrograded amylose (formed when cooked and cooled), as in Boza; RS4, starch chemically or enzymatically modified to resist digestion (Sajilata et al. 2006).

The highest amount of RS is contained in high-amylose corn, oat, and barley (Murphy et al. 2008). RS has beneficial effect when it reaches the large intestine and is consumed by LAB. The RS fermentation produces short-chain fatty acids, which are rapidly absorbed from the colon, then are metabolized in the colonic epithelial cells, liver or other tissues. The high-amylose RS2 (in cereals) was shown to increase hormones and to improve the functions of the gastrointestinal tract (Keenan et al. 2012).

Key studies have demonstrated that RS2 and RS3 consumption raise levels of genera *Bifidobacterium* and *Lactobacillus* (Simpson and Campbell 2015), and specially, the amylolytic LAB. On observing Boza microbial content, several *Lactobacillus* species displaying amylolytic activity were found. They belong to *Lb. plantarum*, *Lb. pentosus*, and *Lb. paracasei* (Gotcheva et al. 2000, Blagoeva and Gotcheva 2013, Petrova et al. 2010, Petrova and Petrov 2012). Amylolytic LAB are able to convert starch directly into lactic acid due to the synthesis enzymes as α-amylases, maltogenic amylases, amylopullanases, pullulanases, neopullulanases, glycogen phosphorylases and 1,6-glucosidases (Petrova et al. 2013, Velikova et al. 2016). Besides lactic acid, in the presence of starch, they are able to produce succinic acid, which contributes to the unique taste and fragrance of the beverage (Petrova et al. 2010).

4.2 Probiotic Bacteria in Boza

The second health-promoting effect of Boza consumption is due to the fact that it is a natural source of prebiotic LAB (Kandylis et al. 2016). The incorporation of probiotics, prebiotics and synbiotics into food provides several health benefits: (1) pathogen interference, exclusion and antagonism; (2) immunostimulation and immunomodulation; (3) anticarcinogenic and antimutagenic activities in animal models; (4) alleviation of symptoms of lactose intolerance; (5) vaginal/urinary tract health; (6) reduction in blood pressure in hypertensive subjects; (7) decreased incidence and duration of diarrhea (antibiotic-associated diarrhea, *Clostridium difficile*, travelers and rotaviral); and (8) maintenance of mucosal integrity (Klaenhammer 2000).

The World Health Organization's 2001 definition of probiotics is "live micro-organisms which, when administered in adequate amounts, confer a health benefit on the host". To exert beneficial effects, the microorganisms must be alive and available in high numbers at the time of consumption. Another requirement of probiotics is to be capable to survive under the harsh conditions of human gastrointestinal tract. Probiotic microorganisms are mostly of human and animal origin, or from fermented foods.

Probiotic bacteria, LAB produce different antimicrobial compounds, including bacteriocins (Table 5), lactic acid, hydrogen peroxide, benzoic acid, fatty acids, diacetyl, and other low molecular weight compounds. During Boza production, the decrease in pH to values around 3.0–4.0 produces optimal conditions for the enzymatic degradation of phytates, helps the product to acquire its typical flavour and prevents the growth of pathogenic and/or spoilage bacteria (Kabak and Dobson 2011).

Bacteriocinogenic probiotic bacteria in cereal-based fermented foods could be beneficial when used as starter cultures, as they may prolong the shelf life of the products and provide the consumer with a healthy dietary component at a considerable low cost (von Mollendorff et al. 2016).

The first, to report the isolation of bacteriocin-producing strains from Boza were Kabadjova et al. (2000) and Ivanova et al. (2000). Of the 80 isolated strains of LAB, a group of 33 showed antibacterial activity against different closely related or pathogenic test microorganisms (*Listeria innocua* F, *Lb. plantarum* 73, *Lc. cremoris* 117 and even against Gram negative bacteria such as *Escherichia coli*). Bacteriocin B14 125 produced by *Lactococcus lactis* subsp. *lactis* 14 was the first reported bacteriocin from Boza (Kabadjova et al. 2000). Mesentericin ST99, produced by *Leuc. mesenteroides* subsp. *dextranicum*, was shown to be a heat- and acid-stable single peptide with bacteriostatic activity on *Listeria innocua* (Todorov et al. 2004). The cell-free supernatant of this strain inhibited the growth of *Bacillus subtilis*, *Ent. faecalis*, several *Lactobacillus* spp., *Lc. lactis* subsp. *cremoris*, *Listeria innocua*, *Listeria monocytogenes*, *Pediococcus pentosaceus*, *Staphylococcus aureus* and *Streptococcus thermophilus*. However, *Clostridium* spp., *Carnobacterium* spp., *Leuconostoc mesenteroides* and Gram-negative bacteria were not inhibited.

A number of bacteriocins, produced by strains of *Pd. pentosaceus*, *Lb. plantarum*, *Lb. paracasei*, *Lb. rhamnosus* and *Lb. pentosus* isolated from Boza have been reported (Table 5). In most cases, the spectrum of activity, the sensitivity to physico-chemical conditions (temperature, pH), and the production kinetics have been determined, along their preliminary purification (von Mollendorff et al. 2006).

The bacteriocin produced by *Lb. lactis* subsp. *lactis* YBD11 (Tuncer and Ozden 2010) inhibited the growth of *Lb. plantarum*, *Lb. sakei*, *Lc. lactis* subsp. *lactis*, *Lc. lactis* subsp. *cremoris*, *Micrococcus luteus*, *Listeria innocua*, *Enterococcus feacalis*, *Staphylococcus aureus*, *Staphylococcus carnosus*, *Pediococcus pentosaceus*, and *Bacillus cereus* but the nisin producer strain *Lactococcus lactis* SIK83 and Gram-negative bacteria were not inhibited. Different enzyme, pH and heat treatments showed that the bacteriocin produced by *Lb. lactis* subsp. *lactis* YBD11 exhibited a similar behavior with control nisin.

Table 5. Bacteriocin-producing strains isolated from Boza.

Species	Strain	Reference
Ent. faecium	ST62BZ	Todorov 2010
Lb. fermentum	JW11BZ	von Mollendorff et al. 2006
	JW15BZ	von Mollendorff et al. 2006
Lb. paracasei	ST242BZ	Todorov and Dicks 2006
	ST284BZ	Todorov and Dicks 2006
Lb. pentosus	ST712BZ	Todorov and Dicks 2006
Lb. plantarum	JW3BZ	von Mollendorff et al. 2006
	JW6BZ	von Mollendorff et al. 2006
	ST194BZ	Todorov and Dicks 2006
	ST414BZ	Todorov and Dicks 2006
	ST69BZ	Todorov 2010
	ST664BZ	Todorov and Dicks 2006
Lb. rhamnosus	ST461BZ	Todorov and Dicks 2006
	ST462BZ	Todorov and Dicks 2006
Lc. lactis	ST63BZ	Todorov 2010
	ST611BZ	Todorov 2010
	ST612BZ	Todorov 2010
Lc. lactis subsp. *lactis*	BZ	Şahingil et al. 2011
	B14	Kabadjova et al. 2000, Ivanova et al. 2000
	YBD11	Tuncer and Ozden 2010
Leuc. mesenteroides subsp. *dextranicum*	ST99	Todorov and Dicks 2004
Leuc. pseudomesenteroides	KM432Bz	Makhloufi et al. 2013
Pd. pentosaceus	ST18	Todorov and Dicks 2005

Another bacteriocin-producing *Leuconostoc pseudomesenteroides* KM432Bz from Boza was reported recently by Makhloufi et al. (2013). The antimicrobial peptide was purified and shown to be identical to other class IIa bacteriocins: leucocin A from *Leuc. gelidum* UAL-187 and *Leuc. pseudomesenteroides* QU15 and leucocin B from *Leuc. carnosum* Ta11a. Gene clusters involved in bacteriocin synthesis generally consist of a gene encoding a bacteriocin precursor, which is further cleaved to yield the active peptide, gene(s) involved in affording immunity to the producing strain, gene(s) encoding the export of the active peptide, and gene(s) involved in regulation and production of the peptide. Leucocin B-KM432Bz gene cluster encodes the bacteriocin precursor (*lcnB*), the immunity protein (*lcnI*) and the dedicated export machinery (*lcnD* and *lcnE*). Leucocin B-KM432Bz did not impair the growth of *Klebsiella pneumoniae*, but inhibited the growth of phylogenetically related strains of lactic acid bacteria belonging to the genera *Lactobacillus*, *Weissella*, and also of several strains of *Leuconostoc*. It was also active against a few enteric pathogens, such as *Ent. faecium* HKLHS and *Ent. faecalis* CIP 103015, and one respiratory tract pathogen, *Streptococcus pneumoniae* CIP 102911. Leucocin B-KM432Bz inhibited the growth of five strains of *Listeria* (*L. innocua*, *L. ivanovii* subsp. *ivanovii*, and *L. monocytogenes*). Bacteriocin ST284BZ revealed high activity against the Herpes Symplex Type 1 virus that causes encephalitis and oro-facial and genital lesions. Growth of *Mycobacterium tuberculosis* was repressed 69% after 5 days of incubation in the presence of bacteriocin ST194BZ (Todorov et al. 2008).

Besides the ability to produce compounds, active against pathogens, the strains isolated from Boza possess additional probiotic properties (LeBlanc and Todorov 2011). The majority of them survives low pH conditions (pH 3.0), grows well at pH 9.0 and are not affected by the presence of 0.3% (w/v) oxbile (Todorov et al. 2008).

Various levels of auto aggregation between the probiotic bacteria and co-aggregation with *Listeria innocua* were observed. The adhesion of probiotics, isolated from Boza to colorectal adenocarcinoma cell lines HT-29 and Caco-2 was similar to that reported for the widely used probiotic strain *Lb. rhamnosus* GG. In addition, the analysis of susceptibility of the strains to substances in commercially available medicaments show that all eight strains tested can withstand high concentrations of N-acetyl-L-cystein, ambroxol, aminophylline, antazoline, aspirin, bisacodyl, cinarizin, heptaminol, codeine, dichlorhidrate hydroxyzine, famotidine, hydrochlorothiaziden, metamizol, metoclopramide hydrochloride, simethicone, methyl-4-hydroxybenzoate, silymarin, and paracetamol and were inhibited by only seven of 24 medicaments tested (Todorov et al. 2008).

Peculiar group of strains, isolated from Boza and possessing significant amylolytic activity were studied for their ability to repress the growth of common foodborne pathogens. *Lb. paracasei* B41, *Lb. plantarum* Bom 816, and *Lb. pentosus* N3 demonstrated antimicrobial activity against *Escherichia coli, Klebsiella pneumoniae, Vibrio cholerae* and *Bacillus subtilis*. The results indicated possible bacteriocin production. None of the tested lactobacilli inhibited the growth of *Saccharomyces cerevisiae* or *Pichia* strains, suggesting the existence of stable microbial community of yeasts and lactobacilli in Boza starters (Petrova and Petrov 2011).

Recently, two different *Lb. paracasei* strains, isolated from Boza (*Lb. paracasei* B41 and LC1) were assessed as meeting almost all criteria for probiotic/prebiotic Boza production: they are amylolytic (Petrova and Petrov 2012, Blagoeva and Gotcheva 2013, Velikova et al. 2016); they are able to ferment prebiotic fibre (Velikova et al. 2014, Petrova et al. 2015); and display antimicrobial activity against pathogens (Petrova and Petrov 2011).

4.3 Other Healthy Effects of Boza

Omega-3 and ω-6 fatty acids. Millet, the traditional grain for Boza production is a source of both ω-3 and ω-6 fatty acids, comprising 0.1 and 2% of its dry weight. Ω-3 fatty acid is known to decrease the risk of major cardiovascular events, such as myocardial infarction, sudden cardiac death, and coronary heart disease (Lavie et al. 2009). Current research suggests that ω-3 may prevent arrhythmias, reduces blood pressure and decreased risk of several cancers such as breast, prostate, colon, and kidney (Anand et al. 2008).

ACE-inhibitory activity of Boza. Angiotensin I-converting enzyme (ACE) plays an important physiological role in regulating blood pressure and fluid–salt balance in human. Compounds that play a role in the regulation of blood pressure by reducing ACE activity are known as ACE inhibitors. Recent study, dedicated to biochemical analysis of the ACE-inhibitory activity of protein hydrolysate and protein fractions obtained from Boza reports that ACE-inhibitory activity of Boza increase 3.5-fold after its digestion in the stomach. All samples, including Boza, protein hydrolysate, fractionated hydrolysates and dialysates obtained after Boza *in vitro* digestion contain bioactive compounds with different ACE inhibitory activities. Based on these results, Boza can be considered as a good source of ACE-inhibitory peptides (Kancabaş and Karakaya 2012).

5. Shelf Life and Safety of Boza

The shelf life of Boza is fairly short, up to 15 days. The selection of cultures is important to obtain product with desirable quality and stability (Yegin and Fernandez-Lahore 2012). Total aerobic bacteria counts detected in Boza are within the range of 1.5×10^4–3.2×10^8 cfu/ml (Altay et al. 2013). However, *E. coli, Salmonella, Staphylococcus aureus* and *Bacillus cereus* have never been found in Boza. The occurrence of coliforms and molds was reported in some of the samples; as well as of several opportunistic human pathogens such as *Candida inconspicua, Rhodotorula mucilaginosa* and *Pichia norvegensis* (Botes et al. 2007). *Aspergillus fumigatus, Penicillum chrysogenum, Fusarium oxysporum, Acremonium* spp., *Geotrichum candidum* and *Geotrichum capitatum* were also isolated from Boza (Uysal et al. 2009).

The most appropriate starter culture combinations using sensory evaluation of Boza was investigated by Zorba et al. (2003). *S. cerevisiae/ Leuc. mesenteroides* subsp. *mesenteroides/Weissella confusa* combination was suggested as a starter culture mixture to obtain desirable sensory characteristics. However, the presence of representatives of genus *Lactobacillus* usually brings the prebiotic effects and increases the shelf life of Boza.

6. Conclusion

Modern eating habits of the human increase the risk of new foodborne illnesses. Due to the rise in awareness of not only food safety, but also of the risk derived from the chemical preservatives, there has been an increasing demand for more "natural" and "health-promoting" food. Boza production has a long history and it which may be evaluated as a non-dairy functional beverage. It is a highly nourishing drink, which is easily digestible, gratifying in taste and fragrance, and exerts beneficial effects on the human health due to its probiotic microbial content. The preservative effect that occurs by the use of lactic acid bacteria in Boza is a result of the antimicrobial action of bacteriocins and other metabolites, produced by LAB. Considering that the production of antimicrobial substances is one of the principles for probiotic strain selection, LAB strains from Boza may have excellent probiotic potential, as they possess remarkable antimicrobial activity both against gram-negative and gram-positive bacteria, and are able to protect the beverage from pathogens' propagation. In addition, the abundance of prebiotic fibre, vitamins, essential amino acids and fatty acids in Boza enriches the daily diet of the consumer and contributes to the health maintenance of people of all ages.

Keywords: Boza, LAB, lactic acid bacteria, Probiotics, Dietary fibres, Yeasts

References

Altay, F., Karbancioglu-Güler, F., Daskaya-Dikmen, C. and Heperkan, D. (2013). A review on traditional Turkish fermented non-alcoholic beverages: Microbiota, fermentation process and quality characteristics. International Journal of Food Microbiology 167: 44–56.

Akpinar-Bayizit, A., Yilmaz-Ersan, L. and Ozcan, T. (2010). Determination of Boza's organic acid composition as it is affected by raw material and fermentation. International Journal of Food Properties 13: 648–656.

Anand, R.G., Alkadri, M., Lavie, C.J. and Milani, R.V. (2008). The role of fish oil in arrhythmia prevention. Journal of Cardiopulmonary Rehabilitation and Prevention 28: 92–98.

Angelov, A., Gotcheva, V., Kuncheva, R. and Hristozova, T. (2006). Development of a new oat-based probiotic drink. International Journal of Food Microbiology 112: 75–80.

Arici, M. and Daglioglu, O. (2002). Boza: a lactic acid fermented cereal beverage as a traditional Turkish food. Food Reviews International 18: 39–48.

Blagoeva, G. and Gotcheva, V. (2013). Occurrence of amylolytic lactic acid bacteria in Bulgarian cereal-based fermented products. Scientific works. Food Science, Engineering and Technologies LX: 965–970.

Blandino, A., Al-Aseeri, M.E., Pandiella, S.S., Cantero, D. and Webb, C. (2003). Cereal-based fermented foods and beverages. Food Research International 36: 527–543.

Borcakli, M., Ozturk, T., Uygun, T., Uygun, N. and Gozum, E. (2014). Boza production method with starter culture. European patent EP2802220.

Botes, A., Todorov, S.D., von Mollendorff, J.W., Botha, A. and Dicks, L.M.T. (2007). Identification of lactic acid bacteria and yeast from boza. Process Biochemistry 42: 267–270.

Caputo, L., Quintieri, L., Baruzzi, F., Borcakli, M. and Morea, M. (2012). Molecular and phenotypic characterization of *Pichia fermentans* strains found among boza yeasts. Food Research International 48: 755–762.

Çelik, I., Işik, F. and Yilmaz, Y. (2016). Effect of roasted yellow chickpea (leblebi) flour addition on chemical, rheological and sensory properties of Boza. Journal of Food Processing and Preservation doi:10.1111/jfpp.12725.

Charalampopoulos, D., Wang, R., Pandella, S.S. and Webb, C. (2002). Application of cereals and cereal components in functional food: a review. International Journal of Food Microbiology 79: 131–141.

Cosansu, S. (2009). Determination of biogenic amines in a fermented beverage, boza. Journal of Food, Agriculture & Environment 7: 54–58.

Enchev, S. (2003). Method for the preparation of a soft drink from cereal crops. Bulgarian patent BG64035.

Estrada, A., Yun, C.H., Van Kessel, A., Li, B., Hauta, S. and Laarveld, B. (1997). Immunomodulatory activities of oat β-D-glucan *in vitro* and *in vivo*. Microbiology and Immunology 41: 991–998.

Gallaher, D. (2006). Dietary Fiber. ILSI Press, Washington, D.C.

Genç, M., Zorba, M. and Ova, G. (2002). Determination of rheological properties of boza by using physical and sensory analysis. Journal of Food Engineering 52: 95–98.

Gibson, G.R. and Roberfroid, M.B. (1995). Dietary modulation of the human colonic microbiota: Introducing the concept of prebiotics. Journal of Nutrition 125: 1401–1412.

Göcmen, D., Korukluoglu, M., Uylaser, V. and Sahin, I. (2000). The yeast flora of bosan put up for consumption in Bursa. Advances in Food Sciences 22: 145–150.

Gotcheva, V., Pandiella, S.S., Angelov, A., Roshkova, Z. and Webb, C. (2001). Monitoring the fermentation of the traditional Bulgarian beverage boza. International Journal of Food Science and Technology 36: 129–134.

Gotcheva, V., Pandiella, S.S., Angelov, A., Roshkova, Z.G. and Webb, C. (2000). Microflora identification of the Bulgarian cereal-based fermented beverage boza. Process Biochemistry 36: 127–130.

Hancioglu, O. and Karapinar, M. (1997). Microflora of boza, a traditional fermented Turkish beverage. International Journal of Food Microbiology 35: 271–274.

Havrlentova, M. and Kraic, J. (2006). Content of beta-d-glucan in cereal grains. Journal of Food Research and Nutrition 45: 97–103.

Ivanova, I., Kabadjova, P., Pantev, A., Danova, S. and Dousset, X. (2000). Detection, purification and partial characterization of a novel bacteriocin substance produced by *Lactococcus lactis* subsp. *lactis* B14 isolated from boza - Bulgarian traditional cereal beverage. Vestnik Moskovskogo Universiteta Khimia 41: 47–53.

Kabadjova, P., Gotcheva, I., Ivanova, I. and Dousset, X. (2000). Investigation of bacteriocin activity of lactic acid bacteria isolated from boza. Biotechnology and Biotechnological Equipment 14: 56–59.

Kabak, B. and Dobson, A.D.W. (2011). An introduction to the traditional fermented foods and beverages of Turkey. Critical Reviews in Food Science and Nutrition 51: 248–260.

Kancabaş, A. and Karakaya, S. (2012). Angiotensin-converting enzyme (ACE)-inhibitory activity of boza, a traditional fermented beverage. Journal of the Science of Food and Agriculture 93: 641–645.

Kandylis, P., Pissaridi, K., Bekatorou, A., Kanellaki, M. and Koutinas, A. (2016). Dairy and non-dairy probiotic beverages. Current Opinion in Food Science 7: 58–63.

Keenan, M.J., Martin, R.J., Raggio, A.M., McCutcheon, K.L., Brown, I.L., Birkett, A., Newman, S.S., Skaf, J., Hegsted, M., Tulley, R.T., Blair, E. and Zhou, J. (2012). High-amylose resistant starch increases hormones and improves structure and function of the gastrointestinal tract: a microarray study. Journal of Nutrigenetics and Nutrigenomics 5: 26–44.

Kivanc, M., Yilmaz, M. and Cakir, E. (2011). Isolation and identification of lactic acid bacteria from boza, and their microbial activity against several reporter strains. Turkish Journal of Biology 35: 313–324.

Klaenhammer, T.R. (2000). Probiotic bacteria: today and tomorrow. Journal of Nutrition 130: 415S–416S.

Köksel, H. and Cetiner, B. (2015). Grain science and industry in Turkey: Past, present, and future. Cereal Foods World 60: 90–96.

Köse, E. and Yücel, U. (2003). Chemical composition of Boza. Journal of Food Technology 1: 191–193.

Lavie, C.J., Milani, R.V., Mehra, M.R. and Ventura, H.O. (2009). Omega-3 polyunsaturated fatty acids and cardiovascular diseases. Journal of the American College of Cardiology 54: 585–594.

LeBlanc, J.G. and Todorov, S.D. (2011). Bacteriocin producing lactic acid bacteria isolated from Boza, a traditional fermented beverage from Balkan Peninsula—from isolation to application. pp. 1311–1320. In: A. Méndez-Vilas (ed.). Science Against Microbial Pathogens: Communicating Current Research and Technological Advances. Formatex, Badajoz.

Mälkki, Y. (2004). Trends in dietary fibre research and development. Acta Alimentaria 33: 39–62.

Makhloufi, K.M., Carré-Mlouka, A., Peduzzi, J., Lombard, C., van Reenen, C.A., Dicks, L.M.T. and Rebuffat, S. (2013). Characterization of leucocin B-KM432Bz from *Leuconostoc pseudomesenteroides* isolated from Boza, and comparison of its efficiency to pediocin PA-1. Plos One 8: e70484.

Murphy, M., Douglass, J.S. and Birkett, A. (2008). Resistant starch intake in the United States. Journal of the American Dietetic Association 108: 67–78.

Osimani, A., Garofalo, C., Aquilanti, L., Milanovic, V. and Clementi, F. (2015). Unpasteurised commercial boza as a source of microbial diversity. International Journal of Food Microbiology 194: 62–70.

Petrova, P., Emanuilova, M. and Petrov, K. (2010). Amylolytic *Lactobacillus* strains from Bulgarian fermented beverage *Boza*. Zeitschrift für Naturforschung C 65: 218–224.

Petrova, P. and Petrov, K. (2012). Direct starch conversion into L(+)-Lactic acid by a novel amylolytic strain of *Lactobacillus paracasei* B41. Starch/Starke 64: 10–17.

Petrova, P. and Petrov, K. (2011). Antimicrobial activity of starch-degrading *Lactobacillus* strains isolated from *boza*. Biotechnology and Biotechnological Equipment 25: 114–116.

Petrova, P., Petrov, K. and Stoyancheva, G. (2013). Starch-modifying enzymes of lactic acid bacteria—structures, properties, and applications. Starch/Starke 65: 34–47.

Petrova, P., Velikova, P., Popova, L. and Petrov, K. (2015). Direct conversion of chicory flour into L(+)-lactic acid by the highly effective inulinase producer *Lactobacillus paracasei* DSM 23505. Bioresource Technology 186: 329–333.

Petrov, M. (1998). Boza composition and method for its production. Bulgarian patent BG101548 (A).

Pizzuti, D., Buda, A., d'Odorico, A., d'Incà, R., Chiarelli, S., Curioni, A. and Martines, D. (2006). Lack of intestinal mucosal toxicity of *Triticum monococcum* in celiac disease patients. Scandinavian Journal of Gastroenterology 41: 1305–1311.

Shahidi, F. and Ambigaipalan, P. (2016). Beverages fortified with omega-3 fatty acids, dietary fiber, minerals, and vitamins. pp. 801–813. *In*: F. Shahidi and C. Alasalvar (eds.). Handbook of Functional Beverages and Human Health. Taylor and Francis Group LLC, UK.

Şahingil, D., Işleroğlu, H., Yildirim, Z., Akçelik, M. and Yildirim, M. (2011). Characterization of lactococcin BZ produced by *Lactococcus lactis* subsp. *lactis* BZ isolated from boza. Turkish Journal of Biology 35: 21–33.

Sajilata, M.G., Singhal, R.S. and Kulkarni, P.R. (2006). Resistant starch—a review. Comprehensive Reviews in Food Science and Food Safety 5: 1–17.

Selçuk, I. (2016). Boza consumption in early-modern Istanbul as an energy drink and a mood-altering substance. Journal of Academic Inquiries 11: 61–81.

Sendov, S. (1996). Installation for the production of Boza and an autoclave for it. Bulgarian patent BG60703 (B1).

Simpson, H.L. and Campbell, B.J. (2015). Review article: dietary fibre-microbiota interactions. Alimentary Pharmacology and Therapeutics 42: 158–179.

Stallknecht, G.F., Gilbertson, K.M. and Ranney, J.E. (1996). Alternative wheat cereals as food grains: inkorn, Emmer, Spelt, Kamut, and Triticale. pp. 156–170. *In*: J. Janick (ed.). Progress in New Crops, ASHA Press, Alexandria.

Tangüler, H. (2014). Traditional Turkish fermented cereal based products: Tarhana, Boza and Chickpea Bread. Turkish Journal of Agriculture - Food Science and Technology 2: 144–149.

Todorov, S.D. (2010). Diversity of bacteriocinogenic lactic acid bacteria isolated from boza, a cereal-based fermented beverage from Bulgaria. Food Control 21: 1011–1021.

Todorov, S.D., Botes, M., Guigas, C., Schillinger, U., Wiid, I., Wachsman, M.B., Holzapfel, W.H. and Dicks, L.M.T. (2008). Boza, a natural source of probiotic lactic acid bacteria. Journal of Applied Microbiology 104: 465–477.

Todorov, S.D. and Dicks, L.M.T. (2006). Screening for bacteriocin producer lactic acid bacteria from boza, a traditional cereal beverage from Bulgaria. Comparison of the bacteriocins. Process Biochemistry 41: 11–19.

Todorov, S.D. and Dicks, L.M.T. (2005). Pediocin ST18, an anti-listerial bacteriocin produced by *Pediococcus pentosaceus* ST18 isolated from Boza, a traditional cereal beverage from Bulgaria. Process Biochemistry 40: 365–370.

Todorov, S.D. and Dicks, L.M.T. (2004). Characterization of mesentericin ST99, a bacteriocin produced by *Leuconostoc mesenteroides* subsp. *dextranicum* ST99 isolated from Boza. Journal of Industrial Microbiology and Biotechnology 31: 323–329.

Tuncer, Y. and Ozden, B. (2010). Partial biochemical characterization of nisin-like bacteriocin produced by *Lactococcus lactis* subsp. *lactis* YBD11 isolated from Boza, a traditional fermented Turkish beverage. Romanian Biotechnological Letters 15: 4940–4948.

Velikova, P., Blagoeva, G., Gotcheva, V. and Petrova, P. (2014). Novel Bulgarian *Lactobacillus* strains ferment prebiotic carbohydrates. Journal of BioScience and Biotechnology SE/Online, pp. 55–60.

Velikova, P., Stoyanov, A., Blagoeva, G., Popova, L., Petrov, K., Gotcheva, V., Angelov, A. and Petrova, P. (2016). Starch utilization routes in lactic acid bacteria: New insight by gene expression assay. Starch/Stärke 68: 953–960.

von Mollendorff, J.W., Todorov, S.D. and Dicks, L.M. (2006). Comparison of bacteriocins produced by lactic-acid bacteria isolated from boza, a cereal-based fermented beverage from the Balkan Peninsula. Current Microbiology 53: 209–216.

von Mollendorff, J.W., Vaz-Velho, M. and Todorov, S.D. (2016). Boza, a traditional cereal-based fermented beverage: a rich source of probiotics and bacteriocin-producing Lactic Acid Bacteria. pp. 157–188. *In*: K. Kristbergsson and S. Ötles (eds.). Functional Properties of Traditional Foods, Springer US, New York.

Uysal, U.D., Oncu, E.M., Berikten, D., Yilmaz, N., Tuncel, N.B., Kivanc, M. and Tuncel, M. (2009). Time and temperature dependent microbiological and mycotoxin (ochratoxin-A) levels in Boza. International Journal of Food Microbiology 130: 43–48.

Yegin, S. and Fernández-Lahore, M. (2012). Boza: A traditional cereal-based, fermented Turkish beverage. pp. 533–542. *In*: Y.H. Hui and E. Özgül Evranuz (eds.). Handbook of Plant-Based Fermented Food and Beverage Technology. CRC Press, Florida.

Yegin, S. and Uren, A. (2008). Biogenic amine content of boza: A traditional cereal-based, fermented Turkish beverage. Food Chemistry 111: 983–987.

Zorba, M., Hancioglu, O., Genc, M., Karapinar, M. and Ova, G. (2003). The use of starter culture in the fermentation of boza, a traditional Turkish beverage. Process Biochemistry 38: 1405–1411.

14

Coffee Fermentation

Fontana Angélique and *Durand Noël* *

1. Introduction

Coffee is one of the important commodities in the world's economy. The total world coffee consumption is estimated to be over 6 million ton/year, with Europe being the largest market, followed by US, and Japan in third position (Sibanda 2006). According to the International Coffee Organization (ICO), the world coffee production reached 145 million bags, representing approximately 8.7 million tons (60 kg/bag) in 2012. The largest producing country was Brazil, followed by Vietnam and Indonesia. Indeed, the coffee trees grow in tropical climate and *Coffea arabica* and *Coffea canephora* species are those used in the preparation of the drink which is one of the most consumed in the world. Depending on the coffee, during the post-harvest treatment used to transform coffee cherries into roasted coffee beans, a fermentation stage can occur. This step is known to contribute to the flavour profile of the coffee.

2. Coffee and Coffee Post-Harvest Treatments

Coffee is a perennial plant belonging to the *Rubiaceae Coffea* family. There are 70 plant species, within Arabica (*Coffea arabica*) and Robusta (*Coffea canephora*) being the two most famous ones. The coffee plant requires a warm and humid weather such as in tropical or sub-tropical regions. The Arabica shrub height reaches 3 to 5 m when cultivated but can reach

UMR Qualisud (CIRAD, Université Montpellier II), 34398 Montpellier Cedex 5, France.
E-mail: afontana@um2.fr
* Corresponding author: noel.durand@cirad.fr

up to 10 m in the wild. This is a delicate shrub, native to the highlands of Ethiopia, which grows at from 800 to 2000 metres altitude (Coste 1989). The Arabica coffee is considered as the top level coffee, with its aroma, smoothness and softness. The Robusta plant usually grows in plains and prefers high temperatures and high levels of humidity. The plant usually reaches 5–8 m of height. *C. canephora* is renowned for its agricultural strength, where its trade name of Robusta originates. The coffee fruit or cherry (e.g., Fig. 1), is classified as drupes, i.e., fleshy mesocarp and endocarp woody fruits.

During ripening, fruit colour changes from green to red (mature) or intense orange depending on the species. Like most drupes, this fruit is composed of the pericarp covering the endosperm (grain). The pericarp consists of an exocarp (skin), a mesocarp (pulp and tissue forming) and an endocarp (parchment). The outer mesocarp, often called pulp, is composed of small parenchymal cells with primary cell walls and increasing diameter toward the interior of the fruit. The inner mesocarp is a tissue organized with fragile primary type cell wall cells that contain high levels of peptic substance (Garcia et al. 1991). "Mucilage" is generally misused to describe the inner mesocarp of coffee as it is slimy, but "mucilaginous tissue" is most appropriate (Fig. 2).

In order to obtain green coffee, it is necessary to remove the membranes that envelop the seed and bring it to a level of humidity such that it enables storage and transport to consumer countries. Two methods are used: (i) wet preparation (Fig. 3), a generic term covering several operations, including a fermentation process designed to remove the mucilaginous tissue and (ii) the dry process (Fig. 4), which consists of sun drying of the whole fruit just after harvest. Arabica coffee typically undergoes a wet preparation (Rolz et al. 1982), with the exception of most of the production from Brazil and Ethiopia, which are obtained by dry process. Robusta coffee is usually subjected to a natural preparation (dry process), with the exception of some of the production of India.

Figure 1. Coffee cherries.

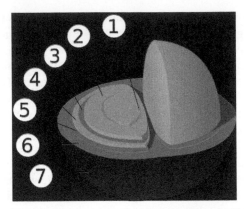

Figure 2. Structure of the fruit and seed of the coffee.

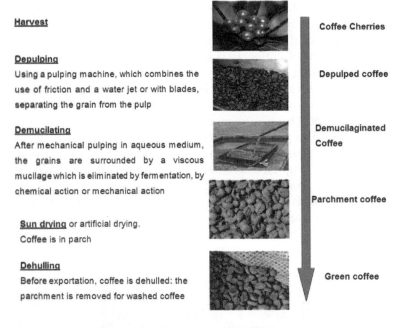

Harvest

Depulping
Using a pulping machine, which combines the use of friction and a water jet or with blades, separating the grain from the pulp

Demucilating
After mechanical pulping in aqueous medium, the grains are surrounded by a viscous mucilage which is eliminated by fermentation, by chemical action or mechanical action

Sun drying or artificial drying.
Coffee is in parch

Dehulling
Before exportation, coffee is dehulled: the parchment is removed for washed coffee

Coffee Cherries

Depulped coffee

Demucilaginated Coffee

Parchment coffee

Green coffee

Figure 3. Post-harvest treatment: Wet process.

Regarding wet preparation of coffee, after harvesting cherries, skin and most of the underlying pectinated mesocarp, which is the pulp, are removed from the ripe harvested cherries by a mechanical process (Fig. 5), called pulping. After de-pulping, a mucilaginous layer (Fig. 6) called mucilage adheres to the grain. Different techniques allow mucilage removal, by mechanical, chemical or biochemical action also called fermentation.

Harvest

Drying
Sun or artificial drying

Dehusking
Before exportation, coffee is dehusked,
husk is removed to obtain green coffee

Coffee Cheries

Husk coffee

Green coffee

Figure 4. Post-harvest treatment: Dry process.

Figure 5. Depulper of coffee in Venezuela.

Mucilage is traditionally removed by natural, solid or liquid fermentation with total immersion of the mass to be fermented. Fermentation reactions are considered "aerobic" when performed in a solid medium with partial oxygen diffusion in interstitial spaces between the coffee grains. Fermentation reactions are considered "anaerobic" when the mass is completely covered with water which limits gas diffusion (Avallone 2001b). Fermented coffee is then washed in water baths that are agitated to

Figure 6. Depulped coffee, with remaining mucilage.

remove remaining mesocarp. At this stage, the wet parchment coffee has a water content of about 55%. The final step is to dry parchment coffee to reach a moisture content of 12%, so it can be stored in good conditions. Green coffee is obtained by removal of the parchment, a process called dehulling. Transformation by dry process includes a step of drying; either natural (sun) or artificial in plant or rotary dryer. At this stage coffee is in shell (shell coffee) and has a water content of about 12%. It can be transported and stored as such or husked and stored in the form of green coffee. The final step to get green coffee is to remove the husk from dry coffee called dehusking. Finally, to ensure good storage conditions, it is necessary to store coffee with relatively low levels of humidity (11–13%).

In the wet process and in some cases, natural fermentation is used, which aims to get rid of the mucilaginous tissue (internal mesocarp) adhering to the parchment (endocarp) (Avallone 2002). This step is known to have a positive impact on coffee quality, but it is very important to control to avoid any secondary fermentation which generates negative flavors (Barel and Jacquet 1994). A fermentation step can also occur during dry processing, if water activity allows it, also in this case, it generates a negative flavour that impacts the organoleptic quality.

Microbial populations associated to coffee vary as a function of the post-harvest treatment. The microflora comes from the saprophytic fruit flora, soil, water and materials used. The different stages of fermentation are crucial regarding the quality of the end product.

3. Coffee Fermentation

"Coffee fermentation" usually refers to the demucilagination process from which the "washed" coffee is obtained. A fermentation step, so called "natural fermentation", can also occur during the dry process as long as the moisture of the coffee beans allows it.

3.1 Coffee Fermentation in the Dry Process

Many studies concerning the fermentation using wet or semi-dry methods (Avallone et al. 2001a, Masoud and Kaltoft 2006, Vilela et al. 2010) are available, but few studies of the fermentation stage in the dry processing of coffee have been published (Silva et al. 2000, Silva et al. 2008a,b). This method involves the fermentation of whole fruit and usually produces coffee with a rather lower cup quality. During the 10–25 days of sun drying, a natural microbial fermentation occurs that can influence the final quality of the product (Schwan and Wheals 2003, Silva et al. 2000). During this stage, enzymes break down the pulp and mucilage and the fermentation of the sugars produces ethanol and acetic, lactic, butyric and other carboxylic acids. The formation of butyric and propionic acids from bacterial fermentation can cause a loss of quality due to diffusion of acids into the beans (Amorim and Amorim 1977). Bacteria, yeasts and filamentous fungi were isolated during dry coffee processing (Silva et al. 2000, 2008a,b). The microbial diversity of the dry process is usually more important than those of the wet process (Silva et al. 2000) but also depends on the hygienic state of the drying areas (Durand et al. 2013). Depending on the study, bacteria are the most abundant group throughout the whole process or dominate at the beginning of the fermentation, when the coffee bean moisture is around 68%. Gram-positive bacteria represent the majority of the all bacteria isolates, and the genera *Bacillus* and *Cellulomonas* are predominant. Gram-negative species of the genera *Pseudomonas, Aeromonas, Serratia, Enterobacter* and *Acinetobacter* are also found. The frequency of yeasts and filamentous fungi increases during the fermentation process when the water activity decreases. Among yeast isolates, almost all are fermentative. The most common genera are *Debaryomyces, Pichia, Candida, Arxula* and *Saccharomycopsis* and the frequency of the three first genera expands as the process advances. There are also some species like *Pichia lynferdii* and *Arxula adeninivorans* which are rarely described. Depending on climatic and geographical conditions, the genus *Aspergillus* is the most abundant or only represents 3% of the fungal isolates. *Penicillium, Fusarium* and *Cladosporium* are the other predominant genera in all cases (Silva et al. 2008b). The dynamics of yeast and filamentous fungi is very close. The evolution of the fungal microbiota is representative of the different stages of drying. According to the delay and thus the decrease in water content, a succession of fungal species more suited to the conditions can be noticed (Durand et al. 2013). The microbial genera and species include members known to have pectinase and cellulase activities. Among the organic acids analyzed and quantified in coffee beans, acetic and lactic acids may have been generated by microbial activity (Silva et al. 2008a).

3.2 Coffee Fermentation in the Wet Process

Coffee fermentation occurs during the degumming step of the wet process or semi-dry process, which is a variation of the wet process in which coffee fruits are also depulped, but the fermentation process is carried out without adding water and thus it is called "dry fermentation". The mucilaginous substrate is mainly constituted of simple sugars and pectic substances (Coleman et al. 1955, Calle 1962, Garcia et al. 1991). The coffee demucilagination is the result of two biological events. The mucilage is hydrolyzed by both endogenous enzymes of the coffee tissues and microorganisms from the cherry skins through the pectin degradation and the microbial production of organic acids from sugars (Sivetz and Foote 1963, Frank and De la Cruz 1964, Van Pee and Castelein 1972, Menchù and Rolz 1973, Avallone et al. 2002, Belitz et al. 2009).

3.2.1 Technical aspects

The aim is to soften the mesocarp (mucilage) that will be later removed by washing. The coffee must be free of all readily fermentable material before being dried because it hinders the rapid drying of beans.

3.2.1.1 Mechanical action

Many alternatives have been proposed to reduce the amount of water used, including mechanical demuculating that eliminates mucilage using abrasion (Bailly et al. 1992). It is usually considered that fermented coffee is of better quality than coffee that was demucilaginated by mechanical means. Apart from the ecological point of view, this alternative is quite interesting.

3.2.1.2 Chemical action

In order to remove the mucilage, various tests were conducted with products such as sodium, potassium or calcium hydroxide, acids or bases. These products facilitate the mucilage separation (Carbonell and Vilanova 1952, Wootton 1963). From a qualitative point of view, the coffee obtained by this method are of lower quality as compared to those obtained by natural fermentation. In addition, this leads to significant higher costs, as well as additional waste to be treated.

3.2.1.3 Biochemical action: fermentation

The aim is to trigger the breakdown of pectic matter of the mucilage. The pectinolysis is accompanied by a dominant lactic fermentation, which is accelerated by the presence of different microorganisms (yeasts, bacteria, fungi). Great care must be taken to not generate secondary fermentations (acetic and butyric) that could be detrimental to the quality of the product. One can observe an increase in the temperature and the acidity of the fermenting mass. But these phenomena should not be too important because to avoid the appearance of pests (fungi, etc.) the dominance of acidogenic species should be favoured. Once the mucilage is degraded, it is important to stop fermentation and to avoid secondary fermentations which, again, may deteriorate the quality of the coffee (Menchú and Rolz 1973).

3.2.1.4 Dry fermentation

Following pulping and draining, coffee (with mucilage) is stored for the time required for the mucilage to reach a stage of fluidity so that it can be removed by washing. It is recommended to regularly stir it in order to keep the fermentation homogeneous.

3.2.1.5 Water fermentation

It consists of covering the mass with water in coffee trays (Fig. 7). The decomposition of the mucilage is slower than in dry fermentation, however, higher quantities of acid are formed. This would allow the exosmosis of

Figure 7. Coffee fermentation under water.

products giving the bitterness and asperity of coffee drink (phenols, caffeine) (Coste 1989).

3.2.1.6 Mixed fermentation

It starts with a dry fermentation followed by fermentation under water. Fermentation in solid medium allows an acidification of the medium, and then the fermentation in liquid medium allows the intrinsic biochemical changes in coffee to occur (Wootton 1963).

3.2.2 Biochemical and microbiological aspects

It is generally assumed that the biological demucilagination of coffee is the result of mixed yeast/bacterial fermentation. However the different studies have shown that the balance between yeasts and bacteria can vary widely. In some cases, yeasts are found to dominate the freshly pulped beans (Velmourougane 2013, Evangelista et al. 2015) followed by bacteria, which sometimes become dominant when the fermentation proceeds. In other cases the microbiota is mainly constituted of bacteria (Avallone et al. 2001a, Vilela 2010). Qualitative and quantitative changes are observed during fermentation. The total microbial population generally increases during the initial hours of fermentation, then stabilizes and decreases in some cases. The best initial growth is of yeasts and lactic acid bacteria and is mainly due to the sugar availability and favourable conditions of pH and temperature. The aerobic mesophilic microflora is quantitatively superior during the first fermentation part but then decreases because of the low pH. It is also shown that the number of strains cultivating on pectin medium remains constant during all the fermentation time and only represents a tiny percentage of the total aerobic microflora. The biodiversity of yeasts and filamentous fungi evolves differently during the wet treatment. However an increase in the number of species during the process was observed in the two cases (Durand et al. 2013). Filamentous fungi are usually found to be significantly less numerous than yeasts and bacteria at the beginning of the fermentation but can either increase or disappear along the process.

The bacterial strains involved in the coffee wet fermentation are lactic acid bacteria, and particularly *Leuconostoc mesenteroides* species, Gram-negative bacilli, especially from *Erwinia* and *Klebsiella* genera, and Gram-positive bacteria from the *Bacillus* genus. Aerobic strains of *Bacillus*, *Pseudomonas*, *Flavobacterium*, *Lactobacillus* and *Leuconostoc* are dominant during the first hours of the fermentation (Velmourougane 2013). *Leuconostoc mesenteroides* species would be involved in the pectin solubilization (Juven et al. 1985). *Erwin herb cola*, *Klebsiella pneumoniae* and *Lactobacillus brevis* are pectinolytic bacteria that are described during the coffee wet process fermentation but these species are not reported as strong pectolytic strains (Perombelon and Kelman 1980, Avallone et al. 2001a). *E. herbicola, E. dissolvens* and *K. pneumonia* produce a pectate lyase which is inefficient in coffee fermentation pH (3.5–5.3) and on highly esterified pectins (Van Pee and Castelein 1972, Avallone et al. 2002). Histological characterization of cell wall polysaccharides of coffee bean mucilage shows many remaining pectins, cellulose, hemicelluloses after fermentation (Avallone et al. 1999). So, despite the general assertion (Calle 1957, Frank and De La Cruz 1964, Frank et al. 1965, Agate and Bhat 1966, Arunga 1973), bacterial participation in the mucilage degradation would depend rather on acidification than on enzyme production (Calle 1965, Wootton 1965). Indeed, the microorganisms produce organic acids (lactic and acetic acids) inducing a pH decrease (Calle 1965, Arunga 1973, Lopez et al. 1989). Like in the cellular response to microbial attack observed in plant/pathogen interactions (d'Auzac 1996), the acidification of the fermentation batch would induce changes in the swelling capacity of the mucilage cell walls in water and these alterations would induce a textural change.

During the dry fermentation, *Enterobacter agglomerans, Escherichia coli, Bacillus cereus, Bacillus megaterium, Bacillus macerans* and *Lactobacillus plantarum* are the predominant bacterial species (Vilela et al. 2010). Isolates of *Erwinia herbicola, Klebsiella pneumonia, Lactobacillus brevis* and *Bacillus subtilis* are also detected. It can be noticed that *B. cereus, Bacillus subtilis, B. macerans, B. megaterium, Acinetobacter* sp. and *Enterobacter agglomerans* are also isolated during the natural fermentation of coffee (Silva et al. 2000, 2008a). Some *Bacillus* species are known to produce cellulases and pectinases that could degrade the mucilage during the fermentation stage.

Many different yeast strains have been isolated from the wet or dry coffee fermentation. *Kluyveromyces marxianus, Saccharomyces cerevisiae, S. bayanus* and *Schizosaccharomyces* sp. (Agate and Bhat 1966, Van Pee and Castelein 1971) are found in fermenting *Coffea robusta* beans. In decreasing frequency, *Kloeckera apis apiculata, Cryptococcus laurentii, Crytococcus albidus* and *Candida guillermondii* are isolated from wet fermenting *Coffea arabica* from Mexico (Avallone et al. 2001a). *Pichia kluyveri* and *Hansenia sporauvarum* are identified as the predominant yeasts found during wet fermentations of *Coffea arabica* in Tanzania. *Pichia anomala, Kluyveromyces marxianus* and *Torulaspora delbrueckii* also occur but in lower concentrations (Masoud et al. 2004). The role of yeasts in the degradation of the pectin-rich mucilage adhering to coffee beans during coffee fermentation are discutable. *K. marxianus, S. cerevisiae* and *Schizosaccharomyces* sp. from *Coffea robusta* are found to have a strong pectinolytic activity (Agate and Bhat 1966). Strains of *Pichia kluyveri, Pichia anomala* and *Kluyveromyces marxianus* possess apolygalacturonase activity (Serrat et al. 2002, Masoud and Jespersen 2006) while *Kloeckera apisapiculata, Cryptococcus laurentii, Crytococcus albidus* and *Candida guillermondii* are non-pectinolytic but are known for their ethanol production through fermentative metabolism of sugars (Avallone et al. 2002).

P. anomala, Hansenia sporauvarum, Torulaspora delbrueckii, Rhodotorula mucilaginosa, Saccharomyces bayanus and *Kloeckera* sp. are present during *Coffea arabica* dry fermentation (Vilela et al. 2010). Because of their ability to tolerate conditions of low water activity (0.6–0.7) (Barnett et al. 2000), yeasts belonging to the genus *Candida* are also found at the end of the fermentation.

Some yeast species appear to be involved whatever the coffee fermentation process (natural, dry or wet) (Masoud et al. 2004, Silva et al. 2008a; Vilela et al. 2010). The predominant genera of filamentous fungi during the wet process are *Aspergillus, Penicillium, Fusarium, Rhizopus, Mucor* and *Cladosporium* in decreasing frequency. The *Aspergillus* representative species are *A. niger, A. terreus, A. nidulans* and *A. tamarii* (Velmourougane 2013). Strains isolated from coffee showed their ability to produce high levels of pectinase (Antier et al. 1993, Boccas et al. 1994).

Among the filamentous fungi isolated during the dry fermentation, the most frequently identified species belong to the *Aspergillus* genus (like *A. tubingensis* and *A. versicolor*) followed by *Cladosporium cladosporioides* and *Penicillium decumbens* (Vilela et al. 2010). *Aspergillus ochraceus,* which is also detected, can cause safety concerns (Batista et al. 2003, 2009). Indeed some *Aspergillus* species are known to produce Ochratoxin A (OTA) which is a toxic secondary metabolite. In tropical zones, OTA is mainly produced in coffee beans by *A. carbonarius, A. niger* (section Nigri), *A. westerdijkiae* and *A. ochraceus* (section Circumdati) (Pitt et al. 2000, O'Callaghan et al. 2003, Frisvad et al. 2004).

However the study of the influence of certain yeasts on growth and sporulation of ochratoxinogenic fungi showed an impact on OTA production. It is observed that the presence of yeasts reduced the incidence of filamentous fungi during the fermentation process (Frank 1999). *Aspergillus westerdijkiae* OTA production is reduced at the transcriptional level by the presence of *Debaryomyces hansenii* (Gil-Serna et al. 2011). The growth of *A. ochraceus* is inhibited during coffee processing by commercial yeasts and *in vitro* by yeasts isolated from coffee samples (Masoud et al. 2005, Masoud and Kaltoft 2006, Velmourougane et al. 2011a). Other yeast strains show their ability to degrade OTA (Molnar et al. 2004, Schatzmayr et al. 2006). An antifungal activity is also observed when confronting lactic acid bacteria isolated from silage coffee pulp and *A. carbonarius* (Djossou et al. 2011). A microbial consortium including yeast strains usually identified in wet coffee processing and lactic acid bacteria is active against *A. carbonarius* (Massawe and Lifa 2010).

3.3 Physicochemical and Sensory Aspects

The coffee fermentation duration varies depending on different factors including the amount of coffee being fermented, temperature and mucilage composition. The fermentation stage can take about 6 hours to more than 72 hours. *Coffea canephora* fermentation usually requires at least one day more due to the sticky structure and thickness of the mucilage and its constituents like tannins (Murthy and Naidu 2011, Velmourougane 2013). The end of the fermentation is ascertained by testing beans with fingers. The pH of the coffee fermenting mass gets down from about 6 to 4–4.5 at the end of the fermentation (Avallone et al. 2001a, Jackels and Jackels 2005, Velmourougane 2013). In the same time the temperature increase ranges from 1 to 5°C and is inversely linked to the stage duration. Exothermic degradation reactions are probably responsible for this increase (Velmourougane 2013).

After pulping during the wet and semi-wet coffee processing, the mucilage which remains on the coffee beans composed of water (84.2%), proteins (8.9%), sugars (4.1%), pectic substances (0.91%) and ash (0.7%) (Belitz et al. 2009). The composition analysis of the alcohol-insoluble residues shows the presence of pectic substances (*ca.* 30%), cellulose (*ca.* 8%) and neutral non-cellulosic polysaccharides (*ca.* 18%). Pectins contain uronic acids (*ca.* 60%) with a high degree of methyl esterification and a moderate degree of acetylation (Avallone et al. 2000). When comparing the cell wall polysaccharides of 20 hour fermented beans and unfermented beans, it is shown that qualitative and quantitative compositions are very similar for alcohol insoluble residues. Only hot-water-soluble crude pectic substances and hot-water-insoluble residues seem to be modified (Avallone et al. 2001b). However the microbial species involved in the coffee fermentation metabolize about 60% sugars of the mucilage. As a result, they produce organic acids like acetic and lactic acids and ethanol which contribute to the pH decrease (Avallone et al. 2001a, Jackels and Jackels 2005).

The sugar concentrations in processed beans depend on the type of coffee processing. After dry and wet fermentation, only low amounts of glucose and fructose remain while they are significantly higher after dry processing. Arabinose and mannose, whose overall concentration is about 100 fold lower than glucose and fructose, also decrease in the wet process. Sucrose which is the major low molecular sugars in coffee beans appears to be not significantly affected by coffee processing (Knopp et al. 2006, Duarte et al. 2010). These decreases of the sugar concentrations can be linked to active seed metabolism under the anoxic conditions which occur due to the microbial activity during the fermentation step (Knopp et al. 2006, Bytom et al. 2007, Tarzia et al. 2010). Compared to the semi-dry method, the wet process yields higher contents of chlorogenic acids (CGA) and trigonelline but the caffeine level is not modified (Duarte et al. 2010). The non-protein γ-aminobutyric acid content seems to be lower in coffee beans from the wet processing (Bytof et al. 2005). The reducing and non-reducing sugars, which come from the same metabolic pathway like the various amino acid compounds responsible for the coffee colour and aroma, are associated with quality (Licciardi et al. 2005). The acidity of the green beans which can vary according to the fermentation time is an indicator of product quality since the coffee brew can have an unpleasant taste without enough acidity (Silva et al. 2008a). Although the composition of green coffee is not basically changed by fermentation, many of these modifications in the chemical composition induced by the postharvest treatments can influence the sensory properties of the final product by involving precursors of flavour and colour of the coffee brew (Arnold and Ludwig 1996, Esquivel and Jimenez 2012).

Indeed it is generally assumed that the coffee brew produced from wet-processed coffee beans is considered to have superior aroma and higher acceptance (Bytof et al. 2005, Knopp et al. 2006, Gonzalez-Rios et al. 2007a,b, Subedi 2011, Velmourougane 2013, Lee et al. 2015). The wet-fermented coffees tend to be brighter and drier in taste than dry-fermented coffees that are more complex and sweet. Semi-dry processing leads to a final beverage of intermediate quality between the natural and wet coffees (Vilela et al. 2010). The green coffee beans from the wet fermentation exhibit more volatile compounds, notably alcohols, aldehydes, ketones and esters than those from dry fermentation. These fermented green coffees are characterized by pleasant and fruity aromatic notes due to esters such as methyl acetate, ethyl acetate and ethyl isovalerate, whereas those obtained after mechanical mucilage removal are characterized by volatile compounds with an unpleasant note like butan-2, 3-diol and acetic acid with a sour note (Gonzales-Rios et al. 2007a). The roasted coffee beans from the wet treatment are characterized by fruity, floral and caramel notes, whereas those from the semi-dry process present more neutral olfactory profile, characterized more by nutty and buttery notes. The light roasted coffee obtained after mechanical mucilage removal is characterized by a more unpleasant note like sour, toasted and bitter almond notes, which are intensified by stronger notes of the burnt and spicy types when the roasting degree is increased. Whatever the degree of roasting, aroma differences for specific aromatic notes, such as bitter almond, fruity, floral and spicy appear depending on the post-harvest treatments (Gonzales-Rios et al. 2007b).

However quality experts are now recognizing that mechanically washed coffee (unfermented) can reach an equal sensory quality since a badly conducted fermentation can result in a very poor flavor quality (Gamble and Wootton 1963, Woelore 1993). The microbial events during the coffee fermentation can have a direct influence on the final quality of the coffee beans (Avallone et al. 2002, Avallone et al. 2001a). Lactic, acetic and propionic acids which are produced by bacteria and yeasts can act in inhibiting the fungal growth and prevent the unwanted "fermented taste" (Velmourougane et al. 2011b). While some filamentous fungal species are associated with a good quality coffee, such as *C. cladosporioides* (Licciardi et al. 2005), others depreciate the quality, like *Fusarium*, *Penicillium* or *Eurotium* which produces 2-methyl-isoborneol and geosmin which are responsible for earthy and mouldy notes, respectively (Batista et al. 2003, Batista et al. 2009). One other coffee taste defect is related to methyl-butanoic and cyclohexanoic acid derivatives and S-containing compounds from *Bacillus brevis* or lactic acid bacteria. When fermentation is stopped before completion, all the mucilage is not eliminated by washing and can generate off-flavors from microbial fermentation during drying and storage. An over-fermentation, with butyric or propionic acids production, can be responsible for an unpleasant flavor of the beverage and results in a quality loss (Avallone et al. 2002). Acetic acid production can also interfere with the organoleptic quality of the beans.

As it can be seen, fermentation is a complex stage in the processing of coffee and is not always well controlled. Until now few studies have been carried out concerning the standardization of the coffee fermentation (Avallone et al. 2002). Electric conductivity appears to be a useful indicator of the demucilagination level during fermentation in order to regulate the process (Lima et al. 2008). The food safety of washed coffee is improved using ozone or ultrasound pre-treatments before fermentation, without influencing the cup quality (do Nascimento et al. 2008). Recently, researches are focusing on microbial or enzymatic starters in order to improve the quality and delay the coffee fermentation (Murthy and Naidu 2011, Silva et al. 2013, Lee et al. 2016, Peirera et al. 2016, Alter et al. 2015).

Keywords: Coffee, fermentation, dry process, wet process, microbial population, biochemistry, sensorial

References

Agate, A.D. and Bhat, J.V. (1966). Role of the pectolytic yeasts in the degradation of mucilage layer of *Coffea robusta* cherries. Applied Microbiology 14: 256–260.

Alter, P., Berthiot, L., Bertrand, B., Durand, N., Meile, J.C., Morel, G., Ortiz-Julien, A. and Sieczkowski, N. (2015). Effects of selected *Saccharomyces cerevisiae* yeast inoculation on coffee fermentation and quality. Elsevier, 1 p. International Congress on Cocoa Coffee and Tea. 3, 2015-06-22/2015-06-24, Aveiro (Portugal).

Amorim, H.V. and Amorim, V.L. (1977). Coffee enzymes and coffee quality. pp. 27–55. *In*: R.L. Ory and A.J. St.Angelo (eds.). Enzymes in Food and Beverage Processing. American Chemical Society, Washington.

Arnold, U. and Ludwig, E. (1996). Analysis of free amino acids in green coffee beans. II. Changes of the amino acid content in arabica coffees in connection with post-harvest model treatment. Z. Lebensm. Unters Forsch 203: 379–84.

Arunga, R.O. (1973). Enzymatic fermentation of coffee. Kenya Coffee 38: 354–357.

d'Auzac, J. (1996). Toxic oxygen: protection against pathogens. Plantations, Recherche, Développement 3: 153–170.

Antier, P., Minjares, A., Roussos, S., Raimbault, M. and Viniegra, G. (1993). Pectinase-hyper-producing mutants of *Aspergillus niger* c28b25 for solid-state fermentation of coffee pulp. Enzyme Microbial Technology 15: 254–260.

Avallone, S., Brillouet, J.M., Guyot, B., Olguin, E. and Guiraud, J.P. (2002). Involvement of pectolytic micro-organisms in coffee fermentation. International Journal of Food Science and Technology 37: 191–198.

Avallone, S., Guiraud, J.P., Guyot, B., Olguin, E. and Brillouet, J.M. (2001b). Fate of mucilage cell wall polysaccharides during coffee fermentation. Journal of Agricultural and Food Chemistry 49: 5556–5559.

Avallone, S., Guyton, B., Bailout, J.M., Holguin, E. and Guiraud, J.P. (2001a). Microbiological and biochemical study of coffee fermentation. Current Microbiology 42: 252–256.

Avallone, S., Guiraud, J.P., Guyot, B., Olguin, E. and Brillouet, J.M. (2000). Polysaccharide constituents of coffee-bean mucilage. Journal of Food Science 65: 1308–1311.

Avallone, S., Guyot, B., Michaux-Ferrière, N., Guiraud, J.P., Olguin, E. and Brillouet, J.M. (1999). Cell wall polysaccharides of coffee bean mucilage. Histological characterization during fermentation. Proceedings of 18th International Scientific Colloquium on Coffee, Helsinki. Association for the Science and Information on Coffee (ASIC), Paris, pp. 463–470.

Barel, M. and Jacquet, M. (1994). La qualité du café: ses causes, son appréciation, son amélioration. Plantations, Recherch e développement 1: 5–13.

Barnett, J.A., Payne, R.W. and Yarrow, D. (2000). Yeasts: Characteristics and Identification (3rd edn.). Cambridge University Press, Cambridge.

Batista, L.R., Chalfoun, S.M., Silva, C.F., Cirillo, M., Varga, E.A., Prado, G. and Schwan, R.F. (2009). Ochratoxin A in coffee beans (*Coffea arabica* L.) processed by dry and wet methods. Food Control 20: 784–790.

Batista, L.R., Chalfont, S.M., Prado, G., Schwann, R.F. and Wheals, A.E. (2003). Oxygenic fungi associated with processed (green) coffee beans (*Coffee Arabica* L.). International Journal Food Microbiology 85: 293–300.

Bailly, H., Sallee, B. and Garcia, S.G. (1992). Improving the Quality of Coffee in the Xalapa-Coatepec Region of Mexico—Evaluation of Yields and Pulp Extraction. Cafe Cacao The. 36: 55–61.

Belitz, H.D., Grosch, W. and Schieberle, P. (2009). Food chemistry (4th edn.). Springer, Heidelberg.

Boccas, F., Roussos, S., Gutierrez, M., Serrano, L. and Viniegra, G. (1994). Production of pectinase from coffee pulp in solid-state fermentation system—selection of wild fungal isolate of high potency by a simple 3-step screening technique. Journal Food Science and Technology 31: 22–26.

Bytof, G., Knopp, S.-E., Schieberle, P., Teutsch, I. and Selmar, D. (2005). Influence of processing on the generation of γ-aminobutyric acid in green coffee beans. European Food Research and Technology 220: 245–250.

Bytom, G., Knopp, S.E., Kramer, D., Breitenstein, B., Bergervoet, J.H.W., Groot, S.P.C. and Selmar, D. (2007). Transient occurrence of seed germination processes during coffee post-harvest treatment. Annals of Botany 100: 61–66.

Calle, V.H. (1965). Algunos métodos de desmucilaginado y sus efectos sobre el café en pergamino. Cenicafé (Colombia) 16: 3–11.

Calle, V.H. (1962). Métodos de extracción de las pectinas del café. Cenicafé (Colombia) 13: 69–74.

Calle, V.H. (1957). Activadores bioquimicos para la fermentation del café. Cenicafé (Colombia) 8: 94–101.

Carbonell, R. and Vilanova, T. (1952). El café de El Salvador 248–249.

Coleman, R.J., Lenney, J.F., Coscia, A.T. and Dicarlo, F.J. (1955). Pectic acid from the mucilage of coffee cherries. Archives of Biochemistry and Biophysics 59: 157–164.

Coste, R. 1989. Caféiers et cafés. Paris: Maisonneuve et Larose. Paris.

Djossou, O., Perraud-Gaime, I., Mirleau, F.L., Rodriguez-Serrano, G., Karou, G., Niamke, S., Ouzari, I., Boudabous, A. and Roussos, S. (2011). Robusta coffee beans post-harvest microflora: *Lactobacillus plantarum* sp. as potential antagonist of *Aspergillus carbonarius*. Anaerobe 17: 267–272.

Duarte, G.S., Pereira, A.A. and Farah, A. (2010). Chlorogenic acids and other relevant compounds in Brazilian coffees processed by semi-dry and wet post-harvesting methods. Food Chemistry 118: 851–855.

Durand, N., El Sheikha, A.F., Suarez-Quiros, M.L., Gonzales-Rios, O., Nganou, D.N., Fontana-Tachon, A. and Montet, D. (2013). Application of PCR-DGGE to the study of dynamics and biodiversity of yeasts and potentially OTA producing fungi during coffee processing. Food Control 34: 466–471.

Esquivel, P. and Jimenez, V.M. (2012). Functional properties of coffee and coffee by-products. Food Research International 46: 488–495.

Evangelista, S.R., Miguel, M.G., Silva, C.F., Pinheiro, A.C. and Schwan, R.F. (2015). HYPERLINK "https://www.ncbi.nlm.nih.gov/pubmed/26119187" Microbiological diversity associated with the spontaneous wet method of coffee fermentation. International Journal of Food Microbiology 210: 102–12.

Frank, H.A. and De la Cruz, A.S. (1964). Role of incidental microflora in natural decomposition of mucilage layer in Kona coffee cherries. Journal of Food Science 29: 850–853.

Frank, H.A., Lum, A.N. and De la Cruz, A.S. (1965). Bacteria responsible for mucilage layer decomposition in Kona coffee cherries. Applied Microbiology 13: 201–207.

Frank, J.M. (1999). HACCP and its mycotoxin management control potential: ochratoxin A in coffee production. Proceedings of 7th Meetings of the International Committee on Food Microbiology and Hygiene. Veldhoven, The Netherlands, pp. 122–125.

Frisvad, J.C., Frank, J.M., Houbraken, J.A.M.P., Kuijpers, A.F.A. and Samson, R.A. (2004). New Ochratoxin A producing species of *Aspergillus* section Circumdati. Studies in Mycology 50: 23–43.

Gamble, F.M. and Wootton, A.E. (1963). A note on a very slow fermentations. Kenya Coffee 28: 481–483.

Garcia, R., Arriola, D., Arriola, M.C., Porrez, E. and Rolz, C. (1991). Characterization of coffee pectin. Lebensmittel-Wissenschaft und Technologie 24: 125–129.

Gil-Serna, J., Patino, B., Cortes, L., Gonzalez-Jaen, M.T. and Vazquez, C. (2011). Mechanisms involved in reduction of ochratoxin A produced by *Aspergillus westerdijkiae* using *Debaryomyces hansenii* CYC 1244. International Journal Food Microbiology 151: 113–118.

Gonzalez-Rios, O., Suarez-Quiroz, M.L., Boulanger, R., Barrel, M., Guyton, B., Giraud J.P. and Schorr-Galindo, S. (2007b). Impact of "ecological" post-harvest processing on coffee aroma: II. Roasted coffee. Journal of Food Composition and Analysis 20: 297–307.

Gonzalez-Rios, O., Suarez-Quiroz, M.L., Boulanger, R., Barel, M., Guyot, B., Guiraud, J.P. and Schorr-Galindo, S. (2007a). Impact of "ecological" post-harvest processing on the volatile fraction of coffee beans: I. Green coffee. Journal of Food Composition and Analysis 20: 289–296.

Jackels, S.C. and Jackels, C.F. (2005). Characterization of the coffee mucilage fermentation process using chemical indicators: A field study in Nicaragua. Journal of Food Science 70: C321–C325.

Juven, B., Lindner, P. and Weisslowicz, H. (1985). Pectin degradation in plant-material by *Leuconostoc mesenteroides*. Journal of Applied Bacteriology 58: 533–538.

Knopp, S., Bytof, G. and Selmar, D. (2006). Influence of processing on the content of sugars in green Arabica coffee beans. European Food Research and Technology 223: 195–201.

Lee, L.W., Cheong, M.W., Curran, P., Yu, B. and Liu, S.Q. (2016). Modulation of coffee aroma *via* the fermentation of green coffee beans with *Rhizopus oligosporus*: II. Effects of different roast levels. Food Chemistry 211: 925–36.

Lee, L.W., Cheong, M.W., Curran, P., Yu, B. and Liu, S.Q. (2015). Coffee fermentation and flavor—An intricate and delicate relationship. Food Chemistry 185: 182–191.

Licciardi, R., Pereira, R.G.F.A., Mendonça, L.M.V.L. and Furtado, E.F. (2005). Avaliação físico-química de cafés torrados e moídos, de diferentes marcas comerciais, da Região Sul de Minas Gerais. Cienciae Tecnologia de Alimentos 25: 425–429.

Lima, M.V., Vieira, H.D. and Martins, M.L.L. (2008). Electric conductivity as indicator of coffee depulped during degumming. Cienca Rural 38: 1765–1768.

Lopez, C.I., Bautista, E., Moreno, E. and Dentan, E. (1989). Factors related to the formation of 'over fermented coffee beans' during the wet processing method and storage of coffee. Proceedings of 13th International Scientific Colloquium on Coffee, Paipa. Association for the Science and Information on Coffee (ASIC), Paris, pp. 373–384.

Masoud, W., Cesar, L.B., Jespersen, L. and Jakobsen, M. (2004). Yeast involved in fermentation of *Coffea arabica* in East Africa determined by genotyping and by direct denaturing gradient gel electrophoresis. Yeast 21: 549–556.

Masoud, W., Polland, L. and Jakobsen, M. (2005). Influence of volatile compounds produced by yeasts predominant during processing of *Coffea arabica* in East Africa on growth and ochratoxin A (OTA) production by *Aspergillus ochraceus*. Yeast 22: 1133–1142.

Masoud, W. and Kaltoft, H.C. (2006). The effects of yeasts involved in the fermentation of *Coffea arabica* in East Africa on growth and ochratoxin A (OTA) production by *Aspergillus ochraceus*. International Journal of Food Microbiology 106: 229–234.

Masoud, W. and Jespersen, L. (2006). Pectin degrading enzymes in yeasts involved in fermentation of *Coffea arabica* in East Africa. International Journal of Food Microbiology 110: 291–296.

Massawe, G.A. and Lifa, S.J. (2010). Yeasts and lactic acid bacteria coffee fermentation starter cultures. International Journal of Postharvest Technology and Innovation 2: 41–82.

Menchù, J.F. and Rolz, C. (1973). Coffee fermentation technology. Thé Café Cacao 17: 53–61.

Molnar, O., Schatzmayr, G., Fuchs, E. and Prillinger, H. (2004). *Trichosporon mycotoxinivorans* sp. nov., a new yeast species useful in biological detoxification of various mycotoxins. Systematic and Applied Microbiology 27: 661–671.

Murthy, P.S. and Naidu, M.M. (2011). Improvement of robusta coffee fermentation with microbial enzymes. European Journal of Applied Sciences 3: 130–139.

Do Nascimento, L.C., Lima, L.C.D., Picolli, R.H., Fiorini, J.E., Duarte, S.M.D., da Silva, J.M.S.F., Oliveira, N.D. and Veiga, S.M.D.M. (2008). Ozone and ultrasound: alternative processes in the treatment of fermented coffee. Cienciae Tecnolgia Alimentos 28: 282–294.

O'Callaghan, J., Caddick, M.X. and Dobson, D.W. (2003). A polyketide synthase gene required for Ochratoxin A biosynthesis in *Aspergillus ochraceus*. Microbiology 149: 3485–3491.

Pereira, G.V.M., Soccol, V.T. and Soccol, C.R. (2016). Current state of research on cocoa and coffee fermentations. Current Opinion in Food Science 7: 50–57.

Perombelon, M.C.M. and Kelman, A. (1980). Ecology of the soft rot *Erwinias*. Annual Review of Phytopathology 18: 361–387.

Pitt, J.I., Basilico, J.C., Abarca, M.L. and Lopez, C. (2000). Mycotoxins and toxinogenic fungi. Medical Mycology 38: 41–46.

Rolz, C., Menchù, J.F., Calzada, F. and De Leon, R. (1982). Biotechnology in washed coffee processing. Process Biochemistry 17: 8–10.

Schatzmayr, G., Schatzmayr, D., Pichler, E., Taubel, M., Loibner, A.R. and Binder, E.M. (2006). A novel approach to deactivate ochratoxin A. Proceedings of the 11th International IUPAC Symposium on Mycotoxins and Phycotoxins, Bethesda, USA, pp. 279–288.

Schwan, R.F. and Wheals, A.E. (2003). Mixed microbial fermentations of chocolate and coffee. pp. 426–459. *In*: T. Boekhout and V. Robert (eds.). Yeasts in Food. Behr's Verlag, Hamburg.

Serrat, M., Bermudez, R.C. and Villa, T.G. (2002). Production, purification, and characterization of a polygalacturonase from a new strain of *Kluyveromyces marxianus* isolated from coffee wet-processing wastewater. Applied Biochemistry and Biotechnology 97: 193–208.

Sibanda, L. (2006). Auto-control of ochratoxin-A levels in green coffee. Tea & Coffee Trade Journal Online 178(1).

Silva, C.F., Schwan, R.F., Dias, E.S. and Wheals, A.E. (2000). Microbial diversity during maturation and natural processing of coffee cherries of *Coffea arabica* in Brazil. International Journal of Food Microbiology 60: 251–260.

Silva, C.F., Batista, L.R., Abreu, L.M., Dias, E.S. and Schwan, R.F. (2008a). Succession of bacterial and fungal communities during natural coffee (*Coffea arabica*) fermentation. Food Microbiology 25: 951–957.

Silva, C.F., Batista, L.R. and Schwan, R.F. (2008b). Incidence and distribution of filamentous fungi during fermentation, drying and storage of coffee (*Coffea arabica* L.) beans. Brazilian Journal of Microbiology 39: 521–526.

Silva, C.F., Vilela, D.M., de Souza Cordeiro, C., Duarte, W.F., Dias, D.R. and Schwan, R.F. (2013). Evaluation of a potential starter culture for enhance quality of coffee fermentation. World Journal of Microbiology and Biotechnology 2: 235-47.

Sivetz, M. and Foote, H.E. (1963). Coffee processing technology, Volume 1. The AVI Publishing Co, West Port.

Subedi, R.N. (2011). Comparative analysis of dry and wet processing of coffee with respect to quality and cost in Kavre district, Nepal: a case of Panchkhal village. International Research Journal of Applied and Basic Sciences 2: 181–193.

Tarzia, A., Scholz, M.B.D. and Petkowicz, C.L.D. (2010). Influence of the postharvest processing method on polysaccharides and coffee beverages. International Journal of Food Science and Technology 45: 2167–2175.

Van Pee, W. and Castelein, J.M. (1971). The yeast flora of fermenting robusta coffee. East African Agricultural and Forestry Journal 36: 308–310.

Van Pee, W. and Castelein, J.M. (1972). Study of the pectinolytic microflora, particularly the *Enterobacteriaceae*, from fermenting coffee in the Congo. Journal of Food Science 37: 171–174.

Velmourougane, K., Bhat, R., Gopinandhan, T.N. and Panneerselvam, P. (2011b). Impact of delay in processing on mold development, ochratoxin A and cup quality in arabica and robusta coffee. World Journal of Microbiologyand Biotechnology 27: 1809–1816.

Velmourougane, K. (2013). Impact of natural fermentation on physicochemical, microbiological and cup quality characteristics of arabica and robusta coffee. The Proceedings of the National Academy of Sciences, India, Section B 83: 233–239.

Velmourougane, K., Bhat, R., Gopinandhan, T.N. and Panneerselvam, P. (2011a). Management of *Aspergillus ochraceus* and Ochratoxin A contamination in coffee during on-farm processing through commercial yeast inoculation. Biological Control 57: 215–221.

Vilela, D.M., Pereira, G.V.D., Silva, C.F., Batista, L.R. and Schwan, R.F. (2010). Molecular ecology and polyphasic characterization of the microbiota associated with semi-dry processed coffee (*Coffea arabica* L.). Food Microbiology 27: 1128–1135.

Woelore, W.M. (1993). Optimum fermentation protocols for Arabica coffee under Ethiopian conditions. Proceedings of the 15th International Scientific Colloquium on Coffee, Montpellier. Association for the Science and Information on Coffee (ASIC). Paris, pp. 727–733.

Wootton, A.E. (1965). The importance of field processing to the quality of east African coffees. Proceedings of the 2th International Scientific Colloquium on Coffee, Nairobi. Association for the Science and Information on Coffee (ASIC), Paris, pp. 247–254.

Wootton, A.E. (1963). The fermentation of coffee. Kenya Coffee 28: 239–249.

15

Cocoa Fermentation

Marisol López-Hernández, Gloria Díaz-Ruiz and
Carmen Wacher*

1. Introduction

The process to obtain chocolate from cocoa (*Theobroma cacao*) pods usually takes 2 to 8 days and includes steps like harvesting, fermentation, drying and roasting. Each of these steps plays a role in the process and depends on the type of cocoa that will be processed, the weather, the kind of chocolate that will be obtained as well as other factors. It is a carefully conducted process that starts with an appropriate cultivation of the tree until roasted cocoa beans with the desired flavor, color and texture are obtained. Each step is important and would need to be controlled in order to obtain good quality chocolate; however, the substrate, as well as the microbiota are complex, and not completely understood.

Studies on this process and the effects of each step on the microbiology, flavor and color precursor formation and quality of the beans, including recent research findings and proposals to improve them are presented in this chapter.

2. Cocoa Tree

Theobroma cacao is a small perennial tree (Fig. 1) that is cultivated in tropical regions of Central America and South America, where a hot and humid

Departamento de Alimentos y Biotecnología, Facultad de Química, UNAM.
Circuito exterior S/N, Ciudad Universitaria, Coyoacán. CP: 04510, Ciudad de México, México.
E-mail: marylohdez@gmail.com, glodr@unam.mx
* Corresponding author: wacher@unam.mx

Images courtesy of: Hacienda Cacaotera "Jesús María", Comalcalco, Tabasco, México.

Figure 1. A. Cocoa flower and unripe cocoa pod (*chilillo*) in the tree, B. Cocoa flower, C. *Criollo* ripe cocoa pod, D. Cocoa pods, E. Worker opening cocoa pods, F. Separating beans from the cocoa pod, G. Cocoa beans fermentation, H. Cocoa beans sun drying and I. Dried cocoa beans (Images courtesy of: Hacienda Cacaotera "Jesús María", Comalcalco, Tabasco, México).

weather prevails, in the rainforests between latitudes 15′S and 18′N. Its primary center of diversity has been found to be the Amazonas basin, but it is known to have been produced naturally from the South of Mexico to the Amazon region. It is not known how it arrived to Mesoamerica, but it is the Mexican cocoa that is appreciated worldwide. There are two classes of *T. cacao*, Criollo and Forastero. Criollo was assumed to be cultivated since prehistoric times in Mesoamerica, it has a weak and special flavor, the beans are rounded and white, but the tree is susceptible to diseases. Forastero varieties were found later and have a higher fat content and stronger flavor. Because of these characteristics, it is the most cultivated, especially the variety Amelonado (Wood and Lass 1985, Vela 2012). The trees are hardy and vigorous, the beans are smaller and flattened and the cotyledons are violet in color.

Earlier, only the pulp was used in the Amazon, which was converted to an alcoholic beverage and the bean was discarded. It was domesticated in a different form in Mexico than in the Amazon. The whole grain, including the pulp, was fermented, so that a different product was obtained, and the Criollo variety, developed in Mexico, was selected for its flavor. The term *Theobroma* comes from the Greek *theos*, that means god, and *broma*, that means food, so it is the food for the gods. In México, it was considered as a special food, an important ritual element and a social status marker (Vela 2012). *Theobroma* genus contains 22 species divided into six sections, but the mostly used species is *T. cacao*. Among other species, *T. grandifluorum* is only used in the Amazon and *T. bicolor*, in Tabasco and Mexico, that is called *pataxte* and considered of lower quality than *T. cacao* (Wood and Lass 1985).

The height of the cocoa tree ranges 3–5 m, and it has a straight trunk that divides into 3 to 4 branches. The cocoa fruits or pods are produced in the trunk and branches and each tree produces 30 to 60 beans. It takes from 4 to 7 mon from pollination to maturity. The cotyledons of the dry seeds are 0.5 to 2.0 g, with a shape ranging from spherical to flat, and color ranging from white to dark purple. It is surrounded by the tegument or testa, and this is covered by white mucilage, called the pulp. These are spongy parenchymatous cells, rich in sugars (10–13%), pentosans (2–3%), citric acid (1–2%) and salts (8–10%). These have two functions: they are the substrate for the microorganisms that perform the fermentation and they inhibit the seed's germination. The tree needs special conditions: it has to grow under the shade of a taller tree that also provides nutrients as organic matter accumulated in the soil. It is considered that this interaction with other trees represents an ecologic advantage: "cocoa cultures occur in the woods, not after their destruction" (Vela 2012).

The tree requires a great amount of water (at least 1500 mm of rainfall per year), a good quality soil, with water retention properties, but with good drainage, pH from 5.0 to 7.0, needed for the solubility of minerals and nutrients, high content of organic matter, ability to absorb and release cations, an optimum total nitrogen/total phosphorous of 1.5 and a temperature 20°C over the ambient. These parameters affect cocoa bean flavor quality (Kongor et al. 2016).

2.1 Diseases

A serious problem is the diseases of the tree, which can result in severe losses (40% of the world production, up to 90% in a farm). The following are some of the most important cocoa's diseases:

- **Cocoa swollen shoot virus.** The virus belongs to the genus *Badna virus* and it is the most common pathogen of cocoa trees in West Africa. It is transmitted by mealybugs (*Planococcus citri, P. njalensis*). The symptom of the disease includes leaf discoloration, roots and shoots swelling and death of the trees. It has been controlled by eradicating the infected trees, isolating the infected areas and controlling the insect vector.

- *Phythophthora* **spp**. Several species of this fungus causes the "black pod rot" disease in most producer countries, causing 30% loss of the total world production. It causes a black or brown lesion of the pods and if it continues, the beans can be lost. Trunk and leaf lesions have also been observed. To control this disease, fungicides are used. A general problem with the control of these diseases is due to the ability of antimicrobials resistance of the pathogens.

- *Moniliophtora perniciosa* **and** *Moniliophtora roreri.* They are fungi that cause problems in cocoa trees from Latin America. *M. perniciosa* causes the disease called "witche's broom" whereas *M. roreri* causes "frosty pod rod". Both results in necrosis in the pods, which may start with a lesion, that advances until the whole pod and beans (and sometimes other parts of the plant) are affected.

These diseases have caused huge losses in Latin American countries; in Brazil they caused 75 to 90% pod losses. They have been controlled removing the infected parts, when it is starting, and/or using fungicides and biological control methods. They can also affect fermentation, as they grow and change the usual fermentation pattern (Lopes and Pires 2016).

2.2 How It is Cultivated

Most cocoas are produced in small farms, where workers belong to the same family and their productivities are low (< 700 kg/ha). On the other hand, large plantations produce 2 to 3 ton kg/ha, with a great amount of workers. Small farms use their own seeds, which are a mixture of varieties that include hybrid seeds or seeds from neighbors. Each variety has its own maturity time and the differences in maturity times results in products with low quality. Cocoa pods are usually harvested by hand, each 15–30 days and it is difficult to recognize the different maturities of the pods (as it is also difficult to decide when they are mature and ready to harvest). Therefore, harvested cocoa beans are of different sizes, colors, pulp contents, and this heterogeneity is a common cause of low quality cocoa.

2.3 Where It is Cultivated

According to Food and Agriculture Organization (FAO), the world major cocoa producers are: Ivory Coast, Ghana, Indonesia, Nigeria, Cameroun, Brazil, Ecuador and Mexico.

Mesoamerica

In Mesoamerica, the region from Central Mexico to Central America, cocoa was first used by the Olmec civilization in Central Mexico, then by the Aztecs and Mayas. It was there where it flourished long before the discovery of the American continent. It was not only consumed as food, but was also used as a medicine. Cocoa was one of the gifts that the gods had given to humans and according to the Popol Vuh, a sacred book of the Mayas, it was one of the four cosmic trees that supported the world and was associated with the maize plant, which is also considered sacred. It was also metaphorically related to blood and sacrifice. With all these symbolisms, cocoa acquired an important role in rituals and was only allowed to be consumed by the royalty. This was so serious that whoever dared to taste it, and did not belong to the royalty, was punished. Its value was such that it was considered a coin, and the cost of merchandises would be given in cocoa grains. It was taken to Europe after the colonial rule, made into a beverage and later chocolate and other products were developed, which are very popular product till date.

México

It was the Mesoamerican people who domesticated it, who learned to grow, to harvest and to process it (Vela et al. 2012); however, now Mexico participates with only 1.1% of the world's production, being the 8th world producer of cocoa beans. Tabasco State, in Southeast Mexico, is Mexico's main cocoa producer contributing to 66% of the total national production followed by Chiapas State with 33% and the rest produced by Oaxaca, Guerrero and Veracruz (www.cacaomexico.org). "Workers produce cocoa with the experience and knowledge of traditional agriculture" (Córdova-Avalos et al. 2001). It is a family business and may constitute their main source of income (42 to 86%). The low productivity is related to the old age method of plantations, a poor control of the tree's pests and diseases, inadequate water draining during rainfall, inappropriate handling of the shade cocoa plantations' need, no technical support or training and insufficient support by the state. However, there is recent research, as that of Hernandez-Hernandez et al. (2016), who determined the best method to ferment cocoa beans in Huimanguillo, Tabasco. There are also foundations, such as Cacao Mexico, which are promoting good quality cocoa production through national and international resources, as chocolate companies, research institutes, and other organizations, contributing to the social and economic welfare of workers from the producing states (www.cacaomexico. org). A national strategic program to rescue this system, which for more than 500 years has shown that ecological and economic sustainability is needed (Córdova-Ávalos et al. 2008).

Brazil

The origin of the cocoa plant has been found to be the Amazon basin in Brazil, where cocoa beans were used to produce beverages, but did not have the importance as in Mesoamerica. In modern times cocoa bean production has had a great importance as it has been the basis of the economy of Bahia and other Brazilian states. However, due to financial crises in the international cocoa market and the disease in the plant caused by *Moniliofthora perniciosa*, called "witche's broom", has reduced production and exports dramatically (da Veiga Moreira et al. 2013). Still Brazil is now the worlds 6th larger producer of cocoa beans.

Africa and Asia

Cocoa was introduced into West Africa during the 18th century. After the 1st world war, African planters were interested in cocoa cultivation. Production increased and then declined, but the interest on cocoa was there due to the government programs focused to increase production.

Cocoa is now produced mainly in the African countries, Ivory Coast and Ghana, where 72% of cocoa beans in the world is produced (Reference). Indonesia contributes 13 percent of the total world production. Higher production in this region may be due to the climatic conditions in the country.

Ghana

According to Ntiamoah and Afrane (2008), Ghana, with 700,000 ton/yr production, is the world's second producer of cocoa, after Ivory Coast, but it is leader in premium quality cocoa beans. It exports raw cocoa beans and also processes some of them into finished and semi-finished cocoa products for national and international markets. Semi-finished products include cocoa liquor, cocoa butter, cake and powder. Cocoa is the major source of revenue in Ghana. This industry employs most (60%) of the national agricultural labor and for these farmers cocoa has become a source of 70 to 100% of their annual incomes. This is why the government, to reduce poverty, initiated a Cocoa disease and Pest Control Project: CODAPEC. This Project provided Ghana's cocoa farms with insecticides and fungicides. After spraying, production increased from 340,562 metric tons in 2001/02 to 736,000 in 2003/2004.

Favorable external conditions and internal reforms were thought as the major factors to renovate traditional export crops. In fact, Ghana has maintained a high position among cocoa producing countries, which is due to land expansion and increasing labor and not due to rise in land productivity (Vigneri 2008). An important agricultural innovation available is the hybrid cocoa which has various advantages. It yields more pods per tree, has more than two harvest seasons, has a shorter gestation period, but has not been adopted by all farmers, as it increases costs by acquisition of information, labor, chemicals and machinery. This could only be possible by the financial help by banka and this could be acquired only by high-income producers (Boahene et al. 1999).

A relevant issue is that ownership right of farming land cannot be easily acquired if there is no family relation. Sharecropping is then used: cultivators, which are paid to do weeding and cleaning and others, who take charge of all the tasks and receive half of the harvest. Cultivators own the land, until it remains unused or until they die and then ownership is returned to the original owner (Boahene et al. 1999). Spraying with pesticides and fertilizers caused a boom in cocoa production; however, it resulted in some negative effects on the environment were produced and this would need to be reviewed.

A Life Cycle Assessment (LCA) was used to determine the effect of pesticides and insecticides in the environment. It determines the potential environmental impacts throughout the product's processing (called life cycle), from raw materials to production, use, end-of-life treatment, recycling and disposal (Ntiamoa and Afrane 2008). The objective was to identify and quantify potential environmental impacts associated with cocoa production in Ghana, then to identify activities that lead to the largest environmental impacts, and to suggest improvement options. This information could be used by COCOABOD (Ghana cocoa Board, the government agency in charge of the cocoa industry), by cocoa scientists, farmers and processors to improve performance of the industry.

Ivory Coast

Ivory Coast is peculiar, as it passed from a poor to a successful country in the 1960–1970s, because of an agriculturally-based program they developed. Cocoa became the major export product in 1970, and in 1979, Ivory Coast Gross reached a national product per capita of 8%, the highest in Sub-Saharan Africa. Capital accumulated, foreign exchange was secured, coffee, cocoa and timber were exported, but among these cocoa was the most exported.

Since 1980, Ivory Coast became the world's largest producer, supplying 33% of the cocoa in the world. The possible reasons are reported to be: para-statal interventions in agriculture, money obtained from agricultural exports, an over-valued exchange rate and subsidies to agriculture for food production (Mathurin and Delgado 1984).

Most of cocoa crops are grown by smallholder farmers, who are in charge of harvesting, separating the beans, fermenting them and then they are dried in a central place. Traitants or buyers buy and collect small farms and keep them in warehouses in larger towns, where exporters buy and take the crops to a ship.

Many people are involved in this process including the owner of the fields to the laborers, who may be family members or people hired to do the jobs; however, a practice that should not exist in this 21st century is that of traffickers, who promise paid work, housing and education to children, who do long hours of forced labor in the farm and undergo severe abuse (US Department of Labor 2015). This has been much criticized.

Indonesia

Indonesia is the third largest producer of cocoa after Ghana and Ivory Coast and the most important supplier in East Asia. The advantages of cocoa beans from Indonesia are their low cost, high production capacity, infrastructure and open trading. However, production may be inconsistent and of low quality (Panlibuton and Lusby 2006), which may be due to the pest infestation problems. Cocoa beans are produced in the island of Sulawesi and are the source of income for more than 400,000 farmers and their families. Also, the lack of involvement of the government in price settings has allowed smallholder farmers to receive more money for their beans (75–85 percent) than farmers from other West African countries (50–60 percent).

3. Postharvest Traditional Process

Postharvest cocoa processing is a complex procedure that is crucial to obtain good quality beans. The following steps are essential.

3.1 Pulp-preconditioning

The properties of the pulp are very prone to changes before it is used for fermentation. If the pulp is removed, mechanically or enzymatically, few acids are produced, which affect the quality of the pulp. Excessive pulp results in higher sour cocoa beans, which affects the flavor, but if a proportion of the pulp is removed, less acid is produced, which improves the flavor, and reduce the fermentation time. Enzymatic depulping involves the use of pectin degrading enzymes that reduce the volume and increase the aeration during fermentation (Kongor et al. 2016).

3.2 Pod Storage

Pods can be stored before fermentation which has shown to improve the chemical composition and flavor of the beans. This effect the acidity significantly reduces in polyphenol content, astringency and bitterness (Kongor et al. 2016).

3.3 Fermentation

This is a crucial step together with drying as flavor precursors are formed. Pods are usually stored before processing and then they are broken by *machete*, to remove the husks. The bean water content can be reduced by 40–50%, which affects the pulp which brings the changes in the fermentation process (Kongor et al. 2016). Beans are separated from broken pods and placed in wooden boxes, in heaps or in trays and are left to ferment spontaneously for 4 to 6 days. Two different phases have been distinguished. Phase 1: The pH value of the pulp is low (3.6) and this favors the growth of yeasts, which initiates fermentation in the pulp and on the surface of the beans. During the first 24 h of yeast growth Ethanol and pectinolytic enzymes are produced. The pulp is hydrolyzed by the action of pectinolytic enzymes and the new conditions favors the growth of lactic acid bacteria (LAB), which reach a maximum concentration at approximately 36 h after fermentation. Acetic acid bacteria (AAB) transform alcohol into acetic acid, an exothermic reaction that results in a rise of temperature to about 50°C. The activity of AAB increases with aeration during fermentation as they are obligate aerobes. After 5 or 6 days their counts decrease and aerobic spore-forming bacteria take over, that is considered undesirable, as they acidify the beans and produce off-flavors.

Due to the production of several metabolites complex reactions occurs in the bean during this stage. Lactic acid, acetic acids and ethanol, together with the heat produced during fermentation, kills the seed embryo that is presumably due to penetration of ethanol and acetic acid into the cotyledons. Proteolytic and peptidase activities during fermentation act on the beans cell wall proteins producing peptides and amino acids, increasing the concentration of the latter by approximately 150 to 200% (Good quality cocoa beans contain 8–14 mg/g dry matter). These are extremely important because they are flavor precursors and together with reducing sugars cause browning Maillard reaction, which is important in the beans color. For proteolysis to occur the pH must not be too low at the beginning of fermentation (Saltini et al. 2013).

Phase 2: The second phase occurs within the cotyledons by the action of hydrolytic enzymes, liberating components such as reducing sugars, amino acids and peptides, that are the precursors of chocolate flavor (da Veiga Moreira et al. 2013, Hamdouche et al. 2015). The process is mediated by a complex group of endopeptidase present in the cotyledons and carboxy-exo-peptidase (Kongor et al. 2016).

Volatile compounds (alcohols, organic acids, esters, aldehydes) are formed after fermentation. Phenolic compounds are oxidized and polymerized to form insoluble tannins contributing to decrease the bitterness and astringency of the grains.

Production of precursor molecules and Maillard browning, by the reaction of sugars and amino acids, are the key compounds of this fermentation process, because if a cocoa batch is not fermented, the final product will not possess the refined taste of chocolate.

3.3.1 Methods of fermentation

Different methods are used to ferment the beans. Saltini et al. (2013) have summarized the differences among them:

- *Boxes*. Large boxes ($1 \times 1 \times 1$ or $7 \times 5 \times 1$ m), are used for processing the beans. These wooden boxes or "sweat boxes" are placed above a drain, where pulp juice or "sweat" are carried (Montville et al. 2012). These juices are formed during the pulp fermentation and liberated through holes in the bottom of the box (Thompson et al. 2013). Cocoa beans fermented in boxes contain low concentrations of sugars, ethanol and acetic acid, which depends on the characteristics of the boxes. Increase in temperature is slower than in the other methods, but it has been noticed that the product is not homogenous after fermentation.

- *Heap method*. Heaps are put on the soil, covered with banana leaves and aerated manually (Montville et al. 2012). Temperature at the beginning of fermentation increases faster than in the box method, the grains are fermented more uniformly, and browner and less purple beans are obtained.

- *Platform method*. It is the cheapest method used for fermentation. Cocoa beans are extended on a surface and left for fermentation and drying during the day. During the night they are placed in heaps to maintain the higher temperature. Short fermentation of 2 to 3 days is achieved with Criollo cocoa, in Central America (Thompson et al. 2013); but because of the low quality of the beans obtained, is not generally used. The fermentation rate is low and molds overgrow the beans, producing off-flavors, if not managed properly.

- *Baskets*. Baskets are used by small-scale producers. Beans are placed in the basket and covered with plantain leaves. Low amounts are processed by this method (Thompson et al. 2013).

3.4 Drying

Drying is done to reduce the moisture content from 60% to approximately 7.5%. This process also contributes to the flavor of chocolate, as fermentation's oxidative stage, which is important in reducing astringency, bitterness and acidity, continues. Drying initiates polyphenol oxidizing reactions, producing new flavor components and loss of membrane integrity, which induces brown color reactions.

During drying, reducing sugars participate in the thermal treatment of non-enzymatic browning reactions, the Maillard reactions to form volatile fractions of pyrazines. Amadori compounds, the first produced from the reaction of amino acids and reducing sugars, have been identified in this stage (Kongor et al. 2016). This process has to be slow, to allow completion of flavor reactions that starts during fermentation. If drying is too fast the beans will be too acidic and if it is slow, the molds will have the opportunity to grow; the right color and acidity will not be produced (Saltini et al. 2013).

Cocoa beans can be sun-dried, which is the common method used because it gives good results. Artificial drying would need to be studied further in order to have comparable results (Saltini et al. 2013).

In Indonesia, the desirable moisture content after drying is 10%. The low quality of cocoa bean produced is assumed to be due to the use of sun-drying method, during which browning reactions and loss of polyphenols by oxidation occurs. Dina et al. (2015) proposed the use of solar drying, which uses solar energy, but the disadvantage is that, by nature, it is intermittent, and depends on the weather. To overcome this, they integrated an absorbent within the chamber that absorbs the humidity that accumulates during the night period. They showed the convenience of using this method to improve the product's quality, as 30 h of total drying and 13.29 MJ/Kg moist specific energy consumption were obtained, compared to 55 h and 60.4 MJ/Kg moist for solar drying.

3.5 Roasting

Roasting brings about the formation of brown color, aroma and texture of the beans and this is why it has been considered as the most important technological operation of the postharvest process. It is in this step where, after the long preparation of the previous steps, the final characteristics of cocoa beans are expressed. An adequate drying process is necessary in order to obtain a good product. It is known that the most important variables for this process are roasting time and roasting temperature.

Sacchetti et al. (2016) studied the effect of processing time and temperature of roasting on non-enzymatic browning of cocoa beans. From the kinetics of moisture loss, they found that final moisture content of cocoa beans roasted at 125°C for 62 and 46 min were 1.9 and 2.3 g /100 g, respectively, and no differences were found when the roasting process were: 30 and 25 min at 135°C and 18 and 13 min at 145°C. These temperatures are not those used in the actual process (5 to 20 min at 120 to 150°C), but were used in order to have a more accurate measure of the kinetics. Roasting temperatures increased the overall rate of water loss. A short period of increasing rate of water loss, followed by another of falling rate is observed. This is associated with hardening of the beans surface because heating was too rapid, causing dehydration of the outer layers, which then act as a barrier to water diffusion. The process at 125°C resulted in darker colored beans in comparison to higher temperatures were used. Color changes showed lower activation energies than those of melanoidins formation, therefore, other browning reactions, as polyphenol oxidation, must contribute significantly to them during roasting, but it is non-enzymatic browning reactions that determine color changes, because of the activation of Maillard reaction with the formation of colored melanoidins. Hydroxymethyl furfural (HMF), which is a toxic compound that is also formed during roasting, is present at lower concentrations in high temperature short time processes. The conclusion of this work was that it is possible to control the roasting process in terms of time and temperature, to optimize cocoa beans characteristics of color, quality of the final products and the content of bioactive compounds.

Proanthocyanidins, brown polymeric compounds, are formed during the beans processing and roasting and also participate in the browning of cocoa beans. The final products of Maillard reactions are melanoidins, responsible for sensorial properties of the beans: color formation, taste, flavor and texture. Its importance has increased, as they exhibit antioxidant activity, which is of great importance nowadays, as a bioactive compound that contributes to consider cocoa as a functional food (Sacchetti et al. 2016).

During roasting of cocoa beans the total pyrazine content increases. It has been reported that roasting of beans with different fermentation qualities, the same pyrazine concentrations are formed, but the relative concentrations of the kind of pyrazine compounds is different. And as each type imparts different flavors to the beans, it would be difficult to control the quality. Flavor-active compounds produced during fermentation are 2-methylbutanoate, tetramethylpyrazine, 3-methylbutanol, phenylacetaldehyde-2-methyl-3-(methyldithio) furan, 2-ethyl-3,5-dimethyl- and 2,3-diethyl-5-methylpyrazine. Bitter notes are given by theobromine and caffeine, with diketopiperazines formed during roasting (Kongor et al. 2016, Afoakwa 2015).

Negative changes can also occur during roasting, as the loss of polyphenols that are considered bioactive compounds. Especially epicatequin and procyanidin are lost both in whole beans as in nibs of different particle size. Degradations were lower with higher relative humidity conditions, and the best were: air-flow rate of 0.5 m/s and relative humidity of 0.3% (Zyzelewicz et al. 2016). Conditions and changes occurring during drying and roasting steps are listed in Table 1. Finally, the roasted cocoa seeds are grinded.

4. Cocoa Microbiology

Cocoa bean fermentation is still carried out as an uncontrolled traditional process (Pereira et al. 2013). It is a spontaneous process, which starts immediately after the bean is removed from the pod (Amoa-Awua 2014) and plays an important role in the production of chocolate. It can have major impacts on product quality and value as it involves complex microbial interactions, which is absolutely essential for flavor development (Pereira et al. 2016). During this process, a diversity of yeast, lactic acid bacteria (LAB) and acetic acid bacteria (AAB) are dominant in a successional manner. Apart from them various species of *Bacillus*, other bacteria and filamentous fungi may also be present throughout fermentation and can affect the bean quality and cocoa flavor (Amoa-Awua 2014). The fermentation takes place in the pulp; the microbiological activity allows various biochemical processes important for the development of taste and flavors precursors inside the beans. This process continues during the drying step and is terminated by complete drying, which does not allow any microbial activity (Nielsen et al. 2013).

Traditionally, the seeds within the ripe pod are considered microbiologically sterile (Schwan and Wheals 2004) but the internal tissues of damaged pods are not likely to be sterile and, most probably, are a main source of microorganisms responsible for fermentation, along with the pod surface microbiota (Amoa-Awua 2014). It is also considered that the first relevant complex interactions between microorganisms are on the surface of cocoa fruits, which are the primary sources of microbial

Table 1. Conditions and reactions that occur during the drying and roasting stages of cocoa postharvest process.

Process	Conditions	Reactions
Drying	Time and temperature control to obtain the water content adequate for roasting and to avoid growth of contaminants.	Flavor and color reactions if the right conditions are applied. Mold growth if too slow. Too acid if too fast.
	Water content reduced from approximately 60% to 7.5%.	Non-enzymatic browning from reducing sugars and amino acids. Polyphenol oxidation.
Roasting	Time and temperature control to give the final characteristics of color, flavor and texture of beans that determine its quality. Water content is lowered down to 2–3%.	Desirable reactions: Non-enzymatic browning reactions with melanoidin formation. Determinant in color formation because activation energy of this reaction is lower than the others. Important for color, flavor and texture. Reaction of reducing sugars and amino acids, carbonyl groups from acids coming from lipolytic activity participate. Also antioxidant activity, considered bioactive compounds.
		Undesirable reactions: Loss of polyphenols, epicatequin and procyanidin are especially lost. Undesirable, as considered bioactive compounds.
		Proanthocyanidin formation, participates in color of the beans.
		Pyrazine content increases. Kind of compound, more than concentration, affect flavor.
		Bitter notes by theobromine, caffeine and by diketopiperazines formed during roasting.
		Removal of acetic acid, volatile esters, and other undesirable aroma components.
		Formation of toxic hydroxymethylfurfural (HMF).

contamination in the fermentation process (Pereira et al. 2016). During stages of pod breaking and bean removal for fermentation, the pulp and beans become contaminated with a variety of microorganisms, many of which contribute to the subsequent fermentation (Amoa-Awua 2014). Other sources of contamination have been described, like hands of workers, knives, and baskets used for transport of seeds and dried mucilage left on the walls of boxes after previous fermentations (Shawn and Wheals 2004). Papalexandratou et al. (2011) reported several strains belonged to *Lactobacillus fermentum*, which is considered as one of the dominant LAB during cocoa fermentation, which comes from leaf, pod, hands, boxes, sack, machetes and shovels. The other LAB found to be associated includes *Lb. plantarum, Lb. pentosus, Lactococcus lactis* subsp. *lactis, Leuconostoc pseudomesenteroides*. Apart from LAB, AAB associated are *Acetobacter pasterianus, A. ghanensis, A. indonesiensis/malorum/cerevisiae, A. senegalensis* and *Gluconoacetobacter saccharivorans*.

As already mentioned, microbial ecology of cocoa bean fermentation is complex and involves the successional growth of various species of yeasts, LAB, AAB and, possibly, species of *Bacillus*, other bacteria and filamentous fungi (Ho et al. 2014). *Enterobacteriaceae* has also been described in cocoa fermentation but little attention has been placed on these microorganisms since they have been considered as producers of undesirable metabolites such as gluconic acid (Papalexandratou et al. 2011) but recently reports have suggested that they may play a more important role in the overall metabolic processes of cocoa bean fermentation (Illeghems et al. 2015a). The initial acidity of the pulp (pH 3.0–4.0) and anaerobic conditions, favor initial growth of yeasts (Amoa-Awua 2014), they utilize the pulp carbohydrates to produce ethanol, which is the primary activity of the yeasts (Nielsen et al. 2013). Also, some of the yeasts contribute to pectin degradation in the pulp, which promotes bean aeration and growth of the AAB (Ho et al. 2014). Jespersen et al. (2005) reported an initial yeast cell count of 10^2 cfu/g of pulp and beans.

The most common yeasts isolated from cocoa fermentation are *Hanseniaspora guilliermondii, Hanseniaspora opuntiae, Pichia fermentans, Pichia kluyveri, Pichia kudriavzevii, Pichia membranifaciens* (Ardhana and Fleet 2003, Schwan and Wheals 2004, Jespersen et al. 2005, Nielsen et al. 2005, 2007, Lagunes-Gálvez et al. 2007, Leal et al. 2008, Daniel et al. 2009, Hamdouche et al. 2015). Among them, *Hanseniaspora guilliermondii* or *Hanseniaspora opuntiae* generally dominate the early part of fermentation followed by *Saccharomyces cerevisiae, Kluyveromyces marxianus, Pichia membranifaciens, Pichia kudriavzevii* and *Candida* spp. (Ardhana and Fleet 2003, Daniel et al. 2009, Galvez et al. 2007, Jespersen et al. 2005, Nielsen et al. 2005, 2007). Ho et al. (2014) found that yeast growth and activity are essential for successful cocoa bean fermentation based on shell content, nib color and chocolate sensory criteria as well as other biochemical parameters like ethanol concentration, higher alcohol and ester productions throughout fermentation and pyrazines presence in the roasted product.

As yeast population decreases during fermentation, bacteria, principally LAB and AAB dominate during the process. Semi-anaerobic conditions favor the development of LAB, reaching a maximum concentration approximately after 36–48 h of fermentation (Amoa-Awua 2014). The importance of LAB to this process remains unclear, but the conversion of lactic acid from pulp sugars during fermentation by LAB can be detrimental to cocoa bean and chocolate quality. Also, ability of LAB to produce lactic acid and potential to utilize the citric acid of the pulp may contribute to the pH balance of the process (Ho et al. 2014).

Then AAB metabolism is stimulated by aeration caused by mass revolving and utilizes the ethanol produced by the yeast for their growth and metabolism of acetic acid (Amoa-Awua 2014). Acetic acid is considered a key metabolite in this process, because it is associated with killing of the bean and the start-up of biochemical reactions inside the cocoa bean forming precursor molecules that are needed for the complete development of flavor and characteristic color of the fermented beans (Camu et al. 2008). The role of *Bacillus* spp. in the fermentation of cocoa has not been fully elucidated; however, they have been present in relatively low numbers during the first day of fermentation where pectin breakdown is most important (Nielsen et al. 2013). It has been also speculated that they may cause off-flavours due to their high enzymatic activity (Schwan and Wheals 2004). Nielsen et al. (2007) identified *Bacillus licheniformis, B. megaterium, B. pumilus, B. subtilis, B. cereus* and *B. sphaericus* after 48–60 h of fermentation.

Ouattara et al. (2008, 2010, 2011) have isolated and identified *Bacillus* spp. during cocoa fermentation, which included *Bacillus subtilis, B. pumilus, B. sphaericus, B. cereus, B. thuringiensis* and *B. fusiformis*. All the isolates showed some degree of pectin lyase activity (PL). Among the isolates, *B. fusiformis, B. subtilis,* and *B. pumilus* had pectin lyase yield of more than 9 U/mg of bacterial dry weight, whereas *B. sphaericus, B. cereus* and *B. thuringiensis* had lower enzyme yield.

Enterobacterial activity may play a role (desirable or not) in pectin degradation, citric acid assimilation, and gluconic acid production. The enterobacterial species *Tatumella punctata* and *T. saanichensis* were among the initial microbiota of all of the Ecuadorian fermentation carried out by Papalexandratou et al. (2011). The role of molds in cocoa fermentation has not been elucidated, some authors underlined the importance of avoid damaging the beans during opening, as damaged beans are more susceptible to mold attacks, leading to formation of mycotoxins such as ochratoxin A (Nielsen et al. 2013).

According to Schwan and Wheals (2004), cocoa beans extracted from intact pods are sterile; however, cocoa beans from damaged pods can be already contaminated by toxin producing fungi from the *Aspergillus* and *Penicillium* genera (Copetti et al. 2012). The occurrence of ochratoxin (OTA) was clearly linked to cocoa pods health status, a delayed pod opening and the use of plastic materials for the fermentation of cocoa beans (Kedjebo et al. 2016). Schwan and Wheals (2004), reported the presence of filamentous fungi especially in the last days of fermentation, on the surface or when the cocoa mass is not turned regularly. Species producing aflatoxin (*Aspergillus flavus* and *A. parasiticus*), and ochratoxin (*A. niger* and *A. arbonarius*) have been isolated in samples from fermentations (Copetti et al. 2014). In general, the concentrations aflatoxins and ochratoxin A were reported to be lower than 0.02–0.05 µg/kg, respectively. Roasting decreases OTA concentrations by 17–40%. Overall, it has been reported that during the process of production of chocolate from unroasted bean, there is more than 90% reduction in OTA concentrations, mostly as the result of shelling (Copetti et al. 2014). Cocoa beans are converted into powdered cocoa or chocolate by a series of processing steps, involving heat treatment or segregation of fractions, which impact on fungal or mycotoxin contamination in finished products (Copetti et al. 2014).

Mycotoxins are chemically stable during processing and storage, thus making it critical to avoid the conditions leading to mycotoxin formation during production, harvesting, transport and storage of cocoa beans (Afsah-Hejri et al. 2013). Recently, attention has been focused on inhibitory roles of natural constituents of cocoa beans in suppressing or reducing the toxicity of mycotoxins. Corcuera et al. (2012) evaluated a polyphenol-enriched cocoa extract as an effective antioxidant agent in a cell free system and in HepG2 cells and its ability to reduce aflatoxin B1 and OTA cytotoxicity and reactive oxygen species induction. An overview of the fermentation process including some of the more important chemical reactions involved is shown in Table 2.

5. Recent Findings on the Microbial Ecology of the Fermentation and on the Possibility to Use a Starter Culture

Due to the recent molecular techniques for detection and identification of un-cultivable microorganisms, it has become easy to study the impact of microbial ecology on biochemical components. Microbial communities in nature have been examined and new insights into their taxonomy and metabolism, interactions and even microbial communication have been discovered.

Fermented foods, especially traditional fermented foods have gained interest among the scientific community and cocoa bean fermentation is one of the examples. Through the use of metagenomics, Illeghems et al. (2012) reported microorganisms those are reported before (*Hanseniaspora uvarum, Saccharomyces cerevisiae, Lb. fermentum, Acetobacter pasteurianum*); however, *Erwinia tasmaniensis, Lb. brevis, Lb. casei, Lb. rhamnosus, Lactococcus lactis, Oenococcus oenus* and viruses associated with LAB were also detected. The work of Illeghems et al. (2015a) on the metagenomics of Brazilian cocoa fermentation, revealed the participation of Enterobacteria with mixed acid fermentation, citrate assimilation and pectinolysis. Bacteria belonging to this family were considered undesirable, but these findings

Table 2. Microorganisms and biochemical reactions that occur during the fermentation stage of cocoa postharvest process.

Fermentation time	Predominant microorganisms	Main chemical reactions
First hours	**Yeasts:** *Hanseniaspora guilliermondii* or *Hanseniaspora opuntiae* in early part; *Saccharomyces cerevisiae, Kluyveromyces marxianus, Pichia membranifaciens, Pichia kudriavzevii* and some *Candida* spp. are more frequently dominant, *Issatchenkia hanoiensis, Saccharomycopsis crataegensis, Schizosaccharomyces pombe Torulaspora delbrueckii, Saccharomycopsis crataegensis, Issatchenkia terricola, Issatchenkia hanoiensis Issatchenkia orientalis* (Ardhana and Fleet 2003, Schwan and Wheals 2004, Jespersen et al. 2005, Nielsen et al. 2005, 2007, Lagunes-Gálvez et al. 2007, Leal et al. 2008, Daniel et al. 2009, Papalexandratou et al. 2011, Arana-Sánchez et al. 2015, Hamdouche et al. 2015).	Yeast activity favored by low pH (3.6). Pectinolytic activity, sugar formation, fermentation with ethanol production. Pulp removal, cocoa bean surface exposed. Oxygen consumption, anaerobic-microaerophlic conditions favor growth of LAB.
Up to 36 hr	**Lactic acid bacteria:** *Lactobacillus fermentum* predominate; also *Lb. plantarum, Lb. pentosus, Fructobacillus tropaeoli-like, Lb. fabifermentans, Fructobacillus tropaeoli-like, Lactococcus lactis* subsp. *lactis, Leuconostoc pseudomesenteroides* (Papalexandratou et al. 2011, Lafeber 2011).	Lactic acid, acetic acid and ethanol production from sugars. Proteolytic and peptidase activities on bean cell wall proteins produce peptides and amino acids that are: flavor precursors and color precursors.

Table 2. contd....

...Table 2. contd.

Fermentation time	Predominant microorganisms	Main chemical reactions
After 5 to 6 days	Acetic acid bacteria: *Acetobacter pasteurianum* *A. ghanensis, A. indonesiensis, A. malorum, A. cerevisiae, A. senegalensis, A. tropicalis* and *Gluconoacetobacter saccharivorans.* (Papalexandratou et al. 2011, Lafeber 2011, Pereira et al. 2012). Enterobacteria: *T. punctata* and *T. saanichensis* *Tatumella ptyseos* and *Pantoea terrea* (Papalexandratou et al. 2011).	Cocoa bean aeration. Aerobic conditions favor growth of acetic acid bacteria. Conversion of ethanol into acetic acid, which is an exothermic reaction; temperature rises to 50°C. High temperature and entry of acetic acid into the cocoa bean and embryo is killed. Within the cotyledons: Action of hydrolytic enzymes, liberating components, reducing sugars, amino acids and peptides that are precursors of chocolate flavor. Volatile compounds (alcohols, organic acids, esters, aldehydes). Insoluble tannins, from oxidation and polymerization of phenolic compounds, that contributes to decrease the bitterness and astringency of the grains. Maillard browning from reducing sugars and amino acid reaction.
	Aerobic spore-forming bacteria. With pectin lyase activity: *Bacillus licheniformis, B. megaterium, B. pumilus, B. subtilis, B. cereus* and *B. sphaericus; B. fusiformis, B. subtilis,* and *B. pumilus* with the best activity (Outtara et al. 2007, 2010, 2011).	Desirable or undesirable? Pectinolytic activity, citrate assimilation, gluconic acid production. When acetic acid bacteria decrease growth. Acidity and unpleasant odors?
After fermentation	Molds: *Absidia corymbifera, Aspergillus* sp. nov.*, A. flavus, Penicillium paneum, A. flavus, A. parasiticus, A. nomius, A. niger* group*, A. carbonarius* and *Aspergillus Ochraceus* (Copetti et al. 2011).	Desired pectinolytic and other hydrolytic activities are proposed. Growth, spoilage, from initial and during process contamination, if conditions allow growth. Bean damage, mycotoxin (ochratoxin) formation (Nielsen et al. 2013).

showed that they may have an important role in the fermentation. On the other hand, Ho et al. (2015) have questioned the importance of LAB in the fermentation, as they found no differences in the fermentation with or without LAB. Other activities were related to stress, competitiveness and potential bacteriocin production. The capacity to relate microbial population and their function could make possible the selection of starter cultures for cocoa bean fermentation.

The effect of inoculation of cocoa beans with pure cultures of yeasts on the profile of volatile compounds produced by the bacterial community was studied (Batista et al. 2016). It resulted in the production of different sensory characteristics of chocolate (more intense fruity notes) than the spontaneous non-inoculated fermentation, which had more astringent property (Batista et al. 2016). Inoculation with *S. cerevisiae, Lb. plantarum* and *Acetobacter aceti* in cocoa beans led to a more consistent product, and the fermentation was potentially shortened from 5–7 to 3 days. Visintin et al. (2016) advocated microbial strains as starter cultures to be used for cocoa fermentation should be highly resistant to environmental stress conditions such as high temperature and the presence of specific enzymatic activities (i.e., endopectinase).

Through whole genome sequence analysis, Illeghems et al. (2015b) found that *Lb. fermentum* 222 and *Lb. plantarum* 80, isolated from a spontaneous Ghanaian cocoa bean fermentation process were able to dominate the process. The reason for this domination was sought by genomic analysis and comparative genome analysis of the strains, which revealed they are good candidates as starter cultures, because they possess genetic potential of activities that would make them functional cultures. These activities include carbon and nitrogen metabolic capabilities, stress factors, absence of virulence factors and production of antimicrobial compounds. Additional sources of energy that would make them competitive and functional starter cultures for cocoa bean fermentation. Moreover, they harbor genes that were not found in bacteria of the same species, e.g., *Lb. fermentum* 222 harbors two unique additional (putative) citrate transporters, possible as an adaptation to consume citrate from cocoa beans more efficiently and a gene cluster for aminoacid transfer and metabolism that would permit it to improve cocoa bean fermentation. As a consequence, *Lb. fermentum* did not group together with other *Lb. fermentum* strains in a phylogenetic tree and this might indicate that the strain isolated from cocoa bean has adapted to its niche (Illeghems et al. 2015b). Tavares-Menezes et al. (2016) performed fermentation of four different Brazilian varieties of cocoa beans using an inoculum of *S. cerevisiae*. In spite of this, the effect of the variety of coca bean was evident in the quality obtained with each of them.

The role of AAB in cocoa bean fermentation is acetate production, which is an important metabolite that imparts desired flavors. Using 13C isotope labeling experiments, Adler et al. (2014) elucidated metabolic fluxes in the fermentation environment. Lactate and ethanol are the primary substrates; acetate derives from ethanol, while acetoin and biomass produced by LAB and yeasts, respectively is needed. AAB produces the largest amounts of acetic acid in mixed culture with LAB and yeasts, so a balanced microbial consortium of the three microorganisms would produce good quality cocoa beans. Despite all the new findings, the use of starter cultures has not been applied in the field. As mentioned by de Melo Pereira et al. (2016), thorough understanding, at the microbial ecology level, of the interactions between the inoculated strains and those belonging to the natural community is needed.

6. Recent Findings on the Biochemical Changes During Cocoa Processing

The proximate composition of fresh non-fermented dry beans is: water (32–39%), fat (30–32%), proteins (10–15%) polyphenols (5–6%), starch (4–6%), pentosans (4–6%), cellulose (2–3%), sucrose (2–3%), theobromine (1–2%), acids (1%) and caffeine (1%) (Kongor et al. 2016). Proximate analysis of dried cocoa beans is shown in Table 3.

The predominant sugars are: sucrose (90% of total sugars), glucose and fructose (6%). Cocoa fat contains, in percentage of total lipids, approximately: 95% triacylglicerols, 2% diacylglycerols, < 1% monoacylglycerols, 1% polar lipids, and 1% fatty acids. The predominant kinds of fatty acids in cocoa butter are: saturated (stearic, 18:0 and palmitic, 16:0), monounsaturated (oleic, 18:1) and polyunsaturated (linoleic) (Reference).

There are three groups of polyphenols in cocoa beans: catechins or flavan-3-ols, anthocyanins and proanthocyanins. The (–)-epicatequin is the dominant catechin whereas, the anthocyanins present are cyanidin-3-α-L-arabinoside and cyanidin-3-β-D-galactoside. Proanthocyanidins are flavan-3, 4-diols, that are 4→subunit. Soluble compounds cause the astringent taste sensation of cocoa beans (Kongor et al. 2016).

There are four main fractions of proteins, representing 95% (w/w) of total seed proteins which includes albumins, which are water soluble and not degraded during fermentation, globulins (salt soluble), prolamins (alcohol-soluble) and glutelins (soluble in diluted acid and alkali), that does not produce aroma. Globulins represent 43% of total protein, made up of three polypeptide subunits of 47, 31 and 16 kDa. Marseglia et al. (2014) investigated the presence of oligopeptides in the fermented beans. They identified 44 different peptides based on molecular masses, mass fragmentation patterns and comparison with vicilin and 21 kDa cocoa seed proteins sequencing.

Cocoa butter is considered as the most important extracted compound from the beans. The beans contain 50–58% (w/w) of cocoa butter, which is

Table 3. Proximate analysis of dried cacao beans.

Origin	Beans condition	Moisture (%)	Protein (%)	Fat (%)	Ash (%)	Fiber (%)	Carbohydrate (%)	Reference
Ghana	Unfermented	4.2±0.02	21.6±0.83	55.2±0.10	3.5±0.11	N.R.	15.5±0.63	Afoakwa et al. 2013
	Fermented	4.0±0.02	18.8±0.56	53.4±0.63	2.8±0.07	N.R.	21±0.08	Afoakwa et al. 2013
Nigeria	Unfermented	5.0±0.50	17.5±0.88	62.9±0.27	4.4±0.31	5.9±0.60	9.3±0.51	Aremu et al. 1995

Results presented are mean values of triplicate analysis ± standard deviation a solid fat that melts at 32 to 35°C. It is responsible for flavor, taste, texture, viscosity, plasticity and glossiness and these properties are dependent on the types of fatty acids. According to Zyzelewicz et al. (2014), the fatty acid composition of cocoa beans is not significantly modified during the roasting step. Trans-fatty acids were formed at the beginning of the heat treatment, but disappeared later; however, peroxide value, conjugated dienes and trienes decreased. Roasting conditions of 150°C, 1 m/s air velocity and 5% relative humidity were recommended by these authors as the best conditions to roast cocoa beans.

Chocolate sensorial properties depend on cocoa beans aroma obtained after post-harvest processing. Flavor depends on its genotype, geographical origin and post-harvest treatments. By using a combination of mass fingerprinting and pattern recognition techniques (MS-fingerprinting), Tran et al. (2015) reported that cocoa bean samples with identical roasting conditions resulted in different patterns, depending on their origin; however, in some cases samples from a geographical area were close to those of another area. Results of MS-fingerprinting technique were useful to predict the aroma potential and quality of roasted beans measuring a wide range of aroma precursors of unroasted beans and classified them according to their extent of processing, as well as discarding un-fermented or low-fermented beans or over-roasted samples. The fermentation index, which is determined by the fermentation degree measured with several quality parameters, is a useful marker, with values above 1, indicative of well fermented cocoa beans.

Kadow et al. (2015) devised a "fermentation-like" incubation of surface-sterilized cocoa beans, adding defined amounts of acetic acid and controlling the temperature, so that in the absence of microorganisms they obtained high quality beans and a reproducible process, proposing the application of their results to standardize the fermentation.

7. Cocoa as a Functional Food

Presently food is not only considered as a source of nutrients, but also as a source of compounds that helps to stay healthy or even to recover or improve from an illness. Such foods are called functional foods. Among all natural foods cocoa is possibly one of the best by having several health benefits.

There are many reports on cocoa and its health benefits, including the details on mechanism of action of the bioactive components. De Araujo et al. (2016) have recently published a review, presenting research on beneficial effects of cocoa on different human body systems.

Regarding the cardiovascular system, effects of cocoa have been shown on blood pressure, endothelium function, lipid profile and platelet function. The mechanism seems to be able to maintain optimal nitric oxide levels, associated with lowering the superoxide anion production in the vasculature, which improves arterial stiffness and endothelial function (Reference). Polyphenol-rich cocoa would reverse endothelial disfunction, so that if it is used early in the sickness, it could prevent or delay progression of cardiovascular events, such as stroke. It possesses anti-thrombotic properties, and improves platelet function (Reference). Cocoa and chocolate polyphenols are also known for their neuroprotective and neurological properties (Reference). Via nitric oxide stimulation and modulation, cocoa flavonols enhance blood flow and perfusion in the brain, increasing blood oxigenation. They could then be used to treat or prevent cerebrovascular diseases as stroke and dementia. Polyphenols reduce the risk of neurodegenerative diseases as Alzheimer's and Parkinson's, that are related to oxidative stress in the brain. To show this, neurons pretreated with a cocoa extract reduced the expression and release of calcitocin–gene related peptide expression, and this peptide is a factor that promotes neural inflammation and migraine development. Cocoa also protects from amyloid β-protein-induced neurotoxicity (Reference). This is relevant in Alzheimer disease, caused by accumulation of amyloid plaques and neurofibrillary tangles in the brain that cause loss of memory and decline in cognitive function. Cocoa also protects against cognitive function due to normal aging. Chocolate and cocoa consumption prevent from neurological disorders, as depression. This has been attributed to the conversion of tryptophan in cocoa to serotonin, whose presence alleviate mood. Chocolate has been found to produce subjective feelings of well-being, to reduce maternal stress and to improve infant temperament. All this evidence would show that cocoa and chocolate are able to maintain cognitive function and through life and prevent against cardiovascular disease age-related development (De Araujo et al. 2016). This subject has also been reviewed by Sokolov et al. (2013).

Pro-anthocyanidins from cocoa have been shown to inhibit diabetes-induced cataract formation (Reference). Cocoa polyphenols prevent diet-induced obesity by modulating lipid metabolism decreasing fatty acid synthesis and transport systems, as well as enhancing thermogenesis in hepatic and white adipose cells. Cocoa without sugar has been shown to inhibit plaque accumulation and caries inhibiting glucan synthesis by *Streptococcus*. Among other studies, it has also been demonstrated that cocoa has anti-influenza virus effect by activating innate immunity (Kamei et al. 2015). All the benefits described are achieved by the consumption of a healthy diet and a small amount of preferably dark chocolate. Minimum, maximum and optimum amounts and consumption frequencies have been determined in different studies and in some cases cocoa enriched with polyphenols or flavonoids are used (De Araujo et al. 2016).

8. Discussion

The method of processing cocoa beans to obtain chocolate was devised by the ancient populations in Mexico, since Pre-hispanic times. It was taken to Europe, where its application was diversified and because of its fine aroma, different products were developed. Chocolate then became very popular and since then is consumed by the whole population, differently from ancient times, that was allowed to be consumed only by the royal community.

Not only cocoa bean consumption as chocolate, but also cultivation of cocoa bean trees and post-harvest processing was extended to areas with the adequate weather and strict conditions that the tree needs to grow and produce the beans. The correct cultivation conditions have been found in African, Asian and South American countries, where cocoa bean production has become the major source of income for populations.

Important findings that have been recently reported were presented here. They represent an important advancement in the knowledge of this complex system; however, in spite of the knowledge gained, most producers still use the traditional post-harvest method, which may result in low-quality products and in great losses.

In general, the biochemical reactions and microorganisms responsible for the changes have been described and found to be similar in processes from different countries or areas. However, it is now recognized that small modifications or differences in tree origins, type of beans, and degree of maturity or process parameters result in beans with different characteristics, such as aroma, color and texture. Recent developments have made possible to know biochemical composition and modifications of the grains with great detail (Marseglia et al. 2014, Zyzelewicz et al. 2014, Tran et al. 2015, Zyzlelwicz et al. 2016). This has resulted in a better understanding of the process, that can make possible its control and standardization. Studies on the factors that affect aroma precursors and compounds formation (Kongor et al. 2016), together with sensitive and faster methods to determine subtle differences in aroma of the final product (Tran et al. 2015) would allow the elaboration of products with specific characteristics and to facilitate the production of constant quality processed cocoa beans.

The actual discovery of the numerous health benefits of cocoa and chocolate justifies its "Food for the Gods" name, but fortunately it can now be appreciated taken advantage of by the whole population.

9. Conclusion

Cocoa (*Theobroma cacao*) is a tree that requires special conditions for its cultivation and a careful post-harvest treatment in order to yield good quality cocoa beans. Much knowledge has been gained on the microbiological, biochemical and sensorial properties of cocoa beans during this treatment and of the final product. This will make possible its production under controlled conditions, with better yields and shorter fermentation times if improvements in the process are made. Strategies to produce beans rich in compounds related to the functional properties attributed to cocoa, such as polyphenols could be devised.

Keywords: Cocoa, post-harvest, fermentation, yeast, lactic acid bacteria, acetic acid bacteria, *Bacillus*, molds, drying, roasting, flavour compounds

References

Adler, P., Frey, L.J., Berger, A., Bolten, C.J., Hansen, C.E. and Wittmann, C. (2014). The key to acetate: metabolic fluxes of acetic acid bacteria under cocoa pulp fermentation – simulating conditions. Applied and Environmental Microbiology 80: 4702–4716.

Afsah-Hejri, L., Jinap, S., Hajeb, P., Radu, S. and Shakibazadeh, S. (2013). A Review on mycotoxins in food and feed: Malaysia case study. Comprehensive Reviews in Food Science and Food Safety 12: 629–651.

Amoa-Awua, W.K. (2014). Methods of cocoa fermentation and drying. pp. 71–128. *In*: R.F. Schwan and G.H. Fleet (eds.). Cocoa and Coffee Fermentations. CRC Press.

Afoakwa, E.O. (2015). Cocoa production and processing technology. Boca Raton: Taylor and Francis Group, 15–21 (109–147).

Afoakwa, E.O., Quao, J., Takrama, K., Simpson, B.A. and Kwesi, S.F. (2013). Chemical composition and physical quality characteristics of Ghanaian cocoa beans as affected by pulp pre-conditioning and fermentation. Journal of Food Science and Technology 50(6): 1097–1105.

Arana-Sanchez, A., Segura-García, L.A., Kirchmayr, M., Orozco-Ávila, I., Lugo-Cervantes, E., and Gschaedler-Mathis, A. (2015). Identification of predominant yeasts associated with artisan Mexican cocoa fermentations using culture-dependent and culture-independent approaches. World Journal of Microbiology and Biotechnology 31: 359–369.

Ardhana, M.M. and Fleet, G.H. (2003). The microbial ecology of cocoa bean fermentations in Indonesia. International Journal of Food Microbiology 86: 87–99.

Aremu, C.Y., Agiang, M.A. and Ayatse, J.O.I. (1995). Nutrient and antinutrient profiles of raw and fermented cocoa beans. Plant Foods for Human Nutrition 48: 217–223.

Boahene, K., Snijders, T.A.B. and Folmeer, H. (1999). An integrated socioeconomic analysis of innovation adoption: the case of hybrid cocoa in Ghana. Journal of Policy Modeling 21(2): 167–184.

Batista, N.N., Lacerda Ramos, C., Ribeiro Dias, D., Marques Pinheiro, A.C. and Freitas Schwan, R. (2016). The impact of yeast starter cultures on the microbial communities and volatile compounds in cocoa fermentation and the resulting sensory attributes of chocolate. Journal of Food Science and Technology 53: 1101–1110.

Camu, N., González, A., De Winter, T., Van Schoor, A., De Bruyne, K., Vandamme P., Takrama, J.S., Addo, S.K. and De Vuyst, L. (2008). Influence of turning and environmental contamination on the dynamics of populations of lactic acid and acetic acid bacteria involved in spontaneous cocoa bean heap fermentation in Ghana. Applied and Environmental Microbiology 74: 86–98.

Copetti, M.V., Iamanaka, B.T., Pitt, J. and Taniwaki, M. (2014). Fungi and mycotoxins in cocoa: From farm to chocolate. International Journal of Food Microbiology 178: 13–20.

Copetti, M.V., Iamanaka, B.T., Pereira, J.L., Frisvad, J.C. and Taniwaki, M.H. (2012). The effect of cocoa fermentation and weak organic acids on growth and ochratoxin A production by *Aspergillus* species. International Journal of Food Microbiology 155: 158–164.

Copetti, M.V., Iamanaka, B.T., Frisvad, J.C., Pereira, J.L. and Taniwaki, M.H. (2011). Mycobiota of cocoa: From farm to chocolate. Food Microbiology 28: 1499–1504.

Corcuera, L.A., Amézqueta, S., Arbillaga, L., Vettorazzi, A., Touriño, S., Torres, J.L. and López de Ceraina, A. (2012). A polyphenol-enriched cocoa extract reduces free radicals produced by mycotoxins. Food and Chemical Toxicology 50: 989–995.

Córdova-Ávalos, V., Sánchez-Hernández, M., Estrella-Chulín, N.G., Macías-Lavalle, A., Sandoval-Castro, E., Martínez-Saldaña, T. and Ortiz-García, C.F. (2001). Factores que afectan la producción de cacao (*Theobroma cacao* L.) en el ejido Francisco I. Madero del Plan Chontalpa, Tabasco, México. Universidad y Ciencia 17(34): 93–100.

Córdova-Avalos, V., Mendoza-Palacios, L., Vargas-Villamil, L., Izquierdo-Reyes, F. and Ortiz-García, C.F. (2008). Participación de las asociaciones campesinas en el acopio y comercialización de cacao (*Theobroma cacao* L.) en Tabasco, México. Universidad y Ciencia 24(2): 147–158.

Da Veiga Moreira, I.M., da Cruz Pedroso, M.G., Ferreira Duarte, W., Ribeiro Dias, D. and Schwan, R.F. (2013). Microbial succession and the dynamics of metabolites and sugars during the fermentation of three different cocoa (*Theobroma cacao* L.) hybrids. Food Research International 54: 9–17.

Daniel, H.M., Vrancken, G., Takrama, J.F., Camu, N., De Vos, P. and De Vuyst, L. (2009). Yeast diversity of Ghanaian cocoa bean heap fermentations. FEMS Yeast Research 9: 774–783.

De Araujo, Q.R., Gattward, J.N., Almoosawi, S., Parada Costa Silva, M.G.C., De Santana Dantas, P.A. and De Araujo Jr., Q.R. (2016). Cocoa and human health: from head to foot-A Review. Critical Reviews in Food Science and Nutrition 56: 1–12.

Dina, S.F., Ambarita, H., Napitupulu, F.H. and Kawai, H. (2015). Study on effectiveness of continuous solar dryer integrated with dessicant thermal storage for drying cocoa beans. Case studies in Thermal Engineering 5: 32–40.

Galvez, S.L., Loiseau, G., Paredes, J.L., Barel, M. and Guiraud, J.P. (2007). Study on the microflora and biochemistry of cocoa fermentation in the Dominican Republic. International Journal Food Microbiology 114: 124–130.

Hamdouche, Y., Guehi, T., Durand, N., Kedjebo, K.B.D., Montet, D. and Meile, J.C. (2015). Dynamics of microbial ecology during cocoa fermentation and drying: Towards the identification of molecular markers. Food Control 48: 117–122.

Hernández-Hernández, C., López-Andrade, P.A., Ramírez-Guillermo, M.A., Guerra-Rampirez, D. and Caballero-Pérez, J.F. (2016). Evaluation of different fermentation processes for use by small cocoa growers in Mexico. Food Science and Nutrition (in press).

Ho, V.T.T., Zhao, J. and Fleet, G. (2015). The effect of lactic acid bacteria on cocoa bean fermentation. International Journal of Food Microbiology 205: 54–67.

Ho, V.T.T., Zhao, J. and Fleet, G. (2014). Yeasts are essential for cocoa bean fermentation. International Journal of Food Microbiology 174: 72–87.

https://www.dol.gov/agencies/ilab/our-work/child-forced-labor-trafficking.

Illeghems, K., Weckx, S. and De Vuyst, L. (2015a). Applying meta-pathway analyses through metagenomics to identify the functional properties of the major bacterial communities of a single spontaneous cocoa bean fermentation process sample. Food Microbiology 50: 54–63.

Illeghems, K., De Vuyst, L. and Weckx, S. (2015b). Comparative genome analysis of the candidate functional starter culture strains *Lactobacillus fermentum* 222 and *Lactobacillus plantarum* 80 for controlled cocoa bean fermentation processes. BMC Genomics 16: 766.

Illeghems, L., De Vuyst, L., Papalexandratou, Z. and Woeckx, S. (2012). Phylogenetic analysis of a spontaneous cocoa bean fermentation metagenomes reveals new insights into its bacterial and fungal diversity. PLoS ONE 7(5): e38040.

Jespersen, L., Nielsen, D.S., Honholt, S. and Jakobsen, M. (2005). Occurrence and diversity of yeasts involved in fermentation of West African cocoa beans. FEMS Yeast Research 5: 441–453.

Kedjebo, K.B.D., Guehi, T.S., Kouakou, B., Durand, N., Aguilar, P., Fontana, A. and Montet, D. (2016). Effect of post-harvest treatments on the occurrence of ochratoxin A in raw cocoa beans. Food Additives and Contaminants 33: 157–166.

Kadow, D., Niemenak, N., Rohn, S. and Lieberei, R. (2015). Fermentation-like incubation of cocoa seeds (*Theobroma cacao* L.)—reconstruction and guidance of the fermentation process. LWT—Food Science and Technology 62: 357–361.

Kamei, M., Nishimura, H., Takahashi, T., Takahashi, K.I., Inocuchi, K., Mato, T. and Takahashi, K. (2015). Anti-influenza virus effects of cocoa. Journal of the Science of Food and Agriculture 96: 1150–1158.

Kongor, J.E., Hinneh, M., Van der Walle, D., Afoakwa, O., Boeckx, P. and Dewettinck, K. (2016). Factors influencing quality variation in cocoa (*Theobroma cacao*) bean flavor profile—A review. Food Research International 82: 44–52.

Lafeber, T., Gobert, W., Vrancken, G., Camu, N. and De Vuyst, L. (2011). Dynamics and species diversity of communities of lactic acid bacteria and acetic acid bacteria during spontaneous cocoa bean fermentation in vessels. Food Microbiology 28: 457–464.

Lagunes-Gálvez, S., Loiseau, G., Paredes, J.L., Barel, M. and Guiraud, J.P. (2007). Study on the microflora and biochemistry of cocoa fermentation in the Dominican Republic. International Journal of Food Microbiology 114: 124–130.

Leal, G.A., Gomes, L.H., Efraim, P., Tavares, F.C.A. and Figueira, A. (2008). Fermentation of cacao (*Theobroma cacao* L.) seeds with a hybrid *Kluyveromyces marxianus* strain improved product quality attributes. FEMS Yeast Research 8: 788–798.

Lopes, U.V. and Pires, J.L. (2016). Botany and production of cocoa. pp. 43–70. *In*: R.F. Schwan and G.H. Fleet (eds.). Cocoa and Coffee Fermentations. CRC Press.

Marseglia, A., Sforza, S., Faccini, A., Bencivenni, M., Palla, G. and Caligiani, A. (2014). Extraction, identification and semi-quantification of oligopeptides in cocoa beans. Food Research International 63: 382–389.

Mathurin, G. and Delgado, C.L. (1984). Lessons and Constraints of Export Crop-Led Growth: Cocoa in Ivory Coast. The Political Economy of Ivory Coast. Ed. I. William Zartman, et al. New York. Praeger. http://krishikosh.egranth.ac.in/bitstream/1/2054816/1/MPKV-1257.pdf.

Montville, T.J., Matthews, K.R. and Kniel, K.R. (2012). Food Microbiology, an Introduction. ASM Press, Washington, D.C.

Nielsen, D.S., Crafack, M., Jespersen, L. and Mogens, J. (2013). The microbiology of cocoa fermentation. pp. 39–60. *In*: R.R. Watson, S. Zibadi and V.R. Preedy (eds.). Chocolate in Health and Nutrition, Nutrition and Health. Springer Science + Business Media. New York.

Nielsen, D.S., Teniola, O.D., Ban-Koffi, L., Owusu, M., Andersson, T.S. and Holzapfel, W.H. (2007). The microbiology of Ghanaian cocoa fermentations analyzed using culture-dependent and culture independent methods. International Journal of Food Microbiology 114: 168–186.

Nielsen, D.S., Honholt, S., Tano-Debrah, K. and Jespersen, L. (2005). Yeast populations associated with Ghanaian cocoa fermentations analyzed using denaturing gradient gel electrophoresis (DGGE). Yeast 2005, 22: 271–284.

Ntiamoah, A. and Afrane, G. (2008). Environmental impacts of cocoa production and processing in Ghana: life cycle assessment approach. Journal of Cleaner Production 16: 1735–1740.

Ouattara, H.G., Reverchon, S., Niamke, S.L. and Nasser W. (2011). Molecular identification and pectate lyase production by *Bacillus* strains involved in cocoa fermentation. Food Microbiology 28: 1–8.

Ouattara, H.G., Reverchon, S., Niamke, S.L. and Nasser, W. (2010). Biochemical Properties of Pectate Lyases Produced by Three Different *Bacillus* Strains Isolated from Fermenting Cocoa Beans and Characterization of Their Cloned Genes. Applied and Environmental Microbiology, pp. 5214–5220.

Ouattara, H.G., Ban, L., Karou, K.G.T., Sangare, A., Niamke, S.L. and Diopoh, J.K. (2008). Implication of *Bacillus* sp. in the production of pectinolytic enzymes during cocoa fermentation. World Journal of Microbiology and Biotechnology 4: 1753–1760.

Panlibuton, H. and Lusby, F. (2006). Cocoa production in Ivory Coast. *In*: Indonesia cocoa bean value chain case study. USAID MicroREPORT # 65. June 2006.

Papalexandratou, Z. and De Vuyst, L. (2011). Assessment of the yeast species composition of cocoa bean fermentations in different cocoa-producing regions using denaturing gradient gel electrophoresis. FEMS Yeast Research 11: 564–574.

Pereira, G.V.M., Soccol, V.T. and Soccol, C.R. (2016). Current state of research on cocoa and coffee fermentations. Current Opinion in Food Science 7: 50–57.

Pereira, G.V.D., Magalhaes, K.T., de Almeida, E.G., Coelho, I.D. and Schwan, R.F. (2013). Spontaneous cocoa bean fermentation carried out in a novel-design stainless steel tank: influence on the dynamics of microbial populations and physical-chemical properties. International Journal of Food Microbiology 161: 121–133.

Pereira, G.V.D., Pedrozo, M.G.D., Lacerda Ramos, C. and Schwan, R.F. (2012). Microbiological and physicochemical characterization of small-scale cocoa fermentations and screening of yeast and bacterial strains to develop a defined starter culture. Applied and Environmental Microbiology 78: 5395–5405.

Sacchetti, G., Ioannone, F., De Gregorio, M., Di Mattia, C., Serafini, M. and Mastrocola, D. (2016). Non-enzymatic browning during cocoa roasting as affected by processing time and temperature. Journal of Food Engineering 169: 44–52.

Saltini, R., Akkerman, R. and Frosh, T. (2013). Optimizing chocolate production through traceability: A review of the influence of farming practices on cocoa bean quality. Food Control 29: 167–187.

Schwan, R.F. and Wheals, A.E. (2004). The microbiology of cocoa fermentation and its role in chocolate quality. Critical Reviews in Food Science and Nutrition 44(4): 205–221.

Sokolov, A.N., Pavlova, M.A., Klosterhalften, S. and Enck, P. (2013). Chocolate and the brain: neurobiological impact of cocoa flavonols on cognition and behavior. Neuroscience and Behavioral Reviews 37: 2445–2453.

Tavares Menezes, A.G., Batista, N.N., Ramos, C.L., Reis de Andrade e Silva, A., Efraim, P., Marques Pinheiro, A.C. and Freitas Schwan, R. (2016). Investigation of chocolate produced from four different Brazilian varieties of cocoa (*Theobroma cacao* L.) inoculated with *Saccaromyces cerevisiae*. Food Research International 81: 83–90.

Thompson, S.S., Miller, K.B., Lopez, A. and Camu, N. (2013). Cocoa and coffee. pp. 881–889. *In*: M.P. Doyle and R.L. Buchanan (eds.). Food Microbiology: Fundamentals ans Frontiers. ASM Press, Washington, DC.

Tran, P.D., Van de Walle, D., De Clercq, N., De Winne, A., Kadow, D., Lieberei, R., Messens, K., Tran, D.N., Dewettinck, K. and Van Durne, J. (2015). Assessing cocoa aroma quality by multiple analytical approaches. Food Research International 77: 657–669.

Vela, E. (2012). El cacao, un fruto asombroso y el chocolate, el sabor mexicano del mundo. Arqueología Mexicana, Edición especial 45, Editorial Raíces, S.A. de C.V., 88 pp.

Vigneri, M. (2008). Drivers of change in Ghana's cocoa sector. Ghana Strategy Support Program (GSSP) Background Paper 13. International Food Policy Research Institute. http://wwwifpri.org/themes/gssp/gssp.html.

Visintin, S., Valente, A., Dolci, P. and Cocolin, L. (2016). Molecular identification and physiological characterization of yeasts, lactic acid bacteria and acetic acid bacteria isolated from heap and box cocoa bean fermentations in West Africa. International Journal of Food Microbiology 216: 69–78.

Wood, G.A.R. and Lass, R.A. (1985). Cocoa. Blackwell Science, Iowa.

Zyzelewicz, D., Krysiac, W., Oracz, J., Sosnowska, D., Budrin, G. and Nebesny, E. (2016). The influence of the roasting processing conditions on the polyphenol content in cocoa beans, nibs and chocolates. Food Research International (in press).

Zyzelewicz, D., Budrin, G., Krysiac, W., Oracz, J., Nebesny, E. and Bojczuk, M. (2014). Influence of the roasting conditions on fatty acid composition and oxidative changes of cocoa butter extracted from cocoa beans of Forastero variety cultivated in Togo. Food Research International 63: 328–343.

16

Kimchi
Microbiology, Biochemistry and Health Benefits

Kun-Young Park[1],* and *Soon-Ah Kang*[2]

1. Introduction

Kimchi is a representative lactic acid bacteria (LAB)-fermented vegetable food in Korea. LAB that occur naturally in the vegetables can be the starters for fermenting kimchi. During the brining of the vegetables with salt, the levels of spoilage microorganisms are lowered; however, the LAB counts are increased and they become the main starter cultures for the kimchi fermentation. Other sub-ingredients, especially condiments such as red pepper powder, garlic, ginger and other nutritious or functional sub-ingredients are added to the brined cabbage, and then fermented.

Fermentation of salted vegetable foods was used as a means of food preservation as far back as 2000 years ago in Korea (Kim 1997). Kimchi foods were described as fermented vegetables eaten by the Korean people in ancient Chinese literature written during the Kokuyro Dynasty of Korea (B.C. 37 to A.D. 668). There are 167–193 kinds of kimchi in Korea (Park et al. 2014, Yoon 2004). However, Baechu kimchi which is mainly prepared

[1] Department of Food Science and Biotechnology, Cha University, Seongnam, Gyeonggi-do 13488, South Korea.
[2] Department of Conversing Technology, Graduate School of Venture, Hoseo University, Seoul 06711, South Korea.
  E-mail: sakang@hoseo.edu
* Corresponding author: kunypark@cha.ac.kr

with Baechu cabbage (Chinese cabbage) is the major kimchi. More than 70% of the kimchi consumed in Korea is Baechu kimchi, thus the term, kimchi, generally refers to Baechu kimchi. Kakdugi (sliced radish kimchi), Chongkak kimchi (ponytail kimchi), Dongchimi (whole radish kimchi with added water), Yeolmu (leafy radish) kimchi and the Baechu kimchi are the five main types of kimchi consumed in Korea.

The LAB level is about 10^4/ml or gram in the brined cabbage at the beginning of the fermentation process, however, the level significantly increases to 10^{8-9}/ml in the optimally fermented kimchi. When 100 g of kimchi is consumed by a person in one day, 10^{10-11}/ml probiotic LAB could be consumed. Kimchi fermentation is governed mainly by hetero-fermentative LAB, producing lactic acid, acetic acid, ethanol, CO_2 and mannitol (Jung et al. 2012, 2014). Kimchi fermentation could be influenced by predominant LAB, fermentation temperatures, pH, anaerobic condition and thus taste and health functionality could be changed by the microbial community and the fermentation conditions.

Baechu cabbage, radish, cabbage and mustard leaves that are used as main ingredients of kimchi are cruciferous vegetables, which contain isothiocyanate, indole-3-canbinal, β-sitosterol and other agents that exhibit anticancer and other functional effects (Park et al. 2014). Garlic, ginger, radish and red pepper powder exhibited anti-cancer effect, anti-obesity and anti-inflammatory effect (Choi and Park 1999, 2000, Choi 2001, Yoon et al. 2005). In addition, LAB also contribute to functional properties of kimchi. Kimchi has various health benefits, including: anti-oxidative, anti-aging, anti-cancer, anti-obesity, anti-atherosclerotic, anti-diabetic effects (Park and Rhee 2005). These physiological effects can be attributed to the bioactive compounds contained in the ingredients of kimchi, and their functionalities increased during the fermentation processes (Kim et al. 2014a, Kim et al. 2015).

Kimchi is brined with salt, so people could be concerned about the salt intake (if higher amount of product are consumed daily). Salt plays an important role during the kimchi fermentation such as improving flavor, balancing Na with K, inhibiting spoilage microorganisms, and accelerating the growth of LAB. The salt content has typically been about 2–5%, however, it becomes lower at 1.2–2.0% in these days. Therefore, when people consume 100g of kimchi per day, the salt intake is only 1.2–2.0g/day from kimchi. In addition, if we use baked salt (e.g., bamboo salt) for kimchi preparation, the taste and health functionality can be increased even more (Park et al. 2014). In this chapter, we will present the kimchi recipe, nutrients and process method, and then the microbiology, biochemistry and health benefits of kimchi will be discussed.

2. Kimchi Recipe, Nutrients and Process Method

The kimchi ingredients, composition and nutrients, and major nutraceuticals for a standardized recipe are shown in Table 1.

The standardized kimchi recipe was prepared from the recipes of kimchi factories, cooking books, research papers and reputable family recipes (Cho 1999). The composition of the kimchi was as follows: 100% of brined Baechu cabbage, 3.5% red pepper powder, 1.4% crushed (chopped) garlic, 0.6% crushed (chopped) ginger, 2.2% fermented anchovy juice, 1.0% sugar, 13.0% sliced radish and 2.0% green onion, and the final salt content is 2.5%. The salt content of kimchi has gradually decreased in recent years to 1.2–2.0% in kimchi factories in Korea (Lee 2015).

The ingredients contain various nutrients, generally sugars, vitamins, minerals, dietary fibers and other nutraceuticals, and especially fermentation metabolites formed by LAB during the fermentation process. As mentioned above, Baechu cabbage and radish, the main ingredients of kimchi, are cruciferous vegetables that have many health benefits. Baechu cabbage contains sugars, vitamin C and other vitamins, dietary fibers, β-sitosterol, indols, isothiocyanate, Ca and other minerals. Red pepper powder, garlic and ginger, which are important condiments that Koreans use in cooking at meals and in making kimchi, contain various phytochemicals to contribute health benefits. Red pepper powder contains sugars, protein, lipids, especially polyunsaturated fatty acid (PUFA), vitamin A (carotenoids), vitamin C, dietary fibers, capsaicinoids, flavonoids, Ca, P and other

Table 1. Standardized kimchi recipe and nutrients.

Composition	Ratio (%)	Nutrients and nutraceuticals
Main Baechu cabbage[1]	100	Sugar, vitamin C, K, DF,[2] β-sitosterol, indoles, isothiocyanate, flavonoids
Sub-ingredients Red pepper power	3.5	PUFA,[3] sugar, vitamin A, vitamin C, Ca, P, K, DF, capsaicinoid, flavonoids
Crushed garlic	1.4	Sugar, allyl compounds, alliin allicin, diallysulfide, methyl linolate
Crushed ginger	0.6	Niacin, K, gingerol
Fermented anchovy juice	2.2	Protein (amino acids), Ca, P, Fe, Na
Sugar	1.0	Sugar
Radish (Sliced)	13.0	Sugar, niacin, Ca, isothiocyanate
Green onion	2.0	Vitamin A, vitamin C, chlorophyll, sulfur compound, flavonoids
Final salt concentration	2.5	

[1] brined cabbage, [2] dietary fiber, [3] polyunsaturated fatty acid

minerals. Garlic contains sugar, allyl compounds, alliin, methyl linolate, and allicin which can combine with vitamin B1 and form allithiamine, and thus vitamin B1 can be stored in the body longer, and give energy and neuronal stability. Ginger has niacin, K, gingerol. Fermented anchovy juice contributes a pleasant savory taste to the kimchi, it has amino acids, Ca, P, Fe, and Na. Green onion contains vitamin A, vitamin C, chlorophyll, flavonoids (Park 1995). When the kimchi is fermented by LAB, various metabolites such as lactate, acetate, CO_2, ethanol, mannitol, etc. are formed with increased levels of LAB (Jung et al. 2011). Thus, taken together, kimchi contains vitamin C, carotenoids, chlorophyll, tocopherol, isothiocyanate, indol-3-carbinol, β-sitosterol, flavonoids, dietary fibers, capsaicinoids, unsaturated fatty acids, allyl sulfur compounds, gingerol, organic acids, LAB and LAB-fermented metabolites.

Other vegetables, fruits and nuts, cereals, fishes or meats are added depending on the family tradition (Cheigh and Park 1994). Recently, functional sub-ingredients for preventing disease have been added to prepare kimchi for specific health functions. Custom-made kimchi could be developed by adding functional ingredients to the standardized kimchi recipe. Other ingredients are added such as Chinese pepper, organically cultivated Baechu cabbage, mustard leaf, Korean mistletoe extract, chitosan, green tea, bamboo salt, starters to enhance anti-cancer and anti-obesity effects of kimchi (Cho 1999, Kil 2004, Choi and Park 1999, 2000, Kong et al. 2010, Bong 2014, Kim et al. 2014b).

Kimchi preparation methods differ, depending on kimchi types and ingredients used. The standardized preparation methods of Baechu kimchi (Tong Baechu kimchi and Cut Baechu kimchi) consists of pretreatment, brining, mixing of ingredients, and then fermentation. Pretreatments include grading, washing and cutting of the cabbage. Other sub-ingredients are also graded, washed, and cut or crushed (chopped) for the mixing and fermentation (Fig. 1, Cheigh and Park 1994).

Tong Baechu kimchi (Whole Baechu cabbage kimchi) preparation method:

1. Pretreated Baechu cabbage is cut lengthwise into 2 or 4 parts which are brined overnight with either dry salt or in brine solution (10% salt concentration) and then rinsed with fresh water.
2. The excess water is drained from the brined Baechu cabbage for 3 h.
3. The pre-mixture of crushed or sliced sub-ingredients, generally includes garlic, ginger, radish, red pepper powder, salt-fermented fishes (anchovy, and other vegetables according to the recipe are mixed and stuffed between the leaves of the cabbage.
4. Wrapping the cabbage with outer leaves of the cabbage.
5. Putting the kimchi in to the pottery or other container, and pile up and press to remove O_2, giving anaerobic condition at low temperature (4–5°C) for 2–3 weeks, it will be optimally ripened.

Main ingredient

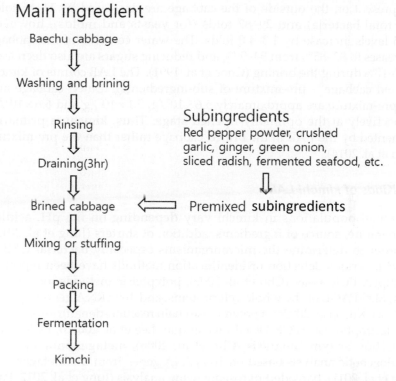

Baechu cabbage

⇩

Washing and brining

⇩

Rinsing

⇩

Draining(3hr)

⇩

Brined cabbage ⟸ Premixed **subingredients**

⇩

Mixing or stuffing

⇩

Packing

⇩

Fermentation

⇩

Kimchi

Subingredients
Red pepper powder, crushed garlic, ginger, green onion, sliced radish, fermented seafood, etc.

⇩

Figure 1. Kimchi (Tong Baechu kimchi) processing method (Cheigh and Park 1994).

Cut Baechu cabbage kimchi preparation method:

1. The pretreated Baechu cabbage is cut into 3–5 cm pieces, brined in salt solution (8–15% concentration) overnight or 2–7 h, rinsed with fresh water, and drained.
2. Crushed garlic and ginger, sliced radish, red pepper powder, green onion, salt-fermented fishes, etc. are combined to make the pre-mixture according to the recipe.
3. This pre-mixture is mixed with the cut brined cabbage.
4. The kimchi placed in the container, pressed down to remove O_2 and container is covered firmly so that the kimchi is fermented in anaerobic conditions by the LAB at low temperature of 4–5°C.

3. Microbiology of Kimchi

Kimchi fermentation is governed by the LAB existing in Baechu cabbage and the sub-ingredients. During the brining of the main ingredient of the cabbage, spoilage bacteria, molds and yeasts, which are aerobes that are present on the outside of the cabbage are decreased by 11–16 folds (for total bacteria) and 29–87 folds (for yeasts and molds), however, LAB levels increase by 1.3–4.0 folds. The water contents of the cabbage decreases to 82–85% from 94–97% and reducing sugars are also decreased by 7–17% during the brining (Choe et al. 1991). The LAB counts of kimchi (brined cabbage + pre-mixture of sub-ingredients), brined cabbage and the pre-mixture are approximately $5.6 \times 10^7/g$, $5.1 \times 10^7/g$ and $6.6 \times 10^6/g$, respectively at the optimally-ripened stage. Thus, kimchi is primarily fermented by the LAB in the brined cabbage rather than the pre-mixture (Yun et al. 2014).

3.1 Kinds of kimchi LAB

Microbial populations in kimchi vary depending on the pH, acidity, temperature, source of ingredients, addition of starters (Jung et al. 2012). In order to determine the microorganisms especially bacterial LAB in kimchi, various detection or identification methods have been reported. Multiplex PCR assay (Cho et al. 2009), polyphasic methods including a PCR, SDS-PAGE of the whole-cell proteins, and 16S rRNA gene sequence analysis (Kim et al. 2002c), a polymerase chain reaction-denaturing gradient gel electrophoresis (PCR-DGGE) technique (Lee et al. 2005b), PCR-based restriction enzyme analysis (Cho et al. 2006), metagenomic analysis (phylogenetic analysis based on 16S rRNA genes from the metagenome; Jung et al. 2011), barcoded pyrosequencing analysis (Jung et al. 2012, Park et al. 2012a), culture-independent 16S rRNA gene clone libraries (Kim and Chun 2005), etc. The kimchi microbiome was dominated by three genera members: *Leuconostoc*, *Lactobacillus* and *Weissella* (Jung et al. 2011). The following species are the dominant LAB identified in kimchi regardless of kimchi type and starter inoculation. *Ln. mesenteroides*, *Ln. citreum*, *Ln. gasicomitatum*, *Ln. geladum*, *Ln. inhae*, *Ln. kimchii*, *Ln. pseudomesenteroides* in *Leuconostoc* species, *Lb. plantarum*, *Lb. sakei*, *Lb. brevis*, *Lb. curvatus*, *Lb. graminis* in the *Lactobacillus* species, *W. koreensis*, *W. cibaria*, *W. confusa*, *W. kimchii*, etc. in the *Weissella* species.

In commercial kimchi, *Weissella*, *Leuconostoc* and *Lactobacillus* are also the main genera detected, with the *Weissella* being the predominant genus. In five of commercial kimchi at pH 4.2 (Kim and Chun 2005), *W. koreensis* was found in all samples and predominated in three of them (43–82%). *Ln. gasicomitatum* and *Lb. sakei* were followed in the remaining kimchi clone libraries (34%) in the species level. In our studies, nine commercial kimchi samples were collected, and hetero-fermentative LAB were found to be the dominant LAB using the pyrosequencing method. pH values of all kimchi samples ranged from 4.3 to 4.7 and total LAB counts were 1.3×10^7–1.6×10^9 CFU/g. *Weissella*, *Leuconostoc* and *Lactobacillus* again were the predominant genera, accounting for 52, 28 and 20% of all genera found, respectively. At the species level, *W. koreensis* (35%) predominated, followed by *Lb. graminis* (13%), *W. cibaria* (11%), *Lb. sakei* (9%) and *Ln. gelidum* (9%) which are all hetero-fermentative LAB.

3.2 Microbial Succession During Kimchi Fermentation

Generally, the process of kimchi fermentation is divided into four stages based on acidity: Initial stage (acidity < 0.2), immature stage (acidity 0.2–0.4), optimally-ripened stage (acidity 0.4–0.9) and over-ripened stage or rancid stage (acidity >0.9) (Codex 2001). *Ln. mesenteroides* appeared at the immature stage and optimally-ripened stage, *Lb. sakei* and *Lb. plantarum* at the initial stage and until over-ripened/rancid stage. *W. koreensis* appeared at the over-ripening (late) stage and increased to the end of fermentation (Cho et al. 2009).

In kimchi fermentation, the hetero-fermentative LAB, *Ln. mesenteroides* and *Ln. citreum*, predominate at the early and middle stages under weaker acidic and less anaerobic conditions, and *Lactobacillus* and *Weissella* species become dominant as fermentation conditions change to more acidic and anaerobic conditions at the later stages (Cho et al. 2006, Chang and Chang 2010, Lee and Lee 2010). Raw materials, especially cabbage and radish, contain high level of *Leuconostoc* (8.1–9.1%) at the first day of fermentation, whereas only small amounts of *Lactobacillus* (at the first day: 0.36–0.59%) were detected. This is another reason for the high level of *Leuconostoc* at the early stage. However, the level of *Lactobacillus* markedly increased. Then *Leuconostoc* levels decreased during the middle stage, and *Lactobacillus* growth continued during the late stage (Jung et al. 2012). When *Ln. mesenteriodes* was used as a starter, the proportion of *Leuconostoc* increased, and *Lactobacillus* decreased, and although the level of *Weissella* in the kimchi raw materials (5.2–10.3%) was high, they were not dominant in the fermentation.

Natural fermentation results in variations in sensory and functional qualities of kimchi. Thus, commercial kimchis that need uniform quality and better taste have begun using starters for kimchi products. When *Ln. mesenteroides* is used as a starter, kimchi products become less acidic and more refreshing in taste due to the production of hetero-fermentative products of ethanol, CO_2, mannitol and reduced production of lactic acid. As indicated, the predominant LAB commercial kimchis in Korea, are mostly rich in hetero-fermentative LAB. Therefore, kimchi hetero-fermentative LAB are most important for achieving a good taste and uniform quality.

Jung et al. (2011) showed the phylogenetic taxonomic composition of the kimchi microbiome during fermentation at 4°C for 29 days. As shown in Fig. 2, about 20% of 16S rRNA gene sequences for all samples

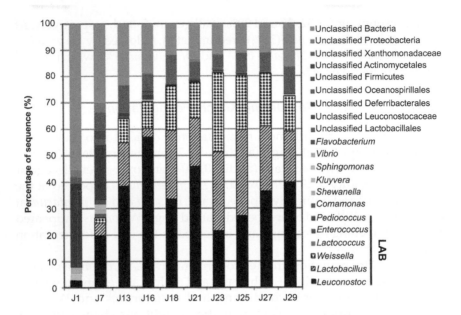

Figure 2. Phylogenetic taxonomic composition of the kimchi microbiome during fermentation. 16S rRNA gene sequences were classified to the genus level using the MG-RAST server based on the RDP II (16S rRNA gene) database (E value, 0.01; minimum alignment length, ≥ 50 bp). (Jung et al. 2011).

were unclassified phylotypic groups during the kimchi fermentation. Kimchi fermentation is governed by three genera, *Leuconostoc*, *Lactobacillus* and *Weissella*. Among them, the genus *Leuconostoc* was most abundant during the fermentation, followed by *Lactobacillus* and *Weissella*. *Leuconostoc* dominated at the early stages, but as the fermentation progressed, the abundance of *Lactobacillus* and *Weissella* increased.

After the middle stages of fermentation, *Leuconostoc*, *Lactobacillus* and *Weissella* became the predominant bacterial groups in the microbial community. The three predominant bacterial groups constituted more than 98% of all bacterial groups when the unclassified phylotypic groups were excluded from the analysis. The abundance of *Leuconostoc* increased again, and that of *Lactobacillus* and *Weissella* gradually decreased after day 23. *Leuconostoc* species dominated at 4°C (low temperature) and even high acidity in this experiment. Although temperature and acidity are important determinants of the microbial community, they suggested that metabolism of free sugars and production of lactate, acetate and mannitol were all closely correlated with the growth of *Leuconostoc*, *Lactobacillus* and *Weissella*.

3.3 Changes in Metabolites During Kimchi Fermentation

Kimchi taste and flavors are mainly from the levels of kimchi metabolites including sugars, amino acids and organic acids, the productions of which are influenced by the microbial community during kimchi fermentation. Ha et al. (1989) indicated that the major sugars found in kimchi (brined Baechu cabbage) were mannose, fructose, glucose and galactose with the levels of 5.2, 3.1, 21.6 and 0.89 mg/100 g, respectively, and the levels decreased with the fermentation, and mannitol is formed during the fermentation. Jung et al. (2012) analyzed kimchi metabolites by using ^{1}H-NMR spectroscopy during fermentation at 4°C. The levels of free sugars, especially fructose and glucose were high at 0 day of fermentation in Baechu kimchi due to the continuous liberation of free sugars from vegetables. After 10 days of fermentation, their concentration began to decrease. However, the levels of lactate and acetate increased inversely with the decrease in the free sugars.

Fructose and glucose were the main free sugars with levels of approximately 68 and 39 mM, respectively (Jung et al. 2011). The levels of free sugars decreased slowly in the early fermentation stage, however, the levels decreased rapidly between days 16 and 23 at 4°C. The sugar levels were relatively constant after day 23. However, lactate, acetate and ethanol were formed as fermentation products accompanied by decreases in free sugars, indicating that hetero-fermentative fermentation occurred by the LAB. It was also observed that considerable amounts of mannitol are formed by *Ln. mesenteroides* (Jung et al. 2011, 2012, McFeeters and Chen 1986, Wisselink et al. 2002) during the fermentation. The mannitol is an effective hydroxyl radical scavenger, and has antioxidant effects. The mannitol in foods gives a sweet, cool and refreshing taste, it has no tooth-decaying properties and is a good sugar source for diabetic foods (Wisselink et al. 2002). The fermentation products inversely correlated with the levels of free sugars. The change in free sugars and fermentation products were also closely correlated with pH changes and growth of the bacterial population. However, sucrose and ethanol levels were very low. The low level of ethanol seemed to be due to the loss during freeze-drying of the kimchi sample.

4. Biochemistry of Kimchi

The fermentation process of kimchi causes a variety of biochemical and microbiological changes in the Baechu cabbage and sub-ingredients in kimchi, which generates the taste and health functionality depending on the fermentation conditions. A series of complex biochemical and microbiological activities obviously occur during the fermentation. Also, the nutritional characteristics of kimchi are determined by the complex relationship of ingredients and biochemical reactions, which is greatly dependent on the initial ingredients, enzymes, and microorganisms in the kimchi. Cheigh and Park (1994) introduced changes in vitamin C and other vitamins of B_1, B_2 and niacin, along with changes in organic acids and acidification, all of which influence the changes in texture and softening in kimchi. We will discuss kimchi LAB metabolism, changes in ingredients and organic acids, taste and flavor compounds, and preservation and packaging condition in the sections below.

4.1 Kimchi LAB Metabolism

Kimchi fermentation is generally carried out by obligatory or facultative hetero-fermentative LAB. Figure 3 shows a proposed pathway for glucose and fructose metabolism of hetero-fermentative LAB (Wisselink et al. 2002, Jung et al. 2014). Glucose and fructose are the major free sugars detected in kimchi. *Leuconostoc* and group III *Lactobacillus* (obligatory hetero-fermentative) use 6-phosphogluconate/phosphoketolase pathway (hexose monophosphate shunt) for hexose fermentation (Kendler and Weiss 1986). They lack the fructose 1, 6-diphosphate aldolase of the EMP pathway, therefore, glucose is converted to equimolar amounts of lactic acid, ethanol and CO_2. Acetyl phosphate can be oxidized to yield acetate or reduced to yield ethanol depending on the O-R potential of the conditions. Pyruvate

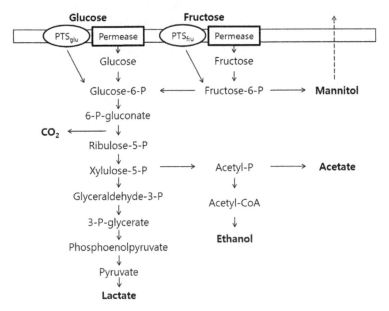

Figure 3. Proposed pathway for hexose metabolism of hetero-fermentative LAB, during kimchi fermentation (modified from Wisselink et al. 2002).
PTS: Phosphotransferase system

and fructose can be reduced under anaerobic conditions. Fructose is reduced to mannitol by mannitol dehydrase. *Ln. mesenteroides* produces mannitol from the mixture of glucose and fructose as the carbohydrate sources (Erten 1998). When 1 mol of glucose is fermented, approximately 2 mol of fructose is reduced to mannitol (Grobben et al. 2001). The conversion equation is as follows: 1 glucose + 2 fructose $\rightarrow$ 1CO_2 + 1 lactate + 1 acetate + 2 mannitol.

Other organic acids such as propionic acid, butanoic acid, 2-methyl propionic acid, fumaric acid, succinic acid and tartaric acid are also produced as minor components (Shim et al. 2012). Amino acids accumulate due to proteolysis of proteins (from raw materials and kimchi LAB) by proteases during fermentation. Kimchi LAB, especially *Lb. sakei* and *Lb. buchneri* produce GABA (γ-aminobutyric acid) during the fermentation (Cho et al. 2007, 2011). The main metabolites are lactate, acetate, CO_2 and mannitol, however, other minor compounds such as diacetyl, acetoin, acetaldehyde, secondary alcohols, esters, and lactones are produced from carbohydrates or fatty acids by kimchi LAB (Chae and Jhon 2007, Kang et al. 2003). These are the compounds that contribute the organoleptic properties of kimchi. Kimchi LAB also produce bacteriocins, which prevent spoilage and over-ripening of kimchi products and inhibit the growth of pathogens.

4.2 Changes in the Ingredients and Organic Acids

Sugars and especially free glucose and fructose in kimchi are fermented rapidly by microorganisms and decrease with conversion to other compounds, such as lactic acid, acetic acid, mannitol, gamma-aminobutyric acid (GABA) and carbon dioxide. The non-fibrous carbohydrates, which are consumed by endogenous LAB, are high in red pepper powder and garlic, even though the main sugar source is Baechu cabbage. Functional health materials are also made from condiments such as garlic, red pepper, ginger, and the cabbage, etc.

Kimchi LAB can affect the composition of bioactive compounds in fermented foods, and the proportion of various forms of bioactive compounds, e.g., flavonoids, its glycosides and aglycones, in kimchi influence its functionality (Chun et al. 2007, Rekha and Vijayalakshmi 2010). Flavonoids in plants are degraded from glycoside to the aglycone and the activity is improved. Therefore the most important fermentation reaction is the biochemical changes to the ingredients.

In general, the contents of B vitamins in kimchi increase gradually with time from the initial fermentation stage, and accumulate to reach their maximum on day 21, and then substantially decrease afterwards (Lee et al. 1960). Kimchi LAB, especially *Ln. mesenteroides* and *Lb. sakei*, produce B vitamins (especially riboflavin and folate). Profiles of health beneficial components in kimchi change during fermentation through diverse biochemical reactions. Also, changes in chlorophyll compounds and other substances occur during the storage of fermented kimchi (Kim 1967, Cheigh and Lee 1991, Cheigh and Park 1994).

The major components of non-volatile organic acids in kimchi are lactic, succinic, fumaric, and malic acids, whereas the major components of volatile organic acids are acetic acid and propionic acid. Their contents and relative compositions change according to the fermentation stage. Ryu et al. (1984) studied changes in organic acids during kimchi (brine cabbage (100%) and brined cabbage (100%) + red pepper powder (4%)) fermentation at 12–16°C for 1, 4 and 7 days as shown in Table 2, the main nonvolatile organic acid was lactic acid, and the level significantly increased from 0.07 to 0.33 in brined cabbage, and from 0.19 to 1.64 meq/100 g in brined cabbage + red pepper powder from day 1 to day 7 of fermentation, respectively. Addition of red pepper powder to the brined cabbage significantly increased the final lactic acid concentrations from 0.33 to 1.64 after seven days of fermentation. Succinic acid, fumaric acid and malic acid levels were high at the beginning, but the levels became trace or low after seven days of fermentation. The final concentration of the volatile organic acid, acetic acid, also increased significantly after fermentation by the addition of red pepper powder to the brined cabbage, from 1.84 to 4.82 meq/100 g. Addition of garlic and green onion also significantly increased lactic acid and acetic acid production compared to brined cabbage only (Ryu et al. 1984). Thus additions of sub-ingredients such as red pepper powder, garlic and green onion are

Table 2. Organic acid contents in brined cabbage and brined cabbage + 4% red pepper powder during fermentation at 12–16°C.

(Unit: meq/100 g)

	Samples	Brined cabbage			Brined cabbage 100 + red pepper powder 4		
	Fermentation period	1	4	7	1	4	7
	Organic acid						
Non-volatile organic acids	Lactic acid	0.07	0.14	0.33	0.19	0.83	1.64
	Succinic acid	0.70	0.35	0.29	0.08	0.83	0.69
	Fumaric acid	0.48	trace	trace	0.04	trace	trace
	Malic acid	3.25	1.24	trace	1.04	0.09	trace
Volatile organic acids	Acetic acid	0.27	0.64	1.84	0.27	0.81	4.82
	Propionic acid	0.16	0.23	0.54	1.51	1.50	1.62
	Butyric acid/ Valeric acid	0.51	0.76	0.82	0.44	0.76	0.68
	Caproic acid	0.03	0.11	0.11	0.07	0.07	0.08
	Heptanoic acid	0.04	0.11	0.11	0.05	0.26	0.08

Source: Ryu et al. (1984)

important to increase the production of the organic acids. Propionic acid levels slightly increased or were the same after the fermentation. Butyric acid, valeric acid caproic acid and heptanoic acid levels were low. Garlic also supports the growth of the hetero-fermentative LAB, *Ln. mesenteroides* and *Lb. brevis*. The CO_2 content was also high in garlic-added kimchi. Chyun and Rhee (1976) reported that low salt levels and low temperatures support increased production of acetic acid.

4.3 Changes in Taste and Flavor Compounds

Kimchi increases the appetite because of its taste, flavor, texture, color, etc. The fresh taste of its vegetables, the tart taste from lactic acid bacteria fermentation, and the flavoring metabolites from the condiments and fermented anchovy, shrimp, etc. all contribute to kimchi's characteristic taste and flavors. The amino acid formation is affected mainly by the addition of ingredients such as fermented sea foods, including anchovy, shrimp, etc. and meats; savory amino acids are formed by microorganisms during the fermentation, and amino acids accumulate due to protein hydrolysis. The activities of amylase and protease increase during the initial fermentation period and then decrease later (Chung 1970, Cheigh and Park 1994). During brining of the cabbage, salt or water-soluble compounds from solid tissue migrate to the liquid portion of kimchi. The eluted soluble compounds accumulate in the kimchi liquid (Choe et al. 1991). The levels of sugars are decreased, with increase in organic acids and a lowering of pH (Mheen et al. 1981). With changes in the organic acid content, the reducing sugars decrease sharply at the initial fermentation stage and then gradually decrease until very little remains. There are several kinds of sugars, especially glucose, fructose and mannitol that affect the sweetness of kimchi.

Titratable acidity (TA) and pH are considered to be the major quality attributes of kimchi because it has the characteristic sour taste (Hong and Park 1997). Ku et al. (2003) reported that commercially available kimchi showed a TA ranging from 0.28% to 1.0%. A longer fermentation period is required to reach this acidity range when kimchi is fermented at low temperatures (−1~5°C), compared to that at 20°C. Acid production has been used as a useful index for evaluating kimchi fermentation. Along with organic acids, sugar and salt are the most important factors affecting the taste of kimchi. The sour and other acidic tastes of kimchi are affected mainly by the amounts of lactic acid and succinic acids, with minor contributions by other organic acids. The volatile organic acids contribute the most to the sour odor of kimchi. However, the nonvolatile organic acids in kimchi are affected by salt content and other fermentation conditions.

The typical flavor of kimchi depends not only on organic acid, carbon dioxide and seasonings, but also on free amino acids. Amino acids are produced by breaking down protein sources in vegetables, condiments, and fermented sea foods, but some of amino acids (lysine, leucine) can be produced during the fermentation. The flavor of kimchi appears to be better in kimchi containing large amounts of lysine, aspartic acid, glutamic acid, valine, methionine, leucine, and isoleucine. Some amino acids (alanine, glycine, serine and threonine) confer a sweet taste. Leucine, phenylalanine, and valine confer a bitter taste (Cho and Rhee 1979). The major aroma components are ethanol, methyl allyl sulfide, acetic acid, dimethyl disulfide, camphene, 1-phellandrene, diallyl disulfide, methyl allyl trisulfide, and zingibirene. These compounds increase during the ripening period and then decrease. The methyl allyl sulfide produced during fermentation seems to be a major volatile flavor compound in well-fermented kimchi. Various biochemical changes occur during fermentation along with the formation of acceptable flavor and texture. However, undesirable fermentation or over-fermentation after ripening may produce a poor quality kimchi, resulting in acidification and cabbage tissue softening (Lee et al. 1992). Because the raw materials contain sufficient sugar to convert to organic acids, excessive acid can be formed with continuous fermentation by the more acid-tolerant microorganisms, especially by homo-fermentative LAB. Softening is associated with decomposition of pectic substances in the tissues of the cabbage or radish (Whang 1960). Polygalacturonase, which is responsible for tissue softening, shows a higher activity during the later fermentation period and is known to be excreted primarily from aerobic and surface-film-forming microorganisms.

4.4 Preservation and Packaging Conditions

When kimchi is fermented at low temperatures, more hetero-fermentative LAB, especially *Ln. mesenteroides* become predominate and gives better taste and quality (Park and Cheigh 2004). The low temperature fermentation and storage is important to preserve kimchi maintain good quality for a longer period. Low temperature by refrigeration or mild freezing of kimchi is a preferred preservation method during storage for marketing purposes. As already shown, kimchi stored at 4°C maintained a high quality for only for 20 days, whereas kimchi stored at −5°C to 0°C keeps its quality for up to three months (Ha et al. 1989). The salinity of kimchi lowers the freezing point, and the texture of kimchi cabbage is not affected by temperatures as low as −5°C even though the kimchi broth is frozen. Thus it is recommended that kimchi should be fermented at a low temperature range (7 to 15°C) and stored at a lower temperature of around −1°C (Kim et al. 1991b, Cheigh and Park 1994). Kimchi refrigerators have been developed for the specific purpose of kimchi fermentation and storage in Korea.

Kimchi fermentation takes place in the container. Thus, the kinds of kimchi container affect biochemical changes differently during the fermentation. Traditionally fermented foods, including kimchi, are stored in Korean traditional earthenware (pottery) called *onggi* (Fig. 4A, Lee and Park 1981, Kim 2007). *Onggi* is made of yellow pulverized clay and is well

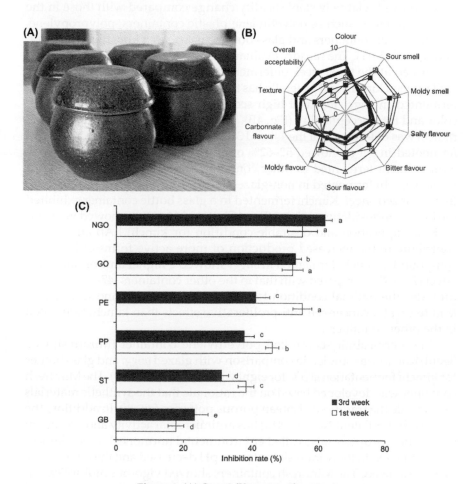

Figure 4. (A) Onggi (Korean earthenware).
(B) Sensory evaluation changes of kimchi fermented in each type of containers after 3 weeks of fermentation. (Jeong et al. 2011)

•, non-glazed onggi; ▲, glazed onggi; ■, polyethylene plastic container; ○, polypropylene container for use in a kimchi refrigerator; △, stainless steel container; □, glass bottle.

(C) DPPH radical-scavenging activity of kimchi fermented in each type of containers.

NGO, non-glazed onggi; GO, glazed onggi; PE, polyethylene plastic container; PP, polypropylene container for use in a kimchi refrigerator; ST, stainless steel container; GB, glass bottle.

[a–e]Means with different letters in the same storage period (weeks) are significantly different ($p < 0.05$) by Duncan's multiple range test.

known for its ability to positively contribute to the quality of fermented foods (Chung et al. 2004, Chung et al. 2005, 2008, Seo et al. 2006, Jeong et al. 2011). During the fermentation at 4°C for 4 weeks, kimchi fermented in *onggi* showed relatively stable acidity change compared with those in the other containers, such as polyethylene plastic containers, polypropylene, stainless steel containers and glass bottles for use in a kimchi refrigerator, (Jeong et al. 2011). The sensory evaluations for each type of container were performed at the third week of fermentation at 5°C which is optimum for well ripened kimchi and generally has the best taste (pH 4.1–4.3). The kimchi fermented in *onggi* received high scores in texture, carbonic acid taste, color and total acceptability (Fig. 4B). The anti-oxidative activity to DPPH radical was determined at the 1st and 3rd week of fermentation, and kimchi fermented in *onggi* showed 52–62% of DPPH radical-scavenging activity while kimchi fermented in other container showed 18–54% at the third week. Kimchi fermented in non-glazed *onggi* showed higher activity than that in glazed *onggi*. Kimchi fermented in a glass bottle container exhibited the lowest antioxidative activity (Fig. 4C). Kimchi in *onggi* showed vigorous LAB multiplication and desirable condition for kimchi LAB and it may contribute to the increased production of more active forms of bioactive compounds. Kimchi fermented in *onggi* showed a higher anti-proliferative effect (74–75%) compared with that in the other containers (47–62%). *Onggi* provides the optimal condition for kimchi fermentation and it may also lead to greater cancer-cell anti-proliferative effects than kimchi fermented in the other containers.

The permeability-controlled polyethylene container (Mirafresh Co., Seoul, Korea) was studied in comparison with glazed *onggi* and glass bottles for kimchi fermentation at 5°C for eight weeks (Lee et al. 2010). The Mirafresh container was developed based on the rationale that the synthetic materials may mimic the traditional Korean porous potteries (*onggi*). In addition, the Mirafresh (US Patent No. 5972815) has antimicrobial activity and no known endocrine disrupters (also called environmental hormones). When kimchi was fermented in the Mirafresh container, pH decreased and the total acidity increase slowed. The Mirafresh containers showed vigorous multiplication of LAB, but inhibited growth of total aerobic bacteria. The texture of kimchi and overall acceptability was excellent in sensory evaluation, The DPPH radical-scavenging activity of kimchi from the Mirafresh container was greater (91%) than those of kimchi in glazed onggi (73%) and glass bottle (63%). The O_2 and CO_2 permeabilities of the Mirafresh container were also higher (458 and 357 nmol h^{-1} m^{-2}atm^{-1}, respectively) than those of the other containers. Glass bottles showed no permeability. This artificial container seems to mimic traditional Korean pottery onggi, thereby helping to extend preservation periods and preserve the traditional better taste of the kimchi.

5. Health Benefits of Kimchi

The secret of kimchi's health benefits lie in fermentation, especially by LAB. Health functionality of kimchi, based on our researches and others, includes a probiotic effect, anticancer, anti-oxidative, anti-obesity, anti-constipation, improving lipid dysfunction, and antidiabetic, enhancing immune function, anti-aging effects (Park and Rhee 2005, Park and Kim 2012, Park et al. 2012b, 2014).

5.1 Probiotic Effects

The LAB found in kimchi are commonly probiotic bacteria that have various beneficial physiological functions. Probiotic strains in the human gastro-intestinal (GI) tract exhibit properties that include colonizing in the human large intestine, anti-oxidant activities, generating anti-viral components, producing anti-cancer compounds, enhancing the immune systems, reducing inflammatory or allergic reactions, and lowering lipids (Vizoso et al. 2006, Ljungh and Wadström 2006). Survival ability of *Lb. plantarum* KCTC 3099 isolated from kimchi was excellent compared to other LAB strains based on the assays determined by binding capacity to Caco-2 intestinal epithelial cells (Lee 2005a). The same strain showed better survival ability against reactive oxygen species such as H_2O_2, hydroxyl radicals, and superoxide anion (Lee et al. 2005b). *Ln. mesenteroides* YML003 isolated from kimchi exerted antiviral activities against anti-H9N2, avian influenza virus (Seo et al. 2012). *Lb. plantarum* KC21 from kimchi showed strong bacteriocin activity against pathogenic bacteria (Lim and Im 2009). *Lb. plantarum* and *Ln. mesenteroides* isolated from kimchi significantly decreased tumor formation in mice by 57% and 39%, respectively. Also, *Lb. plantarum* and *Ln. mesenteroides* from kimchi inhibited cancer growth by 42% and 44% in C57BL/6 mice with Lewis lung carcinoma, respectively (Kim et al. 1991a). Kimchi ingredients, the mixture of microorganisms from kimchi, and especially kimchi *Lb. plantarum* (cell lysate) decreased ascites tumor formation and extend life span by 60% in a Balb/c mouse model system with sarcoma-180 cells implanted (Shin et al. 1998). *Lb. plantarum* CLP-0611 from kimchi strongly inhibited the expression of IL-1β and IL-6, and activation of the NF-κB and AP1 in LPS-stimulated peritoneal macrophage in 2, 4, 6-trinitrobenzene sulfonic acid (TNBS)-induced colitis mouse model (Jang et al. 2014). Also, *Lb. plantarum* CLP-0611 substantially reduced TNBS-induced body weight loss, colon shortening, myeloperoxidase activity, IRAK-1 phosphorylation, NF-κB and MAP kinase (p38, ERK, JNK) activation, and iNOS and COX-2 expression (Jang et al. 2014). Viable or heat-inactivated *Lb. sakei* probio 65 from kimchi could reduce scratching frequency in skin, which could be a good indicator that it modulated immune responses and skin allergies (Kim et al. 2013). Lactobacilli isolated from kimchi were tested for their capacity to modulate the T helper (Th)1/Th2 balance using ovalbumin (OVA)-sensitized mouse splenocytes, and lactobacilli from kimchi might modulate the Th1/Th2 balance via macrophage activation in the hypersensitive reaction caused by Th2 cells (Won et al. 2011). Hypocholesterolemic effects by *Lb. plantarum* KCTC 3928 from kimchi have been demonstrated in C57BL/6 mice because of the induction of fecal bile acid secretion followed by increased degradation of hepatic cholesterol into bile acids (Jeun et al. 2010).

5.2 Anti-oxidant Properties and Slowing Down the Aging Process

Kimchi has lots of powerful antioxidants which are natural free radical scavengers including vitamin C, chlorophyll, phenols, carotenoids, dietary fibers, and other phytochemicals. During kimchi fermentation, metabolites produced from kimchi decrease or eliminate oxidants, pro-oxidants, and other free radicals, which might be direct or indirect causes of aging (Ryu et al. 2004). Ethanolic extracts of kimchi showed strong antioxidant effects in refrigerated cooked pork based on the assays of total phenolic contents, TBARS and peroxide values. The antioxidant effects were significantly different between Baechu and Buchu kimchi in cooked ground pork during storage for 14 days at 4°C (Lee and Kunz 2005).

It is well known that the free radical theory of aging hypothesizes that oxygen-derived free radicals are responsible for the age-related damage at the cellular and tissue levels (Harman 1956). In a normal situation, a balanced-equilibrium exists among oxidants, antioxidants and biomolecules. Kim et al. (2002b) demonstrated that kimchi treatment was able to decrease the high levels of total free radicals, OH radicals, and H_2O_2 in control mouse brains of senescence-accelerated mice (SAM). Kimchi intake over a long period was particularly effective in inhibiting oxidation in the brain. To investigate the effect of kimchi intake on anti-aging properties in liver of SAM, Kim et al. (2002a) conducted another SAM study and reported that the antioxidant enzymes such as Cu, Zn-SOD, Mn-SOD, GSH-px, catalase and GSH/GSSG in liver of kimchi groups increased significantly.

The protective effect of kimchi on mRNA and protein expression of COX-2, iNOS, NF-κB p65, and IκB against SIN-1 treated LLC-PK1 cells was observed (Kim et al. 2014a). The mRNA and protein expressions of COX-2 and iNOS were elevated in SIN-1-treated LLC-PK1 cells. However, treatment with 500 µg/mL of kimchi resulted in markedly decreased mRNA and protein expression of COX-2 and iNOS. In particular, over-ripened (OvR) and optimally-ripened (OptR) kimchi caused more significant downregulation of these expressions than other kimchi extract-treated groups. SIN-1 also increased the protein expression of NF-κB p65, while OvR kimchi caused a marked decrease in its expression. And IκB expression was decreased against the SIN-1-treated group, however, treatment with kimchi resulted in up-regulated expression of IκB. Thus kimchi is protective against oxidative stress related to COX-2, iNOS, NF-κB, p65 and IκB expression.

A study of anti-aging effects of kimchi on human skin cells revealed that ripened kimchi extract remarkably decreased the cytotoxicity induced by H_2O_2 in the keratinocyte cells by protecting skin cells against oxidative damage (Ryu et al. 1997). Kim et al. (2011a) suggested that OptR kimchi helped in regulating and attenuating the inflammation that accelerated the aging process through NF-κB-related gene regulation in an *in vitro* system. The anti-oxidative phytochemicals in kimchi might exert protective effect against the oxidative damage and shield the body from the harmful effects of oxygen free radicals.

The effect of kimchi intake on anti-aging characteristics, free radical production, and anti-oxidative enzyme activities in the brains of senescence accelerated mice have been evaluated (Kim et al. 2002a). Kimchi feeding moderated the increase in free radical production due to aging. Among the kimchi fed groups, both Baechu kimchi with both 30% mustard leaf and mustard leaf kimchi fed groups showed greater inhibition of free radical production than standard Baechu kimchi.

The anti-aging activity of kimchi has also been evaluated using stress-induced premature senescence (SIPS) WI-38 human fibroblasts challenged with H_2O_2. H_2O_2-treated WI-38 cells showed a loss of cell viability, and an increase in lipid peroxidation and shortening of the cell lifespan, indicating the induction of SIPS. However, treatment with kimchi, especially OptR kimchi attenuated cellular oxidative stress by increasing cell viability and inhibiting lipid peroxidation. In addition, the lifespan of WI-38 cells was extended, suggesting a promising role of kimchi as an anti-aging agent (Kim et al. 2011a).

The protective effect of kimchi against Aβ25-35-induced memory impairment was investigated in an *in vivo* Alzheimer's mouse model using ICR mice (Choi et al. 2014). The aggregation of Aβ25-35 was injected into the brains of the mice (5 nmol/mouse), and then OptR kimchi was orally administered at 100 or 200 mg/kg of body weight for two weeks. Objective cognitive ability (old object and new object) and T-maze (space perceptive ability) tests and the Morris water maze test indicated that kimchi exerted strong protective effect against cognitive impairment induced by Aβ25-35 (p < 0.05). Kimchi also showed protective effects on lipid peroxidation in the brain by the treatment of Aβ25-35. The MDA level of the control group was 55.1 mmol/mg protein, whereas the normal group was 45.8, and kimchi treated groups administered 100 or 200 mg/kg were significantly lower at 37.7 and 37.0 mmol/mg protein, respectively. These results suggest that kimchi protected against Aβ-induced impairment of memory and cognition as well as attenuating oxidative stress.

5.3 Prevention of Cardiovascular Disease and Obesity

Cardiovascular Disease: Regular consumption of kimchi lowers the serum concentrations of cholesterol and triglyceride. Diseases related to atherosclerosis such as ischaemic heart disease, stroke and peripheral arterial diseases are associated with higher plasma lipids. Kimchi had been shown to have efficacy for decreasing the levels of plasma cholesterol, triglyceride, and LDL-cholesterol and increasing HDL-cholesterol (Kwon et al. 1997). Kim and Lee (1997) investigated the effects of kimchi on lipid metabolism and immune function using male SD rats fed six kinds of baechu kimchi for 4 weeks and found that kimchi stimulated lipid mobilization to epididymal fat pad and lipid excretion via feces and lowered serum and liver triglyceride concentration. The fermented kimchi stimulated the proliferation of B cells and lowered the lipid accumulation in epididymal fat pad, especially kimchi fermented for 6 weeks at 4°C. Several human studies have shown that daily kimchi consumption provided hypolipidemic effects in middle-aged men (Kwon et al. 1999) and increased HDL-cholesterol level and decreased body fat and BMI in college students (Lee et al. 2012). According to Choi et al. (2013), 100 volunteers were assigned to 2 dietary groups, low (15 g/day, n = 50) and high (210 g/day, n = 50) kimchi intake, and were housed together in a dormitory for 7 days. Lipid lowering effects of kimchi were more profound for total cholesterol and LDL-cholesterol levels in both groups. Fasting blood glucose was significantly decreased in groups with high kimchi intake compared to those with low intake groups.

Obesity: Choi et al. (2002) suggested that the red pepper powder and kimchi diet fed groups showed significantly lower final body weights than high fat diet-fed groups after 4 weeks. Total lipids, triglycerides, and cholesterol in liver showed similar patterns. It is known that capsaicin in red pepper stimulates lipid mobilization from adipose tissue and lowers the perirenal adipose tissue weight and serum triglyceride concentration in lard-fed rats (Kawada et al. 1986). Anti-obesity effects were found not only for kimchi with red pepper powder but also whitish baechu kimchi without red pepper powder (baek-kimchi). Yoon et al. (2004) reported that the whitish kimchi (without red pepper) exerted more effective anti-obesity properties than Baechu kimchi (with red pepper) in a high fat diet induced obesity animal model due to the higher content of radish and pear used in whitish baechu kimchi. Red pepper powder (RPP), garlic and ginger are commonly used spices in kimchi. These spices significantly decreased weight gain compared to HFD, but garlic and ginger showed a greater effect on reducing body weight gain than RPP (Yoon et al. 2005). The weights of liver and epididymal and perirenal fat pads in garlic and ginger diet groups were significantly lower than those of the HFD groups (p < 0.05). The garlic and ginger also decreased triglyceride and cholesterol contents in liver and epididymal and perirenal fat pads, reversing the higher levels seen in HFD.

We confirmed the anti-obesity effect of kimchi in high fat diet-induced obese (DIO) C57BL/6 mice. Mice were fed HFD for 4 weeks to induce obesity, from the 5th to 8th weeks, S-kimchi (standardized kimchi) group, and D-kimchi (Korean commercial D-kimchi) group were fed a HFD containing 10% of S-kimchi and D-kimchi, respectively. Body weight and adipose tissue weights were significantly lower in the kimchi-treated groups than in the HFD group. The D-kimchi group significantly decreased serum levels of TG, TC, LDL-C, insulin and leptin and increased HDL-C and adiponectin. The hepatic histological study conducted after the 8-week experimental period, showed that kimchi treatment significantly reduced HFD-induced lipid formation in liver. Hepatic mRNA expression of adipogenesis-related genes of C/EBPα, PPAR-γ, SREBP-1c and FAS were lower than in the HFD group but fatty acid oxidation related CPT-1 expressions were higher. Histological assays showed that kimchi administration reduced HFD-induced fat accumulation in epididymal fat tissues (Fig. 5A). Mean adipocyte size in the D-kimchi group (1.8±0.7 units) was smaller than in the HFD or S-kimchi

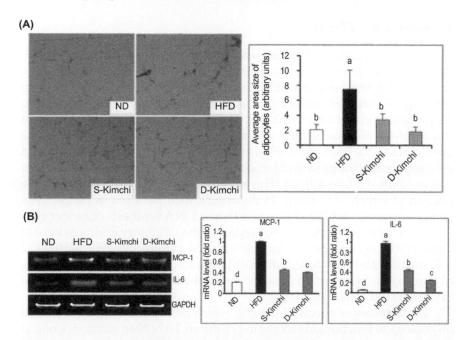

Figure 5. Histological observations (A) and the mRNA expressions of MCP-1 and IL-6 (B) in the epididymal fat tissues of DIO mice (8th week).

[a~d] Means with different letters are significantly different (p<0.05) by Duncan's multiple range tests.

The average sizes of adipocytes were measured on histological slides using Image J (right panel). The intensities of bands were measured by densitometry and are expressed as folds of the control (HFD) divided by GAPDH.

groups (HFD 7.5 ± 2.6 units and S-Kimchi: 3.4 ± 0.8 unit), and was similar to that of the ND group (2.0 ± 0.7 unit). Furthermore, the two kimchi treated groups had significantly lower MCP-1 mRNA expressions (an important mediator of macrophage activity and promoter of inflammatory response in adipose tissues), and IL-6 expressions (an inflammatory cytokine) in epididymal fat tissues than the HFD group after the 8-week experimental period (Fig. 5B). These results suggest the possibility that sub-ingredients, such as, leeks and starter (*Ln. mesenteroides* DRC 0211) enhance the anti-obesity effect of kimchi. Furthermore, our results show that commercial Korean kimchi can reduce body fat as well as standardized kimchi prepared in the laboratory (Cui et al. 2015).

According to Kim et al. (2011b), positive effects of fermented kimchi were found for lowering body mass index and body fat in overweight and obese patients, and for preventing the development of factors implicated in metabolic syndrome such as waist-hip ratio, fasting glucose concentration, fasting insulin, total cholesterol, and blood pressure.

5.4 Anti-cancer Effects

Several studies validated the anti-cancer properties of kimchi (Park 1995, Park et al. 2014, Kim et al. 2014b, Hur et al. 2000, Park et al. 2003). An active component, β-sitosterol, in kimchi and kimchi extract blocked the signaling pathway of oncogenic H-Ras^{v12}-induced DNA synthesis, thus preventing the proliferation of transformed cells (Park et al. 2003). Furthermore, several studies tested anticancer effect of various extracts of kimchi. Treatment of a methanol soluble fraction (MSF) of kimchi in sarcoma-180 cell transplanted mice suppressed the growth of tumors, inhibiting lipid peroxide production and xanthine oxidase activity (Hur et al. 2000). Also, the dichloromethane fraction of kimchi decreased ^{3}H thymidine incorporation in cancer cells and consequently inhibited the growth and DNA synthesis of cancer cells (Hur et al. 1999). Choi et al. (1997) reported the inhibitory effects of kimchi extract (500 μg/mL) on carcinogen-induced cytotoxicity of C3H10T1/2 cells mediated by 7,12-dimethylbenz[a]anthracence and N-methy-N'-nitro-N-nitrosoguanidine and on 3-methylcholanthrene-induced transformation in C3H10T1/2 cells.

The anti-cancer effects of kimchi, which had special ingredients with physiological functionality including chitosan, have been studied. Kong et al. (2010) revealed that chitosan-added kimchi exerted the up-regulation of Bax expression and down-regulation of Bcl-2, cellular inhibitor of apoptosis (cIAP) including cIAP-1 and cIAP-2, cyclooxygenase-2, inducable nitric oxide synthase, and NF-κB expressions when compared to control kimchi (without chitosan addition) in HT-29 human colon carcinoma cells.

(A)

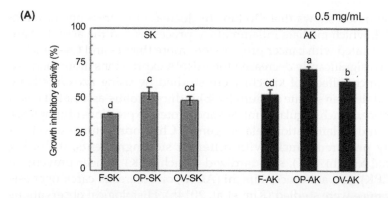

(B)

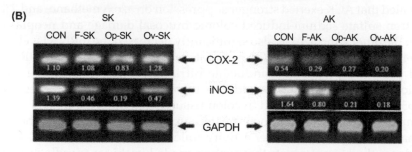

Figure 6. Inhibitory effects of kimchi on growth of HT-29 human colon cancer cells based on an MTT assay (A), and effects of SK and AK on mRNA expressions of COX-2 and iNOS in HT-29 cancer cells (B) (Kim et al. 2015).

SK: standardized kimchi, AK: Anticancer kimchi, F-SK: fresh SK, Op-SK: optimally ripened SK, Ov-SK: over-ripened SK, F-AK: fresh AK, Op-AK: optimally ripened AK, Ov-AK: over-ripened AK.

[a-d]Mean values with different letters on bars are significantly different (p < 0.05) based on Duncan's multiple range test (A). The number below the bands is the band intensity of the factor divided by the band intensity of GAPDH (B).

Kimchi is a valuable food which helps decrease the risk of development of various cancers. Kimchi ingredients, vegetables, LAB and fermentation process help to prevent the colon cancers. We prepared anti-cancer kimchi (Kil 2004), by changing kimchi ingredients.

AK (anticancer kimchi) exhibited superior anti-cancer effects to SK (standardized kimchi) in MTT assay on HT-29 human cancer cells (Kim et al. 2015). Figure 6A shows that Op-SK (optimally ripened-SK) achieved a 54.3% growth inhibition in HT-29 cancer cells, while F-SK (fresh-SK) showed a 39.6% inhibition and Ov-SK (over-ripened SK) was 49.0%. However, Op-AK achieved the highest growth inhibition of 71.2%, F-AK was 53.0% and Ov-AK was 62.3%. Thus the anticancer effect was the highest in Op-SK and AK, and then Ov-kimchi and F-kimchi, and AK also showed higher anticancer effect than SK, thus manipulation of ingredients can increase anticancer effects. Figure 6B shows that Op-kimchi decreased expressions of iNOS and COX-2, which mediate inflammatory processes and the activities are closely associated with cancer progression, more than F- and Ov-kimchis, and also AK significantly decreased the mRNA expressions more than SK.

Anti-cancer effects of kimchi were studied by using azoxymethane (AOM) and dextran sulfate sodium (DSS) induced colitis-associated colon cancer in mice. Kimchi regulated mRNA and protein expressions of apoptosis, cell cycle and inflammation related genes. Chemopreventive effects of differently prepared kimchi with different sub-ingredients, including commercial kimchi (CK), standardized kimchi (SK), cancer-preventive kimchi (CPK), and anticancer kimchi (ACK), on colorectal carcinogenesis in Balb/c mice were studied (Kim et al. 2014b). Histological observations revealed that ACK exerted stronger suppression on azoxymethane- and 2% dextran sulfate sodium-induced colonic mucosal damage and neoplasia than the other kimchi. ACK also significantly decreased the mRNA levels of pro-inflammatory cytokines (TNF-α, IL-6, and IFN-γ) as well as the mRNA and protein expressions of inducible nitric oxide (iNOS) synthase and cyclooxygenase-2 (COX-2). In addition, the mRNA and protein expression of p53 and p21 were elevated in colon tissues from the ACK-treated mice compared with the other kimchi-treated groups (Fig. 7). ACK was shown to have the highest anticancer activity followed by CPK > SK > CK. Our evidence therefore suggests that custom-made kimchi for specific disease prevention is possible. However, more study is needed in this research field.

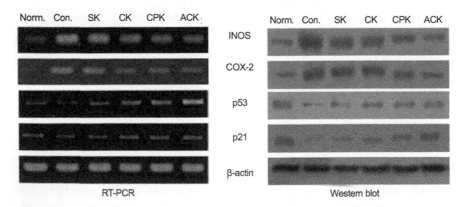

Figure 7. Effects of extracts from different types of kimchi (1.89 g/kg) on mRNA and protein levels of iNOS, COX-2, p53 and p21 in colon tissue from BALB/c mice with AOM- and DSS-induced colitis associated colon cancer (Kim et al. 2014b).

CK: Commercial kimchi; SK: Standardized kimchi; CPK: Cancer preventive kimchi; ACK: Anticancer kimchi.

Helicobacter pylori infection may cause gastroduodenal disorders including acute and chronic gastritis, gastroduodenal ulcer, chronic atrophic gastritis, and gastric malignancy. In our recent studies, oral infection with *H. pylori* at the concentration of 1×10^9 CFU/mL caused gastric inflammation in C57BL/6 mice models for 36 weeks. However, oral administration of 0.63 and 1.83 g/kg kimchi to mice (corresponding to 100 and 300 g/day when converted into kimchi intake of 60-kg humans) significantly reduced the symptoms of gastric inflammation. Especially, mice with 1.83 g/kg kimchi and *H. pylori* had almost healthy stomachs compared to normal mice group. Kimchi treated groups had less lesions such as erosion and ulceration in the stomach of mice infected with *H. pylori* than mice without kimchi treatment based on the results of H&E staining. Mice with the higher concentration of kimchi had less lesions than those with lower concentration of kimchi. Expression of COX-2, NF-κB, p-STAT3 and macrophage infiltration (F4/80 staining) were significantly decreased by the kimchi treatment in the stomach tissue of the mice treated with high concentrations of kimchi (Chung et al. 2014). Thus kimchi may help to reduce the inflammation caused by *H. pylori* and possibly be a new therapeutic food for targeted therapy.

6. Conclusion

Kimchi is a vegetable fermented probiotic food. The main ingredients are cruciferous vegetables, and sub-ingredients which include red pepper powder, garlic, ginger, green onions, fermented fishes, etc. The naturally present microorganisms, especially LAB in the cabbage become the starters for fermentation. Using high quality ingredients, predominant LAB, low temperature fermentation and facultative anaerobic conditions for the fermentation are important factors for making good kimchi. The LAB level significantly increases to 10^{8-9} CFU/mL or gm after the fermentation, thus becoming a functional vegetable probiotic food. The predominant LAB genera are *Leuconostoc*, *Lactobacillus* and *Weissella* in kimchi fermentation. The representative species are *Ln. mesenteroides*, *Ln. citrium*, *Lb. plantarum*, *Lb. sakei* and *W. koreensis*. Although hetero-fermentative and homo-fermentative LAB are involved in kimchi fermentation, the hetero-fermentation become dominant and provide better taste and uniform quality to the kimchi. The biochemical changes of kimchi are mediated by the predominant LAB, taste and flavor compounds from the ingredients, and fermentation conditions such as temperature, containers for fermentation, etc., and their combined effects result in kimchi quality and functionality. Kimchi has various health benefits such as probiotics, anti-obesity, anti-cancer, anti-atherosclerotic, anti-oxidative and anti-aging effects, etc. due to the presence of LAB and various nutraceuticals and metabolites from the ingredients which also impart good taste. Kimchi can also be developed as custom-made foods for different needs, race, age-adjusted kimchi and specific disease prevention such as anticancer, anti-obesity, anti-allergy, etc. by changing the ingredients, LAB starters and other fermentation conditions.

Keywords: Kimchi, recipe, nutrients, process method, lactic acid bacteria, microbiology, biochemistry, health benefits

References

Bong, Y.J. (2014). Probiotic effects of kimchi lactic and bacteria (LAB) and increased health functionality of Baechu kimchi by LAB starters. MS Thesis, Pusan National University, Busan, Korea.

Chae, M.H. and Jhon, D.Y. (2007). Effects of commercial fructooligosaccharides on Bifidobacteria kimchi fermentation. Korean Journal of Food Science and Technology 39: 61–65.

Chang, J.Y. and Chang, H.C. (2010). Improvements in the quality and shelf life of kimchi by fermentation with the induced bacteriocin-producing strain, *Leuconostoc citreum* GJ7 as a starter. Journal of Food Science 75: M103–M110.

Cheigh, H.S. and Park, K.Y. (1994). Biochemical, microbiological and nutritional aspect of kimchi. Critical Review in Food Science and Nutrition 34: 175–203.

Cheigh, H.S. and Lee, J.M. (1991). Classification and review of the literatures on kimchi (I). Bulletin of College of Home Economics, Pusan National University 17: 11–18.

Cho, E.J. (1999). Standardization and cancer chemopreventive activities of Chinese cabbage kimchi. Ph.D. Thesis, Pusan National University, Busan, Korea.

Cho, J.J., Lee, D.Y., Yang, C.N., Jeon, J.I., Kim, J.H. and Han, H.U. (2006). Microbial population dynamics of kimchi, a fermented cabbage product. Federation of European Microbiological Societies Microbiology Letter 257: 262–267.

Cho, K.M., Math, R.K., Islam, S. Md.A., Lim, W.J., Hong, S.Y., Kim, J.M., Yun, M.G., Cho, J.J. and Yun, H.D. (2009). Novel multiplex PCR for the detection for lactic acid bacteria during kimchi fermentation. Molecular and Cellular Probes 23: 90–94.

Cho, S.Y., Park, M.J., Kim, K.M., Ryu, J.H. and Park, H.J. (2011). Production of high γ-aminobutyric acid (GABA) sour kimchi using lactic acid bacteria isolated from mukeunjee kimchi. Food Science and Biotechnology 20: 403–408.

Cho, Y. and Rhee, H.S. (1979). A study in flavorous taste components in kimchi. Korean Journal of Food Science and Technology 11: 26–31.

Cho, Y.R., Chang, J.Y. and Chang, H.C. (2007). Production of gamma-aminobutyric acid (GABA) by *Lactobacillus buchneri* isolated from kimchi and its neuroprotective effect on neuronal cells. Journal of Microbiology and Biotechnology 17; 104–109.

Choe, S.M., Jun, Y.S., Park, K.Y. and Cheigh, H.S. (1991). Changes in the contents of moisture, reducing sugar, microorganisms, NO_2 and NO_3 during salting in various varieties of Chinese cabbage for kimchi fermentation. Research Bulletin College of Home Economics Pusan National University 17: 25–30.

Choi, I.H., Noh, J.S., Han, J.S., Kim, H.J., Han, E.S. and Song, Y.O. (2013). Kimchi, a fermented vegetable, improves serum lipid profiles in healthy young adults: randomized clinical trial. Journal of Medicinal Food 16: 223–229.

Choi, J.M., Lee, S.H., Park, K.Y., Kang, S.A. and Cho, E.J. (2014). Protective effect of kimchi against Aβ25-35-induced impairment of cognition and memory. Journal of Korean Society of Food Science and Nutrition 43: 360–366.

Choi, M.W., Kim, K.H., Kim, S.H. and Park, K.Y. (1997). Inhibitory effects of kimchi extracts on carcinogen-induced cytotoxicity and transformation in C3H10T1/2 cells. Journal of Food Science and Nutrition 2: 241–245.

Choi, S.M., Jeon, Y.S., Rhee, S.H. and Park, K.Y. (2002). Red pepper powder and kimchi reduce body weight and blood and tissue lipids in rats fed a high fat diet. Nutraceuticals and Food 7: 162–167.

Choi, S.M. (2001). Antiobesity and anticancer effects of red pepper powder and kimchi. Ph.D. Thesis, Pusan National University, Busan, Korea.

Choi, W.Y. and Park, K.Y. (2000). Increased preservative and antimutagenic activities of kimchi with addition of green tea leaves. Journal of Food Science and Nutrition 5: 189–193.

Choi, W.Y. and Park, K.Y. (1999). Anticancer effect of organic Chinese cabbage kimchi. Journal of Food Science and Nutrition 4: 113–116.

Codex Alimentarius Commission (Codex) (2001). Codex standard for kimchi. Codex Stan 223. Rome, Italy: Food and Agriculture Organization of the United Nations.

Chun, J., Kim, G.M., Lee, K.W., Choi, I.D., Kwon, G.H., Park, J.Y., Jeong, S.J., Kim, J.S. and Kim, J.H. (2007). Conversion of isoflavone glucosides to aglycones in soymilk by fermentation with lactic acid bacteria. Journal of Food Science 72: M39–M44.

Chung, D.H. (1970). Studies on the composition of kimchi.Ⅲ.Oxidation-reduction potential during kimchi fermentation. Korean Journal of Food Science and Technology 2: 34–37.

Chung, M.K., Kim, H.S., Park, J.M., Kim, E.H., Han, Y.M., Kwon, S.H., Park, K.Y. and Hahm, K.B. (2014). *Helicobacter pylori* infection and food intervention for preventing associated gastric diseases including gastric cancer. The Korean Journal of *Helicobacter* and Upper Gastrointestinal Research 14: 225–232.

Chung, S.K., Lee, K.S. and An, D.S. (2008). Fermentation characteristics of Kochujang in onggi with different porosities. Journal of Korea Society of Packaging Science and Technology 14: 9–14.

Chung, S.K., Kim, Y.S. and Lee, D.S. (2005). Effects of vessel on the quality changes during fermentation of Kochujang. Korean Journal of Food Preservation 12: 292–298.

Chung, S.K., Lee, K.S. and Cho, S.H. (2004). Effect of fermentation vessel on quality of anchovy soy sauce. Korean Journal of Food Preservation 11: 233–239.

Chyun, J.H. and Rhee, H.S. (1976). Studies on the volatile fatty acids and carbon dioxide produced in different kimchis. Korean Journal of Food Science and Technology 8: 90–94.

Cui, M., Kim, H.Y., Lee, K.H., Jeong, J.K., Hwang, J.H., Yeo, K.Y., Ryu, B.H., Choi, J.H. and Park, K.Y. (2015). Antiobesity effects of kimchi in diet-induced obese mice. Journal of Ethnic Foods 2: 137–144.

Erten, H. (1998). Metabolism of fructose as an electron acceptor by *Leuconostoc mesenteroides*. Process Biochemistry 33: 735–739.

Grobben, G.J., Peters, S.W.P.G., Wisselink, H.W., Weusthuis, R.A., Hoefnagel, M.H.N., Hugenholtz, J. and Eggink, G. (2001). Spontaneous formation of a mannitol-producing variant of *Leuconostoc pseudomesenteroides* grown in the presence of fructose. Applied and Environmental Microbiology 67: 7037–7044.

Ha, J.H., Hawer, W.S., Kim, Y.J. and Nam, Y.J. (1989). Changes of free sugars in kimchi during fermentation. Korean Journal of Food Science and Technology 21: 633–638.

Harman, D. (1956). Aging: a theory based on free radical and radiation chemistry. Journal of Gerontology 11: 298–300.

Hong, S.I. and Park, W.S. (1997). Sensitivity of color indicators to fermentation products of kimchi at various temperatures. Korean Journal of Food Science and Technology 29: 21–25.

Hur, Y.M., Kim, S.H., Choi, Y.W. and Park, K.Y. (2000). Inhibition of tumor formation and changes in hepatic enzyme activities by kimchi extracts in sarcoma 180 cell transplanted mice. Journal of Food Science and Nutrition 5: 48–53.

Hur, Y.M., Kim, S.H. and Park, K.Y. (1999). Inhibitory effects of kimchi extracts on the growth and DNA synthesis of human cancer cells. Journal of Food Science and Nutrition 4: 107–112.

Jang, S.E., Han, M.J., Kim, S.Y. and Kim, D.H. (2014). *Lactobacillus plantarum,* CLP-0611 ameliorates colitis in mice by polarizing M1 to M2-like macrophages. International Immunopharmacology 21: 186–192.

Jeong, J.K., Kim, Y.W., Choi, H.S., Lee, D.S., Kang, S.A. and Park, K.Y. (2011). Increased quality and functionality of kimchi when fermented in Korean earthenware (onggi). International Journal of Food Science and Biotechnology 46: 2015–2021.

Jeun, J., Kim, S., Cho, S.Y., Jun, H.J., Park, H.J., Seo, J.G., Chung, M.J. and Lee, S.J. (2010). Hypocholesterolemic effects of *Lactobacillus plantarum* KCTC3928 by increased bile acid excretion in C57BL/6 mice. Nutrition 26: 321–330.

Jung, J.Y., Lee, S.H. and Jeon, C.O. (2014) Kimchi microflora: history, current status, and perspectives for industrial kimchi production. Applied Microbiology and Biotechnology 98: 2385–2393.

Jung, J.Y., Lee, S.H., Lee, H.J., Seo, H.Y., Park, W.S. and Jeon, C.O. (2012). Effects of *Leuconostoc mesenteroides* starter cultures on microbial communities and metabolites during kimchi fermentation. International Journal of Microbiology 153: 378–387.

Jung, J.Y., Lee, S.H., Kim, J.M., Park, M.S., Bae, J.W., Hahn, Y., Madson, E.L. and Jeon, C.O. (2011). Metagenomic analysis of kimchi, a traditional Korean fermented food. Applied Environmental Microbiololgy 77: 2264–2274.

Kang, J.H., Lee, J.H., Min, S. and Min, D.B. (2003). Changes of volatile compounds, lactic acid bacteria, pH, and headspace gases in kimchi, a traditional Korean fermented vegetable product. Journal of Food Science 68: 849–854.

Kawada, T., Hagihara, K. and Iwai, K. (1986). Effects of capsaicin on lipid metabolism in rats fed high fat diet. Journal of Nutrition 116: 1272–1278.

Kendler, O. and Weiss, N. (1986). Regular, nonsporing gram-possitive rods. *In*: P.H.A. Sneath, N.S. Mair, M.E. Sharpe and J.G. Holt (eds.). Bergeys Manual of Systematic Bacteriology, Vol. 2 (9th ed), Williams and Wilkins Company, Baltimore, MD, USA.

Kil, J.H. (2004). Studies on development of cancer preventive and anticancer kimchi and its anticancer mechanism. Ph.D. Thesis, Pusan National University, Busan, Korea.

Kim, B.K., Song, J.L., Ju, J.H., Kang, S.A. and Park, K.Y. (2015). Anticancer effect of kimchi fermented for different times and with added ingredients in human HT-29 colon cancer cells. Food Science and Biotechnology 24: 629–633.

Kim, B.K., Choi, J.M., Kang, S.A., Park, K.Y. and Cho, E.J. (2014a). Antioxidative effects of kimchi under different fermentation stage on radical-induced oxidative stress. Nutrition Research and Practice 8: 638–643.

Kim, B.K., Park, K.Y., Kim, H.Y., Ahn, S.C. and Cho, E.J. (2011a). Anti-aging effects and mechanisms of kimchi during fermentation under stress-induced premature senescence cellular system. Food Science and Biotechnology 20: 643–649.

Kim, E.K., An, S.Y., Lee, M.S., Kim, T.H., Lee, H.K., Hwang, W.S., Choe, S.J., Kim, T.Y., Han, S.J., Kim, H.J., Kim, D.J. and Lee, K.W. (2011b). Fermented kimchi reduced body weight and improves metabolic parameters in overweight and obese patients. Nutrition Research 31: 436–443.

Kim, H.Y., Song, J.L., Chang, H.K., Kang, S.A. and Park, K.Y. (2014b). Kimchi protects against azoxymethane/dextran sulfate sodium-induced colorectal carcinogenesis in mice. Journal of Medicinal Food 17: 833–841.

Kim, H.Y., Bae, H.S. and Baek, Y.J. (1991a). *In vivo* antitumor effects of lactic acid bacteria on Sarcoma 180 and mouse Lewis lung carcinoma. Journal of Korean Cancer Association 23: 188–196.

Kim, J.H., Kwon, M.J., Lee, S.Y., Ryu, J.D., Moon, G.S., Cheigh, H.S. and Song, Y.O. (2002a). The effect of kimchi intake on production of free radicals and anti-oxidative enzyme activities in the liver of SAM. Journal of the Korean Society of Food Science and Nutrition 31: 109–116.

Kim, J.H., Kwon, M.J., Lee, S.Y., Ryu, J.D., Moon, G.S., Cheigh, H.S. and Song, Y.O. (2002b). The effect of kimchi on production of free radicals and antioxidative enzyme activities in the brain of SAM. Journal of the Korean Society of Food Science and Nutrition 31: 117–123.

Kim, J.Y. and Lee, Y.S. (1997). The effects of kimchi intake on lipid contents of body and mitogen response of spleen lymphocytes in rats. Journal of the Korean Society of Food Science and Nutrition 26: 1200–1207.

Kim, J.Y., Park, B.K., Park, H.J., Park, Y.H., Kim, B.O. and Pyo, S. (2013). Atopic dermatitis-mitigating effects of new *Lactobacillus* strain, *Lactobacillus sakei* probio 65 isolated from kimchi. Journal of Applied Microbiology 115: 517–526.

Kim, M.J. (1967). Fermentation and preservation of Korean kimchi. M.S. thesis, University of Leeds, Leeds, UK.

Kim, M.J. and Chun, J.S. (2005). Bacterial community structure in kimchi, a Korean fermented vegetable food, as revealed by 16S rRNA gene analysis. International Journal of Food Microbiology 103: 91–96.

Kim, S.B. (1997). Culture; History of Korea food life. Kwangmoonkak Publishing Co., Seoul, Korea, pp. 146–260.

Kim, S.H. (2007). Porous and pottery with dark brown glaze. Journal of Contents Association 7: 157–164.

Kim, T.W., Lee, J.Y., Jung, S.H., Kim, Y.M., Jo, J.S., Chung, D.K., Lee, H.J. and Kim, H.Y. (2002c). Identification and distribution of predominant lactic acid bacteria in kimchi, a Korean traditional fermented food. Journal of Microbiology and Biotechnology 12: 635–642.

Kim, W.J., Kang, K.O., Kyung, K.H. and Shin, J.I. (1991b). Addition of salts and their mixtures of storage stability of kimchi. Korean Journal of Food Science and Technology 23: 188–191.

Kong, C.S., Bahn, Y.E., Kim, B.K., Lee, K.Y. and Park, K.Y. (2010). Antiproliferative effect of chitosan-added kimchi in HT-29 human colon carcinoma cells. Journal of Medicinal Food 13: 6–12.

Ku, K.H., Cho, M.H. and Park, W.S. (2003). Characteristics analysis for the standardization of commercial kimchi. Korean Journal of Food Science and Technology 35: 316–319.

Kwon, M.J., Chun, J.H., Song, Y.S. and Song, Y.O. (1999). Daily kimchi consumption and its hypolipidemic effect in middle-aged men. Journal of the Korean Society of Food Science and Nutrition 28: 1144–1150.

Kwon, M.J., Song, Y.O. and Song, Y.S. (1997). Effects of kimchi on tissue and fecal lipid composition and apoprotein and thyroxine levels in rats. Journal of the Korean Society of Food Science and Nutrition 26: 507–513.

Lee, C.W., Ko, C.Y. and Ha, D.M. (1992). Microfloral changes of the lactic acid bacteria during kimchi fermentation and identification of the isolates. Korean Journal of Applied Microbiology and Biotechnology 20: 102–107.

Lee, E.J., Park, S.E., Choi, H.S., Han, G.J., Kang, S.A. and Park, K.Y. (2010). Quality characteristics of kimchi fermented in permeability controlled polyethylene containers. Korean Journal of Food Preservation 17: 793–799.

Lee, G.J. and Park, C.K. (1981). Lead content leached out from glazed potteries bull. Korean Fish Society 4: 158–164.

Lee, G.Y. (2015). Survey on middle and high school students' kimchi intake patterns and studies on the development of kimchi for adolescent. MS Thesis, Pusan National University, Busan, Korea.

Lee, J., Hwang, K.T., Heo, M.S., Lee, J.H. and Park, K.Y. (2005a). Resistance of *Lactobacillus plantarum* KCTC 3099 from kimchi to oxidative stress. Journal of Medicinal Food 8: 299–304.

Lee, J.M. (2005). Adhesion of kimchi *Lactobacillus* strains to Caco-2 cell membrane and sequestration of aflatoxin B_1. Journal of Korean Society of Food Science and Nutrition 34: 581–585.

Lee, J.S., Heo, G.Y., Lee, J.W., Oh, Y.J., Park, J.A., Park, Y.H., Pyun, Y.R. and Ahn, J.S. (2005b). Analysis of kimchi microflora using denaturing gradient gel electrophoresis. International Journal of Food Microbiology 102: 143–150.

Lee, J.Y. and Kunz, B. (2005). The antioxidant properties of baechu-kimchi and freeze-dried kimchi-powder in fermented sausages. Meat Science 69: 741–747.

Lee, K. and Lee, Y. (2010). Effect of *Lactobacillus plantarum* as a starter on the food quality and microbiota of kimchi. Food Science and Biotechnology 19: 641–646.

Lee, S.Y., Song, Y.O., Han, E.S. and Han, J.S. (2012). Comparative study on dietary habits, food intakes, and serum lipid levels according to kimchi consumption in college students. Journal of the Korean Society of Food Science and Nutrition 41: 351–361.

Lee, T.Y., Kim, J.S., Chung, D.H. and Kim, H.S. (1960). Studies on kimchi. II. Variations of vitamins during kimchi fermentation. Bulletin of Science Research Institute of Korea 5: 43–50.

Lim, S.M. and Im, D.S. (2009). Screening and characterization of probiotic lactic acid bacteria isolated from Korean fermented foods. Journal of Microbiology and Biotechnology 19: 178–186.

Ljungh, A. and Wadström, T. (2006). Lactic acid bacteria as probiotics. Current Issues in Intestinal Microbiology 7: 73–89.

Mheen, T.I., Kwon, T.W. and Lee, C.H. (1981). Traditional fermented food products in Korea. Korean Journal of Applied Microbiology and Biotechnology 9: 253–261.

McFeeters, R.F. and Chen, K.H. (1986). Utilization of electron acceptors for anaerobic mannitol metabolism by *Lactobacillus plantarum*; compounds which serve as electron acceptors. Food Science 3: 73–81.

Park, E.J., Chun, J.S., Cha, C.J., Park, W.S., Jeon, C.O. and Bae, J.W. (2012a). Bacterial community analysis during fermentation of ten representative kinds of kimchi with barcoded pyrosequencing. Food Microbiology 30: 197–204.

Park, J.A., Tirupathi Pichiah, P.B., Yu, J.J., Oh, S.H., Daily, J.W. 3rd and Cha, Y.S. (2012b). Anti-obesity effect of kimchi fermented with *Weissella koreensis* OK1-6 as starter in high-fat diet-induced obese C57BL/6J mice. Journal of Applied Microbiology 113: 1507–1516.

Park, K.Y. (1995). The nutritional evaluation and anti-mutagenic and anti-cancer effects of kimchi. Journal of the Korean Society of Food and Nutrition 24: 169–182.

Park, K.Y., Jeong, J.K., Lee, Y.E. and Daily III, J.W. (2014). Health benefits of kimchi (Korean fermented vegetables) as a probiotic food. Journal of Medicinal Food 17: 6–20.

Park, K.Y. and Kim, B.K. (2012). Lactic acid bacteria in vegetable fermentations. pp. 187–211. *In*: S. Lahtinen, S. Salminen, A. Ouweh and von A. Wright (eds.). Lactic Acid Bacteria. CRC Press Inc., Boca Raton, FL, USA.

Park, K.Y. and Rhee, S.H. (2005). Functional foods from fermented vegetable products: Kimchi (Korean fermented vegetables) and functionality. pp. 341–380. *In*: J. Shi, C.T. Ho and F. Shahidi (eds.). Asian Functional Foods. CRC Press Inc., Boca Raton, FL, USA.

Park, K.Y. and Cheigh, H.S. (2004). Kimchi. pp. 621–655. *In*: Y.H. Hui, L.M. Goddik, A.S. Hansen, J. Josephsen, W.K. Nip, P.S. Stanfield and F. Toldre (eds.). Handbook of Food and Beverage Fermentation Technology. Marcel Dekker Inc., New York, NY, USA.

Park, K.Y., Cho, E.j., Rhee, S.H., Jung, K.O., Yi, S.J. and Jhun, B.H. (2003). Kimchi and an active component, beta-sitosterol, reduce oncogenic H-Ras(v12)-induced DNA synthesis. Journal of Medicinal Food 6: 151–156.

Rekha, C.R. and Vijayalakshmi, G. (2010). Bioconversion of isoflavone glycosides to aglycones, mineral bioavailability and vitamin B complex in fermented soymilk by probiotic bacteria and yeast. Journal of Applied Microbiology 109: 1198–1208.

Ryu, B.M., Ryu, S.H., Lee, Y.S., Jeon, Y.S. and Moon, G.S. (2004). Effect of different kimchi diets on oxidation and photooxidation in liver and skin of hairless mice. Journal of the Korean Society of Food Science and Nutrition 33: 291–298.

Ryu, J.Y., Lee, J.H. and Rhee, H.S. (1984). Changes of organic acids and volatile flavor compounds in kimchis fermented with different ingredients. Korean Journal of Food Science and Technology 16: 169–174.

Ryu, S.H., Jeon, Y.S., Kwon, M.J., Moon, J.W., Lee, Y.S. and Moon, G.S. (1997). Effect of kimchi extracts to reactive oxygen species in skin cell cytotoxicity. Journal of the Korean Society of Food Science and Nutrition 26: 814–821.

Seo, B.J., Rather, I.A., Kumar, V.J., Choi, U.H., Moon, M.R. Lim, J.H. and Park, Y.H. (2012). Evaluation of *Leuconostoc mesenteroides* YML003 as a probiotic against low-pathogenic avian influenza (H9N2) virus in chickens. Journal of Applied Microbiology 113: 163–171.

Seo, G.H., Song, B.S., An, D.S. and Chung, S.K. (2006). Physical properties of Korean earthenware (Onggi) as food container. Journal of Korea Society of Packaging Science and Technology 12: 87–90.

Shim, S.M., Kim, J.Y., Lee, S.M., Park, J.B., Oh, S.K. and Kim, Y.S. (2012). Profiling of fermentative metabolites in kimchi: Volatile and non-volatile organic acids. Journal of the Korean Society for Applied Biological Chemistry 55: 463–469.

Shin, K., Chae, O., Park, I., Hong, S. and Choe, T. (1998). Antitumor effects of mice fed with cell lysate of *Lactobacillus plantarum* isolated from kimchi. Korean Journal of Biotechnology and Bioengineering 13: 357–363.

Vizoso, P.M.G., Franz, C.M., Schillinger, U. and Holzapfel, W.H. (2006). *Lactobacillus* spp. with *in vitro* probiotic properties from human faeces and traditional fermented products. International Journal of Food Microbiology 109: 205–214.

Whang, K.C., Chung, Y.S. and Kim, H.S. (1960). Microbiological studies on kimchi. II. Isolation and identification of aerobic bacteria. Bulletin of Science Research Insight 5: 51.

Won, T.J., Kim, B., Song, D.S., Lim, Y.T., Oh, E.S., Lee, D.I., Park, E.S., Min, H., Park, S.T. and Hwang, K.W. (2011). Modulation of Th1/Th2 balance by *Lactobacillus* strains isolated from kimchi *via* stimulation of macrophage cell line J774A.1 *in vitro*. Journal of Food Science 76: H5–H61.

Yoon, J.Y., Jung, K.O., Kil, J.H. and Park, K.Y. (2005). Anti-obesity effect of major Korean spices (red pepper powder, garlic and ginger) in rats fed high-fat diet. Journal of Food Science and Nutrition 10: 58–63.

Yoon, J.Y., Jung, K.O., Kim, S.H. and Park, K.Y. (2004). Anti-obesity effect of baek-kimchi (whitish baechu kimchi) in rats fed high fat diet. Journal of Food Science 9: 259–264.

Yoon, S.J. (2004). Korean preservative fermented food. Shinkwang Publishing Co., Seoul, Korea, pp. 87–127.

Yun, J.Y., Jeong, J.K., Moon, S.H. and Park, K.Y. (2014). Effects of brined Baechu cabbage and seasoning on fermentation of kimchi. Journal of Korean Society of Food Science and Nutrition 43: 1081–1087.

Wisselink, H.W., Weusthuis, R.A., Eggink, G., Hugenholtz, J. and Grobben, G.J. (2002). Mannitol production by lactic acid bacteria: a review. International Dairy Journal 2: 151–161.

17

Amasi and *Mageu*

Expedition from Ethnic Southern African Foods to Cosmopolitan Markets

Eugenie Kayitesi,[1,*] *Sunil K. Behera,*[2] *Sandeep K. Panda,*[1,*]
Dlamini Bheki[1] *and A.F. Mulaba-Bafubiandi*[2]

1. Introduction

Fermentation is understood as the culture of microorganisms on a growth medium with a goal of producing a specific biochemical product. Fermented foods are advantageous as fermentation helps in preservation, flavour improvement, toxicity reduction and enhancement of the nutritional quality of the final product (Panda et al. 2014a, 2014b, Rolle and Satin 2002). The science of the involvement of microorganisms in fermentation of foods and beverages came to limelight with the discovery of live yeast cells in the production of wine from grape juice in the year 1857 by Louis Pasteur (Alba-Lois and Segal-Kischinevzky 2010). Traditional foods have specific nutritional and organoleptic properties depending upon the region of origin and the raw materials used (Nout 2003). However, out of the several traditional fermented food products, alcoholic beverages and lactic acid fermented food and beverages have been accepted worldwide

---

[1] Department of Biotechnology and Food Technology, Faculty of Science, University of Johannesburg, P.O. Box 17011, Doornfontein Campus, Johannesburg, South Africa.
[2] Department of Metallurgy, Faculty of Engineering and the Built Environment, University of Johannesburg, P.O. Box 17911, Doornfontein Campus, 2028, Johannesburg, South Africa.
* Corresponding authors: eugeniek@uj.ac.za; sandeeppanda2212@gmail.com

in the modern times. The African continent has a wide range of traditional indigenous fermented foods embedded within the socio-cultural linkage. Several studies have been conducted to derive the mechanism of the development of indigenous fermented foods of Africa. Studies reveal that except for alcoholic beverages, the rest of the fermented foods of Africa contain probiotic bacterial strains—live microorganisms which when administered in adequate amounts confer a health benefit on the host (FAO/WHO 2002). Mostly the probiotic fermented products of Africa are non-alcoholic fermented cereals, fermented vegetables, fermented milk. Even within the continent, the indigenous products are prepared in different technologies and named differently.

In South Africa and Zimbabwe, the traditional fermented milk is known as *amasi*, in Ethiopia it is named as *ergo* and *ititu*. Similarly, fermented maize products are *ogi* in Nigeria, *uji* in Kenya, *mageu* in South Africa and *aflata* of Ghana. Although several articles have described the indigenous fermented food diversity of Africa but there is rarely any study describing the successful commercialization of the ethnic fermented foods of Africa. The current research work focuses on *amasi* and *mageu*. *Mageu* is known in different spellings as per the pronunciations in different regions such as *magou, amahewu* or *aramrewu* (Fritz 2005). This chapter covers the origin of *amasi* and *mageu*, along with biochemical and, microbiological features and products' successful commercial transformation.

2. History, Origin and Traditional Preparation of *Amasi* and *Mageu*

Amasi and *mageu* are indigenous fermented non-alcoholic beverages originating from Southern Africa. Both are considered as weaning foods for infants in the African continent and are believed to protect against food borne diseases (Gadaga et al. 2004). *Amasi* is a traditional fermented milk associated with the *Zulu* community (Franz et al. 2014). Indigenously, it is prepared by fermenting the unpasteurised milk in a calabash (bottle gourd, *Lagenaria siceraria*) container. Prior to the inoculation of milk in calabash, the inner part of the calabash is smoked to prevent the contamination due to undesired fungi. After inoculation, the calabash is sealed and is kept indoors in a warm place or near a source of mild heat. Earthen and metal pots are also used instead of the calabash container. Fermentation is facilitated by the natural bacterial microflora of the milk inside the close container. The fermentation process is allowed for 3–5 days. With due course of fermentation, the milk medium is separated into two layers, a thin liquid layer (*umlaza*) and a thick coagulant (*amasi*). The liquid layer is discarded and the thick coagulant or *amasi* is harvested (Osvik et al. 2013). Fresh milk is rarely consumed by Zulus, rather they believe that *amasi* consumption makes a person healthy and strong, except for some taboos (such as menstruation or physically affected person) (Russel 2007). The preparation steps of *amasi* in traditional process are presented in Fig. 1.

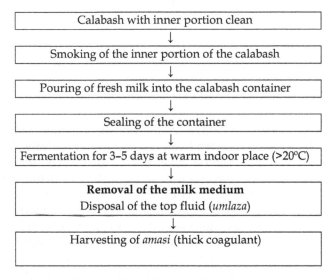

Figure 1. Flow chart for making *amasi* in indigenous process.

Similarly, *mageu* is a popular South African indigenous refreshing sour beverage and is prepared by the fermentation of maize flour (Byaruhanga et al. 1999). Maize flour is boiled with water and stirred to make porridge (Hesseltine 1979). Generally one part of maize flour is boiled with nine parts of water and care is taken to avoid formation of lumps. Then, the slurry is cooled down to around 25°C and a small amount of wheat flour (5% of the maize flour used) is added to it as a source of natural microflora. Further, the inoculated medium is allowed to ferment up to 3 days for preparation of *mageu* (Nyanzi et al. 2010). The traditional preparation technology of *mageu* is demonstrated in Fig. 2.

3. Challenges in Indigenous Processing of *Amasi* and *Mageu*

Preparation of *amasi* and *mageu* is practiced in common households of South Africa and Zimbabwe. There is no uniform statute regarding the ingredients, physicochemical conditions and microbial inoculum concentration for the indigenous preparation of *amasi* and *mageu*. Hence, the biochemical as well as sensory quality of *amasi* and *mageu* is dissimilar when observed in different places. Several studies have been conducted to examine the quality of the final product (*amasi* and *mageu*) as in the indigenous fermentation

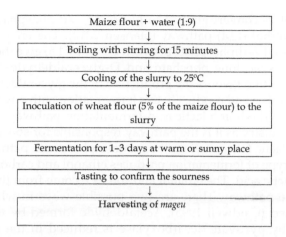

Figure 2. Flow chart for making *mageu* in indigenous processing.

process, no specific microorganism is harnessed. Hence, there is always a risk of microbial contamination or even food poisoning. Coliforms and *E. coli* population density in food materials are regarded as an indicator of its quality. A study was conducted by Gran et al. (2002) on the unwanted microbiological interference in naturally fermented sour milk (*amasi*) in Zimbabwe. It was observed that the mean number of coliforms and *E. coli* in naturally soured milk, collected from different small holder diary processing units were 9.3×10^4 and 5.7×10^4 CFU/ml, respectively. Coliforms count of more than 10^4 (per ml) and *E. coli* count of more than 10^3 (per ml) in food products are regarded as grade C products (Zimbabwe Dairy Regulations (RGN 886) 1977, Norwegian Food Control Authority (SNT) 1994). Furthermore, it is reported that lethal food borne pathogens such as *Bacillus cereus, E. coli, Salmonella* sp., *Staphylococcus aureus, Vibrio cholerae, Aeromonas, Klebsiella, Campylobacter* and *Shigella* sp. are common contaminants of traditional African fermented foods (Gadaga et al. 2004).

4. Science of The Fermentation of *Amasi* and *Mageu*

The fermentation of *amasi* and *mageu* was earlier an art, and later with the knowledge of microbiology and biochemistry, the process of fermentation became well understood.

4.1 Biochemistry of Lactic Acid Fermentation of Amasi and Mageu

Lactic acid fermentations can be divided into two broad categories distinguishable by the products formed from glucose (Stanier et al. 1979). These are referred to as homo-fermentation and hetero-fermentation.

Homofermenters convert glucose to glucose-1, 6-diphosphate using the Embden Meyerhof (EM) pathway (Hansen 2012). The enzyme aldolase cleaves fructose-l, 6-diphosphate between C3 and C4 to give the phosphate esters dihydroxyacetone phoshate and D-glyceraldehyde-3-phosphate (Fig. 3). The reaction favours the production of the glyceraldehyde isomer at equilibrium. The end product in this fermentation pathway is lactic acid (Hansen 2012). The homo lactic acid fermentation pathway is important in the dairy industry as it is the pathway responsible for souring of milk.

In heterolactic fermentation, the pentose phosphate pathway is used (Fig. 3). This type of fermentation produces ethanol and carbon dioxide in addition to lactic acid. The ethanol and the CO_2 come from the glycolytic portion of the pathway. There are two possible ways in which ethanol may be formed (Caldwell 1995). Acetaldehyde formed by the cleavage of pyruvate by pyruvate decarboxylase is reduced in the presence of alcohol dehydrogenase to form ethanol. Ethanol can also be formed by a combination of acetyl-CoA reduction to acetaldehyde followed by reduction of acetaldehyde by ethanol dehydrogenase. Lactic acid is formed by direct pyruvate reduction with lactate dehydrogenase. Formic acid and acetyl-CoA are produced by the action of pyruvate-ferredoxin on pyruvate. Acetyl-CoA is converted to free acetic acid.

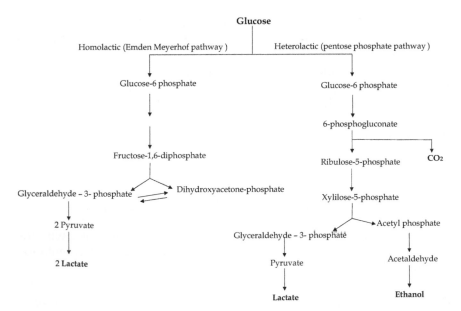

Figure 3. Generalised scheme for the fermentation of glucose by lactic acid bacteria (Adopted from Stanier et al. 1979).

4.2 Microorganisms Involved in Lactic Acid Fermentation of Amasi and Mageu

The name Lactic acid bacteria (LAB) is derived from the fact that ATP is synthesised through fermentations of carbohydrates, which yield lactic acid as a major and sometimes the only end-product (Stanier et al. 1979). Taxonomically these bacteria belong to the heterogeneous group of LAB. They are gram-positive; catalase-negative; acid tolerant; devoid of cytochromes; aerotolerant; non-sporulating; and they are strictly fermentative rods or cocci which produce lactic acid as the major product from the energy-yielding fermentation of sugars (Stiles and Holzapfel 1997, Temmerman et al. 2004, Wessel et al. 2004). The LAB genera generally associated with the fermentation of a variety of foods are *Lactobacillus, Lactococcus, Leuconostoc, Streptococcus, Enterococcus, Pediococcus, Oenococcus, Teragenococcus, Cronobacterium* and *Weissella* (Adams and Martau 1995, Stiles and Holzapfel 1997, Temmerman et al. 2004, Wessel et al. 2004). Lactic acid bacteria commonly used for food fermentation are presented in Table 1.

Table 1. Summary of lactic acid bacteria commonly used for food fermentation.

Bacteria (genus)	Characteristics
Lactobacillus	Gram-positive, rod-shaped, strictly fermentative, aciduric or acidophilic, non-endospore forming bacteria which grow well in anaerobic environments.
Streptococcus	Gram-positive, catalase-negative, anaerobic, aerotolerant, coccus-shaped cells grouped in linear chains.
Leuconostoc	Gram positive, catalase negative cocci that are heterofermentative, facultative anaerobes and ferment glucose to produce D(–)-lactate, ethanol and gas.
Carnobacteria	Gram positive, catalase negative, straight, slender rods, occur singly or in pairs and sometimes in short chains, non-sporing chemoorganotrophs that are heterofermentative and produce mainly L(+)-lactate from glucose.
Enterococci	Gram positive, cells are spherical or ovoid and occur in pairs or short chains, sometimes mobile by means of scanty flagella. Lack obvious capsules, are facultative anaerobic and are chemoorganotrophs with fermentative metabolism. Ferment a wide range of carbohydrates, including lactose, to produce of mainly L(+) lactic acid.
Lactococci	Gram positive, catalase negative, non-motile spherical or ovoid cells that occur singly, in pairs or in chains. Facultative anaerobes which can grow at 10°C but not at 45°C (optimum temperature 30°C). They are chemoorganotrophs with fermentative metabolism.
Vagococci	Gram positive, non-sporing, spheres, ovals or short rods which occur singly, in pairs or in short chains. They are catalase negative and chemoorganotrophic with a fermentative metabolism. They produce acid but no gas from a number of carbohydrates. Glucose fermentation yields mainly L(+)-lactate.

Genus *Lactobacillus* consists of most of the acetyl-CoA species (Giraffa et al. 2010). This heterogeneous group is Gram-positive, rod-shaped, strictly fermentative, aciduric or acidophilic, non-endospore forming bacteria which grow well in anaerobic environments, although they are aerotolerant (Stiles and Holzapfel 1997, Bernardeau et al. 2008, Giraffa et al. 2010). *Lactobacilli* have been isolated from a variety of habitats including the intestinal tract of mammals, plant material, raw milk and sewerage (Stiles and Holzapfel 1997, Giraffa et al. 2010). The genus *Lactobacillus* can be sub-divided into three groups based on sugar fermentation, namely facultative hetero-fermentative (*Group I*), obligated hetero-fermentative (*Group II*) and obligated homo-fermentative (*Group III*) (Stiles and Holzapfel 1997, Bernardeau et al. 2008). *Lactobacilli* from *Group I* ferment hexoses to lactic acid and pentoses to lactic acid and acetic acid, and gas is produced from gluconate but not from glucose. *Group II* bacteria produce carbon dioxide, lactic acid, acetic acid and/or ethanol from hexoses, and produce gas from glucose. *Lactobacilli* from *Group III* do not ferment gluconate or pentoses, but ferment glucose to lactic acid. *Lactobacillus* sp. from all three of these groups can take part in food fermentation.

LAB that belongs to the genus *Streptococcus* is Gram-positive, catalase-negative, anaerobic, aerotolerant, coccus-shaped cells grouped in linear chains (Stiles and Holzapfel 1997, Delorme 2008). In this genus only one species, *Streptococcus thermophilus* (synonym: *Sc. salivarius* subsp. *thermophilus*) is "generally recognised as safe" (GRAS) and found in dairy environments (Delorme 2008, De Vuyst and Tsakalidou 2008). Thus *Streptococcus thermophilus* is the only *Streptococcus* spp. that is used as a starter along with one or more LAB strains from the genus *Lactobacillus*.

LAB present in dairy fermentations can generally be classified in terms of optimum growth temperatures, namely mesophilic and thermophilic LAB. Mesophilic LAB grows optimally between 20°–30°C and thermophilic LAB between 30°–45°C. Therefore, mesophilic LAB is often isolated from traditional fermented dairy products from the colder Northern and Western European countries and thermophilic LAB from traditional fermented dairy products from hotter sub-tropical regions (Wouters et al. 2002). Mixed strain starter cultures are used in various dairy fermentations, including the production of yoghurt and fermented milks (Stiles and Holzapfel 1997, Delorme 2008). *Lactobacillus delbrueckii* subsp. *bilgaricus* and *Sc. thermophilus* have been isolated form commercially produced fermented milk (Chammas et al. 2006, Velez et al. 2007, Giraffa et al. 2010). *Lactobacilli* including *Lb. plantarum* and *Lb. delbrueckii* subsp. *lactis* have been isolated from traditionally prepared *amasi* from South Africa and Zimbabwe. A wide variety of other *lactobacilli* were also isolated from Zimbabwean *amasi* including *Lb. helveticus*, *Lb. casei* subsp. *casei* and *Lb. casei* subsp. *pseudoplantarum* (Gadaga et al. 1999, 2000, McMastera et al. 2005, Todorov et al. 2007). The main microorganisms in native *mageu* are *Leuconostoc mesenteroides* and *Lactobacillus brevis*. Pure cultures of *Lactobacillus acidophilus*, *L. bulgaricus*, and *Streptococcus lactis* have been adapted to the maize meal substrate at 51°C production of *mageu*.

5. Industrial Production of *Amasi* and *Mageu*

5.1 Amasi

During the commercial production of *amasi*, skim milk powder (3.0%) and gelatine (0.5%) are added to raw milk prior to pasteurization to stabilize the product (Fig. 4). The skim milk powder enhances the nutritional and functional properties of *amasi* and prevents syneresis, while gelatine prevents the latter and also gives a smooth uniformly textured product (Karam et al. 2013, Modler et al. 1983). The milk is then pasteurized at 72°C for 15 sec to destroy pathogenic and spoilage microorganism followed by rapid cooling to about 30°C to prevent the outgrowth of spores. LAB starter cultures such as *Lactococcus lactis* subsp. *lactis* and *Lactococcus lactis* subsp. *cremonis* are added to give an initial inoculum level of 10⁶ cfu/ml. The product is then incubated at 30°C for 24 h and subsequently stored at low temperatures (7°C).

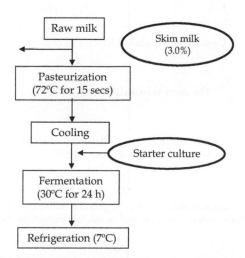

Figure 4. Summary for commercial *amasi* production process.

5.2 Mageu

In South Africa, the making of *mageu* has been industrialised to meet the demand of the increasing urban market. A flow diagram of a modem processing plant is shown in Fig. 5. The porridge is prepared by first mixing 8% of maize meal with water at ambient temperature and then pumping the mixture into stainless steel pots where it is heated at 85 to 90°C for 20 to 30 min. Heating not only gelatinises the starch, makes it more susceptible to enzymatic hydrolysis, it also kills the unwanted micro-organisms in the maize meal that may compete with the starter culture and spoil the product. The porridge is cooled to 50°C before it is transferred to the bioreactors where 1 to 2% sugar and 0.1 to 0.2% wheat flour are added followed by 7 to 12% of the thermophilic *Lactobacillus* starter culture (an adapted pure culture of *Lactobacillus delbrueckii*) (Mc Carrol 2015). The sugar provides a readily hydrolysable source of energy for the starter culture. Starch is broken down into a mixture of maltose and dextrins by bacterial amylases and then into glucose which means the energy supplied to the bacteria is slow. During the next 17 to 29 h, the temperature gradually falls to about 30°C and the pH reaches 3.4 to 3.8. When the desired pH has been attained, about 2% sugar (based on the total volume) is added as a sweetener. Initial starter culture preparation occurs in the pre-fermenter where porridge at 50°C is inoculated with 10% of an active culture. Within 18 to 24 h at 30°C, the pH gradually drops and reaches a range of 3 to 3.4.

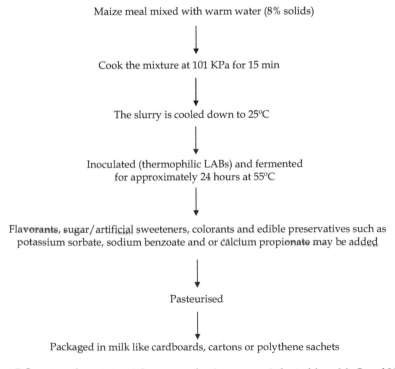

Figure 5. Summary for commercial *mageu* production process (adopted from Mc Carrol 2015).

Lactobacillus delbrueckii is the microorganism of choice because its optimum temperature is high and at this temperature the growth of undesirable microorganisms is suppressed. This starter culture is also known to produce large quantities of lactic acid with few by-products, thus giving a relatively pure lactic flavour. It remains active until a low pH has been attained and it produces lactic acid by the fermentation of glucose, maltose, sucrose, fructose, galactose and dextrins. Fermentation is improved by the addition of bran and high quality proteins also improve the activity of lactic-acid bacteria. To maintain the high rate of acid production, the addition of buffering salts such as $CaHPO_4$ is preferred. The degree of sourness required depends on individual taste but is usually on average 0.4–0.5% titratable acidity, calculated as lactic acid, at which the average pH is 3.5 (Schweigart and Fellingham 1963). The main product of homofermentative *Lactobacillus* cultures is lactic acid which has very little flavour and aroma but has a distinct, refreshing sour taste. In traditionally processed *mageu* where heterofermentation occurs, in addition to lactic acid, acetic acid and butyric acids are also formed which contribute to the flavour and aroma of the product (Schweigart et al. 1960). Maize meal has very little buffering capacity and the pH quickly drops to less than 3.5. The addition of buffering salts such as $CaHPO_4$ and protein rich supplements means that the buffering capacity of *mageu* will be improved and the microorganisms can produce more acid (Schweigart and Fellingham 1963). Sugar is added at a rate of between 1 and 2% in the modem procedure for making *mageu*. The sugar is readily fermented and provides energy for the starter culture. In the traditionally processed product, fermentation sugars are derived from the wheat flour enzymes hydrolysing the starch in the maize porridge (Schweigart et al. 1960). Over the years, there has been a decline in consumption of *mageu* in South Africa possibly due to a change in dietary patterns from that of a traditional diet to a westernised diet (Mc Carrol 2015). A study by Mc Carrol (2015) indicated that *mageu* could be used as a novel ingredient that contributes to an improvement in quality aspects of gluten-free bread. Mc Carrol (2015) observed inhibitory actions of lactic acid bacteria present in the *mageu* against spoilage bacteria and mould growth thus improving gluten-free bread shelf life, higher loaf volume and decreased crumb firmness when compared to the control without *mageu*.

6. Current Market Scenario

The annual milk production in South Africa was approximately 2700 million litres in 2013, which indicates some increase (7%) in milk production when compared to the year 2009 (Milk SA 2014). The South Africa dairy market is divided into 60% liquid and 40% concentrated products. Pasteurized (51%) and UHT (29%) milk are the major liquid products, while *amasi* and buttermilk constantly constituted 5%, in total, over the past five years (MPO, LACTO Data 2012, 2013, 2014). The remainder is made up of yoghurt (13%) and flavoured milk (2%). The demand for *amasi* has been fluctuating over the past few years (Table 2). Although the change in price somehow influences the demand for the product, as observed in the year 2012/2013, other factors such as other dairy products and non-dairy products are competing against the demand. The non-dairy foods, which mainly include luxurious foods such as pizzas and other fast foods, are more preferred by the major target market of *amasi* which constitute of medium-to-low income consumers. Awareness campaigns spearheaded by the government and the dairy industry on the health benefits of dairy and fermented milk products are expected to boost the demand for *amasi* in future.

Non-alcoholic beverages in South Africa constitute about 40% of the total beverage industry by volume, of which dairy products make up only 13% with alcoholic beverages dominating (47%) the sector. In the retail market in South Africa *mageu* is sold in cartons in liquid form in unflavoured and flavoured variants. Popular flavours include cream, banana, strawberry and ginger. Smooth *mageu* variants produced with finer maize flour were introduced to the market in 2009 (Dairy Mail Africa 2009). *Mageu* is also available in a spray dried or roller dried form and is reconstituted by adding a small amount of water to mix thoroughly before adding the balance of the required water. *Mageu* production in South African saw a major drop (≈15%) in volumes in 2009 compared to 2008, possibly due to economic recession (Fig. 6). Thereafter, production volumes recovered steadily to about 85.8 million litres in 2012. The recovery in sales after the recession was, in part, attributed to the small increase in price during 2010. The value of the product substantially increased from approximately R550 million in 2008 to R760 million in 2012. Although the product has been traditionally targeted at a specific demographic, current innovations such as the introduction of smooth and flavoured variants seems to have increased the customer market base. In addition, the availability of larger package sizes in the

Table 2. Year-on-year change in demand and prices of *amasi*.

Year	Change in demand (%)	Change in retail Price (%)
Jan–Dec 2011/Jan–Dec 2012	1.7	7.27
Jan–Dec 2012/Jan–Dec 2013	6.9	3.4
Jan–Dec 2013/Jan–Dec 2014	5.3	13.1

Source: MPO (2012, 2013, 2014) - Lacto Data

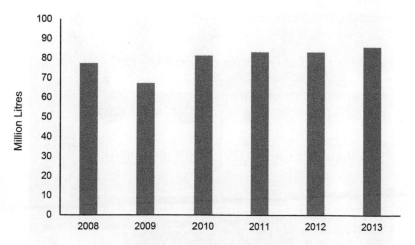

Figure 6. *Mageu* production volumes.

market compared to the common small packs (less than 500 mL) indicates that consumers are willing to make bulk purchases.

7. Conclusion and Critical Opinion

This chapter has described the origin and indigenous preparation of *amasi* and *mageu*. Microbiology, biochemical as well as commercialization of both the fermented products have been elaborated. This chapter has portrayed the successful transformation of *amasi* and *mageu* from indigenous South African fermented beverages to cosmopolitan supermarket products (Figs. 7, 8). But the commercialization of these two products is confined to only the African continent. Although the *amasi* and *mageu* are sold under different brand names across the continent of Africa by leading dairy processing brands but due to various reasons these two products are unable to penetrate the other continents. Indigenous fermented foods are considered as foods related to the culture and tradition. Hence, such indigenous foods represent the culture and tradition of a particular region, country or continent. Foods like *amasi* and *mageu* are representatives of South Africa as well as the whole African continent, so initiatives should be taken by international cultural forums as well as leading food marketers to promote such indigenous food products outside their native continents. Industrialization and promotion of such indigenous food products can be helpful in socio-cultural exchange between different countries and continents.

Figure 7. *Amasi* in the shelf of supermarkets of Johannesburg, South Africa.

Figure 8. *Mageu* of different flavours displayed in the shelf of supermarkets of Johannesburg, South Africa.

Keywords: *Amasi, Mageu,* Probiotics, Lactic acid fermentation, Indigenous processing, Industrial production

References

Adams, M.R. and Marteau, P. (1995). Letter to the Editor on the safety of lactic acid bacteria from food. International Journal of Food Microbiology 27: 263–264.

Alba-Lois, L. and Segal-Kischinevzky, C. (2010). Beer & Wine Makers. Nature Education 3(9): 17.

Bernardeau, M., Vernoux, J.P., Henri-Dubernet, S. and Gueguen, M. (2008). Safety assessment of dairy microorganisms: the *Lactobacillus* genus. International Journal of Food Microbiology 126: 278–285.

Byaruhanga, Y.B., Bester, B.H. and Watson, T.G. (1999). Growth and survival of *Bacillus cereus* in *mageu*, a sour maize beverage. World Journal of Microbiology and Biotechnology 15: 329–333.

Caldwell, D.R. (1995). Microbial physiology and metabolism. WCB Publishers, Dubuque.

Chammas, G.I., Saliba, R., Corrieu, G. and Beal, C. (2006). Characterisation of lactic acid bacteria isolated from fermented milk "laban". International Journal of Food Microbiology 110: 52–61.

Dairy Mail Africa. (2009). Get Smooth with Mageu Number 1. Dairy Industry in Africa 4(2): 26–27.

Delorme, C. (2008). Safety assessment of dairy microorganisms: *Streptococcus thermophilus.* International Journal of Food Microbiology 126: 274–277.

De Vuyst, L. and Tsakalidou, E. (2008). *Streptococcus macedonicus,* a multi-functional and promising species for dairy fermentations. International Dairy Journal 18: 476–485.

FAO/WHO. (2002). Guidelines for the evaluation of probiotics in food. Food and Agriculture Organization of the United Nations and World Health Organization Working Group Report. Available at ftp://ftp.fao.org/docrep/fao/009/a0512e/a0512e00.pdf.

Franz, C.M.A.P., Huch, M., Mathara, J.M., Abriouel, H., Benomar, N., Reid, G., Galvez, A. and Holzapfel, W.H. (2014). African fermented foods and probiotics. International Journal of Food Microbiology 190: 84–96.

Fritz, K. (2005). Incorrect information on mageu. South African Journal of Clinical Nutrition 18(3): 265.

Gadaga, T.H., Mutukumira, A.N. and Narvhus, J.A. (2000). Enumeration and identification of yeasts isolated from Zimbabwean traditional fermented milk. International Dairy Journal 10: 459–466.

Gadaga, T.H., Mutukumira, A.N., Narvhus, J.A. and Feresu, S.B. (1999). A review of traditional fermented foods and beverages of Zimbabwe. International Journal of Food Microbiology 53: 1–11.

Gadaga, T.H., Nyanga, L.K. and Mutukumira, A.N. (2004). The occurrence, growth and control of pathogens in African fermented foods. African Journal of Food Agriculture Nutrition and Development 4(1), http://bioline.org.br/request?nd04009.

Giraffa, G., Chanishvili, N. and Widyastuti, Y. (2010). Importance of *lactobacilli* in food and feed biotechnology. Research in Microbiology 161: 480–487.

Gran, H.M., Mutukumira, A.N., Wetlesen, A. and Narvhus, J.A. (2002). Smallholder dairy processing in Zimbabwe: the production of fermented milk products with particular emphasis on sanitation and microbiological quality. Food Control 13: 161–168.

Hansen, A.S. (2012). Sourdough bread. pp. 498–499. *In*: Y.H. Hui (ed.). Handbook of Plant-Based Fermented Food and Beverage Technology. CRC Press, Boca Raton.

Hesseltine, C.W. (1979). Some important fermented foods of Mid-Asia, the Middle East and Africa. Journal of the American Oil Chemists Society 56: 367–374.

Karam, M.C., Gaiani, C., Hosri, C., Burgai, J. and Scher, J. (2013). Effect of dairy powders fortification on yogurt textural and sensorial properties: a review. Journal of Dairy Research 80: 400–409.

Mc Carrol, L.A. (2015). Evaluation of Mageu-based gluten-free bread in South Africa. M.Tech (Food Technology) Dissertation, University of Johannesburg. South Africa.

McMaster, L.D., Kokott, S.A., Reid, S.J. and Abratt, V.R. (2005). Use of traditional African fermented beverages as delivery vehicles for *Bifidobacterium lactis* DSM 10140. International Journal of Food Microbiology 102: 231–237.

Milk Producer's Organization (MPO). (2015). LACTO DATA: Statistics 18(1): 1–31.

Milk Producer's Organization (MPO). (2014). LACTO DATA: Statistics 17(2): 1–31.

Milk Producer's Organization (MPO). (2013). LACTO DATA: Statistics 16(2) 1–31.

Milk Producer's Organization (MPO). (2012). LACTO DATA: Statistics 15(1): 1–16.

Milk South Africa. (2014). Dairy industry review. Available at: http://www.milksa.co.za/reports/dairy-industry-review.

Modler, H.W., Larmond, M.E., Lin, C.S., Froehlich, D. and Emmons, D.B. (1983). Physical and sensory properties of yoghurt stabilized with milk proteins. Journal of Dairy Research 66(3): 422–429.

Norwegian Food Control Authority (SNT). (1994). Microbiological guidelines for food. (Mikrobiologiskeretningslinjer for næringsmiddel) (2nd ed.). SNT, Oslo, Norway.

Nout, M.J.R. (2003). Traditional fermented product from Africa, Latin America and Asia. pp. 451–500. *In:* T. Boekhout and V. Robert (eds.). Yeasts in Food, Beneficial and Detrimental Aspects. Behr's Verlag, Hamburg.

Nyanzi, R., Jooste, P.J., Abu, J.O. and Beukes, E.M. (2010). Consumer acceptability of a symbiotic version of the maize beverage *mageu*. Development Southern Africa 27(3): 447–463.

Osvik, R.D., Sperstad, S., Hareide, E., Godfroid, J., Zhou, Z., Ren, P., Geoghegan, C., Holzapfel, W. and Ringo, E. (2013). Bacterial diversity of a Masi, a South African fermented milk product, determined by clone library and denaturing gradient gel electrophoresis analysis. African Journal of Microbiology Research 7(32): 4146–4158.

Panda, S.K., Sahu, U.C., Behera, S.K. and Ray, R.C. (2014b). Fermentation of sapota (*Achras sapota* Linn.) fruits to functional wine. Nutrafoods doi:10. 1007/s13749-014-0034-1.

Panda, S.K., Sahu, U.C., Behera, S.K. and Ray, R.C. (2014a). Fermentation of bael (*Aegle marmelos* L.) fruits into wine with antioxidants. Food Bioscience 5: 34–41.

Rolle, R. and Satin, M. (2002). Basic requirements for the transfer of fermentation technologies to developing countries. International Journal of Food Microbiology 75: 181–187.

Russell, M., Armstrong, T. and Dawson, S. (2007). Diet of the Zulu people. Thinkquest. Archived from the original on 2 January 2007. Retrieved 2007–01–18.

Schweigart, F. and Fellingham, S.A. (1963). At study of fermentation in the production of Mahewu, an indigenous sour maize beverage of Southern Africa. Milchwissenschaft 18: 241–246.

Schweigart, F., Van Bergen, W.E.L., Wiechers, S.G. and DeWit, J.P. (1960). The Production of Magewu. CSIR Research Report No. 167. National Nutrition Research Institute Bulletin No. 3 CSIR, Pretoria.

Stanier, R.Y., Adelberg, E.A. and Ingraham, J.L. (1979). Gram-Positive bacteria: The Actinomycete Line, p. 683 *In:* General Microbiology. 4th ed., The Macmillan Press Ltd., London, UK.

Stiles, M.E. and Holzapfel, W.H. (1997). Lactic acid bacteria of foods and their current taxonomy. International Journal of Food Microbiology 36: 1–29.

Temmerman, R., Huys, G. and Swings, J. (2004). Identification of lactic acid bacteria: culture-dependent and culture-independent methods. Trends in Food Science and Technology 15: 348–359.

Todorov, S.D., Nyati, H., Meincken, M. and Dicks, L.M.T. (2007). Partial characterization of bacteriocin AMA-K, produced by *Lactobacillus plantarum* AMA-K isolated from naturally fermented milk from Zimbabwe. Food Control 18: 656–664.

Velez, M.P., Hermans, K., Verhoeven, T.L.A., Lebeer, S.E., Vanderleyden, J. and De Keersmaecker, S.C.J. (2007). Identification and characterization of starter lactic acid bacteria and probiotics from Columbian dairy products. Journal of Applied Microbiology 103: 666–674.

Wessels, S., Axelsson, L., Hansen, E.B., De Vuyst, L., Laulund, S., Lahteenmaki, L., Lindgren, S., Mollet, B., Salminen, S. and von Wright, A. (2004). The lactic acid bacteria, the food chain, and their regulation. Trends in Food Science and Technology 15: 498–505.

Wouters, J.T.M., Ayad, E.H.E., Hugenholtz, J. and Smit, G. (2002). Microbes from raw milk for fermented dairy products. International Dairy Journal 12: 91–109.

Zimbabwe Dairy Regulations (RGN 886). (1977). Government Printer, Harare.

18

Kefir and Koumiss

Origin, Health Benefits and Current Status of Knowledge

Sunil K. Behera,[1,]* Sandeep K. Panda,[2] Eugenie Kayitesi[2] and Antoine F. Mulaba-Bafubiandi[1]

1. Introduction

Milk constitutes an important ingredient of healthy balanced diet of our daily life. It is an important source of vitamins, minerals, and proteins for human nutritional requirement, hence regarded as a complete food. Throughout the world, milk and milk products are valued as natural and traditional food. However, milk is extremely perishable and many means have been adapted to preserve it. The earliest one which has been used for many thousands of years is the process of fermentation (Parveen and Hafiz 2003). Milk can be fermented by inoculating fresh milk with the appropriate bacteria and incubating it at a temperature that favors their growth. As the bacteria grow, they convert milk sugar (lactose) to lactic acid through the process of fermentation. The lactic acid generated during milk fermentation decreases the pH of milk and as a result it prevents the growth of putrefactive and/or pathogenic microorganisms that do not survive in acidic environment. Since the time immemorial the process of

[1] Department of Metallurgy, Faculty of Engineering and the Built Environment, University of Johannesburg, P.O. Box 17911, Doornfontein Campus, 2028, Johannesburg, South Africa.
[2] Department of Biotechnology and Food Technology, Faculty of Science, University of Johannesburg, Doornfontein Campus, Johannesburg, South Africa, P.O. Box 17011.
* Corresponding author: skbehera2020@gmail.com

fermentation has been adapted as a tool for food preservation. With due course of time, it has been noticed that many fermented foods have better nutritional and functional values when compared to their unfermented counterparts (Hasan et al. 2014). Hence the fermentation processes have become the most popular food processing techniques for preservation of foods along and for the addition of better nutritional value (Panda et al. 2014a, 2014b). Worldwide, the known fermented milk products are yogurt, kefir, koumiss, sour cream, cheeses, etc.

The primary function of fermenting milk was, originally, to extend its shelf life. Further the fermentation process brought numerous changes to the nutritional property of milk, such as an improved taste and flavor and enhanced digestibility of the milk. Historically, the fermentation of milk can be traced back to around 10,000 B.C. (Dhewa et al. 2015). Fermentation takes place through the natural microflora present in milk. With the advent of scientific methods the different classes of microorganisms present in the fermented dairy products have been detected.

The most common microorganisms observed in fermented milk products belong to the strains of lactic acid bacteria (LAB), *Lactobacillus, Leuconostoc, Lactococcus,* etc. (Liu et al. 2014). These microorganisms prevent the spoilage of milk and inhibit the growth of other pathogenic microorganisms. Today the fermentation processes are controlled with specific starter cultures and conditions to obtain a wide range of milk products like milk cream, cultured buttermilk, kefir, koumiss, yogurt and amasi. Different starter cultures are used for each fermented dairy product. They consist of microorganisms added to the milk to provide specific characteristics in the final fermented milk product with desired properties. The vital function of lactic acid starters is for fermentation of lactose into lactic acid. In addition, they also contribute to flavor, aroma and alcohol production, while inhibiting interference of spoilage microorganisms. A single strain of bacteria may be added, or a mixture of several microorganisms may be introduced. Bacteria, yeasts and molds perform the process of fermentation at specific range of optimum temperatures. The optimum temperature for thermophilic lactic acid fermentation is about 40 to 45°C, while mesophilic lactic acid fermentation occurs at moderate temperatures ranging from 30 to 40°C (Carminati et al. 2010).

Microbial fermentation is the biochemical conversion of carbohydrates into alcohols or acids (Panda et al. 2013, Panda et al. 2016). In fermented milk products both alcohol and lactic acid may be produced, e.g., kefir and koumiss, or only lactic acid, e.g., sour milk cream. The bacteria convert the lactose to lactic acid and raise the acidity of the milk. The rise in acidity causes denature of milk proteins thus inhibiting the growth of other organisms that are not acid tolerant. Following the completion of fermentation process, the fermented dairy products are marketed with added flavors.

In addition to extending the shelf life of milk products, the fermentation process gives probiotic properties to the milk products (Amara and Shibl 2015). After invention of the microscope, the microbiological studies revealed that the fermented milk products contain live microorganisms and the microbial metabolites that are highly beneficial for human health (Fernandez et al. 2015). Further development of scientific knowledge has made it clear that the human intestinal microflora consists of trillions of microbial cells. These microorganisms play a vital role in several physiological activities, metabolic activities and immune functions (Guinane and Cotter 2013). Historically the fermented milk products have health benefits and have a good taste which enables their consumption, hence milk was the first probiotics food adapted by the men (Amara and Shibl 2015). Men knew how to prepare different types of fermented milk products even before the invention of the microscope (Amara and Shibl 2015). The different types of the microorganisms used as starter culture induce different reactions and as a result it produces different types of fermented products like yogurt, kefir, koumiss, sour cream, etc. The knowledge of such traditional processes for food preservation is transferred from generation to generation. Looking to the importance of the beneficial properties of kefir and koumiss, this chapter describes the nutritional importance and health promoting properties of kefir and koumiss.

2. Kefir

Kefir is a fermented milk drink which is traditionally produced by fermenting cow, goat and sheep milk by using "kefir grains" as starter culture (Farnworth 2005). The kefir grains contain active microorganisms and when added to fresh milk, they produce kefir. Kefir grains consist of protein and polysaccharide matrix containing different species of yeasts, LAB, acetic acid bacteria, and mycelial fungi (Witthuhn et al. 2005). Kefir produced through the fermentation of milk by the microorganisms is a white or yellowish colored, sour, carbonated and a mild alcoholic beverage. It has yeasty aroma with acidic taste (Irigoyen et al. 2005). Kefir is sometimes commercially available without carbonation and alcohol (when yeast is not added to the starter culture), resulting in a product that is very similar to yogurt. The flavor, taste, nutritional composition of kefir varies with the type of milk and microbial strains used for kefir production.

Kefir originated from the Caucasian mountains and then it became popular in central and Eastern Europe (Assadi et al. 2000). It is important food stuff in Russia. Traditionally kefir is produced in the households of the Caucasian Mountains and Tibetian region of China. They used the kefir grains which were inherited from their ancestors for kefir preparation.

Generally, kefir is prepared from cow, sheep and goat milk; however soya milk is also known to be used for kefir preparation (Farnworth 2005).

In the later part of the nineteenth century, the benefit of kefir was highlighted when it helped for effective treatment of tuberculosis, intestinal and chronic diseases in the Caucasus region. The benefits of kefir were first reported by the Russian doctors working in this region. Kefir has been reported for its healing effects on the high blood pressure, anaemia, obesity control, gall bladder problem, etc. (Bellamy and MacLean 2005). Due to its abundant health benefits, kefir has gained popularity as a functional healthy probiotic food throughout the world.

2.1 Kefir Production

Kefir production started from the indigenous process by incubating the kefir grains with milk of cow, sheep or goat. Later the process was refined scientifically to produce it on a commercial scale.

2.1.1 Kefir grains

Microorganisms present in kefir grains are responsible for the fermentation of milk. The original kefir grains are slightly yellowish in colour and they resemble a cauliflower (Nielsen et al. 2014). There is no scientific evidence about the origin of kefir grains. Kefir grains are yellowish-white, cauliflower shaped, semi hard granules containing different yeast and bacterial stains, which exist in symbiotic association (Fig. 1). When the grains are added to sterilized milk and incubated the microorganisms are activated to ferment milk. Kefir grains are made up of a complex microbial biomass matrix composed of polysaccharide, fat and protein of kefir microorganism

Figure 1. Kefir grain.
(source: https://julietwhev.files.wordpress.com/2013/03/kefir-grains.jpg).

origin (Rea et al. 1996). Microorganisms involved in kefir production secrete exopolysaccharides that accumulate along with proteins and fat molecules to form kefir grains. Further, growth of kefir grains occur by the accumulation of microbial biomass on the pre-existing kefir grains during kefir production. The major constituent of kefir grain matrix is composed of polysaccharide "kefiran" (La Riviere et al. 1967). The kefiran is a hetero-polysaccharide made from glucose and galactose. Lactic acid bacteria are the main exopolysaccharide producing microorganisms in kefir that give the rheological and texture properties to the kefir formed from the fermented milk (Frengova et al. 2002).

2.1.2 Traditional/Indigenous process of kefir production

The traditional process for kefir preparation is described as follows: (a) Incubation of milk (cow, sheep or goat) with kefir grains at room temperature for 24 hr; (b) At the end of the incubation period kefir grains are separated from milk by filtration process; (c) the fermented milk, i.e., the kefir is preserved for further consumption. The kefir grains separated from the kefir can be further reused as a starter culture for preparation of kefir from fresh batch of milk.

2.1.3 Modern process of kefir production

The kefir can be prepared from the milk of sheep, cow or goat. For large scale production of kefir, kefir grains are not used but rather sterilized milk is incubated with selected microorganisms directly. At the industrial scale, at first the milk is sterilized by the process of homogenization and pasteurization. After sterilization the milk is kept for cooling down up to 20°C and the milk is incubated with specific strains of microorganisms for 24 hr further to prepare kefir (Assadi et al. 2000, Otles and Cagindi 2003). The nutritional and sensory qualities of kefir vary with the type of milk and microbial strain used. The taste, flavor and aroma of kefir differ from other milk products because it is a produced through a combination of eukaryotic and prokaryotic (yeast and bacteria) fermentation process. A typical process of kefir production is graphically presented in Fig. 2.

Kefir resembles yogurt to some extent. Many people believe that kefir and yogurt are similar, but in reality they have many significant differences based on biochemical and organoleptic properties. Both kefir and yogurt are cultured milk products but they contain different strains of microorganisms. Generally, yogurt contains bacterial strains that belong to the genera of *Lactobacillus* and *Streptococcus*, while kefir contains several other bacterial

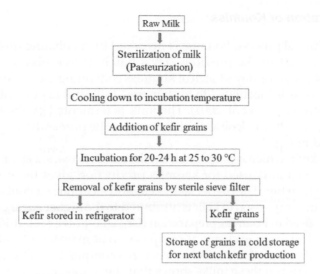

Figure 2. A typical process of Kefir production.

strains such as *Lactobacillus, Leuconostoc, Acetobacter, Streptococcus* and *Pseudomonas* spp. Apart from the bacterial strains, kefir contains different yeast strains belonging to *Saccharomyces, Candida* and *Kluyveromyces* genus, hence kefir is a fermented milk product of combined action of yeast and bacterial fermentation process. Appearance of kefir is not as creamy as yogurt. Generally, kefir has a sourer or tart taste due to lactic acid content. It has a little effervescence, due to the carbon dioxide and alcohol content released by alcoholic fermentation. The acidity generated drops the pH of milk up to 4 to 4.5, and it varies depending on the type of milk and fermentation conditions.

3. Koumiss

Koumiss is a traditional milk beverage produced from fermentation of mares' milk by indigenous microorganisms (Montanari et al. 1996). Koumiss is also known by other names like *koumiss, kumiss, kumis, kymis, kymmyz*. It is a fermented drink traditionally made from the milk of horses by people in Central Asia and China, where it is one of the most important basic foodstuffs. Koumiss is similar to the kefir; however it is prepared by a liquid starter culture in contrast to the solid kefir grains used in kefir production. Koumiss is also widely produced in Russia, Kazakhstan in Western Asia. In Mongolia, it has been adapted as the national drink and is known as *Airag* (Uniacke-Lowe et al. 2010).

3.1 Preparation of Koumiss

In the traditional process, koumiss is prepared by incubating fresh mare's milk with a part of the previous day's batch of koumiss as a starter culture. The previous day's batch of koumiss containing indigenous native microorganisms is inoculated to fresh mares' milk and kept for about 8 hr of incubation (Cagno et al. 2004). The milk is fermented by the LAB and yeast to produce lacto-alcoholic rich beverage. The preparation of koumiss is presented in Fig. 3.

Unlike kefir, which is prepared from the milk of cow, sheep or goat, the mare and camel milk used for koumiss production gives the product its distinctive character. The higher lactic acid and alcoholic content observed in koumiss in comparison to kefir is attributed to the higher sugar content of mare's milk used in koumiss preparation (Uniacke-Lowe et al. 2010, Bornaz et al. 2010). The average chemical composition of mare's milk and bovine milk shows that the mare's milk has a distinct composition. The noticeable differences between these milks shows that the mares milk contain lower fat (12.1 g/kg) and higher lactose (63.7 g/kg) content compared to bovine milk (Uniacke-Lowe et al. 2010). The predominant microbial strains used for koumiss are LAB and yeast strains of *Saccharomyces* (Wouters et al. 2002). The lactic acid produced by the lactic acid bacterial strains gives the koumiss an acidic characteristic and the yeasts generate alcohol in koumiss through alcoholic fermentation process. Thus the koumiss obtained by the fermentation of mare's milk is a milky grey, fizzy liquid with a sharp alcohol and acidic taste. Since the mare's milk has higher lactose content, the final fermented product, i.e., koumiss has comparatively higher lactic acid and

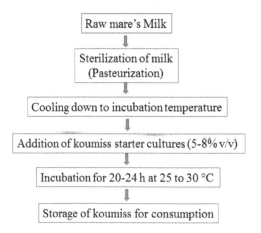

Figure 3. A typical process of Koumiss production.

alcoholic content when compared with kefir. Koumiss contains alcohol up to 2%, and therefore, it is also called milk wine. Although the koumiss is a popular fermented milk beverage, the availability of the mare's milk is the limiting factor for its large scale production. Presently, the large scale production of koumiss is performed from cow's milk supplemented with the additional sugar to make it approximate with the composition of mare's milk (Zhang and Zhang 2012).

4. Microorganisms Associated With Kefir and Koumiss

The microbial diversity of kefir and koumiss is very complex. However, several reports have described the microbial diversity with these two products. The microorganisms entrapped in the protein and polysaccharide matrix of the kefir grains live symbiotically (Lopitz-Otsoa et al. 2006). The microbial populations presented in the grains mostly belong to LAB, acetic acid bacteria (AAB), yeasts, and mycelia fungi (Marshall 1984, Witthuhn et al. 2005). A brief list of microorganisms detected in different kefirs is described in the Table 1.

The notable bacterial strains found in kefir are *Lactobacillus acidophilus, Lb. brevis, Lb. casei, Lb. fermentum, Lb. helveticus, Lb. kefir, Lb. parakefiri, Lactococcus lactis, Leuconostoc mesenteroides, Lb. delbrueckii, Acetobacter sicerae* sp. nov, *Streptococcus*, etc. (Witthuhn et al. 2005, Simova et al. 2002, Li et al. 2014).

Several strains of yeasts have also been reported in kefir such as *Kluyveromyces marxianus, Torula kefir, Saccharomyces exiguus, Candida lambica, Candida kefir, Saccharomyces cerevisiae, Candida krusei* and *Candida famata* (Lopitz-Otsoa et al. 2006, Witthuhn et al. 2005). Yeast cells provide flavor and aroma to kefir (Simova et al. 2002). In addition, they provide essential nutrients (vitamins and amino acids) for bacteria growth and thus assist growth of bacterial strains present in kefir (Farnworth 2005, Irigoyen et al. 2005). Furthermore, they inhibit growth of pathogenic microorganisms by decreasing the pH of the medium through the production of ethanol, carbon dioxide and organic acids.

The notable microorganisms present in the starter culture used for koumiss preparation belong to the bacterial strains of *Lactobacillus* and *Lactococcus*, and fungal strains of *Kluyveromyces* and *Saccharomyces* (Uniacke-Lowe et al. 2010). Cagno et al. (2004) reported the use of bacterial strains *Lactobacillus delbrueckii* and *Streptococcus thermophillus* in starter culture for the preparation of koumiss from mare's milk. Brief lists of microorganisms detected in different koumiss are described in the Table 2.

Table 1. Microorganisms associated with Kefir.

Microorganisms	References
Lactococcus spp., *Lactobacillus* spp.	Cui et al. 2013, Chen et al. 2008, Simova et al. 2002
Lactobacillus keranofaciens	Chen et al. 2008
Lactobacillus buchneri	Garofalo et al. 2015
Lactobacillus plantarum	Wang et al. 2015
Lactobacillus kefiri	Chen et al. 2008
Acetobacter aceti	Li et al. 2014
Kluyveromyces marxianus	Wang et al. 2012 Chang et al. 2014
Saccharomyces turicensis	Wang et al. 2012
Pichia fermentans	Wang et al. 2012
Kazachstaniaunispora	Garofalo et al. 2015
Dekkeraanomala	Garofalo et al. 2015
Yeast	Soupioni et al. 2013

Table 2. Microorganisms associated with Koumiss.

Microorganisms	References
Lactobacillus spp.	Guo et al. 2015
Lb. acidophilus	El-Ghaish et al. 2011
Lb. helveticus	Miyamoto et al. 2015
Lb. salivarius	Danova et al. 2005
Lb. buchneri	Danova et al. 2005
Lb. plantarum	Danova et al. 2005
Lb. delbrueckii	Cagno et al. 2004
Streptococcus spp.	Kozhahmetova et al. 2013
Str. thermophiles	Cagno et al. 2004
Leuconostoc	Guzel-Seydim et al. 2009
Yeasts	El-Ghaish et al. 2011
Torula kumiss	Kosikowski 1982
Saccharomyces lactis	Kosikowski 1982
Saccharomyces unisporus	Montanari et al. 1996
Kluyveromyces lactis	Kucukcetin et al. 2003

5. Biochemical Mechanisms Involved in The Production of Kefir and Koumiss

The biochemistry of kefir and koumiss production depends upon the type of lactic acid fermentation occurred, namely homo fermentation and hetero fermentation. The homo fermentative microorganisms produce only lactic acid as the final product whereas the hetero fermentative microorganisms produce ethanol and CO_2 along with lactic acid. The details of the homo and hetero lacto-fermentation have been described in the previous chapter of this book (see Chapter 17).

The characteristic flavor of kefir and koumiss are due to the presence of various biochemical compounds formed during fermentation process, such as organic acids (lactic, propionic, citric, acetic, pyruvic and succinic acids, etc.). These acids are produced due to the metabolism of milk by the microorganisms involved in milk fermentation (Guzel-Seydim et al. 2000).

The microorganisms present in kefir, metabolise the milk sugar to produce energy through the biochemical process of glycolysis and other metabolic pathways (alcoholic and lactic acid fermentation and Krebs cycle). The final product formed in glycolysis is pyruvic acid. Further, metabolic process depends on the availability of the oxygen for the microorganisms present in kefir. In presence of oxygen, the aerobic microbes metabolize the pyruvate through the Krebs cycle. In the Krebs cycle the pyruvate is converted into different organic acids like citric, oxalic, succinic acid, etc. However, in absence of oxygen the pyruvic acid is fermented to either alcohol or lactic acid.

The organic acids produced by microbial metabolism have several functional properties. Acetic acid is a weak acid used as a preservative and a food additive. It is soluble in lipids and therefore it can diffuse through the plasma membrane of a microbial cell and affect its internal pH, causing the death of food spoiling or pathogenic microorganisms (Giannattasio et al. 2013). Citric acid is a preservative and flavoring agent generally used in food and pharmaceutical industries. This acid acts as a chelating agent for the metal ions present in the medium and inhibit microbial growth. Pyruvic acid is also applied as a flavoring agent and as a preservative agent (Nielsen et al. 2014, Theron and Lues 2011). Lactic acid is also a preservative and a pH regulating agent (Theron and Lues 2011). Propionic acid is used as a preservative and as a flavoring agent (Nielsen 2004). Hence, due to the production of the above organic metabolites during milk fermentation, the fermented milk products have long shelf life. In addition to the above metabolites, exopolysaccharides are also formed by the microorganisms found in kefir and koumiss. The microorganisms belonging to the genera of *Lactobacillus* and *Lactococcus* are the mostly exopolysaccharide producing bacterial strains in kefir (Welman and Maddox 2003, Irigoyen et al. 2005, Miguel et al. 2010).

Kefir and koumiss contain complex microbial community which include bacterial strains dominated by LAB, acetic acid bacteria and yeast strains. The bacteria and yeast symbiotically exist in kefir, and the yeast strains play the major role in the development of specific aroma and flavor. Lactic acid bacteria are the prevalent bacterial strain found in the kefir grains.

The yeast cells do not grow efficiently when the bacteria are separated from the kefir grain (Leite et al. 2013). Since some strains of yeasts such as *Saccharomyces cerevisiae*, *Saccharomyces turicensis*, *Torulaspora delbrueckii*, *Kazachstania unispora*, etc. found in kefir are unable to metabolize lactose (Leite et al. 2013). This inability makes them dependent on the lactic acid bacteria, which are capable of metabolizing the milk sugar lactose.

Kefir is usually made from partially skimmed cow's milk. The final product contains live bacteria and yeasts that produce carbon dioxide gas. This gas production gives kefir a "sparkling" sensation on the tongue when consumed. Kefir has been referred to as the champagne of fermented dairy products.

Koumiss is a milk drink with a sharp alcohol and acidic taste (Salimei and Fantuz 2012). The lactic acid content of koumiss varies from 0.7 to 1.8% and the ethanol content varies between 0.6 to 2.5%. Koumiss is categorized into mild, medium and strong depending upon the degree of lactic acid and ethanol content. Due to higher alcoholic content the koumiss is referred as milk wine.

6. Health Beneficial Properties of Kefir and Koumiss

Kefir and koumiss are microbial fermented milk products. The fermentation process induces changes in the nutritional value, flavor, aroma and color, etc. of the milk. The intake of kefir and koumiss promote wide range of health benefits. Primarily, they are rich probiotic food for human consumption. In addition to their probiotic nature they possess a wide range of health benefits such as anti-bacterial and anti-fungal properties, regulate immunity, maintain healthy gastrointestinal system, regulate cholesterol and sugar levels, regulate blood pressure, help to get rid over the lactose intolerance, induce production of some essential vitamins, etc. (Bakir et al. 2015, Apostolidis et al. 2007). Recent studies that have focused on the importance of probiotic food materials have enumerated the following points to describe the health benefits achieved from kefir and koumiss consumption.

6.1 Kefir and Koumiss as Potential Source of Probiotics

According to The Food and Agriculture Organization of the United Nations/World Health Organization (FAO/WHO 2001), the term probiotics can be defined as "live microorganisms administered in an adequate quantity that continue to exist in the intestinal environment, to perform a health positive effect on the host" (Reid et al. 2003). The kefir and koumiss are excellent source of probiotic microorganisms.

The microbes present in the kefir and koumiss live symbiotically, yet the microbial population composition in them may differ due to the origins, methods and substrates used for preparation of these products. However, there are common species of microorganisms such as bacterial strains of *Lactobacillus acidophilus, Lb. brevis, Lb. casei, Lb. fermentum, Bifidobacterium bifidum, B. adolescentis, Streptococcus lactis, Str. alivarius, Str. thermophilus, Bacillus, Enterococcus* and yeast and mold strains of *Saccharomyces cerevisiae, S. bourlardii, Aspergillus niger, A. oryzae, Candida pintolopesii,* etc.

Most of the probiotic bacterial strains colonize in the digestive tract of our body. The widely known probiotic microorganisms belong to bacterial strains of *Lactobacillus* and *Bifidobacterium*, however, bacterial strains belonging to *Pediococcus, Lactococcus, Bacillus* and several strains of yeasts are also reported for their probiotic nature (Soccol et al. 2010, Blaiotta et al. 2013). The bacterial strains of *Lactobacillus*, i.e., LAB strains are dominant microorganisms distributed throughout the gastrointestinal and genital tracts of man and higher animals. They are non-pathogenic and produce lactic acid by their metabolism; as a result they lower the pH of the gastric and genital tract of human body. This class of microorganisms also produce hydrogen peroxide, ethanol and/or acetic acid, thus inhibiting the proliferation of unwanted pathogenic microorganisms in the gastric and genital tract of the body. That is why these microorganisms are called natural living protectors of the human body.

6.2 Kefir and Koumiss has Potent Antimicrobial Properties

The nutritional and organoleptic properties of the probiotic milk kefir and koumiss offer the potential to combat against pathogenic microbial infections (Franco et al. 2013, Carasi et al. 2014, Silva et al. 2009). Recent scientific studies have confirmed the antimicrobial properties of traditional kefir and koumiss. In a study, Chifiriuc et al. (2011) evaluated the antimicrobial properties of kefir by *in vitro* analysis. The authors investigated the antimicrobial activity of kefir against the bacterial strains of the *Bacillus subtilis, Staphylococcus aureus, Enterococcus faecalis, Escherichia coli, Salmonella enteritidis, Pseudomonas aeruginosa* and *Candida albicans*. They compared the antimicrobial properties of the kefir fermented for 24 hr and 48 hr, as well as 7 day old preserved kefir by *in vitro* disk diffusion method. The authors compared the antimicrobial activities of the kefir with those of the antibiotics ampicillin and neomycin and observed higher antimicrobial potential of kefirs against the bacterial strains used for the study. The authors claim that the kefir has higher antibacterial properties when compared to control antibiotics (ampicillin and neomycin) taken for the study.

A recent study of the antimicrobial properties of the koumiss was conducted by Chen et al. (2015). The authors used koumiss from Inner Mongolia, China and evaluated the anti-bacterial properties of the mycotoxin secreted by the yeast cell of koumiss by *in vitro* and *in vivo* analysis. Through genomic analysis, the authors identified three strains of *S. cerevisiae*, and two strains of *Kluyveromyces marxianus* producing mycocin in the traditional Koumiss from Inner Mongolia. The *in vitro* and *in vivo* study on mice confirmed the anti-bacterial properties of the mycotoxin isolated from the yeast cells of koumiss against the pathogenic *E. coli* bacteria. Hence, the study shows that the traditional fermented milk beverages have potential antimicrobial properties.

6.3 Kefir and Koumiss as a Substitute for Lactose Intolerant People

Lactose is a naturally occurring sugar in milk. Most of the adult populations of the world are unable to digest the lactose content of the milk properly. Such condition of lactose indigestion is called lactose intolerance. The LAB present in kefir and koumiss ferment lactose to lactic acid, and as a result these dairy foods are much lower in lactose content than raw milk. Hence, in general kefir and koumiss are well tolerated by the people with lactose intolerance in comparison to regular raw milk products (Fox et al. 2015, Zubillaga et al. 2001).

6.4 Kefir and Koumiss Stimulate Expression of Growth Factors

In the recent *in vivo* study conducted on mice model by Bakir et al. (2014) it was revealed that the probiotic dairy products positively affect the release of growth factors and stimulates the increase in body weight of mice. The authors conducted immune-histochemical studies on the liver and kidney cells on the mice nourished with kefir and koumiss and observed the expression of the platelet derived growth factor-c (PDGF-C) and platelet derived growth factor receptor-alpha (PDGFR-α). The authors found that PDGF-C and PDGFR-α were expressed more in the kidney cells and hepatic cells and stimulated increase in live weights of the mice models used for the study. The platelet derived growth factor receptor-alpha (PDGFR-α) is a member of the PDGF receptor and they play a vital role for activation of platelet derived growth factor (Andrae et al. 2008). The platelet derived growth factors (PDGF) are a class of growth factors that influence growth and development of body. From the above report it may be concluded that the probiotic dairy products kefir and koumiss act as growth and development promoters.

6.5 Kefir May be Protective Against Cancer

Cancer is one of the world's leading causes of death. It occurs when there is an uncontrolled growth of abnormal cells in the body such as a tumor. The probiotics in fermented dairy products are believed to inhibit tumor growth by reducing formation of carcinogenic compounds, as well as by stimulating the immune system (Leite et al. 2013).

6.6 Trends in Transformation of Traditional Kefir and Koumiss to Commercialization

According to a market survey conducted by Transparency Market Research, the global sale of probiotic products is $15.9 billion in 2019 as compared to $11.6 billion in 2012. Although kefir is not known in all corners of the world, still it has a stake among the probiotic products. A company named Lifeway Foods, Inc, recently named one of Fortune Small Business Fastest Growing Companies, is known to sell of $100 million annually. In 2013 the company introduced the products to the UK and subsequently to Canada. The products are available in different flavors such as vanilla, raspberry, strawberry and mango flavors. Some of the kefir products of Lifeway Foods have been displayed in Fig. 4. More than 1000 leading stores such as Kroger, Wegmans and Whole Foods are known to sell kefir in the US. However the

Figure 4. Kefir with different flavours marketed by Lifeway industries.
(source: http://lifewaykefir.com/).

commercialization of koumiss is not clearly documented. Unlike kefir the commercialization of koumiss is difficult as the substrate for the koumiss, i.e., mare's milk is rarely available.

7. Conclusion

Fermented dairy products such as kefir and koumiss have numerous functional properties. It is a convenient and traditional process of food preservation. Fermentation lowers down the pH of milk and inhibits the growth of food spoilage microorganisms. The microbial metabolites like different organic acids, exopolysaccharides produced during kefir and koumiss production enhances their flavor, aroma and texture, which are attractive for the consumer. This chapter has described the history, microbiology and biochemistry of kefir and koumiss. The current market status of the two products has been mentioned. Popularization of kefir and koumiss can be expedited through deliberation and discussion of the health beneficial properties of the products in different forums such as social media, internet, etc. Future research should be directed to upscale and for the commercialization of kefir and koumiss, keeping in view the health beneficial properties and unique organoleptic properties of the products.

Keywords: Kefir, koumiss, milk, fermentation, probiotics

References

Amara, A.A. and Shibl, A. (2015). Role of Probiotics in health improvement, infection control and disease treatment and management. Saudi Pharmaceutical Journal 23: 107–114.

Andrae, J., Gallini, R. and Betsholtz, C. (2008). Role of platelet-derived growth factors in physiology and medicine. Genes & Development 22: 1276–1312.

Apostolidis, E., Kwon, Y.I., Ghaedian, R. and Shetty, K. (2007). Fermentation of milk and soymilk by *Lactobacillus bulgaricus* and *Lactobacillus acidophilus* enhances functionality for potential dietary management of hyperglycemia and hypertension. Food Biotechnology 21: 217–236.

Assadi, M.M., Pourahmad, R. and Moazami, N. (2000). Use of isolated kefir starter cultures in kefir production. World Journal of Microbiology & Biotechnology 16: 541–543.

Bakir, B., Sari, E.K., Aydin, B.D. and Yil, S.E. (2015). Immunohistochemical examination of effects of kefir, koumiss and commercial probiotic capsules on platelet derived growth factor-c and platelet derived growth factor receptor-alpha expression in mouse liver and kidney. Biotechnic & Histochemistry 90: 190–196.

Bellamy, I. and MacLean, D. (2005). Radiant Healing: The many paths to personal harmony and planetary wholeness: Joshua Books, Queensland, Australia, pp. 272.

Blaiotta, G., Gatta, B.L., Capua, M.D., Luccia, A.D., Coppola, R. and Aponte, M. (2013). Effect of chestnut extract and chestnut fiber on viability of potential probiotic *Lactobacillus* strains under gastrointestinal tract conditions. Food Microbiology 36: 161–169.

Bornaz, S., Guizani, N., Sammari, J., Allouch, W., Sahli, A. and Attia, H. (2010). Physicochemical properties of fermented Arabian mares' milk. International Dairy Journal 20: 500–505.

Cagno, R.D., Tamborrino, A., Gallo, G., Leone, C., Angelis, M.D., Faccia, M., Amirante, P. and Gobbetti, M. (2004). Uses of mares' milk in manufacture of fermented milks. International Dairy Journal 14: 767–775.

Carasi, P., Diaz, M., Racedo, S.M., Antoni, G.D., Urdaci, M.C. and Serradell, M.A. (2014). Safety characterization and antimicrobial properties of kefir-isolated *Lactobacillus kefiri*. BioMed Research International 2014: 1–7.

Carminati, D., Giraffa, G., Quiberoni, A., Binetti, A., Suarez, V. and Reinheimer, J. (2010). Advances and trends in starter cultures for dairy fermentations. *In*: F. Mozzi, R.R. Raya and G.M. Vignolo (eds.). Biotechnology of Lactic Acid Bacteria: Novel Applications, Wiley-Black well, USA.

Chang, J., Ho, C., Mao, C., Barham, N., Huang, Y., Ho, F., Wu, Y., Hou, Y., Shih, M., Li, W. and Huang, C. (2014). A thermo-and toxin-tolerant kefir yeast for biorefinery and biofuel production. Applied Energy 132: 465–474.

Chen, Y., Aorigele, C., Wang, C., Simujide, H. and Yang, S. (2015). Screening and extracting mycocin secreted by yeast isolated from koumiss and their antibacterial effect. Journal of Food and Nutrition Research 3: 52–56.

Chen, H., Wang, S. and Chen, M. (2008). Microbiological study of lactic acid bacteria in kefir grains by culture-dependent and culture-independent methods. Food Microbiology 25: 492–501.

Chifiriuc, M.C., Cioaca, A.B. and Lazar, V. (2011). *In vitro* assay of the antimicrobial activity of kephir against bacterial and fungal strains. Anaerobe 17: 433–435.

Cui, X., Chen, S., Wang, Y. and Han, J. (2013). Fermentation conditions of walnut milk beverage inoculated with kefir grains. LWT - Food Science and Technology 50: 349–352.

Danova, S., Petrov, K., Pavlov, P. and Petrova, P. (2005). Isolation and characterization of *Lactobacillus* strains involved in koumiss fermentation. International Journal of Dairy Technology 58: 100–105.

Dhewa, T., Mishra, V., Kumar, N. and Sangu, K.P.S. (2015). Koumiss nutritional and therapeutic values. *In*: A.K. Puniya (ed.). Fermented Milk and Dairy Products. CRC Press, Boca Raton.

El-Ghaish, S., Ahmadova, A., Hadji-Sfaxi, I., Mecherfi, K.E.E., Bazukyan, I., Choiset, Y., Rabesona, H., Sitohy, M., Popov, Y.G., Kuliev, A.A., Mozzi, F., Chobert, J. and Haertle, T. (2011). Potential use of lactic acid bacteria for reduction of allergenicity and for longer conservation of fermented foods. Trends in Food Science & Technology 22: 509–516.

Farnworth, E.R. (2005). Kefir-a complex probiotic. Food Science and Technology Bulletin: Functional Foods 2: 1–17.

Fernandez, M., Hudson, J.A., Korpela, R. and de los Reyes-Gavilan, C.G. (2015). Impact on human health of microorganisms present in fermented dairy products: an overview. BioMed Research International 412714.

Fox, P.F., Uniacke-Lowe, T., Mc Sweeney, P.L.H. and O'Mahony, J.A. (2015). Dairy Chemistry and Biochemistry, Springer, Heidelberg, New York.

Franco, M.C., Golowczyc, M.A., de Antoni, G.L., Perez, P.F., Humen, M. and Serradell, M.D.L.A. (2013). Administration of kefir-fermented milk protects mice against Giardia intestinalis infection. Journal of Medical Microbiology 62: 1815–1822.

Frengova, G.I., Simova, E.D., Beshkova, D.M. and Simov, Z.I. (2002). Exopolysaccharides produced by Lactic Acid Bacteria of kefir grains. Z. Naturforsch 57c: 805–810.

Garofalo, C., Osimani, A., Milanovi, V., Aquilanti, L., Filippis, F.D., Stellato, G., Mauro, S.D., Turchetti, B., Buzzini, P., Ercolini, D. and Clementi, F. (2015). Bacteria and yeast microbiota in milk kefir grains from different Italian regions. Food Microbiology 49: 123–133.

Giannattasio, S., Guaragnella, N., Zdralevic, M. and Marra, E. (2013). Molecular mechanisms of *Saccharomyces cerevisiae* stress adaptation and programmed cell death in response to acetic acid. Frontiers in Microbiology 4: 33–42.

Guinane, C.M. and Cotter, P.D. (2013). Role of the gut microbiota in health and chronic gastrointestinal disease: understanding a hidden metabolic organ. Therapeutic Advances in Gastroenterology 6: 295–308.

Guo, C., Zhang, S., Yuan, Y., Yue, T. and Li. J. (2015). Comparison of lactobacilli isolated from Chinese suan-tsai and koumiss for their probiotic and functional properties. Journal of Functional Foods 12: 294–302.

Guzel-Seydim, Z., Koktas, T. and Greene, A.K. (2009). Kefir and Koumiss: microbiology and technology. Development and manufacture of yogurt and other functional dairy products. Fatih Yildiz (ed.). CRC Press, Boca Raton.

Hasan, M.N., Sultan, M.Z. and Mar-E-Um, M. (2014). Significance of fermented food in nutrition and food science. Journal of Scientific Research 6: 373–386.

Irigoyen, A., Arana, I., Castiella, M., Torre, P. and Ibanez, F.C. (2005). Microbiological, physicochemical, and sensory characteristics of kefir during storage. Food Chemistry 90: 613–620.

Kosikowski, F. (1982). Cheese and Fermented Milk Foods, F. V. Koskiowski and Associates, New York, U.S.A.

Kozhahmetova, Z. and Kasenova, G. (2013). Selection of lactic acid bacteria and yeast for koumiss starter and its impact on quality of koumiss. ATI—Applied Technologies & Innovations 9: 138–142.

Kucukcetin, A., Yaygin, H., Hinrichs, J. and Kulozik, U. (2003). Adaptation of bovine milk towards mares' milk composition by means of membrane technology for koumiss manufacture. International Dairy Journal 13: 945–951.

La Riviere, J.W., Kooiman, P. and Schmidt, K. (1967). Kefiran, a novel polysaccharide produced in the kefir grain by *Lactobacillus brevis*. Archives of Microbiology 59: 269–278.

Leite, A.M.O., Miguel, M.A.L., Peixoto, R.S., Rosado, A.S., Silva, J.T. and Paschoalin, V. M. F. (2013). Microbiological, technological and therapeutic properties of kefir: a natural probiotic beverage. Brazilian Journal of Microbiology 44: 341–349.

Li, L., Wieme, A., Spitaels, F., Balzarini, T., Nunes, O.C., Manaia, C.M., Landschoot, A.V., Vuyst, L.D., Cleenwerck, I. and Vandamme, P. (2014). *Acetobacter sicerae* sp. nov., isolated from cider and kefir, and identification of species of the genus *Acetobacter* by *dnaK*, *groEL* and *rpoB* sequence analysis. International Journal of Systematic and Evolutionary Microbiology 64: 2407–2415.

Liu, S., Holland, R. and Crow, V.L. (2004). Esters and their biosynthesis in fermented dairy products: a review. International Dairy Journal 14: 923–945.

Lopitz-Otsoa, F., Rementeria, A., Elguezabal, N. and Garaizar, J. (2006). Kefir: A symbiotic yeast-bacteria community with alleged healthy capabilities. Revista Iberoamericana DeMicologia 23: 67–74.

Marshall, V.M.R.E. (1984). The microflora and production of fermented milks. *In*: M.R. Adams (ed.). Progress in Industrial Microbiology. Vol. 23. Microorganisms in the Production of Food. Elsevier, Amsterdam.

Miguel, M.G.C.P., Cardoso, P.G., Lago, L.A. and Schwan, R.F. (2010). Diversity of bacteria present in milk kefir grains using culture-dependent and culture-independent methods. Food Research International 43: 1523–1528.

Miyamoto, M., Ueno, H.M., Watanabe, M., Tatsuma, Y., Seto, Y., Miyamoto, T. and Nakajima, H. (2015). Distinctive proteolytic activity of cell envelope proteinase of *Lactobacillus helveticus* isolated from airag, a traditional Mongolian fermented mare's milk. International Journal of Food Microbiology 197: 65–71.

Montanari, G., Zambonelli, C., Grazia, L., Kamesheva, G.K. and Shigaeva, M.K. (1996). *Saccharomyces unisporaas* the principle alcoholic fermentation microorganism of traditional koumiss. Journal of Dairy Research 63: 327–331.

Nielsen, B., Gurakan, G.C. and Unlu, G. (2014). Kefir: A Multifaceted Fermented Dairy Product. Probiotics & Antimicrobial Protection 6: 123–135.

Otles, S. and Cagindi, O. (2003). Kefir: A Probiotic Dairy-Composition, Nutritional and Therapeutic Aspects. Pakistan Journal of Nutrition 2: 54–59.

Panda, S.K., Behera, S.K., Sahu, U.C., Ray, R.C., Kayitesi, E. and Mulaba-Bafubiandi, A.F. (2016). Bioprocessing of jackfruit (*Artocarpus heterophyllus* L.) pulp into wine: Technology, proximate composition and sensory evaluation. African Journal of Science, Technology, Innovation and Development 8: 27–32.

Panda, S.K., Sahu, U.C., Behera, S.K. and Ray, R.C. (2014b). Bio-processing of bael (*Aegle marmelos* L.) fruits into wine with antioxidants. Food Bioscience 5: 34–41.

Panda, S.K., Sahu, U.C., Behera, S.K. and Ray, R.C. (2014a). Fermentation of sapota (*Achras sapota* linn.) fruits to functional wine. Nutrafoods 13: 179–186.

Panda, S.K., Swain, M.R., Singh, S. and Ray, R.C. (2013). Proximate compositions of a herbal purple sweet potato (*Ipomoea batatas* L.) wine. Journal of Food Processing and Preservation 37: 596–604.

Parveen, S. and Hafiz, F. (2003). Fermented cereal from indigenous raw materials. Pakistan Journal of Nutrition 2: 289–291.

Rea, M.C., Lennartsson, T., Dillon, P., Drina, F.D., Reville, W.J., Heapes, M. and Cogan, T.M. (1996). Irish kefir-like grains: their structure, microbial composition and fermentation kinetics. J. Appl. Microbiol. 81: 83–94.

Reid, G., Sanders, M.E., Gaskins, H.R., Gibson, G.R., Mercenier, A., Rastall, R., Roberfroid, M., Rowland, I., Cherbut, C. and Klaenhammer, T.R. (2003). New scientific paradigms for probiotics and prebiotics. Journal of Clinical Gastroenterology 37: 105–118.

Salimei, E. and Fantuz, F. (2012). Equid milk for human consumption. International Dairy Journal 24: 130–142.

Silva, K.R., Rodrigues, S.A., Filho, L.X. and Lima, A.S. (2009). Antimicrobial activity of broth fermented with kefir grains. Appl. Biochem. Biotechnol. 152: 316–325.

Simova, E., Beshkova, D., Angelov, A., Hristozova, T., Frengova, G. and Spasov, Z. (2002). Lactic acid bacteria and yeasts in kefir grains and kefir made from them. Journal of Industrial Microbiology & Biotechnology 28: 1–6.

Soccol, C.R., Vandenberghe, L.P.S., Spier, M.R., Medeiros, A.B.P., Yamaguishi, C.T., Lindner, J.D.D., Pandey, A. and Thomaz-Soccol, V. (2010). The potential of probiotics: A review. Food Technology and Biotechnology 48: 413–434.

Soupioni, M., Golfinopoulos, A., Kanellaki, M. and Koutinas, A.A. (2013). Study of whey fermentation by kefir immobilized on low cost supports using 14C-labelled lactose. Bioresource Technology 145: 326–330.

Theron, M.M. and Lues, J.F. R. (2011). Organic acids and food preservation, 1st edition, Chapter 1 and 2, CRC Press, Taylor & Francis Group, Boca Raton.

Uniacke-Lowe, T., Huppertz, T. and Fox, P.F. (2010). Equine milk proteins: chemistry, structure and nutritional significance. International Dairy Journal 20: 609–629.

Wang, J., Zhao, X., Tian, Z., Yang, Y. and Yang, Z. (2015). Characterization of an exopolysaccharide produced by *Lactobacillus plantarum* YW11 isolated from Tibet Kefir. Carbohydrate Polymers 125: 16–25.

Wang, S., Chen, K., Lo, Y., Chiang, M., Chen, H., Liu, J. and Chen, M. (2012). Investigation of microorganisms involved in biosynthesis of the kefir grain. Food Microbiology 32: 274–285.

Welman, A.D. and Maddox, I.S. (2003). Exopolysaccharides from lactic acid bacteria: perspectives and challenges. Trends in Biotechnology 21: 269–274.

Witthuhn, R.C., Schoeman, T. and Britz, T.J. (2005). Characterisation of the microbial population at different stages of kefir production and kefir grain mass cultivation. International Dairy Journal 15: 383–389.

Wouters, J.T.M., Ayad, E.H.E., Hugenholtz, J. and Smit, G. (2002). Microbes from raw milk for fermented dairy products. International Dairy Journal 12: 91–109.

Zhang, W. and Zhang, H. (2012). Fermentation and koumiss. *In*: Y.H. Hui (ed.). Handbook of Animal-Based Fermented Food and Beverage Technology, Second Edition, CRC press, Boca Raton.

Zubillaga, M., Weill, R., Postaire, E., Goldman, C., Caro, R. and Boccio, J. (2001). Effect of probiotics and functional foods and their use in different diseases. Nutrition Research 21: 569–579.

19

Yogurt
Microbiology, Organoleptic Properties and Probiotic Potential
Françoise Rul

1. Introduction: An Ancestral Fermented Food with an Expanding Contemporary Market

Fermentation has been used for thousands of years to preserve food. Thanks to the acidifying activity of bacteria, the shelf life of milk is increased because the growth of undesirable microorganisms is prevented.

Traces of fermented milk products apppear rather quickly after the emergence of agriculture, as early as 8,000 B.C. in Turkey and Eastern Europe. Based on the presence of milk lipids recently discovered on pottery shards, the inhibitants of what is now modern-day Libya were consuming fermented dairy products around 7,000 B.C. (Dunne et al. 2012). Traces of kefir have also been detected on a Bronze Age mummy in China (Yang et al. 2014). Yogurt seems to make an appearance around 5,000 B.C. and was discovered by nomadic peoples living in the Middle East. It has been consumed for thousands of years by different civilizations. "Yogurt" comes from the Turkish word "yogurmak," which means to thicken, coagulate, or curdle.

Micalis Institute, INRA, AgroParisTech, Université Paris-Saclay, 78350 Jouy-en-Josas, France.
E-mail: francoise.rul@inra.fr

In France, yogurt appears around 1542. King Francis I, who suffered from chronic diarrhea, was cured by eating yogurt. In 1905, Stamen Grigorov, a Bulgarian medical student studying in Geneva, Switzerland, was the first to describe the spherical and rod-shaped lactic acid bacterium that is found in Bulgarian yogurt; the species was named *Bacillus bulgaricus*. Then, in the 20th century, Russian Nobel laureate Elie Metchnikoff, a scientist at the Pasteur Institute in Paris, hypothesized that Bulgarians lived unusually long lives because they regularly consumed yogurt; his research helped make yogurt popular in Europe and served as the foundation for the field of probiotics, which is still growing a century later.

Another major event in yogurt's history was the food's transformation into a commercial product by Isaac Carasso in 1919, in Barcelona, Spain. Yogurt's commercialization was taken further by Danone, a private company, and the food was industrialized and spread throughout Europe starting in the 1960s.

A traditional food that is consumed on a daily basis in the Middle East and Europe, yogurt is currently expanding its market across the globe. Demand has grown dramatically in North and South America, as well as in Asia (> 100% between 2000 and 2010 for yogurt and fermented dairy products; Mikkelsen 2013). At present, more than 30% of the world's population eats yogurt, and worldwide yogurt consumption has hit around 15 million tons per year. The global yogurt market was projected to surpass $65 billion in 2015 (www.strategyr.com). Traditional yogurts—produced at small scales—currently coexist with industrially produced yogurts, and we are seeing renewed interest in homemade foods. Yogurt's nutritional value and healthful properties are universally recognized. The food has a positive market image, attributable to its specific organoleptic properties (fresh taste, sourness, unique aroma), which has improved following the discovery of its probiotic properties and society's movement toward greater health consciousness.

Producing yogurt requires milk to acidify, whereupon curds are formed. This acidification process, which has to be rapid in industrial settings, largely depends on the growth and activity of bacteria that produce lactic acid by fermenting lactose. The association between the two yogurt lactic acid bacteria (LAB) *Streptococcus thermophilus* (*S. thermophilus*) and *Lactobacillus delbrueckii* ssp. *bulgaricus* (*Lb. bulgaricus*) is regarded as a protocooperation because it is beneficial for both species, but each bacterium can grow alone in milk (Tamime and Robinson 1999). This protocooperation has industrial importance because it can improve yogurt's properties, such as the texture (*via* exopolysaccharide production; Bouzar et al. 1997), the acidification rate (Pette 1950c, Moon and Reinbold 1976, El-Soda et al. 1986, Amoroso et al. 1988, Beal and Corrieu 1991, Bautista et al. 1996), and the flavor (*via* the production of aromic compounds; Hamdan et al. 1971, Bottazzi et al.

1973, El-Abbassy and Sitohy 1993, Courtin and Rul 2004). This association at least partly relies on metabolite exchanges and involves elements of competition (for the nutrients in the milk) and mutualism (the fellow bacterium synthesizes and hydrolyzes metabolites).

In 1984, the FAO/WHO defined yogurt as "the coagulated milk product obtained by lactic acid fermentation through the action of *Lactobacillus delbrueckii* ssp. *bulgaricus* (*Lb. bulgaricus*) and *Streptococcus thermophilus* from milk and milk products. The microorganisms in the final product must be viable and abundant." If other bacteria are added, such as probiotics (e.g., *Bidobacteria, Lactobacilli* spp.), the product must be called "fermented milk" and cannot carry the yogurt label. The Codex Alimentarius entry for fermented milk (Codex STAN 243-2003) specifies that yogurt should contain a minimum of 2.7% (m/m) milk proteins, a maximum of 15% milk fate, a minimum of 0.6% titratable acidity (expressed as % of lactic acid), and a minimum of 10^7 CFU/g of microorganisms (total microorganisms in the starter culture). Yogurt has highly attractive nutritional properties—it is low in calories (around 90 kcal per serving) but contains enough macro- and micronutrients (proteins, fatty acids, calcium, phosphorus, and vitamins) to cover a person's daily needs.

The purpose of this chapter is to provide an overview of the nature of such bacterial associations in yogurts, describe the interactions among the bacteria involved, and detail how bacterial metabolic activities impact the properties of the end product. We will focus on traditional yogurts; yogurts or fermented milk products that contain probiotic species (such as *Bifidobacterium* or *Lactobacillus* spp.), stabilizers, added aromas, or other additives will not be discussed.

2. A Fermented Food Originating in a Mutually Beneficial Association between Two Thermophilic LAB Species

2.1 How Yogurt Bacteria Grow Together in Milk

Typically, yogurts are produced at temperatures around 42°C, which promotes the optimal growth of both *S. thermophilus* and *Lb. bulgaricus*. When milk is inoculated with these two bacteria, they usually grow in succession and *S. thermophilus* presents diauxic growth (Tamime and Robinson 1999, Courtin et al. 2002, Letort et al. 2002, Courtin and Rul 2004, Sieuwerts et al. 2010; Fig. 1). *S. thermophilus* first grows exponentially (for the first 90 to 120 min) and then experiences a short latency period. *Lb. bulgaricus* stays at inoculation levels. *S. thermophilus* subsequently resumes growth, albeit at a reduced rate, and *Lb. bulgaricus* starts to grow exponentially. As the milk becomes acidified, reaching a pH of around 5.2, *S. thermophilus* stops growing. In contrast, the growth of *Lb. bulgaricus* continues until

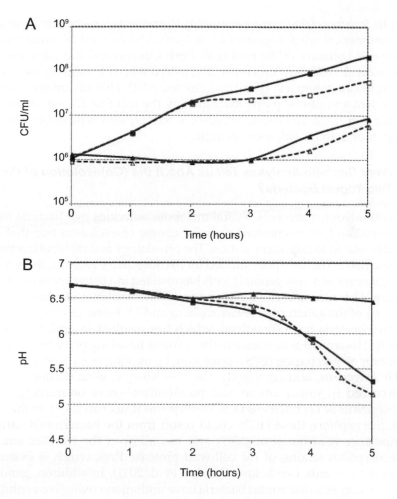

Figure 1. (A) Growth curves of *S. thermophilus* CNRZ385 in pure cultures (□) or co-cultures (■) and of *Lb. bulgaricus* CNRZ398 in pure cultures (▲) or co-cultures (△); cultures were grown in microfiltered milk (Marguerite®).

(B) Acidification curves of milk inoculated with *S. thermophilus* CNRZ385 (■), *Lb. bulgaricus* CNRZ398 (▲), or both bacteria (△).

pH levels drop to around 4.4 (Beal and Corrieu 1991). The acidification curves usually correlate well with these growth patterns (Fig. 1). The first round of acidification corresponds to *S. thermophilus'* period of exponential growth; during the latency period, there is a slowdown in acidification. Then, acidification accelerates with the tandem growth of both bacteria.

Lb. bulgaricus is better than *S. thermophilus* at handling the acidic environment, possibly partly because it can transform ornithine into putrescine, which raises the intracellular pH (Azcarate-Peril et al. 2004).

Despite *S. thermophilus'* greater sensitivity to acidity, this bacterium generally has a numerical advantage over *Lb. bulgaricus* by the end of fermentation (Pette and Lolkema 1950a, Beal et al. 1994, Courtin and Rul 2004, Herve-Jimenez et al. 2008, Ben-Yahia et al. 2012), even when *Lb. bulgaricus* starts off at a higher inoculum level (Béal and Corrieu 1991). This advantage is strain dependent and can be partly explained by the fact that *Lb. bulgaricus* has stricter nutritional requirements. Also, *S. thermophilus* is probably a better competitor than *Lb. bulgaricus* in milk.

2.2 What Genome Analyses Tell us About the (Co)evolution of the Two Yogurt Bacteria?

Growth in yogurt involves several metabolic activities that bacteria have conserved and/or re-enforced over the course of evolution and that are directly related to milk composition. The physiology and metabolic activity of these two LABs have been studied for decades. More recently, the advent of sequencing and post-genomic tools has resulted in a better, more complete picture of how these bacteria evolved and how they have adapted to milk. Analysis of the genomes of *S. thermophilus* and *Lb. bulgaricus* suggests that the two bacteria have coevolved, which has resulted in optimized joint growth. Horizontal gene transfers (HGTs) may be taking place between the two: exopolysaccharide (EPS) genes may be moving from *S. thermophilus* to *Lb. bulgaricus*, and conversely, the *cbs-cblB-cysE* gene cluster—which is involved in sulfur amino acid metabolism—may be moving from *Lb. bulgaricus* or *Lb. helveticus* to *S. thermophilus* (Liu et al. 2009). In the case of *S. thermophilus*, these HGTs could result from the bacterium's natural competence (Gardan et al. 2009) and has allowed the transfer among *S. thermophilus* strains, of the cell-wall protease PrtS, which is essential for growth in milk (see below; Dandoy et al. 2011). In addition, genome analysis suggests that yogurt bacteria have undergone reductive evolution (Bolotin et al. 2004, Makarova et al. 2006): their "domestication" in milk has led to metabolic simplification and specialization. More specifically, among the Lactobacillaceae, the two yogurt LABs have the highest number of pseudogenes, frameshift mutations, nonsense mutations, and deletions (around 10%; Bolotin et al. 2004, Makarova et al. 2006, Goh et al. 2011), leading to a loss of functional genes.

2.3 Yogurt Bacteria are Metabolically Well Adapted to the Composition of Milk

The growth of *S. thermophilus* and *Lb. bulgaricus* in milk largely depends on their ability to efficiently use the medium's major carbon and nitrogen sources (lactose and caseins, respectively), as well as to synthesize any growth-limiting nucleotide bases that are lacking. These metabolic traits are key for the bacteria's associated growth and thus have a major impact on the properties of the resulting yogurt; they will be discussed further below. Furthermore, *S. thermophilus* produces CO_2 (Driessen et al. 1982, Tinson et al. 1982, Spinnler et al. 1987, Ascon-Reyes et al. 1995), which stimulates the growth of *Lb. bulgaricus*. The CO_2 comes from the decarboxylation of urea—present in milk—by urease (Tinson et al. 1982), which most *S. thermophilus* strains possess (Juillard et al. 1988). Other factors also influence the specifics of this association, such as the production of deleterious compounds like lactate, H_2O_2, or bacteriocins; however, they will not be discussed here.

2.3.1 Lactose metabolism

The main source of carbohydrates in milk is lactose. Because it is still present at high concentrations (around 40 g/L) at the end of fermentation, it is not growth limiting for yogurt bacteria and thus does not directly fuel competition. Yogurt bacteria prefer lactose over other simple sugars, such as glucose or sucrose, as a carbon source (Chervaux et al. 2000, Goh et al. 2011, Thomas et al. 2011), probably because both organisms possess an efficient lactose/galactose antiporter LacS. Lactose is imported into the bacteria and hydrolyzed by β-galactosidase (LacZ) into two compounds: galactose, which is largely exported by LacS permease, and glucose, which feeds the glycolysis pathway. The galactose moiety of lactose is not used by most *S. thermophilus* strains, mainly because they have low galactokinase activity (Vaughan et al. 2001, Vaillancourt et al. 2004) or low levels of induction of the galactose promotor (Van den Bogaard et al. 2004). However, when growth conditions become difficult (e.g., lactose is limited and galactose is present at high concentrations), galactose can be used (Terence and Vaughan 1984, Hutkins et al. 1985, Levander et al. 2002).

Lactose utilization is chromosomally encoded in both yogurt bacteria, which ensures that this trait is maintained. In contrast, it is plasmid encoded in other LABs, such as *Lactococcus lactis*, and is thus less stable. Pyruvate, the end product of glycolysis, is then converted into (L- and D-) lactate, which is excreted, leading to milk acidification. This process is mediated by L-lactate dehydrogenase (L-Ldh) in *S. thermophilus* and by D-lactate dehydrogenase (D-Ldh) in *Lb. bulgaricus*. Even if L-Ldh genes are present in the *Lb. bulgaricus* genome, 90% of pyruvate is nonetheless converted into D-lactate.

2.3.2 Nitrogen metabolism

Optimal bacterial growth depends on efficient protein synthesis and, as a result, on the availability of amino acids (AAs). LABs are auxotrophic for amino acids: one to several in *S. thermophilus* strains (Hols et al. 2005, Pastink et al. 2009) and 15–20 in *Lb. bulgaricus* strains, which are only able to synthesize 3–4 AAs (Asp, Asn, Thr, +/– Lys; Van de Guchte et al. 2006, Hao et al. 2011).

Milk is poor in nitrogen compounds that can be directly assimilated by LABs (free amino acids and short peptides), but yogurt bacteria possess a complex and efficient proteolytic system that provides them with exogenous nitrogen sources stemming from milk proteins. This multiprotein system has been extensively studied (for reviews, see Christensen et al. 1999, Savijoki et al. 2006, Liu et al. 2010). It is able to hydrolyze caseins—the major proteins in milk. It transports the resulting oligopeptides into the cells and then degrades them into smaller oligopeptides and AAs. The system is composed of a cell-wall protease (Prt), various AA and peptide transporters, and several peptidases, mostly intracellular, the coordinated action of which leads to the recovery of free AAs for protein synthesis (Fig. 2).

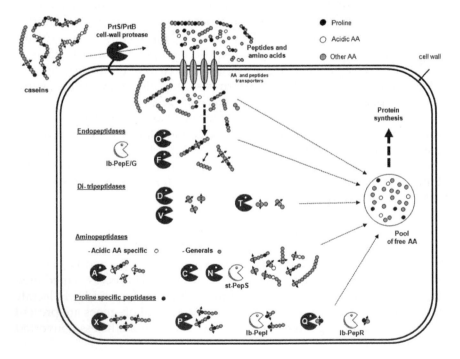

Figure 2. The proteolytic system of *S. thermophilus* and *Lb. bulgaricus*.
🌑: protease/peptidase common to the two bacteria
ℭ: bacterium-specific peptidase; "st" stands for *S. thermophilus* and
   "lb" for *Lb. bulgaricus*

Lb. bulgaricus has higher overall levels of proteolytic activity than *S. thermophilus* (Shankar and Davis 1978, Rajagopal and Sandine 1990, Courtin and Rul 2004). It liberates most of the AAs (Courtin and Rul 2004) that stimulate *S. thermophilus* growth, including valine, histidine, glycine, leucine, isoleucine, methionine, and various dipeptides (Pette and Lolkema 1950b, Bautista et al. 1966, Accolas et al. 1971, Bracquart et al. 1978, 1979, Shankar and Davies 1978, Radke-Mitchell and Sandine 1984, El-Soda et al. 1986, Rajagopal and Sandine 1990, Courtin and Rul 2004). Levels of overall proteolytic activity in *Lb. bulgaricus* vary among strains and differentially promote the growth of *S. thermophilus* when the two bacteria are associated in milk (Courtin and Rul 2004).

Cell-wall protease—PrtB in *Lb. bulgaricus* and PrtS in *S. thermophilus*—plays a major role in stimulating growth in milk because it initiates the breakdown of caseins into various oligopeptides. PrtS is often absent from older strains of *S. thermophilus* (Shahbal et al. 1991) but is frequently found in more recent industrial strains. Prt is particularly important in allowing *Lb. bulgaricus*, which is poorly equipped for AA synthesis (see above), to grow in milk; PrtS- or PrtB-negative mutants develop more slowly than do wild-type strains (Gilbert et al. 1997, Courtin et al. 2002). The PrtS-encoding gene was probably acquired *via* HGT from a species related to *S. suis* (Delorme et al. 2010), and it can be transferred by natural competence to other *S. thermophilus* strains (Dandoy et al. 2011). The expression of PrtS is induced during the latency period (Letort et al. 2002), most probably because available peptides are lacking. In co-cultures, the following are true: (i) PrtB gene expression is probably induced by the presence of *S. thermophilus*, which reduces the peptides available to *Lb. bulgaricus* (as compared to pure cultures; Sieuwerts et al. 2010), and (ii) PrtS is no longer essential to *S. thermophilus* growth if *Lb. bulgaricus* PrtB is present. Indeed, PrtB may be more efficient than PrtS in making nitrogen available because when PrtB-positive *Lb. bulgaricus* strains co-occur with PrtS-negative *S. thermophilus* strains, *S. thermophilus* populations are larger than when PrtS-positive *S. thermophilus* strains co-occur with PrtB-negative *Lb. bulgaricus* strains (Courtin et al. 2002). We also cannot rule out the possibility that the casein-hydrolyzation specificities of PrtB and PrtS are different. For example, the substrate-binding region, which influences Prt specificity, differs between PrtS (Fernandez-Espla et al. 2000) and PrtB (Gilbert et al. 1996).

Yogurt bacteria also possess different transport systems—for AAs, dipeptides, tripeptides, and oligopeptides (the latter have been extensively studied in LABs; for a review, see Savijoki et al. 2006)—that can efficiently target nitrogenous compounds in the medium. *Lb. bulgaricus* lacks the general DtpT di/tri-peptide transporter found in *S. thermophilus*, but its absence may be compensated for by a Dpp transporter that preferentially takes up hydrophobic di/tripeptides. Some *S. thermophilus* transport systems are upregulated in co-cultures, including the oligopeptide carrier

Ami/Opp (Sieuwerts et al. 2010) and potential polar AA transporters (Hervé-Jimenez et al. 2009).

S. *thermophilus* and *Lb. bulgaricus* have a similar number of protease/ peptidase genes (44 [Hols et al. 2005] vs. 45–49 [Hao et al. 2011, Zheng et al. 2012], respectively). Around a dozen have been studied to determine their roles in nitrogen metabolism (Fig. 2). They were found to have various peptide hydrolysis specificity and, *a priori*, are sufficient to meet bacterial needs for AAs. There are even peptidases dedicated to hydrolyzing proline-containing peptide bonds, which are difficult to break down but essential for casein degradation as caseins are rich in proline.

One group of AAs is particularly important in allowing yogurt bacteria to grow in milk: branched-chain AAs (BCAAs), arginine, and cystein (Bracquart and Lorient 1977, Garault et al. 2000). These AAs are predicted to be among the most common in proteins of *S. thermophilus* (Hervé-Jimenez et al. 2009) and *Lb. bulgaricus* (Sieuwerts et al. 2010); in contrast, they are largely lacking from caseins and the two bacteria probably compete for these AAs. This fact may explain why *S. thermophilus* and *Lb. bulgaircus* increases BCAA and arginine biosynthesis, and BCAA permease activity, respectively, in co-cultures as compared to in pure cultures. In addition, when the two yogurt bacteria co-occur in milk, there is an increase in activity along the serine-to-methionine and cysteine-conversion pathways, as compared to in pure cultures (Sieuwerts et al. 2010). Finally, *S. thermophilus* and *Lb. bulgaricus* possess the necessary peptidolytic and transport pathways for casein exploitation; they hydrolyze caseins into free amino acids, thus fulfilling their protein synthesis needs. When the two bacteria co-occur in milk, they both compete and complement each other in terms of their nitrogen metabolisms.

2.3.3 Formate, folate, and purine metabolism

Some of the first metabolic exchanges described in *S. thermophilus* and *Lb. bulgaricus* associations in milk were interactions involving folic acid (Rao et al. 1984, Sybesma et al. 2003), pyruvic acid (Higashio et al. 1978), formic acid (Veringa et al. 1968), and CO_2 (Driessen et al. 1982). These compounds all fed, directly or indirectly, into the purine biosynthesis pathway (Fig. 3). Formate is necessary for the synthesis of purine bases (i.e., xanthine, adenine, and guanine) that are nucleic acid precursors (Suzuki et al. 1986); both yogurt bacteria require it to grow (Galesloot et al. 1968, Suzuki et al. 1986, Derzelle et al. 2005, Horiuchi and Sasaki 2012, Nishimura et al. 2013).

Depending on the strain, *Lb. bulgaricus* gains a boost in growth at formate concentrations ranging from 0.5 to 27 mM (Galesloot et al. 1968, El-Abbassy et al. 1993, Horiuchi and Sasaki 2012). In addition, Courtin and Rul (2004) showed that formate concentrations decreased more strongly in

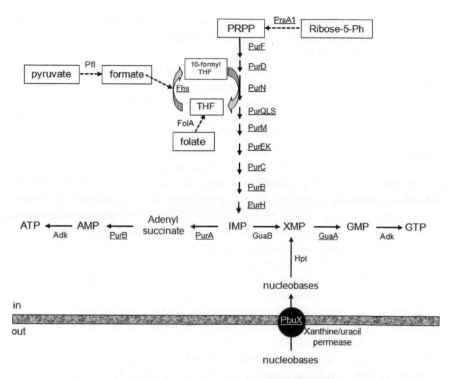

Figure 3. Purine pathway in *S. thermophilus*.
Underlined proteins are proteins or their corresponding genes that were less abundant or repressed during the growth of *S. thermophilus* in co-culture in milk with *Lb. bulgaricus* (data from Hervé-Jimenez et al. 2009).

milk co-cultures than in pure *S. thermophilus* cultures; they differed by a factor of 4 to 8, depending on the *Lb. bulgaricus* strain present and probably because of the latter's consumption of formate.

Pyruvate formate lyase (Pfl), which converts pyruvate into formate, is abundantly expressed in *S. thermophilus* and is induced when the species is grown in milk (as compared to when it is grown in M17-rich medium; Derzelle et al. 2005). If formate (5 mM) or purines (adenine and guanine, 50 µM) are added to the milk, Pfl is no longer overexpressed, and *S. thermophilus* experiences a boost in growth (Derzelle et al. 2005). In contrast, Pfl is absent from the *Lb. bulgaricus* genome (Van de Guchte et al. 2006). In several bacterial species, Pfl activity is oxygen sensitive (Knappe et al. 1974, Yamada et al. 1985, Sawers and Watson 1998). It has recently been suggested that, in *S. thermophilus*, oxygen conversion by NADH oxidase (NOX) could improve growing conditions, by promoting Pfl activity and, as a consequence, formate synthesis (Horiuchi and Sasaki 2012). Interestingly, when added to milk, formate stimulates EPS production in *Lb. bulgaricus* by a factor of 4, which may be a mechanism contributing to improved growth as well as to enhanced cell-wall synthesis and bacterial division (Nishuimura et al. 2013).

Thanks to post-genomic approaches, it has been discovered that the purine biosynthesis pathway in *S. thermophilus* slows down in co-cultures; almost all the enzymes involved become less abundant or are expressed at lower levels (Hervé-Jimenez et al. 2009) (Fig. 3). This result was unexpected because the pathway is ramped up when the bacterium is grown in pure cultures (Hervé-Jimenez et al. 2008), confirming that purines are essential for *S. thermophilus'* growth in milk. One might hypothesize that, in co-cultures, purines or purine precursors are supplied by *Lb. bulgaricus*; indeed, a potential xanthine/uracil permease (a transporter of purine precursors) was expressed at higher levels when *S. thermophilus* was associated with *Lb. bulgaricus* (Hervé-Jimenez et al. 2009). However, a more recent post-genomic study of yogurt bacteria associations (Sieuwerts et al. 2010) showed that, when fermentation times were similar (around 5 h), purine synthesis and folate cycling pathways were upregulated in *S. thermophilus*, while folic acid and purine synthesis were downregulated in *Lb. bulgaricus*. These contradictory results underscore the importance of milk type and strain identity, as they differed in the two studies (skim milk vs. μ-filtered milk and strains CNRZ1066-ATCC BAA-65 vs. LMG18311-ATCC11842, respectively). However, yogurt bacteria need to utilize purines to grow in milk, and the process can be modulated. It is assumed that *S. thermophilus* supplies *Lb. bulgaricus* with the compounds needed for purine biosynthesis, such as formate, a precursor, and folic acid, which is a co-factor and produced in co-cultures (Crittenden et al. 2002). In turn, *Lb. bulgaricus* may provide *S. thermophilus* with other purine precursors.

The protocooperative association of the two yogurt bacteria has been studied for years (for a review see Sieuwerts et al. 2008), but new genomic and post-genomic approaches have made it possible to gather more detailed knowledge about the general and specific metabolic mechanisms involved (Fig. 4). They have revealed new, entirely unexpected interactions and exchanges. For instance, iron metabolism in *S. thermophilus* (Hervé-Jimenez et al. 2009) and fatty acid metabolism in *Lb. bulgaricus* (Sieuwerts et al. 2010) are modulated when the bacteria are grown in milk co-cultures, but not when they are grown in pure cultures. It has become clear that the association is the sum of a variety of bacterial interactions, both positive ones, such as mutualism and commensalism, as well as negative ones, such as competition and amensalism.

These new findings are crucial when it comes to designing and selecting novel compatible and complementary strains and strain cocktails with specific properties, with a view to creating tailored dairy products or developing entirely new foods.

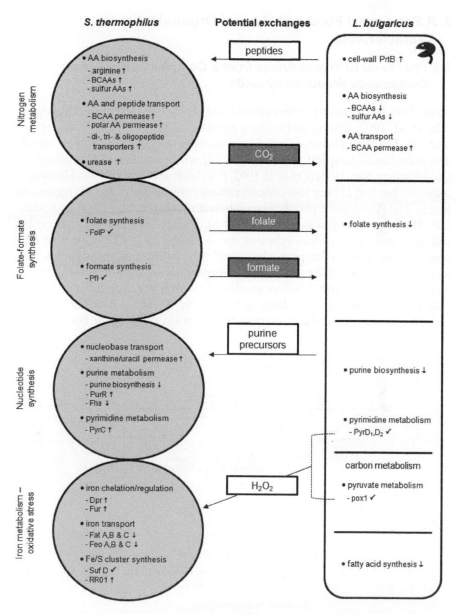

Figure 4. Metabolite exchanges between the two yoghurt bacteria, in particular regarding the metabolism pathways that were modulated during their co-culture in milk.

↑,↓: Proteins or genes that, respectively, were more or less abundant or expressed during the co-culture.

✓: Proteins or genes that were present or expressed without quantitative variation during the co-culture.

3. A Fermented Food with Typical Organoleptic Characteristics

3.1 The Flavor of Yogurt Arises from a Complex Mix of Aroma Compounds Produced by LABs

More than 100 different aroma compounds have been identified in yogurt (Ott et al. 1997, Cheng 2010), as a result of GC (gas chromatography) and GC-MS (gas chromatography-mass spectrometry) analyses, which are sometimes coupled with human olfactory assays (GC-sniffing or GC-olfactory detection) (Ott et al. 1997, Friedrich and Acree 1998). However, most are present at very low concentrations; only a few occur at significant levels. The flavor we typically associate with yogurt comes from its acidity (i.e., the presence of lactic acid) (Ott et al. 2000b). It is also influenced by different carbonyl compounds that were identified quite some time ago and that result from the proteolysis and degradation of amino acids into alcohol, aldehydes, and esters. These compounds are mainly acetaldehyde, acetoin, diacetyl, and 2,3-pentanedione, which are, for the most part, produced by bacterial metabolic activity (Fig. 5). Their production and/ or accumulation in milk in co-cultures is strain dependent because, for

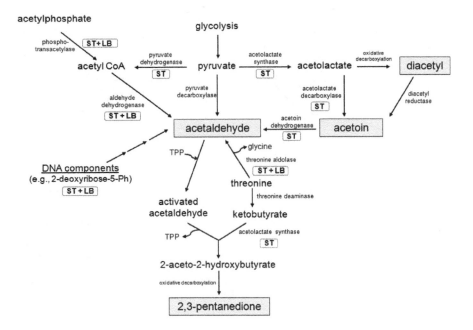

Figure 5. Production pathways of the main aroma compounds found in yogurt bacteria. ST, LB, and ST + LB indicate that the gene or enzyme is present in *S. thermophilus*, *Lb. bulgaricus*, and *S. thermophilus* and *Lb. bulgaricus* together, respectively.

example, when a given strain of *S. thermophilus* co-occurs with different strains of *Lb. bulgaricus*, different levels of acetaldehyde, acetoin, or diacetyl are generated (Courtin and Rul 2004). Their levels do not vary significantly when yogurt is stored at 4°C.

Acetaldehyde is the most typical yogurt flavor component (Pette and Lolkema 1950c, Dumont and Adda 1973, Law 1981) and is responsible for the food's fresh and fruity notes (e.g., hints of green apple and nuts). It is usually found at concentrations of 1 to 25 mg/L (Hamdan et al. 1971, Bottazzi et al. 1973, Rysstad and Abrahamsen 1987, Kneifel et al. 1992, Beshkova et al. 1998). The accumulation of acetaldehyde as yogurt fermentation progresses could be related to the bacteria's limited ability to use it (as hypothesized by Manca de Nadra et al. [1988]) and convert it into ethanol, as other LABs do (Lees and Jago 1976, Chaves et al. 2002).

Acetaldehyde can be produced in several ways: via pyruvate decarboxylation (from DNA) and via alcohol dehydrogenation (Fig. 5). However, the acetaldehyde found in yogurt bacteria probably mainly stems from the transformation of the amino acid threonine (Thr) into glycine (Gly) and acetaldehyde by threonine aldolase (Lees and Jago 1976, Raya et al. 1986, Ott et al. 2000a, Chaves et al. 2002). A threonine aldolase has been purified and characterized in *Lb. bulgaricus* (Manca de Nadra et al. 1987), and there is evidence for its involvement in flavor development (Marshall and Cole 1983). It is inhibited by the presence of glycine, via pH dependent way (Manca de Nadra et al. 1987), similarly to the threonine aldolase found in *S. thermophilus* (Marranzini et al. 1989). When threonine is added to milk, acetaldehyde production in *S. thermophilus* increases (Chaves et al. 2002, Ozer and Atasoy 2002); additionally, the higher Thr to Gly ratio generally enhances acetaldehyde production by both bacteria (Marranzini et al. 1989). Depending on the strain, decarboxylase or aldehyde dehydrogenase activity, which results in acetaldehyde production (Fig. 5), may or may not be present (Lees and Jago 1976 and Raya et al. 1986, respectively).

Apart from acetaldehyde, the other essential aroma compounds are diacetyl, acetoine, and 2–3 pentanedione, which give yogurt its buttery note. Diacetyl is also responsible for yogurt's full, delicate flavor (Rasic and Kurmann 1978); it occurs at concentrations of 0.2 to 3 mg/L (Cheng 2010) and is produced by both yogurt bacteria (Dutta et al. 1973, Rasic and Kurmann 1978). Production of 2–3 pentanedione is ramped up 3 to 5 fold in co-cultures, as compared to its combined production by each bacterium considered separately (1 mg/kg; Imhof et al. 1995).

While diacetyl and acetadehyde are the main compounds that define yogurt's flavor (Ott et al. 1997, Friedrich and Acree 1998), the ratio between the different aromatic compounds matters more than their individual concentrations (Pette and Lolkelma 1950c, Bottazzi and Vescoso 1969). For example, Bottazzi and Vescoso (1969) observed that a stronger flavor was obtained when the acetaldehyde-to-acetone ratio was 2.8:1 but not when it was 0.4:1. Diacetyl takes on greater importance when the acetaldehyde concentration is low (Groux 1973, Rysstad and Abrahamsen 1987). Although a 1:1 ratio of acetaldehyde to diacetyl yields the preferred, typical yogurt's flavor associated with yogurt (Bottazzi and Dellaglio 1967, Zourari and Desmazeaud 1991), ratios of 7–10:1 can nonetheless create a "good" flavor (Beshkova et al. 1988). In addition, some low-acetaldehyde yogurts present a typical aroma (Hamdan et al. 1971, Groux 1973), possibly thanks to the presence of diacetyl (Kneifel et al. 1992).

Genome analysis has revealed that *Lb. bulgaricus* is probably poorly equipped to produce the aroma compounds mentioned above (Hao et al. 2011). *S. thermophilus* possesses several aminotransferases that convert Asp, aromatic AAs, or BCAAs into flavor compounds and produces alpha ketoglutarate from glutamate dehydrogenase, thus helping generate different volatiles (Pastink et al. 2009).

Finally, yogurt flavor results from a subtle balance between the main flavor compounds, which stem from bacterial proteolytic processes, and from fatty-acid derivatives (Turcic et al. 1969, Dumont and Adda 1973, Beshkova et al. 1998).

3.2 The Texture of Yogurt Results from the Acidification Capacity and Exopolysaccharide Production of LABs

In addition to its unique flavor, yogurt also has a very specific texture, which plays an important role in gustative quality. This texture is largely a function of bacterial activities: the acidification of milk leads to the formation of a coagulated gel and the production of EPSs creates a matrix which participates in shaping yogurt texture.

During milk fermentation, pH drops, and when it reaches the isoelectric point of the caseins (pH of 4.6), the latter precipitate and then aggregate, generating a gel network in which water and fat are embedded. The firmness and viscosity of the curd ultimately depend on the final pH, as well as on bacterial proteolytic activity, which can result in syneresis (Marshall 1987) and change the structure or microstructure of the yogurt.

S. thermophilus and *Lb. bulgaricus* both produce EPSs (Cerning et al. 1986, 1988, Cerning 1990, Laws et al. 2001). These polysaccharides are key players in the development of yogurt's rheological properties and help determine yogurt firmness, unctuosity, stickiness, and mouthfeel (Broadbent et al. 2003, Vaninelgem et al. 2004, Purwandari et al. 2007, Qin et al. 2011). Polysaccharides can bind to casein micelles and thus both increase water retention in the curd and reduce whey exudation at the yogurt's surface (Amatayakul et al. 2006a,b, Purohit et al. 2009). *S. thermophilus* EPSs also act to protect *Lb. bulgaricus* (Ramchandran and Shah 2009).

Fermentation conditions (e.g., temperature, time of incubation), the composition of the medium (e.g., carbon and nitrogen sources, the carbon-to-nitrogen ratio), the level of acidity, and the identity of the strain affect EPS quantity and sugar composition (De Vuyst et al. 1998, Degeest and De Vuyst 1999, Tamime and Robinson 1999, Zisu and Shah 2003, Vaningelgem et al. 2004, Zhang et al. 2014). Depending on the strain and culture conditions, *S. thermophilus* and *Lb. bulgaricus* produce EPSs either during the exponential growth phase (De Vuyst et al. 1998) or late fermentation (Petry et al. 2000, Broadbent et al. 2003, Sieuwerts et al. 2010). Production levels are higher in co-cultures than in pure cultures (Cerning 1990, Frengova et al. 2000, Sieuwerts et al. 2010), possibly because the drop in pH favors EPS production by *Lb. bulgaricus*. EPS production by *Lb. bulgaricus* can be stimulated by formate and other compounds such as vitamins or nucleobases.

EPSs are synthesized in the cytoplasm by polymerization of repeating sugar units that are attached to a lipid carrier. They are translocated to the membrane before being secreted, which is what distinguishes them from capsular polysaccharides (which are permanently attached to the cellular surface). LAB EPSs comprise multiple copies of an oligosaccharide that contains several residues linked in different patterns. All the EPSs characterized in yogurt bacteria up until now are mainly composed of galactose. Galactose's omnipresence is probably a consequence of the ubiquity of lactose in milk—it is the main carbon source and, when hydrolyzed, forms (i) glucose, which is preferentially used for glycolysis-based energy production and (ii) galactose, which can be used to synthesize nucleotide sugars for EPS production.

In *S. thermophilus*, EPSs are most commonly composed of galactose and glucose (De Vuyst et al. 1988, Petit et al. 1991, Laws et al. 2001, Marshall et al. 2001, Nordmark et al. 2005, Sawen et al. 2010, Qin et al. 2011). In some strains, additional mannose (Cerning et al. 1986) or rhamnose (Cerning et al. 1986, Escalante et al. 1998) are also included; in others, it is small amounts of xylose, arabinose, and mannose (Cerning et al. 1988). Alternatively, EPSs can also be solely composed of rhamnose and galactose (Ariga et al. 1992, Faber et al. 1998, 2001) or contain sugar derivatives such as acetylgalactosamine and/or fucose (Doco et al. 1990, Stingele et al. 1996, Laws et al. 2001), D-galactopyranose and L-rhamnopyranose residues (Bubb et al. 1997), or N-acetylglucosamine and glucuronic acid, which are components of hyaluronic acid (Izawa et al. 2009). In *Lb. bulgaricus*, most of EPSs described to date contain galactose, glucose, and rhamnose (Cernning et al. 1986, Zourari et al. 1992, Gruter et al. 1993, Grobben et al. 1995, Petry et al. 2000, Marshall et al. 2001, Lamothe et al. 2002) and occasionally traces of mannose (Petry et al. 2000). However, some *Lb. bulgaricus* EPSs are composed of galactose and glucose (Petry et al. 2000, Faber et al. 2001); galactose, glucose, and traces of mannose (Bouzar et al. 1996); or xylose and arabinose (Cerning et al. 1988). Even if both bacteria produce EPSs containing similar sugars, the proportions of these sugars differ, generating EPS diversity and shaping yogurt viscosity in different ways (Faber et al. 1998). In addition, EPS composition can vary during fermentation (Bouzar et al. 1997).

Several EPS gene clusters have been described in different yogurt bacterial strains, which often possess two of said clusters. The *eps* clusters usually exhibit a modular organization and a chimeric structure; they are highly diverse in terms of sequence and genetic context across clusters, both within and among strains (Stingele et al. 1996, Bourgoin et al. 1999, Jolly and Stingele 2001, Broadbent et al. 2003, Rasmussen et al. 2008, Goh et al. 2011, Hao et al. 2011). Sequence divergence ranges from 10 to 50% for *S. thermophilus* EPSs (Bourgoin et al. 1999). The diversity of these genes suggests that HGTs are being acquired from other bacteria in the environment. For instance, *S. thermophilus* strain LMD-9 contains some *eps* genes that are very similar to those found in *Lactococcus lactis* (Bourgoin et al. 1999, Goh et al. 2011). In addition, Wu et al. (2014) has proposed that *S. thermophilus* may produce EPSs of different molecular sizes because the researchers detected the presence of two gene pairs that are involved in chain-length determination.

4. A Fermented Food with Probiotic Potential

Yogurt can also be considered to be a probiotic food (Guarner et al. 2005). Indeed, yogurt starters clearly meet the definition of probiotics—"Live microorganisms [that] when administered in adequate amounts confer a health benefit on the host"—proposed by the Joint Food and Agriculture Organization/World Health Organization Working Group (2002) and adopted by the International Scientific Association for Probiotics and Prebiotics (Reid et al. 2003). In addition, yogurt possesses well-documented healthful properties, including the ability to help alleviate lactose intolerance. This latter characteristic is the basis of a health claim recently accepted by EFSA (Section 4.1, 2010).

There are numerous examples in the literature of the probiotic effects of LAB, when delivered in capsules, sprays, or via yogurt. However, the goal here is to focus exclusively on the probiotic effects of traditional yogurt, which includes both and only *S. thermophilus* and *Lb. bulgaricus*.

4.1 Bacterial β-galactosidase Participates in Lactose Digestion in the Gastrointestinal Tract

Lactose maldigestion/intolerance is the main cause of milk intolerance in adults. Yogurt's ability to alleviate the symptoms of lactose maldigestion is well documented and recognized at the regulatory level (FAO/WHO 2001, 2002, EFSA 2010). Lactose maldigestion results from reduced or absent lactase activity (the hydrolysis of lactose into glucose and galactose) in the brush border membrane of the small intestine. Undigested lactose travels to the colon, where it is subject to fermentation by resident microbiota, leading to excessive gas production (i.e., methane, carbonic gas, hydrogen) and, consequently, symptoms such as abdominal pain, bloating, cramps, or diarrhea. Lactose intolerance/maldigestion can be diagnosed by the breath-hydrogen concentration test. It measures the quantity of exhaled hydrogen, which is proportional to the quantity of ingested lactose reaching the colon and is thus inversely proportional to the level of lactose digestion in the intestine (Savaiano 2014). Hypolactasia (lactase deficiency) is a physiological condition and is affected by age, sex, and ethnic origin. For instance, lactase activity rapidly decreases after weaning in the majority of children. Hypolactasia prevalence is over 50% in adult Africans, American Hispanics, and American Indians, and close to 100% in some Asian populations (Wilt et al. 2010, Lember 2012), because of a genetically programmed loss of lactase after weaning.

Though the lactose concentration in yogurt is similar to that in milk (around 40 g/L), yogurt consumption is recommended for people suffering from lactose intolerance because it alleviates the symptoms of the disorder. For example, lactose intolerance/maldigestion is more severe in young children (e.g., causing acidic diarrhea), but yogurt consumption appears to help (i.e., is associated with a decrease in acidic feces occurrence and volume) (Dewit et al. 1987, Shermak et al. 1995). These benefits can only be obtained if the product contains live yogurt bacteria; they are lost if the yogurt is thermized and the bacteria are killed (Goodenough and Kelyn 1975, Gilliland and Kim 1984, Savaiano et al. 1984, Lerebours et al. 1989, Pochart et al. 1989, de Vrees et al. 2001).

Older work by Alm et al. (1982) suggested that the lactase produced by yogurt bacteria (β-galactosidase, see above)—which all strains of yogurt bacteria have—could promote lactose hydrolysis in the digestive tract. This hypothesis was later experimentally supported in mice by Douault et al. (2002). In lactase-deficient subjects, more than 90% of the lactose found in the small intestine was hydrolyzed by the β-gactactosidase of yogurt starters (Marteau et al. 1990). The results were dependent on the size of the bacterial population present in the yogurt; total hydrolysis occurred when 10^8 UFC/g yogurt were present but was limited when only 10^6 UFC/g were present (Pelletier et al. 2001).

It has been suggested that lactose could enter yogurt bacteria as a result of the permeabilization of their envelopes by bile (Noh and Gilliland 1994), or that bacteria lyze and release their β-galactosidase into the lumen (Marteau et al. 1997). Yogurt's texture—which is more viscous and thicker than that of milk—could also slow down gastric emptying and gastrointestinal transit times (Marteau et al. 1990), thus favoring the action of residual intestinal lactase on enterocytes by favoring contact with lactose in the lumen. In rats, intestinal lactase activity was higher in animals fed yogurt than in animals fed pasteurized yogurt or milk (Goodenough and Kleyn 1975, Besnier et al. 1983, Garvie et al. 1984).

4.2 Yogurt Consumption Can Have Beneficial Effects on Infections, Inflammatory Diseases, and Cancers

Infections, antibiotic treatments, and tube feeding are major causes of diarrhea. In developing countries, bacterial enteropathogens, such as enterotoxigenic *E. coli*, are frequently responsible for diarrhea in children and travellers (Narayan et al. 2010). The WHO (1995) has recommended using yogurt to treat acute diarrhea because yogurt is associated with the production of antimicrobial compounds (H_2O_2, bacteriocins, or organic acids such as lactic acid), acts via immunomodulation, and inhibits pathogen adhesion to the intestinal epithelium. In particular, yogurt helps limit chronic diarrhea in children (Boudraa et al. 1990, Touhami et al. 1992), by reducing both its frequency and duration (Boudraa et al. 2001).

Maintaining microbial equilibrium is an important part of intestinal physiological health and homeostasis; it also helps eliminate pathogenic enteric bacteria. For patients with chronic liver disease, consuming yogurt can decrease microbiota imbalances (Liu et al. 2010), and in patients with inflammatory bowel disease (IBS), yogurt consumption has been linked to a decline in the pathogen *Bilophila wadworthia* (Veiga et al. 2014). The feces of subjects that have consumed yogurt have higher microbial densities and contain larger amounts of LABs as compared to *Bacteroides* species (Garcia-Albiach et al. 2008). Yogurt consumption resulted in lower *E. coli* counts in patients suffering from chronic liver disease (Liu et al. 2010). It also led to lower levels of *Clostridium* in the elderly, which is beneficial because Clostridia generate putrefactive products that are potentially toxic for the colic mucosa (Canzi et al. 2002). Interestingly, a recent meta-analysis looking at the influence of diet on tooth erosion in children and adolescents has indicated that yogurt has protective effects (Salas et al. 2015). *In vitro* experiments suggest they may stem from bactericidal effects on cariogenic *S. mutans* species (Petti et al. 2008).

In the 1990s, research began to suggest that the administration of live LABs could modify immune responses. The effects of each of the two yogurt bacteria on inflammatory and immune responses have been documented, but a few studies have also shown that yogurt containing both bacteria can have beneficial impacts. Yogurt consumption stimulates the immune system and decreases allergies in young adults (20–40 years old) and older adults (55–70 years old) (Trapp et al. 1993, Van de Water et al. 1999) and is recommended for the immunocompromised (Meydani and Ha 2000). Based on the results of *in vitro* and *in vivo* studies, one could hypothesize that the immunomodulatory and anti-inflammatory effects of yogurt could stem from the induction of cytokines (γ-interferon, TNFα, IL-12), the production of immunoglobulins (IgA in particular), and an improved barrier effect due to increased mucosa thickness.

Furthermore, in animal models with induced cancers, yogurt consumption reduces tumor number (Narushima et al. 2010); it also inhibits tumor progression and spread of colon cancer by increasing apoptosis (De Moreno de Leblanc and Perdigon 2004), by decreasing the inflammatory immune response (mediated by increases in IgA), and increasing IL-10 expression (Perdigon et al. 2002). In humans, yogurt consumption is associated with a reduction in colorectal cancer (Pala et al. 2011) and may also reduce the risk of breast cancer (Le et al. 1986, Van't Veer et al. 1989). Again, yogurt's cancer-fighting properties are lost if the food is thermized (Pool-Zabel et al. 1993), underscoring that the benefits are due to the presence of live bacteria.

4.3 The Proteolytic Activity of Yogurt Bacteria Contributes to Yogurt's Antihypertensive Potential

The proteolytic activity of yogurt starters is essential for bacterial growth in milk (see above), as well as for the production of peptides that possess varying degrees of functional activity. Several casein-derived peptides produced by yogurt LABs have been identified, and their biological activity has been demonstrated *in vitro*. *In vivo*, following their ingestion, these yogurt peptides must confront the proteolytic enzymes of the gastrointestinal tract (i.e., pepsin, trypsin, chymotrypsin, as well as carbo-, amino- and membrane endopeptidases), which can hydrolyze them and thus prevent them from reaching their target within the host. Nevertheless, proline-rich peptides can withstand the gauntlet of gastrointestinal hydrolysis (Korhonen and Pihlanto 2003) because peptide bonds containing proline can only be hydrolyzed with specific enzymes. Consequently, the tripeptides IPP (Ile-Pro-Pro) and VPP (Val-Pro-Pro), which are produced in yogurt (Donkor et al. 2007) and have antihypertensive properties (Hirota et al. 2007), resist *in vivo* degradation; a product containing these two peptides has been commercialized. The antihypertensive effects of such peptides results from their inhibition of angiotensin-converting enzyme (ACE), which plays a crucial role in regulating blood pressure by modulating the levels of the vasoconstricting peptide angiotensin II and those of the vasodilatory peptide bradykinin. More recently, β-casein(94–123)-derived peptides, which are present in yogurt, have been shown to enhance the expression of mucin-encoding genes and the number of goblet and Paneth cells in the small intestine (Plaisancié et al. 2013); they may thus play a role in protecting the intestinal epithelium and maintaining its homeostasis. More generally, bioactive peptides with diverse functional properties have been isolated from dairy products (Madureira et al. 2010, Muro Urista et al. 2011).

Apart from the extensive data provided above that demonstrate yogurt's probiotic effects, other findings indicate that yogurt could have promise in treating prevalent pathologies—such as obesity, diabetes II (Margolis et al. 2011), and cardiovascular diseases (Sonestedt et al. 2011)—and maintaining better general health in aging populations (El-Abbadi et al. 2014). Furthermore, the lactate present in yogurt not only acts as an antibacterial agent, but also helps define the food's organoleptic properties (see above) and has bioactive properties (Garrote et al. 2015). It functions as a signaling molecule between bacteria and the host and, in particular, modulates the physiology of the colon epithelium (Rul et al. 2011, Thomas et al. 2011). Recently, Tsilingiri and Rescigno (2013) have proposed the concept of "postbiotic" factors, which are "soluble factors produced by probiotics [that] are sufficient to elicit the desired response" and that "could be a safe alternative for clinical applications, especially in chronic inflammatory conditions like inflammatory bowel disease." However, clinical trials in humans are still lacking, and most of the potential mechanisms involved remain to be specifically elucidated.

5. Conclusions

Yogurt, a healthful, traditional food that has been consumed for millenia, has modern relevance because its combination of nutritional and probiotic proprieties result in unique benefits. It is an interesting food ecosystem which has recently been rediscovered using post-genomic tools. We now have a better understanding of yogurt microbiology, and especially of the metabolic processes that are essential for yogurt production. Nevertheless, data are lacking on some key technical aspects (e.g., the nature of optimal bacterial combinations, the regulation of EPS production) and the factors responsible for yogurt's healthful effects. Research has yet to fully clarify yogurt's probiotic potential and the underlying mechanisms. Indeed, the list of yogurt's possible probiotic properties in the face of various pathologies continues to grow as various animal models are explored. However, findings in humans, particularly in healthy populations, are still needed.

The emergence of new food consumption patterns and of health consciousness on the part of consumers, as well as the global aging of the population, are reasons for continuing to promote fermented foods, such as yogurt, and to develop new ones. For example, vitamin- and/or iron-fortified yogurts, thermized yogurts, and dairy snacks are now available, and it is easy to imagine that these products will continue to be developed and adapted to different populations (e.g., varying in age, sex, ethnic origin, and geographical location) and different nutritional habits. Traditional yogurt probably has benefits that remain to be identified, and exploring its characteristics in greater detail may yield new probiotic applications. Yogurt bacteria EPSs provide a good example: older studies focused on describing these texturing agents of yogurts, and more recent studies have suggested that texturing agents could be involved in inflammatory diseases (Sengul et al. 2006) and have immunostimulating (Makino et al. 2006), antiviral (Nagai et al. 2011), and antibacterial (Aslim et al. 2007) effects.

Acknowledgements

We are very grateful to Jessica Pearce-Duvet for language editing in the manuscript.

Keywords: Acidification, aroma, carbohydrate metabolism, exopolysaccharide, flavor, lactic acid bacteria, *Lactobacillus bulgaricus*, lactose, nitrogen metabolism, probiotics, proteolysis, protocooperation, *Streptococcus thermophilus*, texture, yogurt

References

Accolas, J.P., Veaux, M. and Auclair, J. (1971). Etude des interactions entre diverses bactéries lactiques thermophiles et mésophiles, en relation avec la fabrication des fromages à pâte cuite. Lait 505-506: 249–272.

Alm, L. (1982). Effect of fermentation on lactose, glucose, and galactose content in milk and suitability of fermented milk products for lactose intolerant individuals. Journal of Dairy Science 65: 346–352.

Amatayakul, T., Sherkat, F. and Shah, N.P. (2006b). Syneresis in set yogurt as affected by EPS starter cultures and levels of solids. International Journal of Dairy Technology 59: 216–221.

Amatayakul, T., Sherkat, F. and Shah, N.P. (2006a). Physical characteristics of set yoghurt made with altered casein to whey protein ratios and EPS-producing starter cultures at 9 and 14% total solids. Food Hydrocolloids 20: 314–324.

Amoroso, M.J., Manca de Nadra, M.C. and Oliver, G. (1988). Glucose, galactose, fructose, lactose and sucrose utilization by Lactobacillus bulgaricus and Streptococcus thermophilus isolated from commercial yoghurt. Milchwissenshaft 43: 626–631.

Ariga, H., Urashima, T., Michihata, E., Ito, M., Morizono, N., Kimura, T. and Takahashi, S. (1992). Extracellular polysaccharide from encapsulated Streptococcus salivarius subsp. thermophilus OR-901 isolated from commercial yogurt. Journal of Food Science 57: 625–628.

Ascon-Reyes, D.B., Ascon-Cabrera, M.A, Cochet, N. and Lebeault, J.M. (1995). Indirect conductance for measurements of carbon dioxide produced by Streptococcus salivarius ssp. thermophilus TJ 160 in pure and mixed cultures. Journal of Dairy Science 48: 8–16.

Aslim, B., Onal, D. and Beyatli, Y. (2007). Factors influencing autoaggregation and aggregation of Lactobacillus delbrueckii subsp. bulgaricus isolated from handmade yogurt. Journal of Food Protection 70(1): 223–237.

Azcarate-Peril, M.A., Altermann, E., Hoover-Fitzula, R.L., Cano, R.C. and Klaenhammer, T.R. (2004). Identification and inactivation of genetic loci involved with Lactobacillus acidophilus Acid tolerance. Applied and Environmental Microbiology 70: 5315–5322.

Bautista, E.S., Dahiya, R.S. and Speck, M.L. (1966). Identification of compounds causing symbiotic growth of Streptococcus thermophilus and Lactobacillus bulgaricus in milk. Journal of Dairy Research 33: 299–307.

Beal, C., Spinnler, H.E. and Corrieu, G. (1994). Comparison of growth, acidification and productivity of pure and mixed cultures of Streptococcus salivarius subsp. thermophilus 404 and Lactobacillus delbrueckii subsp. bulgaricus 398. Microbiology and Biotechnology 41: 95–98.

Beal, C. and Corrieu, G. (1991). Influence of pH, temperature, and inoculum composition on mixed cultures of Streptococcus thermophilus 404 and Lactobacillus 398. Biotechnology and Bioengineering 38: 90–98.

Ben-Yahia, L., Mayeur, C., Rul, F. and Thomas, M. (2012). Growth advantage of Streptococcus thermophilus over Lactobacillus bulgaricus in vitro and in the gastrointestinal tract of gnotobiotic rats. Beneficial Microbes 3: 211–219.

Beshkova, D., Simova, E., Frengova, G. and Simov, Z. (1998). Production of flavour compounds by yogurt starter cultures. Journal of Industrial Microbiology and Biotechnology 20: 180–186.

Besnier, M.O., Bourlioux, P., Fourniat, J., Ducluzeau, R. and Aumaitre, A. (1983). Effect of yogurt ingestion on the intestinal lactase activity in axenic or holoxenic mice. Annals of Microbiology (Paris) 134A: 219–230.

Bolotin, A., Quinquis, B., Renault, P., Sorokin, A., Ehrlich, S.D., Kulakauskas, S., Lapidus, A., Goltsman, E., Mazur, M., Pusch, G.D., Fonstein, M., Overbeek, R., Kyprides, N., Purnelle, B., Prozzi, D., Ngui, K., Masuy, D., Hancy, F., Burteau, S., Boutry, M., Delcour, J., Goffeau, A. and Hols, P. (2004). Complete sequence and comparative genome analysis of the dairy bacterium Streptococcus thermophilus. Nature Biotechnology 22: 1554–1558.

Bottazzi, V., Battistotti, B. and Montescani, G. (1973). Influence des souches seules et associées de Lb. bulgaricus et Str. thermophilus ainsi que des traitements du lait sur la production d'aldéhyde acétique dans la yaourt. Lait 525-526: 295–308.

Bottazzi, V. and Vescovo, M. (1969). Carbonyl compounds produced by yoghurt bacteria. Netherland. Milk and Dairy Journal 23: 71–78.

Bottazzi, V. and Dellaglio, F. (1967). Acetaldehyde and diacetyl production by Streptococcus thermophilus and other lactic streptococci. Journal of Dairy Research 34: 109–113.

Boudraa, G., Benbouabdellah, M., Hachelaf, W., Boisset, M., Desjeux, J.F. and Touhami, M. (2001). Effect of feeding yogurt versus milk in children with acute diarrhea and carbohydrate malabsorption. Journal of Pediatric Gastroenterology and Nutrition 33: 307–313.

Boudraa, G., Touhami, M., Pochart, P., Soltana, R., Mary, J.Y. and Desjeux, J.F. (1990). Effect of feeding yogurt versus milk in children with persistent diarrhea. Journal of Pediatric Gastroenterology Nutrition 11: 509–512.

Bourgoin, F., Pluvinet, A., Gintz, B., Decaris, B. and Guédon, G. (1999). Are horizontal transfers involved in the evolution of the Streptococcus thermophilus exopolysaccharide synthesis loci? Gene 233: 151–61.

Bouzar, F., Cerning, J. and Desmazeaud, M. (1997). Exopolysaccharide production and texture-promoting abilities of mixed-strain starter cultures in yogurt production. Journal of Dairy Science 80: 2310–2317.

Bouzar, F., Cerning, J. and Desmazeaud, M. (1996). Exopolysaccharide production in milk by Lactobacillus delbrueckii ssp. bulgaricus CNRZ 1187 and by two colonial variants. Journal of Dairy Science 79: 205–211.

Bracquart, P. and Lorient, D. (1979). Effet des acides aminés et peptides sur la croissance de Streptococcus thermophilus. III. Peptides comportant Glu, His et Met. Milschwissenschaft 34: 679–679.

Bracquart, P., Lorient, D. and Alais, C. (1978). Effet des acides aminés sur la croissance de Streptococcus thermophilus. II: Etude sur cinq souches. Milchwissenshaft 33: 341–344.

Bracquart, P. and Lorient, D. (1977). Effet des acides aminés sur la croissance de Streptococcus thermophilus. Milchwissenshaft 32: 221–224.

Broadbent, J.R., Mc Mahon, D.J., Oberg, W.C. and Moineau, S. (2003). Biochemistry, genetics, and applications of exopolysaccharide production in *Streptococcus thermophilus*: a review. Journal of Dairy Science 86: 407–423.

Bubb, W.A., Urashima, T., Fujiwara, R., Shinnai, T. and Ariga, H. (1997). Structural characterisation of the exocellular polysaccharide produced by *Streptococcus thermophilus* OR 901. Carbohydrate Research 301(1-2): 41–50.

Canzi, E., Casiraghi, M.C., Zanchi, R., Gandolfi, R., Ferrari, A., Brighenti, F., Bosia, R., Crippa, A., Maestri, P., Vesely, R. and Bianchi Salvadori, B. (2002). Yogurt in the diet of the elderly: a preliminary investigation into its effect on the gut ecosystem and lipid metabolism. Lait 82: 713–723.

Cerning, J. (1990). Exocellular polysaccharides produced by lactic acid bacteria. FEMS Microbiological Reviews 7: 113–130.

Cerning, J., Bouillanne, C., Desmazeaud, M.J. and Landon, M. (1988). Exocellular polysaccharide production by *Streptococcus thermophilus*. Biotechnology Letters 10: 255–260.

Cerning, J., Bouillanne, C., Desmazeaud, M.J. and Landon, M. (1986). Isolation and characterization of exopolysaccharide produced by *Lactobacillus bulgaricus*. Biotechnology Letters 8: 625–628.

Chaves, A.C.S.D., Fernandez, M., Lerayer, A.L.S., Mierau, I., Kleerebezem, M. and Hugenholz, J. (2002). Metabolic engineering of acetaldehyde production by *Streptococcus thermophilus*. Applied and Environmental Microbiology 68: 5656–5662.

Cheng, H. (2010). Volatile flavor compounds in yogurt: a review. Critical Reviews in Food Science and Nutrition 50(10): 938–950.

Chervaux, C., Ehrlich, S.D. and Maguin, E. (2000). Physiological study of *Lactobacillus delbrueckii* subsp. *bulgaricus* strains in a novel chemically defined medium. Applied Environmental. Microbiology 66: 5306–5311.

Christensen, J.E., Dudley, E.G., Pederson, J.A. and Steele, J.L. (1999). Peptidases and amino acid catabolism in lactic acid bacteria. Antonie van Leeuwenhoek 76: 217–246.

Courtin, P. and Rul, F. (2004). Interactions between microorganisms in a simple ecosystem: yogurt bacteria as a study model. Lait 84(1-2): 125–134.

Courtin, P., Monnet, V. and Rul, F. (2002). Cell-wall proteinases PrtS and PrtB have a different role in *Streptococcus thermophilus*/*Lactobacillus bulgaricus* mixed cultures in milk. Microbiology 148: 3413–3421.

Crittenden, R.G., Martinez, N.R. and Playne, M.J. (2002). Synthesis and utilisation of folate by yoghurt starter cultures and probiotic bacteria. International Journal of Food Microbiology 80: 217–222.

Dandoy, D., Fremaux, C., Henry de Frahan, M. and Horvath, P., Boyaval, P., Hols, P. and Fontaine, L. (2011). The fast milk acidifying phenotype of *Streeptococcus thermophilus* can be acquired by natural transformation of the genomic island encoding the cell-envelope proteinase PrtS. Microbial Cell Factories 10(suppl 1): 521.

De Moreno de Leblanc, A. and Perdigón, G. (2004). Yogurt feeding inhibits promotion and progression of experimental colorectal cancer. Medical Science Monitor Journal 10: BR96–104.

de Vrese, M., Stegelmann, A., Richter, B., Fenselau, S., Laue, C. and Schrezenmeir, J. (2001). Probiotics—compensation for lactase insufficiency. The American Journal of Clinical Nutrition 73(2): 421s–429s.

De Vuyst, L., Vanderveken, F., Van de Ven, S. and Degeest, B. (1998). Production by and isolation of exopolysaccharides from *Streptococcus thermophilus* grown in a milk medium and evidence for their growth-associated biosynthesis. Journal of Applied Microbiology 84: 1059–1068.

Degeest, B. and De Vuyst, L. (1999). Indication that the nitrogen source influences both amount and size of exopolysaccharides produced by *Streptococcus thermophilus* LY03 and modelling of the bacterial growth and exopolysaccharide production in a complex medium. Applied and Environmental Microbiology 65(7): 2863–2870.

Delorme, C., Bartholini, C., Bolotine, A., Ehrlich, S.D. and Renault, P. (2010). Emergence of a Cell Wall Protease in the *Streptococcus thermophilus* Population. Applied and Environmental Microbiology 76: 451–460.

Derzelle, S., Bolotin, A., Mistou, M.-Y. and Rul, F. (2005). Proteome analysis of *Streptococcus thermophilus* grown in milk reveals pyruvate formate-lyase as the major upregulated protein. Applied and Environmental Microbiology 71: 8597–8605.

Dewit, O., Boudraa, G., Touhami, M. and Desjeux, J.F. (1987). Breath hydrogen test and stools characteristics after ingestion of milk and yogurt in malnourished children with chronic diarrhoea and lactase deficiency. Journal of Tropical Pediatrics 33: 177–180.

Doco, T., Wieruszeski, J.-M., Fournet, B., Carcano, D., Ramos, P. and Loones, A. (1990). Structure of an exocellular polysaccharide produced by *Streptococcus thermophilus*. Carbohydrate Research 198: 313–321.

Donkor, O.N., Henriksson, A., Singh, T.K., Vasiljevic, T. and Shah, N.P. (2007). ACE-inhibitory activity of probiotic yoghurt. International Dairy Journal 17: 1321–1331.

Driessen, F.M., Kingma, F. and Stadhouders, J. (1982). Evidence that *Lactobacillus bulgaricus* in yogurt is stimulated by carbon dioxide produced by *Streptococcus thermophilus*. Netherland Milk and Dairy Journal 36: 135–144.

Drouault, S., Anba, J. and Corthier, G. (2002). *Streptococcus thermophilus* is able to produce a b-galactosidase active during its transit in the digestive tract of germ-free mice. Applied and Environmental Microbiology 68: 938–941.

Dumont, J. and Adda, J. (1973). Méthode rapide d'étude des composés très volatils de l'arôme des produits laitiers. Application au yoghourt. Dairy Science and Technology 53: 12–22.

Dunne, J., Evershed, R.P., Salque, M., Cramp, L., Bruni, S., Ryan, K., Biagetti, S. and di Lernia, S. (2012). First dairying in green Saharan Africa in the fifth millennium bc. Nature 486: 390–394.

Dutta, S., Kuila, R. and Ranganathan, B. (1973). Effect of different heat treatments of milk on acid and flavour production by five single strain cultures. Milchwissenschaft 28: 321–323.

El-Abbadi, N.H., Dao, M.C. and Meydani, S.N. (2014). Yogurt: role in healthy and active aging. The American Journal of Clinical Nutrition 99: 1263S–1270S.

El-Abbassy, M.Z. and Sitohy, M. (1993). Metabolic interaction between *Streptococcus thermophilus* and *Lactobacillus bulgaricus* in single and mixed starter yogurts. Nahrung 37(1): 53–58.

El-Soda, M.A., Abou-Donia, S.A., El-Shafy, H.K., Mashaly, R. and Ismail, A.A. (1986). Metabolic activities and symbiosis in zabady isolated cultures. Egyptian Journal of Dairy Science 14: 1–10.

Escalante, A., WacherRodarte, C., GarciaGaribay, M. and Farres, A. (1998). Enzymes involved in carbohydrate metabolism and their role on exopolysaccharide production in *Streptococcus thermophilus*. Journal of Applied Microbiology 84: 108–114.

European and Food Safety Authority. (2010). Scientific opinion on the substantation of health claims related to live yogurt cultures and improved lactose digestion. EFSA Journal 8: 1763.

Faber, E.J., Kamerling, J.P. and Vliegenthart, J.F. (2001). Structure of the extracellular polysaccharide produced by *Lactobacillus delbrueckii* subsp. *bulgaricus* 291. Carbohydrate Research 331: 183–194.

Faber, E.J., Zoon, P.J., Kamerling, P. and Vliegenthart, J.F.G. (1998). The exopolysaccharides produced by *Streptococcus thermophilus* Rs and Sts have the same repeating unit but differ in viscosity of their milk cultures. Carbohydrate Research 310(4): 269–276.

FAO/WHO. (2001). Report of a joint FAO/WHO Expert Consultation on evaluation of health and nutritional properties of probiotics in food including power milk with lactic acid bacteria. ftp.fao.org/es/esn/food/probio_report_en.pdf.

FAO/WHO. (2002). Report of a joint FAO/WHO Working Group on drafting guidelines for the evaluation of probiotics in food. ftp.fao.org/es/esn/food/wgreport2.pdf.

Fernandez-Espla, M.-D., Garault, P., Monnet, V. and Rul, F. (2000). *Streptococcus thermophilus* cell wall-anchored proteinase: release, purification, and biochemical and genetic characterization. Applied and Environmental Microbiology 66: 4772–4778.

Frengova, G.I., Simova, E.D., Beshkova, D.M. and Simov, Z.I. (2000). Production and monomer composition of exopolysaccharides by yogurt starter cultures. Canadian Journal of Microbiology 46: 1123–1127.

Friedrich, J.E. and Acree, T.E. (1998). Gas chromatography olfactometry (GC/O) of dairy products. International Dairy Journal 8: 235–241.

Galesloot, T.E., Hassing, F. and Veringa, H.A. (1968). Symbiosis in yoghurt(I). Stimulation of *Lactobacillus bulgaricus* by a factor produced by *Streptococcus thermophilus*. Netherland Milk Dairy Journal 22: 50–63.

Garault, P., Letort, C., Juillard, V. and Monnet, V. (2000). Branched-chain amino acid biosynthesis is essential for optimal growth of *Streptococcus thermophilus* in milk. Applied and Environmental Microbiology 66: 5128–5133.

García-Albiach, R., José, M., de Felipe, P., Angulo, S., Morosini, M.I., Bravo, D., Baquero, F. and del Campo, R. (2008). Molecular analysis of yogurt containing *Lactobacillus delbrueckii* subsp. *bulgaricus* and *Streptococcus thermophilus* in human intestinal microbiota. American Journal of Clinical Nutrition 87: 91–6.

Gardan, R., Besset, C., Guillot, A., Gitton, C. and Monnet, V. (2009). The oligopeptide transport system is essential for the development of natural competence in *Streptococcus thermophilus* Strain LMD-9. Journal of Bacteriology 191: 4647–4655.

Garrote, G., Abraham, A.G. and Rumbo, M. (2015). Is lactate an undervalued functional component of lactic acid bacteria-fermented food products? Frontiers in Microbiology 6.

Garvie, E.I., Cole, C.B., Fuller, R. and Hewitt, D. (1984). The effect of yoghurt on some components of the gut microflora and on the metabolism of lactose in the rat. Journal of Applied Bacteriology 56: 237–245.

Gilbert, C., Blanc, B., Frot-Coutaz, J., Portalier, R. and Atlan, D. (1997). Comparison of cell surface proteinase activities within the *Lactobacillus* genus. Journal of Dairy Research 64: 561–571.

Gilbert, C., Atlan, D., Blanc, B., Portalier, R., Germond, J.E., Lapierre, L. and Mollet, B. (1996). A new cell surface proteinase: sequencing and analysis of the *prtB* gene from *Lactobacillus delbrueckii* subsp. *bulgaricus*. Journal of Bacteriology 178: 3059–3065.

Gilliland, S.E. and Kim, H.S. (1984). Effect of viable starter culture bacteria in yogurt on lactose utilization in humans. Journal of Dairy Science 67: 1–6.

Goh, Y., Goin, C., O'Flaherty, S., Altermann, E. and Hutkins, R. (2011). Specialized adaptation of a lactic acid bacterium to the milk environment: the comparative genomics of *Streptococcus thermophilus* LMD-9. Microbial Cell Factories 10(Suppl 1): S22.

Goodenough, E.R. and Kleyn, D.H. (1975). Influence of viable yoghurt microflora on digestion of lactose by the rat. Journal of Dairy Science 59: 601–606.

Grobben, G.J., Sikkema, J., Smith, M.R. and de Bont, J.A.M. (1995). Production of extracellular polysaccharides by *Lactobacillus delbrueckii* ssp. *bulgaricus* NCFB 2772 grown in a chemically defined medium. Journal of Applied Bacteriology 79: 103–107.

Groux, M. (1973). Etude des composants de la flaveur du yoghourt. Lait 523-524: 146–153.

Gruter, M., Leeflang, B.R., Kuiper, J., Kamerling, J.P. and Vliegenthart, J.E. (1993). Structural characterization of the exopolysaccharide produced by *Lactobacillus delbrueckii* subspecies *bulgaricus* rr grown in skimmed milk. Carbohydrate Research 239: 209–226.

Guarner, F., Perdigon, G., Corthier, G., Salminen, S., Koletzjo, B. and Morelli, L. (2005). Should yoghurt cultures be considered probiotic? British Journal of Nutrition 93: 783–786.

Hamdan, I.Y., Kunsman, J.E. and Deane, D.D. (1971). Acetaldehyde production by combined yogurt cultures. Journal of Dairy Science 54(7): 1080–1082.

Hao, P., Zheng, H., Yu, Y., Ding, G., Gu, W., Chen, S., Yu, Z., Ren, S., Oda, M., Konno, T., Wang, S., Li, X., Ji, Z.S. and Zhao, G. (2011). Complete sequencing and pan-genomic analysis of *Lactobacillus delbrueckii* subsp. *bulgaricus* reveal its genetic basis for industrial yogurt production. PLoS One 6: e15964.

Herve-Jimenez, L., Guillouard, I., Guedon, E., Boudebbouze, S., Hols, P., Monnet, V., Maguin, E. and Rul, F. (2009). Postgenomic analysis of *Streptococcus thermophilus* cocultivated in milk with *Lactobacillus delbrueckii* subsp. *bulgaricus*: involvement of nitrogen, purine, and iron metabolism. Applied and Environmental Microbiology 75: 2062–2073.

Herve-Jimenez, L., Guillouard, I., Guedon, E., Gautier, C., Boudebbouze, S., Hols, P., Monnet, V., Rul, F. and Maguin, E. (2008). Physiology of *Streptococcus thermophilus* during the late stage of milk fermentation with special regard to sulfur amino-acid metabolism. Proteomics 8: 4273–4286.

Higashio, K., Kikuchi, T. and Furuichi, E. (1978). Symbiose entre *Lactobacillus bulgaricus* et *Streptococcus thermophilus* dans le yoghourt. XX congrès international de laiterie, Paris.

Hirota, T., Ohki, K., Kawagishi, R., Kajimoto, Y., Mizuno, S., Nakamura, Y. and Kitakaze, M. (2007). Casein hydrolysate containing the antihypertensive tripeptides Val-Pro-Pro and Ile-Pro-Pro improves vascular endothelial function independent of blood pressure-lowering effects: contribution of the inhibitory action of angiotensin-converting enzyme. Hypertension Research 30: 489–496.

Hols, P., Hancy, F., Fontaine, L., Grossiord, B., Prozzi, D., Leblond-Bourget, N., Decaris, B., Bolotin, A., Delorme, C., Ehrlich, S.D., Guédon, E., Monnet, V., Renault, P. and Kleerebezem, M. (2005). New insights in the molecular biology and physiology of *Streptococcus thermophilus* revealed by comparative genomics. FEMS Microbiology Reviews 29: 435–463.

Horiuchi, H. and Sasaki, Y. (2012). Effect of oxygen on symbiosis between *Lactobacillus bulgaricus* and *Streptococcus thermophilus*. Journal of Dairy Science 95: 2904–2909.

Hutkins, R., Morris, H.A. and McKay, L.L. (1985). Galactokinase activity in *Streptococcus thermophilus*. Applied and Environmental Microbiology 50: 777–780.

Imhof, R., Glättli, H. and Bosset, J.O. (1995). Volatile organic compounds produced by thermophilic and mesophilic single strain dairy starter cultures. Lebensmittel-Wissenschaft und Technologie 28: 78–86.

Izawa, N., Hanamizu, T., Iizuka, R., Sone, T., Mizukoshi, H., Kimura, K. and Chiba, K. (2009). *Streptococcus thermophilus* produces exopolysaccharides including hyaluronic acid. Journal of Bioscience and Bioengineering 107: 119–123.

Jolly, L. and Stingele, F. (2001). Molecular organization and functionality of exopolysaccharide gene clusters in lactic acid bacteria. International Dairy Journal 11: 733–745.

Juillard, V., Desmazeaud, M.J. and Spinnler, H.E. (1988). Mise en évidence d'une activité uréasique chez *Streptococcus thermophilus*. Canadian Journal of Microbiology 34: 818–822.

Korhonen, H. and Pihlanto, A. (2003). Food-derived bioactive peptides—opportunities for designing future foods. Current Pharmaceutical Design 9: 1297–1308.

Knappe, J., Blaschkowski, H.P., GröBner, P. and Schmitt, T. (1974). Pyruvate Formate-Lyase of *Escherichia coli*: the Acetyl-Enzyme Intermediate. European Journal of Biochemistry 50: 253–263.

Kneifel, W., Ulberth, F., Erhard, F. and Jaros, D. (1992). Aroma profiles and sensory properties of yogurt and yogurt-related products. I Screening of commercially available stater cultures. Milchwissenschaft 47: 362–365.

Lamothe, G.T., Jolly, L., Mollet, B. and Stingele, F. (2002). Genetic and biochemical characterization of exopolysaccharide biosynthesis by *Lactobacillus delbrueckii* subsp. *bulgaricus*. Archives of Microbiology 178: 218–228.

Law, B. (1981). The formation of aroma and flavour compounds in fermented dairy products. Dairy Science Abstracts 43: 143–154.

Laws, A., Gu, Y. and Marshall, V. (2001). Biosynthesis, characterization, and design of bacterial exopolysaccharides from lactic acid bacteria. Biotechnol Advances 19: 597–625.

Lê, M.G., Moulton, L.H., Hill, C. and Kramar, A. (1986). Consumption of dairy produce and alcohol in a case-control study of breast cancer. Journal of the National Cancer Institute 77: 633–636.

Lees, G.J. and Jago, G.R. (1976). Formation of acetaldehyde from threonine by lactic acid bacteria. Journal of Dairy Research 43: 75–83.

Lember, M. (2012). Hypolactasia: a common enzyme deficiency leading to lactose malabsorption and intolerance. Polish Archives of Internal Medecine 122: 60–64.

Lerebours, E., N'Djitoyap Ndam, N.C., Lavoine, A., Hellot, M.F., Antoine, J.M. and Colin, R. (1989). Yogurt and fermented-then-pasteurized milk: effects of short-term and long-term ingestion on lactose absorption and mucosal lactase activity in lactase-deficient subjects. The American Journal of Clinical Nutrition 49: 823–827.

Letort, C., Nardi, M., Garault, P., Monnet, V. and Juillard, V. (2002). Casein utilization by *Streptococcus thermophilus* results in a diauxic growth in milk. Applied and Environmental Microbiology 68: 3162–3165.

Levander, F., Svensson, M. and Rådström, P. (2002). Enhanced exopolysaccharide production by metabolic engineering of *Streptococcus thermophilus*. Applied and Environmental Microbiology 68: 784–790.

Liu, M., Siezen, R.J. and Nauta, A. (2009). *In silico* prediction of horizontal gene transfer events in *Lactobacillus bulgaricus* and *Streptococcus thermophilus* reveals protocooperation in yogurt manufacture. Applied and Environmental Microbiology 75: 4120–4129.

Liu, J.E., Zhang, Y., Zhang, J., Dong, P.L., Chen, M. and Duan, Z.P. (2010). Probiotic yogurt effects on intestinal flora of patients with chronic liver disease. Nursing Research 59: 426–432.

Madureira, A.R., Tavares, T.A., Gomes, M.P., Pintado, M.E. and Malcata, F.X. (2010) Invited review: Physiological properties of bioactive peptides obtained from whey proteins. Journal of Dairy Science 93: 437–455.

Makarova, K., Slesarev, A., Wolf, Y., Sorokin, A., Mirkin, B., Koonin, E., Pavlov, A., Pavlova, N., Karamychev, V., Polouchine, N., Shakhova, V., Grigoriev, I., Lou, Y., Rohksar, D., Lucas, S., Huang, K., Goodstein, D.M., Hawkins, T., Plengvidhya, V., Welker, D., Hughes, J., Goh, Y., Benson, A., Baldwin, K., Lee, J.H., Díaz-Muñiz, I., Dosti, B., Smeianov, V., Wechter, W., Barabote, R., Lorca, G., Altermann, E., Barrangou, R., Ganesan, B., Xie, Y., Rawsthorne, H., Tamir, D., Parker, C., Breidt, F., Broadbent, J., Hutkins, R., O'Sullivan, D., Steele, J., Unlu, G., Saier, M., Klaenhammer, T., Richardson, P., Kozyavkin, S., Weimer, B. and Mills, D. (2006). Comparative genomics of the lactic acid bacteria. Proceedings of the National Academy of Sciences 103: 15611–15616.

Makino, S., Ikegami, S., Kano, H., Sashihara, T., Sugano, H., Horiuchi, H., Saito, T. and Oda, M. (2006). Immunomodulatory effects of polysaccharides produced by *Lactobacillus delbrueckii* ssp. *bulgaricus* OLL1073R-1. Journal of Dairy Science 89: 2873–2881.

Manca de Nadra, M.C., Amoroso, M.J. and Oliver, G. (1988). Acetaldehyde metabolism in *Lactobacillus bulgaricus* and *Streptococcus thermophilus* isolated from market yogurt. Microbiologie, Aliments Nutrition 6: 269–272.

Manca de Nadra, M.C., Raya, R.R., Pesce de Ruiz Holgadi, A. and Oliver, G. (1987). Isolation and properties of threonine aldolase of *Lactobacillus bulgaricus* YOP12. Milchwissenschaft 42: 92–94.

Margolis, K.L., Wei, F., de Boer, I.H., Howard, B.V., Liu, S., Manson, J.E., Mossavar-Rahmani, Y., Phillips, L.S., Shikany, J.M., Tinker, L.F. and for the Women's Health Initiative Investigators. (2011). A diet high in low-fat dairy products lowers diabetes risk in postmenopausal women. The Journal of Nutrition 141: 1969–1974.

Marranzini, R.M., Schmidt, R.H., Shireman, R.B., Marshall, M.R. and Cornell, J.A. (1989). Effect of threonine and glycine concentrations on threonine aldolase activity of yogurt Microorganisms during growth in a modified milk prepared by ultrafiltration. Journal of Dairy Science 72: 1142–1148.

Marshall, V.M. (1987). Lactic acid bacteria: starters for flavour. FEMS Microbiological Reviews 46: 327–336.

Marshall, V.M., Laws, A.P., Gu, F., Levander, Y., Radstrom, P., DeVuyst, L., Degeest, B., Vaningelgem, F., Dunn, H. and Elvin, M. (2001). Exopolysaccharide-producing strains of thermophilic lactic acid bacteria cluster into groups according to their EPS structure. Letters in Applied Microbiology 32: 433–437.

Marshall, V.M. and Cole, W.M. (1983). Threonine aldolase and alcohol dehydrogenase activities in *Lactobacillus bulgaricus* and *Lactobacillus acidophilus* and their contribution to flavour production in fermented milks. Journal of Dairy Research 50: 375–379.

Marteau, P., Minekus, M., Havenaar, R. and Huis In't Veld, J.H.J. (1997). Survival of lactic acid bacteria in a dynamic model of the stomach and small intestine: Validation and the effects of bile. Journal of Dairy Science 80: 1031–1037.

Marteau, P., Flourie, B., Pochart, P., Chastang, C., Desjeux, J.F. and Rambaud, J.C. (1990). Effect of the microbial lactase (EC 3.2.1.23) activity in yoghurt on the intestinal absorption of lactose: an *in vivo* study in lactase-deficient humans. British Journal of Nutrition 64: 71–79.

Meydani, S.N. and Ha, W.-K. (2000). Immunologic effects of yogurt. The American Journal of Clinical Nutrition 71: 861–872.

Mikkelsen, P. (2013). The world demands more yoghurts. International dairy magazine 11/12.

Moon, N.J. and Reinbold, G.W. (1976). Commensalism and competition in mixed cultures of *Lactobacillus bulgaricus* and thermophilus. Journal of Milk and Food Technolology 39: 337–341.

Muro Urista, C., Álvarez Fernández, R., Riera Rodriguez, F., Arana Cuenca, A. and Téllez Jurado, A. (2011). Review: Production and functionality of active peptides from milk. Food Science and Technology International 17: 293–317.

Nagai, T., Makino, S., Ikegami, S., Itoh, H. and Yamada, H. (2011). Effects of oral administration of yogurt fermented with *Lactobacillus delbrueckii* ssp. *bulgaricus* OLL1073R-1 and its exopolysaccharides against influenza virus infection in mice. International Immunopharmacology 11: 2246–2250.

Narayan, S.S., Jalgaonkar, S., Shahani, S. and Kulkarni, V.N. (2010). Probiotics: current trends in the treatment of diarrhoea. Hong Kong Medical Journal 16: 213–218.

Narushima, S., Sakata, T., Hioki, K., Itoh, T., Nomura, T. and Itoh, K. (2010). Inhibitory effect of yogurt on aberrant crypt foci formation in the rat colon and colorectal tumorigenesis in RasH2 mice. Experimental Animals 59: 487–94.

Nishimura, J., Kawai, Y., Aritomo, R., Ito, Y., Makino, S., Ikegami, S., Isogai, E. and Saito, T. (2013). Effect of formic acid on exopolysaccharide production in skim milk fermentation by *Lactobacillus delbrueckii* subsp. *bulgaricus* OLL1073R-1. Bioscience of Microbiota, Food and Health 32: 23–32.

Noh, D.O. and Gilliland, S.E. (1994). Influence of Bile on β-Galactosidase Activity of Component Species of Yogurt Starter Cultures. Journal of Dairy Science 77: 3532–3537.

Nordmark, E.L., Yang, Z., Huttunen, E. and Radstrom, P. (2005). Structural studies of an exopolysaccharide produced by *Streptococcus thermophilus* THS. Biomacromolecules 6: 105–108.

Ott, A., Hugi, A., Baumgartner, M. and Chaintreau, A. (2000b). Sensory investigation of yogurt flavor perception: mutual influence of volatiles and acidity. Journal of Agricultural and Food Chemistry 48: 441–450.

Ott, A., Germond, J.E. and Chaintreau, A. (2000a). Origin of acetaldehyde during milk fermentation using C-13-labeled precursors. Journal of Agricultural and Food Chemistry 48: 1512–1517.

Ott, A., Fay, L.B. and Chaintreau, A. (1997). Determination and origin of the aroma impact compounds of yogurt flavor. Journal of Agricultural and Food Chemistry 45: 850–858.

Ozer, B. and Atasoy, F. (2002). Effect of addition of amino acids, treatment with b-galactosidase and use of heat-shocked cultures on the acetaldehyde level in yoghurt. International Journal of Dairy Technology 55: 166–170.

Pala, V., Sieri, S., Berrino, F., Vineis, P., Sacerdote, C., Palli, D., Masala, G., Panico, S., Mattiello, A., Tumino, R., Giurdanella, M.C., Agnoli, C., Grioni, S. and Vittorio, K. (2011). Yogurt consumption and risk of colorectal cancer in the Italian European prospective investigation into cancer and nutrition cohort. International Journal of Cancer 129: 2712–2719.

Pastink, M.I., Teusink, B., Hols, P., Visser, S., de Vos, W.M. and Hugenholtz, J. (2009). Genome-scale model of *Streptococcus thermophilus* LMG18311 for metabolic comparison of lactic acid bacteria. Applied and Environmental Microbiology 75: 3627–3633.

Pelletier, X., Laure-Boussage, S. and Donazzolo, Y. (2001). Hydrogen excretion upon ingestion of dairy products in lactose-intolerant male subjects: importance of the live flora. European Journal of Clinical Nutrition 55: 509–512.

Perdigón, G., de Moreno de LeBlanc, A., Valdez, J. and Rachid, M. (2002). Role of yoghurt in the prevention of colon cancer. European Journal of Clinical Nutrition 56(Suppl 3): S65–S68.

Petit, C., Grill, J.P., Maazouzo, N. and Marczak, R. (1991). Regulation of polysaccharide formation by *Streptococcus thermophilus* in batch and in fed-batch cultures. Applied Microbiology and Biotechnology 36: 216–221.

Petry, S., Furlan, S., Crepeau, M.-J., Cerning, J. and Desmazeaud, M. (2000). Factors affecting exocellular polysaccharide production by *Lactobacillus delbrueckii* subsp. *bulgaricus* grown in a chemically defined medium. Applied and Environmental Microbiology 66: 3427–3431.

Pette, J.W. and Lolkema, H. (1950c). Acid formation and aroma formation in yoghurt. Netherland milk and Dairy Journal 4: 261–273.

Pette, J.W. and Lolkema, H. (1950b). Growth stimulating factors for *Sc. thermophilus*. Netherland Milk and Dairy Journal 4: 209–224.

Pette, J.W. and Lolkema, H. (1950a). Symbiosis and antibiosis in mixed cultures *Lb. bulgaricus* and *S. thermophilus*. Netherland Milk and Dairy Journal 4: 197–208.

Petti, S., Tarsitani, G. and Simonetti D'Arca, A. (2008). Antibacterial activity of yoghurt against viridans streptococci *in vitro*. Archives of Oral Biology 53: 985–990.

Plaisancié, P., Claustre, J., Estienne, M., Henry, G., Boutrou, R., Paquet, A. and Léonil, J. (2013). A novel bioactive peptide from yoghurts modulates expression of the gel-forming MUC2 mucin as well as population of goblet cells and Paneth cells along the small intestine. Journal of Nutritional Biochemistry 24: 213–221.

Pochart, P., Dewit, O., Desjeux, J. and Bourlioux, P. (1989). Viable starter culture, beta-galactosidase activity, and lactose in duodenum after yogurt ingestion in lactase-deficient humans. American Journal of Clinical Nutrition 49: 828–831.

Pool-Zobel, B.L., Münzner, R. and Holzapfel, W.H. (1993). Antigenotoxic properties of lactic acid bacteria in the *S. typhimurium* mutagenicity assay. Nutrition and Cancer 20: 261–270.

Purohit, D.H., Hassan, A.N., Bhatia, E., Zhang, X. and Dwivedi, C. (2009). Rheological, sensorial, and chemopreventive properties of milk fermented with exopolysaccharide-producing lactic cultures. Journal of Dairy Science 92: 847–856.

Purwandari, U., Shah, N.P. and Vasiljevic, T. (2007). Effects of exopolysaccharide-producing strains of *Streptococcus thermophilus* on technological and rheological properties of set-type yoghurt. International Dairy Journal 17: 1344–1352.

Qin, Q.Q., Xia, B.S., Xiong, Y., Zhang, S.X., Lou, Y.B. and Hao, Y.L. (2011). Structural characterization of the exopolysacchairde produced by *Streptococcus thermophilus* 05-34 and its application in yogurt. Journal of Food Science 76: C1226–1230.

Radke-Mitchell, L. and Sandine, W.E. (1984). Associative growth and differential enumeration of *Streptococcus thermophilus* and *Lactobacillus bulgaricus*: a review. Journal of Food Protection 47(3): 245–248.

Rajagopal, S.N. and Sandine, W.E. (1990). Associatice growth and proteolysis of *Streptococcus thermophilus* and *Lactobacillus bulgaricus* in skim milk. Journal of Dairy Science 73: 894–899.

Ramchandran, L. and Shah, N.P. (2009). Effect of exopolysaccharides on the proteolytic and angiotensin-I converting enzyme-inhibitory activities and textural and rheological properties of low-fat yogurt during refrigerated storage. Journal of Dairy Science 92: 895–906.

Rao, D.R., Reddy, A.V., Pulusani, S.R. and Cornwell, P.E. (1984). Biosynthesis and utilization of folic acid and vitamin B by lactic cultures in skim milk. Journal of Dairy Science 67: 1169–1174.

Rasic, J. and Kurmann, J.A. (1978). Flavour and aroma in yoghurt. pp. 90–98. *In*: Rasic and Kurmann (eds.). Yoghurt Scientific Grounds, Technology, Manufacture and Preparations. Copenhagen.

Rasmussen, T.B., Danielsen, M., Valina, O., Garrigues, C., Johansen, E. and Pedersen, M.B. (2008). *Streptococcus thermophilus* core genome: comparative genome hybridization study of 47 strains. Applied and Enviromental Microbiology 74: 4703–10.

Raya, R.R., Manca de Nadra, M.C., Pesce de Ruiz Holgado, A. and Oliver, G. (1986). Acetaldehyde metabolism in lactic acid bacteria. Milchwissenschaft 41: 397–399.

Reid, G., Sanders, M.E., Gaskins, H.R., Gibson, G.R., Mercenier, A., Rastall, R., Roberfroid, M., Rowland, I., Cherbut, C. and Klaenhammer, T.R. (2003). New scientific paradigms for probiotics and prebiotics. Journal of Clinical Gastroenterololgy 37: 105–108.

Rul, F., Ben-Yahia, L., Chegdani, F., Wrzosek, L., Thomas, S., Noordine, M.-L., Gitton, C., Cherbuy, C., Langella, P. and Thomas, M. (2011). Impact of the Metabolic Activity of *Streptococcus thermophilus* on the colon epithelium of gnotobiotic rats. Journal of Biological Chemistry 286: 10288–10296.

Rysstad, G. and Abrahamsen, R.K. (1987). Formation of volatile aroma compounds and carbon dioxide in yogurt starter grown in cow's and goat's milk. Journal of Dairy Research 54: 257–266.

Salas, M.M., Nascimento, G.G., Vargas-Ferreira, F., Tarquinio, S.B., Huysmans, M.C. and Demarco, F.F. (2015). Diet influenced tooth erosion prevalence in children and adolescents: Results of a meta-analysis and meta-regression. Journal of Dentistry 43: 865–875.

Savaiano, D.A. (2014). Lactose digestion from yogurt: mechanism and relevance. The American Journal of Clinical Nutrition 99: 1251S–1255S.

Savaiano, D.A., Abou El Anouar, A., Smith, D.E. and Levitt, M.D. (1984). Lactose malabsorption from yogurt, pasteurized yogurt, sweet acidophilus milk, and cultured milk in lactase-deficient individuals. The American Journal of Clinical Nutrition 40: 1219–1223.

Savijoki, K., Ingmer, H. and Varmanen, P. (2006). Proteolytic systems of lactic acid bacteria. Applied Microbiology and Biotechnology 71: 394–406.

Säwén, E., Huttunen, E., Zhang, X., Yang, Z. and Widmalm, G. (2010). Structural analysis of the exopolysaccharide produced by *Streptococcus thermophilus* ST1 solely by NMR spectroscopy. Journal of Biomolecular NMR 47: 125–134.

Sawers, G. and Watson, G. (1998). A glycyl radical solution: oxygen-dependent interconversion of pyruvate formate-lyase. Molecular Microbiology 29: 945–954.

Sengül, N., Aslím, B., Uçar, G., Yücel, N., Işik, S., Bozkurt, H., Sakaoğullari, Z. and Atalay, F. (2006). Effects of exopolysaccharide-producing probiotic strains on experimental colitis in rats. Diseases of the Colon and Rectum 49(2): 250–258.

Shahbal, S., Hemme, D. and Desmazeaud, M. (1991). High cell wall-associated proteinase acticity of some *Streptococcus thermophilus* strains (H-strains) correlated with a high acidification rate in milk. Lait 71: 351–357.

Shankar, P.A. and Davies, F.L. (1978). Proteinase and peptidase activities of yogurt starter bacteria. XX Intern. Dairy Congr., Paris.

Shermak, M.A., Saavedra, J.M., Jackson, T.L., Huang, S.S., Bayless, T.M. and Perman, J.A. (1995). Effect of yogurt on symptoms and kinetics of hydrogen production in lactose-malabsorbing children. American Journal of Clinical Nutrition 62: 1003–1006.

Sieuwerts, S., Molenaar, D., van Hijum, S.A.F.T., Beerthuyzen, M., Stevens, M.J.A., Janssen, P.W.M., Ingham, C.J., de Bok, F.A.M., de Vos, W.M. and van Hylckama Vlieg, J.E.T. (2010). Mixed culture transcriptome analysis reveals molecular basis of mixed culture growth in *Streptococcus thermophilus* and *Lactobacillus bulgaricus*. Applied and Environmental Microbiology 76: 7775–7784.

Sieuwerts, S., de Bok, F.A., Hugenholtz, J. and van Hylckama Vlieg, J.E. (2008). Unraveling microbial interactions in food fermentations: from classical to genomics approaches. Applied and Environmental Microbiology 74(16): 4997–5007.

Sonestedt, E., Wirfält, E., Wallström, P., Gullberg, B., Orho-Melander, M. and Hedblad, B. (2011). Dairy products and its association with incidence of cardiovascular disease: the Malmö diet and cancer cohort. European Journal of Epidemiology 26: 609–618.

Spinnler, H.E., Bouillanne, C., Desmazeaud, M.J. and Corrieu, G. (1987). Measurement of the partial-pressure of dissolved CO_2 for estimating the concentration of *Streptococcus thermophilus* in coculture with *Lactobacillus bulgaricus*. Applied Microbiology and Biotechnology 25: 464–470.

Stingele, F., Neeser, J.R. and Mollet, B. (1996). Identification and characterization of the eps (Exopolysaccharide) gene cluster from *Streptococcus thermophilus* Sfi6. Journal of Bacteriology 178: 1680–1690.

Suzuki, I., Kato, S., Kitada, T., Yano, N. and Morichi, T. (1986). Growth of *Lactobacillus bulgaricus* in milk. 1. Cell elongation and the role of formic acid in boiled milk. Journal of Dairy Science 69: 311–320.

Sybesma, W., Starrenburg, M., Kleerebezem, M., Mierau, I., de Vos, W.M. and Hugenholtz, J. (2003). Increased production of folate by metabolic engineering of *Lactococcus lactis*. Applied and Environmental Microbiology 69: 3069–3076.

Tamime, A.Y. and Robinson, R.K. (1999). Yoghurt Science and technology. Tamime and Robinson (eds.). Woodhead Publishing, Cambridge England.

Terence, D.T. and Vaughan, L.C. (1984). Selection of galactose-fermenting *Streptococcus thermophilus* in lactose-limited chemostat cultures. Applied and Environmental Microbiology 48: 186–191.

Thomas, M., Wrzosek, L., Ben-Yahia, L., Noordine, M.-L., Gitton, C., Chevret, D., Langella, P., Mayeur, C., Cherbuy, C. and Rul, F. (2011). Carbohydrate metabolism is essential for the colonization of *Streptococcus thermophilus* in the digestive tract of gnotobiotic rats. PloS one 6(12).

Tinson, W., Broome, M.C., Hillier, A.J. and Jago, G.R. (1982). Metabolism of *Streptococcus thermophilus*. 2. Production of CO2 and NH3 from urea. Australian J Dairy Technol 37: 14–16.

Touhami, M., Boudraa, G., Mary, J.Y., Soltana, R. and Desjeux, J.F. (1992). Clinical consequences of replacing milk with yogurt in persistent infantile diarrhea. Annales de Pediatrie (Paris) 39: 79–86.

Trapp, C.L., Chang, C.C., Halpern, G.M., Keen, C.L. and Gershwin, M.E. (1993). The influence of chronic yogurt consumption on population of young and elderly adults. International Journal of Immunotherapy 9: 53–64.

Tsilingiri, K. and Rescigno, M. (2013). Postbiotics: what else? Beneficial Microbes 4: 101–107.

Turcic, M., Rasic, J. and Canic, V. (1969). Influence of *Str. thermophilus* and *Lb. bulgaricus* culture on volatile acids content in the flavour components of yoghurt. Milchwissenschaft 5: 277–280.

Vaillancourt, K., LeMay, J.-D., Lamoureux, M., Frenette, M., Moineau, S. and Vadeboncoeur, C. (2004). Characterization of a galactokinase-positive recombinant strain of *Streptococcus thermophilus*. Applied and Environmental Microbiology 70: 4596–4603.

Van de Guchte, M., Penaud, S., Grimaldi, C., Barbe, V., Bryson, K., Nicolas, P., Robert, C., Oztas, S., Mangenot, S., Couloux, A., Loux, V., Dervyn, R., Bossy, R., Bolotin, A., Batto, J.-M., Walunas, T., Gibrat, J.-F., Bessières, P., Weissenbach, J., Ehrlich, S.D. and Maguin, E. (2006). The complete genome sequence of *Lactobacillus bulgaricus* reveals extensive and ongoing reductive evolution. Proceedings of the National Academy of Sciences USA 103: 9274–9279.

Van de Water, J., Keen, C.L. and Gershwin, M.E. (1999). The influence of chronic yogurt consumption on immunity. Journal of Nutrition 129(7 Suppl): 1492S–5S.

Van den Bogaard, P.T.C., Hols, P., Kuipers, O.P., Kleerebezem, M. and de Vos, W.M. (2004). Sugar utilisation and conservation of the gal-lac gene cluster in *Streptococcus thermophilus*. Systematic and Applied Microbiology 27: 10–17.

Vaningelgem, F., Zamfir, M., Adriany, T. and De Vuyst, L. (2004). Fermentation conditions affecting the bacterial growth and exopolysaccharide production by *Streptococcus thermophilus* ST 111 in milk-based medium. Journal of Applied Microbiology 97: 1257–1273.

Van't Veer, P., Dekker, J.M., Lamers, Jos W.J., Kok, F.J., Schouten, E.G., Brants, H.A.M., Sturmans, F. and Hermus, R.J.J. (1989). Consumption of fermented milk products and breast cancer: a case-control study in the Netherlands. Cancer Research 49: 4020–4023.

Vaughan, E.E., vandenBogaard, P.T.C., Catzeddu, P., Kuipers, O.P. and deVos, W.M. (2001). Activation of silent *gal* genes in the *lac-gal* regulon of *Streptococcus thermophilus*. Journal of Bacteriology 183: 1184–1194.

Veiga, P., Pons, N., Agrawal, A., Oozeer, R., Guyonnet, D., Brazeilles, R., Faurie, J.M., van Hylckama Vlieg, J.E., Houghton, L.A., Whorwell, P.J., Ehrlich, S.D. and Kennedy, S.P.

(2014). Changes of the human gut microbiome induced by a fermented milk product. Scientific Reports 4: 6328.

Veringa, H.A., Galesloot, T.E. and Davelaar, H. (1968). Symbiosis in yoghourt (II). Isolation and identification of a growth factor for *Lactobacillus bulgaricus* produced by *Streptococcus thermophilus*. Netherland Milk and Dairy Journal 22: 114–120.

Wilt, T.J., Shaukat, A., Shamliyan, T., Taylor, B.C., MacDonald, R., Tacklind, J., Rutks, I., Schwarzenberg, S.J., Kane, R.L. and Levitt, M. (2010). Lactose intolerance and health. Evidence report/technology *assessment* (Full rep.) 192: 1–410.

WHO-World Health Organization (1995). The treatment of diarrhoea. A manual for physicians and other senior health workers. WHO/CDR/95.3 10/95. Geneva: WHO.

Wu, Q., Tun, H.M., Leung, F.C.C. and Shah, N.P. (2014). Genomic insights into high exopolysaccharide-producing dairy starter bacterium *Streptococcus thermophilus*. Scientific Reports 4(4974).

Yamada, T., Takahashi-Abbe, S. and Abbe, K. (1985). Effects of oxygen on pyruvate formate-lyase *in situ* and sugar metabolism of *Streptococcus mutans* and *Streptococcus sanguis*. Infection and Immunity 47: 29–134.

Yang, Y., Shevchenko, A., Knaust, A., Abuduresule, I., Li, W., Hu, X., Wang, C. and Shevchenko, A. (2014). Proteomics evidence for kefir dairy in Early Bronze Age China. Journal of Archaeological Science 45: 178–186.

Zhang, Q., Yang, B., Brashears, M.M., Yu, Z., Zhao, M., Liu, N. and Li, Y. (2014). Influence of casein hydrolysates on exopolysaccharide synthesis by *Streptococcus thermophilus* and *Lactobacillus delbrueckii* ssp. *bulgaricus*. Journal of Food Science and Agriculture 94: 1366–1372.

Zheng, H., Liu, E., Hao, P., Konno, T., Oda, M. and Ji, Z.S. (2012). *In silico* analysis of amino acid biosynthesis and proteolysis in *Lactobacillus delbrueckii* subsp. *bulgaricus* 2038 and the implications for bovine milk fermentation. Biotechnology Letters 34: 1545–1551.

Zisu, B. and Shah, N.P. (2003). Effects of pH, Temperature, Supplementation with Whey Protein Concentrate, and Adjunct Cultures on the Production of Exopolysaccharides by *Streptococcus thermophilus* 1275. Journal of Dairy Science 86: 3405–3415.

Zourari, A. and Desmazeaud, M.J. (1991). Caractérisation de bactéries lactiques thermophiles isolées de yaourts artisanaux grecs. I. Souches de *Lactobacillus delbrueckii* subsp. *bulgaricus* et cultures mixtes avec *Streptococcus salivarius* subsp. *thermophilus*. Lait 71: 463–482.

Zourari, A., Accolas, J.P. and Desmazeaud, M.J. (1992). Metabolism and biochemical characteristics of yogurt bacteria. A review. Lait 72: 1–34.

20

Fermentations Generating the Cheese Diversity

Sophie Landaud* and Henry-Eric Spinnler

1. Introduction

The cheese making process involves the growth and expressions of a large diversity of species. It starts by the growth of lactic acid bacteria (LAB) in the curd before or during the curd drainage and it is followed by ripening steps which will determine the cheese properties. In large cheeses, if a phase of ripening is made at high temperature (e.g., 23°C), the development of *Propionibacterium* spp. generate propionic acid and CO_2. These species are responsible for the formation of eyes in Emmental (Emmental is a yellow, medium-hard Swiss cheese that originated in the area around Emmental, Canton Bern). The development of *Penicillium camemberti* at the cheese surface of soft cheeses will produce the white coverage. If the rind is washed, the moulds cannot grow but the growth of yeasts and corynebacteria will produce cheeses rich in sulphur flavours and with an orange rind (wash rind cheeses). The use of molecular methods permits the description of the microbial ecosystem which appears now more complex than 15 years ago but it is still difficult to understand the interactions between the different species and the curd and also of the interactions between the species at the cheese surface. This emerging knowledge will definitely help to control the sensory properties of the cheese produced.

UMR de Génie et Microbiologie des Procédés Alimentaires, AgroParisTech, INRA, Université Paris-Saclay, Site de Grignon, Avenue Lucien Bretignières, 78850 Thiverval Grignon, France. E-mail: henry_eric.spinnler@agroparistech.fr
* Corresponding author: sophie.landaud@agroparistech.fr

2. Principles of Coagulation, Chemical Barriers to Microbial Growth

There are three principles of milk coagulation each one leading to gels with different structure properties and consequently allowing different levels of draining, different microbial growth and finally different cheese characteristics. These different systems of coagulation are usually combined in cheese making processes. The three principles of coagulation are the following:

- by acidification, usually due to the production of lactic acid by LAB,
- by the action of protease such as the chymosin present in the rennet (extract of the calf abomasum), the one produced by moulds like *Mucor pusillus* or *Mucor miehi* or plant proteases such as the one contained in the fig sap,
- by the action of the temperature associated to both of the previous systems like in mozzarella and other Italian "pasta filata" type cheese.

2.1 Coagulation by Lactic Acid

The production of lactic acid by LAB has two main effects:

- The caseins which are grouped in micelles stabilized by the casein κ (Fig. 1) are negatively charged at the surface. These charges provoke repulsion between the micelles and disperse them in the milk. The growth of the LAB during the milk fermentation is associated to an increase of the lactic acid concentration raising the proton concentration shown by the drop in pH. The negative charges of the caseins are neutralized by these protons and the micelles are joining together to build a fragile gel, like in yoghurts.
- The acidification has another effect. The calcium and phosphate fixed to the caseins are released because of a displacement from their corresponding salts by the H^+ concentration. The release of these minerals will reduce the possibility of ionic bounds. Mainly hydrogen bounds, Van der Walls or hydrophobic bounds are possible, all with low energies, are binding together the caseins. The resultant hydrophobic gel will be fragile and should be drained with caution.

In terms of composition, where this principle is the major system of coagulation, cheeses have a lower calcium and phosphate content, they stay more humid and soft, and only small cheeses can be made. The acidification stops the growth of the LAB and so, some lactose is remaining in the curd, at the end of the draining.

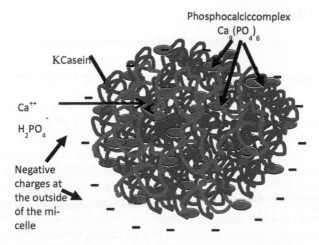

Figure 1. Scheme of the structure of a milk casein micelle. κ-casein is represented in yellow. This amphiphilic casein is at the interface between the hydrophobic κ and κ caseins (in green) and the whey and permits the dispersion of the micelles in the milk. The negative charges of the caseins are grouped outside the micelles and are also contributing to the stability of the dispersion of the micelles in the milk.

2.2 Coagulation by the Rennet

The second principle of coagulation is based on the action of the chymosin on the kappa casein. The micelles becomes less polar and join together to form a gel. In that case the calcium and phosphate remain inside the micelles and the ionic bounds between the caseins permit to make firmer gel with more mechanical strength. That permits to get firm gels after draining, to apply pressure without losing significant amount of caseins in the whey. It is also possible to cook the curd in the whey like in Emmental or Comté types of cheese. This principle permits to get cheeses with big size. It is used to make Holland type cheese (i.e., Gouda), Cheddar, Italian type hard cheeses (Parmiggiano Regianno, Grana Padano), Comté, Cantal, Emmental, etc. Almost every cheese is made with at least a small amount of rennet.

2.3 Coagulation Including Heat Treatment

The temperature increases the speed of destabilization of the micelles by the two previous principles but it gives a curd that is resistant to stretching. It gives a family of cheese called *pasta filata* and is mainly made in Italy like Mozzarella, *Burrata* for fresh cheeses and *Provolone* or *Caciocavallo* for ripened cheeses. The stretching has the effect to give an orientation of the casein fibers as well as the fat along these fibers (Rowney et al. 2004).

3. LAB in Cheeses

There are two types of LAB present in cheeses: mesophilic and thermophilic.

3.1 Mesophilic LAB Involved in Cheese Making Processes

In almost all cheeses, LAB are present in large concentration (over 10^9 CFU/mL) at the beginning of the ripening but the diversity of these LAB depend on the milk treatment and on the technology used. In soft cheeses the mesophilic LAB are the more common. It includes *Lactococcus lactis* ssp. *lactis* or ssp.*cremoris*, in the ssp. *lactis* the *biovar diacetyl lactis* is producing diacetyl which gives a buttery flavour especially in non ripened soft cheeses like *"fromagefrais"*, "cottage cheese", "quark" or "chakka". These homolactic bacteria could be associated with *Leuconostoc* sp. such as *Ln. cremoris* or *Ln. mesenteroides* species that are also producing flavours. These species are taxonomically close to the hetero-fermentative *Lactobacillus* species that are common and important in pressed cheeses. In this group, the species *Lb. plantarum, Lb. casei* and *Lb. rhamnosus* are particularly common.

The LAB are all very efficient on the hydrolysis of curd proteins. Some are able to produce flavour compounds from citrate or amino-acids but their flavour capabilities are slower and of limited impact as compared to the microorganisms found at the surface of cheese during ripening. This leads to a slow evolution of large cheeses, with a small surface/volume ratio, where LAB are one of the major families of microorganisms involved.

3.2 Thermophilic LAB Involved in Cheese Making Processes

Thermophilic LAB are mainly represented in yoghurt and hard cooked cheeses like *swiss* type cheeses. *Streptococcus thermophilus* is a species that grows and convert the lactose into lactic acid very quickly in milk at 40–44°C but it is also able to grow at lower temperature; consequently this species is now also used in different soft cheese technologies. However, *St. thermophilus* is inhibited by pH of 5.6 and it is usually used in combinations with *Lactobacillus* that are more resistant to low pH such as *Lb. helveticus* and *Lb. delbruecki* subsp. *bulgaricus*.

4. Diversity of Cheese Microbial Ecosystems in Ripening Processes

4.1 Mould Ripened Cheeses

The technology of mould ripened cheeses starts with the production of a curd with a pH between 6.3 and 5.8 at the beginning of moulding. The more acidic it will be at the beginning of moulding, the chalkier will be the

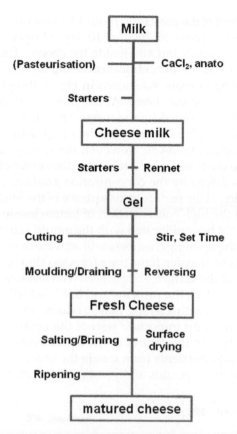

Figure 2. General Technological diagram used to make mould ripened soft cheeses.

curd at the demoulding step and the longer will be the complete ripening. However, the ones which are the chalkier after draining will be the creamier at the end of ripening. The technology of stabilized curds permits to keep a quite high pH of the curd by washing the curd with water or using starter sensitive to low pH (such as *St. thermophilus*). In that case, the texture is soft (not chalky) right after the moulding step but this texture will not change much during the ripening.

Spores of *P. camemberti* (10^4 spores/ml) are usually added to the milk with very low level of *Geotrichum candidum* (20 to 100 spores/ml). *Brevibacterium linens* is often added to give more flavours but is seems that the typical flavours of several mould ripened cheeses is related to the occurrence of Gram negative bacteria. For example, in *camemberts* made from pasteurized milks, *Hafnia alvei* is frequently added to improve the flavour typicality of the mould ripened cheeses.

The development of the yeasts and mould, by consumption of the lactate and production of ammonia, for the first 10 days of ripening leads to a rise in pH (Fig. 3) at the surface but also inside the cheese. This is a major step in the ripening process. The deacidification changes the texture and allows the ripening bacteria to grow. As shown in Fig. 3, the pH rise is quicker at the surface than inside the cheese. At the cheese surface, the lactic acid available is quickly consumed by the fungi. This surface consumption is only related to the metabolic activity of the fungi, which itself is mainly driven by the temperature level. However, the change in pH inside the cheese is related to the transfer of lactate from the center of the cheese onto the surface and is driven by the concentration gradient between the low lactate concentration at the surface, consequence of the fungal consumption of this solute, and the high concentration of lactate inside the cheese. This mass transfer should be equilibrated with the mould activity. Any increase in temperature of the ripening chambers will activate the mould physiology quicker than the mass transfer. If the transfer is too slow as compared to the biological activity at the surface, *P. camemberti* can exhaust the lactate at the surface and start to attack the proteins and lipids which could lead to the production of styrene (with a strong plastic odour) and 1-octene 3-ol with a mushroomy flavour. An important part of the control of the quality of soft cheese ripening is in the control in the equilibrium between the mass transfer of the mould nutrients from inside the cheese to the surface and the biological activity of moulds and yeasts at the surface.

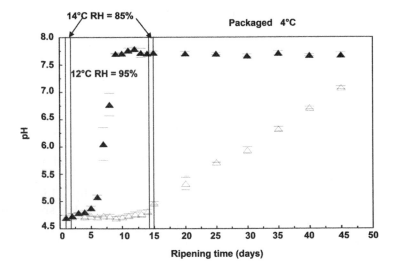

Figure 3. Changes in cheese pH versus ripening time (45 days). Solid symbols = pH of cheese rind and open symbols = core pH. Each evolution represents the mean of two trials and '_' Std deviation (from Leclercq-Perlat et al. 2004a).

4.2 Washed Rind or Smear Cheeses

The microbial ecosystem of washed rind cheeses is probably the ecosystem that has been studied the most in the recent years. It has been checked in Munster, Livarot, Tilsit, Limburger, Taleggio, Gubbeen, Reblochon, etc. In this variety of cheese where some are made from quite lactic curd (*Epoisses*) and where the curd of others are more rennet curds and even pressed curd (*Reblochon*), the common treatment is the wash of the surface with brushes or cloth regularly during ripening. This mechanical treatment, often completed with the addition of salt onto the surface and of ripening starters (such as *Brevibacterium linens*), will prevent the development of moulds. It leads to the formation of a smear which is a quite complex ecosystem made of yeasts and ripening bacteria (Feurer et al. 2004). In *Livarot* for example, up to 80 different types of bacteria (Gram + and Gram –) and yeasts have been evidenced through molecular methods (Mounier et al. 2009).

The yeasts (*Kluyveromyces lactis, Debaryomyces hansenii, Geotrichum candidum, Yarrowia lipolytica*) selectively grow just after the beginning of the ripening due to the curd acidity (Larpin et al. 2006). Their metabolisms lead to the consumption of the lactic acid and to the production of ammonia from the curd amino acids and consequently to rise in pH, as in mould ripened cheese, but slower. This increase in pH will be favourable to the growth of different bacteria, mainly coryneform bacteria (*Arthrobacter arilaitensis, Corynebacterium casei, Microbacter iumgubeenense, Brevibacterium linens*) but also some Staphylococci (e.g., *St. equorum, St. xylosus, St. saprophyticus*) or Gram negative bacteria such as *Proteus vulgaris* (Larpin-Laborde et al. 2011). It has recently been evidenced from the total DNA sequence of several of these strains that iron limitation in the curd is a significant limitation growth factor that probably lead to the selection of microorganisms adapted to a low iron environment (Monnet et al. 2012).

This microflora will generate the typical yellow to brown colour of this cheese type, the typical strong garlic, sulphur flavours and a texture that could be chalky on young cheeses but that will move to very creamy textures.

4.3 Blue Cheeses

In the blue veined cheeses, a main objective is to create a curd with apertures in which *P. roquefortii* will be able to grow. To open the curd, diverse strategies have been developed like the use of a large proportion of hetero -fermentative LAB, such as *Leuconostoc* sp., that produce bubbles of CO_2 in the curd. The stirring of the curd before and during draining is also a means used to obtain a coating of the curd grains that will produce apertures in the cheese during moulding. After moulding, the cheese is pierced with holes (40/cheese for a typical Roquefort), this piercing will create some kind of chimney by which the air will fill up the inside these openings and lead to the development of *P. roquefortii*. These types of cheeses are usually rather salty (up to 6.6% of dry matter) (Prieto et al. 1999), considering the moisture level of these cheeses, it leads to a ratio salt/water which is about 14% that is selective for the growth of secondary flora in the Blue cheese ecosystem. Recent study shows that *Yarrowia lipolytica* is common yeast in the secondary flora of blue cheese and is an important yeast for the flavour of this type of cheese (Gkatzionis et al. 2013).

4.4 Pressed Cheeses

In this family, depending on the coagulation, on the surface treatment and on the geometry of the cheese, a large diversity of cheeses is obtained. The curd is mainly made from the rennet activity and the LAB are growing late during the moulding. When the cheeses are small or of a limited thickness like in Reblochon or Saint Nectaire, the ripening is rather quick. The fungal surface microflora is made of moulds, if the surface is not washed or only made of yeasts if washed, which have strong activities and are able to change the texture and to develop a significant lipolysis and proteolysis. The activities of the bacteria inside and outside are probably less important for the texture. However, the bacterial flora is important for the flavour in these cheeses.

A constant characteristic of the bigger cheeses like Gouda, Cantal or Cheddar is the much lower level of lipolysis occurring during ripening. The ripening is mainly the fact of the activity of LAB and of yeasts in anaerobiosis and is so with a much slower metabolism than in the previous category. In these cheeses the subtle flavours are related to the proteolysis and mainly to amino acid catabolism.

4.5 Hard Cooked Cheeses

In this family are included most of the swiss type cheeses like Emmental, Gruyere and Comté but also the hard cheeses from Italy such as Parmiggiano Reggiano or the Grana family cheeses. The swiss cheeses have an even lower level of lipolysis than the pressed cheeses, when lipolysis is over 2%, taints are observed. After moulding and salting, the cheeses are kept at a temperature of 12°C for at least two weeks in order to permit the salt to diffuse homogeneously inside the cheese. The target level of salt is around 1.7% (w/w) of the humid cheese. After this period, depending on the opening desired, the transfer of the cheese in a warm cellar can be done for three weeks to a month (temperatures up to 33°C). This warm room ripening will increase the speed of the propionic fermentation happening in these cheeses. The high level of propionic fermentation not only produces propionic acid but also CO_2. *Propionibacterium* like *Pr. freundenreichii* or *Pr. acidilactici* are able to produce large amounts of CO_2 and are responsible of the large eyes present in Emmental. After this period the cheese is cooled down to 12°C until the cut of the wheel and should be maintained at 4°C after cutting. For blind cheeses (without eyes) like Beaufort or Comté, cheese ripening proceeds at low temperature (8–12°C) from the beginning to the end and, in that case, propionic fermentation is slow. The CO_2 diffuses from inside the cheese to outside at a quicker speed than the CO_2 is produced and so the CO_2 concentration never exceeds the saturation in the cheese curd and no openings are formed. Dominant flavours for swiss cheeses will be fruity and nutty flavours. They are developed from amino acid catabolism. In terms of texture the proteolysis will permit, after a long time (more than six months) the development of a crumbly texture and permit the precipitation of crystals that are easily perceived when tasted and that are the witness of long ripening.

On the opposite side, in the hard cheeses from Italy that are often made from lamb rennet, the lipolysis level is rather high due to the lipase activities present in this type of rennet. This lipolytic activity will be responsible of the piquant notes in these cheeses, which are probably due to the level of free fatty acids in the cheese.

5. Functionalities Developed By the Microbial Ecosystems Important For the Cheese Sensory Perception

A large number of interactions occur among the microorganisms at the origin of the diversity of sensorial characteristics perceived in cheese (Irlinger and Mounier 2009).

5.1 Texture

Several parameters have an impact on the texture. The major one is probably the pH of the curd. Drained acidic curds have a chalky texture like in a young traditional camembert cheese. But as the pH rises during the ripening, in soft cheese the texture gets more and creamier with the age. In mould ripened cheeses the strong proteolytic and lipolytic activities present at the surface will be a lever on the change in texture. These activities emphasize the changes in texture.

In the big cheeses the texture changes slowly. This change is mainly related to the proteolysis. Progressively the texture will be less elastic and will become crumbly with longer aging. Longer aging can lead to the formation of crystals that are perceived when the cheeses are tasted.

5.2 Colour

In mould surface ripened cheeses the white colour of the mycelium of *P. camemberti* will mainly determine the cheese colour. However the development of *Brevibacterium linens* may lead to the occurrence of yellow to orange spots corresponding to the growth of colonies of *B. linens*. When the surface of these mould ripened cheeses become brownish, it can be related to the surface humidity of the cheese that provokes the cell lysis of the mycelium. This aspect is a characterized defect.

In washed rind cheeses, the colour is related to the development of the yeasts and bacteria mixture at the cheese surface. The interaction between yeasts and bacteria on the cheese colour has been shown to be related to the strains used. For example, the association of *Debaryomyces hansenii* with *B. linens* will give a colour that will depend on the *D. hansenii* strain used (Leclercq-Perlat et al. 2004b). *Geotrichum candidum* will usually decrease the orange intensity of *B. linens* colour. Apart from *B. linens*, several other bacteria species will contribute to the cheese colour (e.g., *Arthrobacter arilaitensis*).

5.3 Flavours

Cheese ripening is the main step where flavour formation occurs. Considering the four pathways involved in cheese volatiles biosynthesis (glycolysis and the utilization of citrate, proteolysis and lipolysis), the identification of key aroma components shows that amino acid degradation is a major process for aroma formation in various cheeses (Curioni et al. 2002). In addition, enzymatic hydrolysis of triglycerides to fatty acids and related volatiles (e.g., methylketones) is also essential to a cheese flavour development (Spinnler 2011).

Production of the aroma components relies on the milk-degrading enzymes of each strain, as well as on the complementation of metabolic pathways between strains, which may lead to an enhancement in the quantity and the variety of flavours. So the complexity of the cheese microbiota, and its functional diversity, plays a crucial role in the final flavour of cheeses (Irlinger et al. 2009).

Here, we will describe the main steps leading to the formation of key aroma compounds, with a focus on metabolic complementarities between members of the microbial ecosystem. Considering that amino acid degradation, branched-chain amino acids (leucine, isoleucine and valine—BcAAs), methionine and aromatic amino acids (phenylalanine, tyrosine, tryptophan—ArAAs) are the major precursors of cheese aroma compounds (Yvon et al. 2001). Their catabolism proceeds by two different pathways. The first one, initiated by elimination reactions, has been observed for methionine and aromatic amino acids and leads by a single step to phenol, indol and methanethiol, respectively. The second pathway, observed for ArAAs and methionine, is initiated by a transamination reaction and the resulting α-keto acids are then degraded to aldehydes, alcohols, carboxylic acids, hydoxy acids or methanethiol for methionine *via* 1 or 2 additional steps. If the variety of aromatic compounds is potentially huge, specific pathways will be involved depending on the specific environment (pH, a_w, redox) and the microflora of each cheese. For example, 3-methylbutanal (leucine) was identified as a potent odorant in Camembert, aged Cheddar, together with 2-methylbutanal (isoleucine) in Emmental and Gruyere types of cheeses (Smit et al. 2009). Nevertheless, aldehydes are transitory compounds in cheese because they are rapidly reduced to primary alcohols or even oxidized to the corresponding acids. If the primary alcohols (3-methyl-1-butanol and 2-methyl butanol) confer a pleasant aroma of fresh cheese, the corresponding acids are important flavour compounds in many cheeses and are associated with typical "old cheese", "sweaty socks" notes. *Pr. freudenreichii* is by far the main agent of the formation of methylbutanoic acids, key aromatic compounds in Swiss-type cheese, where this bacteria prevails (Thierry et al. 2011).

Concerning methionine catabolism, the α-elimination (direct or following the transamination step) producing methanethiol seems to be a major pathway especially in *B. linens*, *P. camemberti* and *G. candidum*. Methanethiol can be further chemically oxidized in polysulfides; dimethyldisulfide (DMDS), dimethyl trisulfide (DMTS) and even dimethyl quadrisulfide (DMQS). Coryneform bacteria, especially *B. linens*, are considered to be the key agents of sulphur-compound production in smear ripened and mould-ripened cheeses such as Munster, Livarot or Camembert (Landaud et al. 2008). Sulphur compounds are of main importance in the basic flavour of cheeses. For example DMDS is present in all ripened cheeses.

During cheese ripening, free fatty acids containing four or more carbon atoms may originate from the lipolysis of milk fat or the breakdown of amino acids, as stated above. Shorter ones are directly linked to lactate fermentation, as for example by *Pr. freudenreichii* resulting in the formation of propionate and acetate (and CO_2), both acids being considered as flavour compounds in Swiss-type cheese (Thierry et al. 2011). However, most of the fatty acids have between 4 and 20 carbon atoms come from the lipolysis of triglycerides by moulds. For example, hexanoic acid is a characteristic flavour component of specific hard cheeses (Grana Padano and Roncal). More significantly, fatty acids are the precursors of important aromatic compounds, in particular methylketones, alcohols, lactones and esters (Spinnler 2011). Indeed, the terminal oxidation of free fatty acids produces methylketones, compounds that are primarily known for their contribution to the aroma of surface-mould ripened and blue-veined cheeses, like sulphur compounds they seem to be present in all ripened cheeses. In surface-ripened cheeses and blue-veined cheeses, the synthesis of methyl ketones is related to the enzymatic activity of *P. roqueforti, P. camemberti* or *G. candidum*. Fruity, floral and musty notes are associated with various methylketones such as 2-octanone, 2-nonanone, 2-decanone and 2-undecanone, while Blue cheese notes are attributed to 2-heptanone, key compounds of these cheeses. In anaerobiosis, these methyl ketones are usually reduced in secondary alcohols having lower olfactive thresholds and fruity notes. For this reason blue cheeses are kept for a while in anaerobiosis to get rounder flavour notes. Another important odorant is 1-octen-3-one, especially in Camembert or aged Cheddar, and has been commonly associated with mushroom odours. The production of this compound by *P. camemberti* probably derives from the linoleic and linolenic pathways.

Finally, esters are common cheese volatiles whose production is at the cross-road of lactose fermentation, amino acid and lipid catabolism (Liu et al. 2004, Sourabié et al. 2012). Their precursors activate short- to medium-chain fatty acids (acetyl or acyl CoA) and primary and secondary alcohols. Most esters encountered in cheese are described as having sweet, fruity and floral notes, especially ethyl esters known for their important role in the formation of a fruity character in cheese. The thiols like alcohols can react with acyl CoA to produce thioesters. Among them, S-methyl thioesters are noteworthy. Their very powerful odours are often associated with cheese flavours because of their low perception thresholds. They have been reported in numerous cheeses including camembert and smear-ripened cheeses (Martin et al. 2004).

Different approaches have been proposed to enhance cheese flavour, including the selection of adjunct cultures based on metabolic complementation. For example, S-methylthioacetate production was enhanced when a yeast *Kluyveromyces lactis*, able to produce esters and hence to accumulate acyl CoA, was co-cultivated with *B. linens*, producing méthanethiol (Arfi et al. 2006). There has also been a recent interest in studying the flavouring capacities of Gram-negative bacteria such as *Proteus vulgaris* or *Psychrobacter* sp. that have been isolated from different traditional cheeses (Deetae et al. 2009).

The recent identification of new key compounds in cheeses, such as thiols (Sourabié et al. 2008) shows that understanding the flavour of cheese is still a vast and attractive field of research.

6. Conclusions

Diversity of cheeses occur due to the diversity of milks, coagulation methods, curds obtained and resulting microbial ecosystems. The type of curd and its consequence on the possible size of cheeses will be a defining step that will determine the balance between the ripening microorganism that will develop at the surface with high metabolic activities and the ripening less active microorganisms that will develop in anaerobiosis. This balance will, on the one hand, determine the conservation of the cheese and on the other hand the elaboration of colour and aspect, texture, taste and flavours of the diversity of the cheeses. In soft cheeses research works of the last 10 years have shown how the equilibrium between the microbial activities at the surface and the diffusion of substrates like lactic acid from the cheese center is important. Flavours defects like styrene flavours have been shown to be related with a lack of balance between the diffusion and the *Penicillium* activity.

Complete genomes are available for more and more cheese microorganisms. These genomes give new insight into the adaptation of the microorganisms to dairy product and cheese technology. Among them it appears that iron limitation of the milk is a selective substrate on which a strong competition between the strains of the ecosystem is observed (Monnet et al. 2012). The recent access to RNA sequences expressed by the micro-organisms in the cheese ecosystem is expected to give new and interesting results on the role of each species in the elaboration of cheese organoleptic characteristics (Dugat-Bony et al. 2015, Monnet et al. 2016).

Keywords: Diversity, cheese, microbial ecosystem, methylketone, sulphur compound, ripening microorganism

References

Arfi, K., Landaud, S. and Bonnarme, P. (2006). Evidence for distinct L-methionine catabolic pathways in the yeast *Geotrichum candidum* and the bacterium *Brevibacterium linens*. Applied Environmental Microbiology 72: 2155–2162.

Curioni, P.M.G. and Bosset, J.O. (2002). Key odorants in various cheese types as determined by gas chromatography-olfactometry. International Dairy Journal 12: 959–984.

Deetae, P., Spinnler, H.E., Bonnarme, P. and Helinck, S. (2009). Growth and aroma contribution of *Microbacterium foliorum*, *Proteus vulgaris* and *Psychrobacter* sp. during ripening in a cheese model medium. Applied Microbiology and Biotechnology 82: 169–177.

Dugat-Bony, E., Straub, C., Teissandier, A., Onésime, D., Loux, V., Monnet, C., Irlinger, I., Landaud, S., Leclercq-Perlat, M.N., Bento, P., Fraud, S., Gibrat, J.F., Aubert, J., Fer, F., Guédon, E., Pons, N., Kennedy, S., Beckerich, J.M., Swennen, D. and Bonnarme, P. (2015). Overview of a surface-ripened cheese community functioning by meta-omics analyses. PLoS One 10: 1–25.

Feurer, C., Irlinger, F., Spinnler, H.E., Glaser, P. and Vallaeys, T. (2004). Assessment of the rind microbial diversity in a farmhouse-produced *vs* a pasteurized industrially produced soft red-smear cheese using cultivation and rDNA-based methods. Journal of Applied Microbiology 97: 546–556.

Gkatzionis, K., Hewson, L., Hollowood, T., Hort, J., Dodd, C.E.R. and Linforth, R.S.T. (2013). Effect of *Yarrowia lipolytica* Lon blue cheese odour development: flash profile sensory evaluation of microbiological models and cheeses. International Dairy Journal 30: 8–13.

Irlinger, F. and Mournier, J. (2009). Microbial interactions in cheese: implications for cheese quality and safety. Current Opinion in Biotechnology 20: 142–148.

Landaud, S., Helinck, S. and Bonnarme, P. (2008). Formation of volatile sulfur compounds and metabolism of methionine and other sulfur compounds in fermented food. Applied Microbiology and Biotechnology 77: 1191–1205.

Larpin-Laborde, S., Imran, M., Bonaiti, C., Bora, N., Gelsomino, R., Goerges, S., Irlinger, F., Ward, A.C., Vancanneyt, M., Swings, J., Scherer, S., Gueguen, M. and Desmasures, N. (2011). Surface microbial consortia from Livarot, a French smear-ripened cheese. Canadian Journal of Microbiology 57: 651–660.

Larpin, S., Mondoloni, C., Goerges, S., Vernoux, J.P., Guéguen, M. and Desmasures, N. (2006). *Geotrichum candidum* dominates in yeast population dynamics in Livarot, a French red-smear cheese. FEMS Yeast Research 6: 1243–1253.

Leclercq-Perlat, M.N., Buono, F., Lambert, D., latrille, E., Spinnler, H.E. and Corrieu, G. (2004a). Controlled production of Camembert type cheeses. Part I: Microbial and physicochemical evolution. Journal of Dairy Research 71: 346–354.

Leclercq-Perlat, M.N., Corrieu, G. and Spinnler, H.E. (2004b). The color of *Brevibacterium linens* depends on the yeast used for cheese deacidification. Journal of Dairy Science 87: 1536–1544.

Liu, S.Q., Holland, R.V. and Crow, L. (2004). Esters and their biosynthesis in fermented dairy products: a review. International Dairy Journal 14: 923–945.

Martin, N., Neelz, V. and Spinnler, H.E. (2004). Supra-threshold intensity and odour quality of sulphides and thioesters. Food Quality Preference 15: 247–257.

Monnet, C., Back, A. and Irlinger, F. (2012). Growth of the aerobic bacteria at the cheese surface is limited by the availability of iron. Applied and Environmental Microbiology 78: 3185–3192.

Monnet, C., Dugat-Bony, E., Swennen, D., Beckerich, J.M., Irlinger, F., Fraud, S. and Bonnarme, P. (2016). Investigation of the activity of the microorganisms in a Reblochon-style cheese by metatranscriptomic analysis, Frontier in Microbiology 7: 536 doi:10.3389/fmicb.2016.00536.

Mounier, J., Monnet, C., Jacques, N., Antoinette, A. and Irlinger, F. (2009). Assessment of the microbial diversity at the surface of Livarot cheese using culture-dependent and independent approaches. International Journal of Food Microbiology 133: 31–37.

Prieto, B., Urdiales, R., Francoa, I., Tornadijob, M.E., Fresnoa, J.M. and Carballo, J. (1999). Biochemical changes in PiconBejes-Tresviso cheese, a Spanish blue-veined variety, during ripening. Food Chemistry 67: 415–421.

Rowney, M.K., Roupas, P., Hickey, M.W. and Everett, D.W. (2004). Salt-induced structural changes in 1-day old Mozzarella cheese and the impact upon free oil formation. International Dairy Journal 14: 809–816.

Smit, B.A., Engels, W.J. and Smit, G. (2009). Branched chain aldehydes: production and breakdown pathways and relevance for flavour in foods. Applied Microbiology and Biotechnology 81: 987–999.

Sourabié, A.M., Spinnler, H.E., Bonnarme, P., Saint-Eve, A. and Landaud, S. (2008). Identification of a powerful aroma compound in Munster and Camembert cheeses: Ethyl 3-mercaptopropionate. Journal of Agriculture and Food Chemistry 56: 4674–4680.

Sourabié, A.M., Spinnler, H.-E., Bourdat-Deschamps, M., Tallon, R., Landaud, S. and Bonnarme, P. (2012). *S*-methyl thioesters are produced from fatty acids and branched-chain amino acids by brevibacteria: focus on L-leucine catabolic pathway and identification of acyl-CoAs intermediates. Applied Microbiology and Biotechnology 93: 1673–1683.

Spinnler, H.E. (2011). Role of lipids in the olfactive perception of dairy products. Sciences des Aliments 30: 105–121.

Thierry, A., Deutsch, S.M., Falentin, H., Dalmasso, M., Cousin, F.J. and Jan, G. (2011). New insights into physiology and metabolism of *Propionibacterium freudenreichii*. International Journal of Food Microbiology 149: 19–27.

Yvon, M. and Rijnen, L. (2001). Cheese flavour formation by amino acid catabolism. International Dairy Journal 11: 185–201.

21

Current Trends in Microbiological, Technological and Nutritional Aspects of Fermented Sausages

Eleftherios H. Drosinos* and Spiros Paramithiotis

1. Introduction

The production of fermented sausages includes mixing of minced meat and fatty tissue with curing agents, carbohydrates and spices, stuffing into casings, fermentation and ripening. A wide range of products results from the variations in the type and amount of raw materials, fermentation as well as drying conditions. The meat employed is usually pork or a mixture of pork and beef and the fatty tissue usually pork back fat. The role of the curing agents is multifunctional; sodium chloride lowers the water activity, interacts with the myofibrillar protein fraction and affects taste. Sodium nitrate functions as a reservoir to provide the ecosystem with nitrites *via* its reduction by Gram-positive catalase-positive microbiota. Sodium nitrite contributes to the development of curing color and the establishment of the desired Gram-positive microbiota and at the same time inhibits the autoxidative processes leading to rancidity. Sodium ascorbate is very frequently added with nitrite to enhance the development of curing color

Department of Food Science and Human Nutrition, Agricultural University of Athens, Athens, Greece.
* Corresponding author: ehd@aua.gr

and flavor. Carbohydrates are used as fermentation substrates to allow sufficient pH value reduction and spices to enhance taste.

Fermentation and ripening of spontaneously fermented sausages may take place in three, two or one step. The first has been reported for the production of salciccia sotto sugna (Gardini et al. 2001a, Parente et al. 2001b); the initial drying stage at 24°C, 80% relative humidity (RH) for 24 hr is followed by fermentation stages at 18–22°C, 75–89% RH for 5 days and 15–20°C, 80–85% RH for 2 mon. Quite often, the second stage of fermentation is not performed in fixed conditions, for example the fermented sausages prepared in Greece. In the latter case, the fermentation temperature gradually decreases from 18–24°C to 11–19°C with an alteration to relative humidity from 60–94% to 60–90% within 5–7 days, according to the occasion (Coppola et al. 1998, Metaxoloupos et al. 2001, Papamanoli et al. 2003, Greco et al. 2005, Comi et al. 2005, Drosinos et al. 2005a, 2007). However, in the case of chorizo, salchichon or the Naples-type salami, this step may as well be omitted (Lizaso et al. 1999, Garcia-Varona et al. 2000, Mauriello et al. 2004). Finally, fermentation may be carried out in only one step, such as in the case of sobrasada or Salame di Senise (Rosselo et al. 1995, Baruzzi et al. 2006).

Fermented meat products have been extensively studied over the last decades. Reformulation attempts and design of effective starter cultures have drawn special attention due to their importance, especially from a technological point of view. Regarding the former, the formulations employed have been subjected to intensive criticism. More accurately, increased fat content as well as sodium and nitrite utilization have been correlated with a number of adverse health effects. Moreover, a discussion on carcinogenicity of consumption of red and processed meat was reinitiated (Bouvard et al. 2015). Thus, several attempts have been made to decrease the amount or even replace these technologically important ingredients with ones that could possibly provide some health benefits. Additionally, the technological properties that an effective starter culture should fulfill are continuously complemented with several functional ones, increasing their significance regarding both quality and safety of the final product. In the following pages the recent advances in these fields have been critically discussed.

2. The Microbiota of Meat Fermentation

The ecosystem of fermented meat products has been studied throughout the world; lactic acid bacteria (LAB) dominate while Gram-positive catalase-positive cocci and yeasts form a secondary microbiota. The main effect of LAB is acidification due to lactic acid production. This has multiple effects on the product: improvement of firmness, cohesiveness and sliceability due to the loss of meat water holding capacity, enhancement of color development and growth inhibition of spoilage and pathogenic microorganisms. Gram-positive catalase-positive cocci mainly contribute to color stability through their nitrate reductase and catalase activities and secondarily to flavor development through their proteolytic and lipolytic activities. The latter is also the main contribution of yeasts.

A wide variety of LAB, Gram-positive catalase-positive cocci and yeasts have been isolated from fermented meat products. *Lactobacillus sakei* is the most abundant while presence of *Lb. curvatus* and *Lb. plantarum* is nearly always reported. Similarly, *Staphylococcus xylosus, St. saprophyticus, St. carnosus* and *St. simulans* seem to prevail in the respective microcommunity. Regarding yeasts, *Debaryomyces hansenii* is most frequently isolated.

Dominance of *Lb. sakei* may be attributed to the quite efficient utilization of the available substrates, namely nucleosides, arginine and N-acetyl-D-neuraminic acid. On the other hand, *Lb. plantarum* and *Lb. curvatus* seem to lack the so-called 'meat specialization' of *Lb. sakei* as this is defined by the ability to utilize additional energy sources that happen to be abundantly present in meat (Hebert et al. 2012, Siezen et al. 2012).

Nucleosides such as adenosine and inosine result from ATP breakdown that takes place in meat *post mortem*. Upon depletion of glucose, nucleoside catabolism may simultaneously commence once a critical biomass level has been reached. Then, nucleosides may enter the bacterial cells and be cleaved by nucleoside phosphorylases or hydrolases to the respective pentose moieties and nucleobases. The former are catabolized to a mixture of acetic acid, formic acid and ethanol while the latter is stoichiometrically excreted to the growth medium. This conversion is very efficient and proceeds towards the depletion of the nucleosides throughout a pH range that is relevant to sausage fermentation. Moreover, at pH 5.0, lactic acid may also be produced by inosine (Rimaux et al. 2011a).

The arginine deiminase (ADI) pathway consists of three enzymes: arginine deiminase, catabolic ornithine transcarbamoyl transferase and carbamate kinase that catalyzes the conversion of arginine to citrulline and ammonia, citrulline to carbamoyl phosphate and ornithine and carbamoyl phosphate to ammonia, carbon dioxide and ATP, respectively (Zuniga et al. 2002). These enzymes are encoded by *arcA, arcB* and *arcC*, respectively. These genes are organized in the *arcABCTDR* operon; *arcD, arcR* and *arcT* encode an arginine/ornithine antiporter, a regulatory gene and a putative transaminase gene, respectively. Moreover, a putative citrulline/ornithine antiporter seems to be encoded by *PTP*, a gene located downstream from the *arcABCTDR* operon (Rimaux et al. 2013). The ADI pathway is induced by arginine, an amino acid that is present in meat in significant quantities due to the action of endogenous proteolytic enzymes. Moreover, ADI pathway operation seems to be subjected to catabolite repression, it is stimulated by anaerobiosis and is affected by environmental pH and growth phase, in a strain-dependent manner (Champomier-Verges et al. 1999, Rimaux et al. 2012). Kinetic analysis of the ADI pathway revealed that citrulline and ornithine excretion took place over a wide pH range (4.50–7.75). However, their ratio was affected by the pH value. More accurately, upon arginine depletion, citrulline was taken up by the cells and converted to ornithine but only when pH value was between 5.0 and 6.5 (Rimaux et al. 2011b). Moreover, transcriptomic analysis of the ADI pathway genes revealed that their expression was higher at the optimal growth pH (pH 6.0) but decreased away from this value (Rimaux et al. 2012). Thus, improved survival was assigned to the generation of extra ATP rather than a protective effect against acidity.

N-acetyl-D-neuraminic acid (Neu5Ac), a nine carbon amino sugar that is a component of glycoproteins and glycolipids, in both eukaryotes and prokaryotes (Tao et al. 2010) is another carbon and nitrogen source that *Lb. sakei* may utilize. Indeed, genome sequencing of *Lb. sakei* 23 K (Chaillou et al. 2005) revealed the presence of three loci putatively involved in Neu5Ac catabolism. Anba-Mondoloni et al. (2013) demonstrated that the *nan* gene cluster was functional and involved in Neu5Ac catabolism. Moreover, a catabolic pathway similar to the one described for *Escherichia coli* (Plumbridge and Vimr 1999) was proposed. However, transport and regulation systems seemed to be specific for the *Lb. sakei* strain (Anba-Mondoloni et al. 2013). A functional and mutational analysis of N-acetylneuraminate lyases (NALs) from *Lb. sakei* 23 K and *Lb. antri* DSMZ 16041 was presented by Garcia-Garcia et al. (2014). NALs catalyze the first step of the Neu5Ac catabolism, i.e., the cleavage of Neu5Ac into pyruvate and N-acetyl mannosamine. The functional characterization of the enzymes from the two bacterial species revealed differences in their optimal temperatures and pH values as well as in the stability at basic and acidic pH values. Moreover, lower catalytic activity was detected compared to the counterparts belonging to the phylogenetically divergent groups that were assigned to their relatively small sugar-binding pocket.

Similarly, genomic analysis of *St. carnosus* strain TM300 was performed in order to justify the omnipresence of this species in meat fermentations. The genetic potential of several properties that may promote growth in this microenvironment, such as the ability to grow at low temperatures and water activity and to catabolize carbohydrates occurring in meat, has been revealed (Rosenstein et al. 2009).

3. Generation of Flavor Compounds

Generation of flavor compounds during manufacture of dry fermented sausages has been extensively studied and the importance of factors such as product formulation and processing parameters has been adequately highlighted (Leroy et al. 2006). Proteolysis and lipolysis are considered as key biochemical reactions for the generation of flavor compounds (Toldra 1998) and thus the contribution of microbial proteolytic and lipolytic activities has been the epicenter of intensive research. Generally, these activities are strain-dependent and their accurate control is necessary to ensure the quality of the final product.

Muscle proteinases, especially cathepsin D, may be held responsible for the initial protein breakdown (Luecke 2000). Then, free amino acids may result from the action of endogenous as well as microbial peptidases (Flores and Toldra 2011). Similarly, muscle lipases along with microbial ones contribute to the lipolytic activities with the former estimated at approximately upto 60%–80% of the total (Molly et al. 1996, 1997). The proteolytic and lipolytic capacity of several LAB, staphylococci, yeast and mold isolates from fermented meat products has been thoroughly investigated. The strain-dependence of this property has been adequately highlighted. Studies on the proteolytic activities of several dry fermented sausage-associated microorganisms such as *Lb. sakei*, *Lb. casei*, *Lb. plantarum*, *Lb. curvatus*, *Pediococcus pentosaceus*, *Pd. acidilactici*, *St. xylosus*, *St. saprophyticus*, *St. simulans*, *St. gallinarum*, *St. cohnii*, *St. capitis*, *St. carnosus*, *St. griseus*, *St. vitulinus*, *D. hansenii* and *Saccharomyces cerevisiae* revealed significant differences regarding the ability to hydrolyze the sarcoplasmic and/or the myofibrillar protein fractions and the mode of this action as well.

Lb. plantarum CRL 681 was able to hydrolyze both protein fractions; significant increase in lysine, arginine and leucine content was assigned to the lytic action on myofibrillar proteins whereas an increase in alanine to the respective sarcoplasmic proteins (Fadda et al. 1999). The proteolytic activities of *Lb. sakei* CRL 1862 was studied by Castellano et al. (2012) and a more pronounced hydrolysis of sarcoplasmic proteins compared to the myofibrillar ones was reported that opposed the results presented by Drosinos et al. (2007) and Villani et al. (2007), in which no action on the sarcoplasmic proteins was detected. More accurately, in the former study, six *Lb. sakei* strains were able to completely hydrolyze myosin, actin and all myofibrillar proteins with molecular weights ranging from 12 to 200 kDa, which in turn was different from the partial hydrolysis of only actin and myosin that was observed by *Lb. plantarum* strain CRL 681 (Fadda et al. 1999). Additionally, two *Lb. sakei* and two *Lb. curvatus* strains exhibited a reduction in the 11.5 and 22.7 kDa myofibrillar protein bands with a concomitant increase in the 13 kDa band (Villani et al. 2007).

Proteolysis seems to be a common property among staphylococci; decomposition of both sarcoplasmic and myofibrillar fractions has been reported. Regarding the former, complete hydrolysis by one *St. xylosus* strain or at least a decrease in the intensity of protein bands at approximately 20.3, 22.4, 41.6 and 48.4 by two other *St. xylosus* strains has been reported by Mauriello et al. (2002). Comparable alteration of the profile obtained by SDS-PAGE was also reported by Drosinos et al. (2007). However, a different decomposition mode was reported for the myofibrillar protein fraction; complete hydrolysis of myosin and actin with the concomitant appearance of bands at about 25 and 100 kDa was reported by Mauriello et al. (2002), whereas complete hydrolysis of the myofibrillar fraction to polypeptides with a molecular weight less than 12 kDa was reported by Drosinos et al. (2007).

The lipolytic activities of LAB and staphylococci have also been studied to some extent. In general, the former seem to possess a weaker lipolytic system compared to the latter (El Soda et al. 1986, Miralles et al. 1996, Coppola et al. 1997, Kenneally et al. 1998, Sanz et al. 1998, Montel et al. 1998, Mauriello et al. 2004, Drosinos et al. 2007, Casaburi et al. 2007). Lipolysis results in release of fatty acids that are concomitantly subjected to oxidative reactions to form aliphatic hydrocarbons, alcohols, aldehydes, ketones, and esters. Excessive oxidation may result in the formation of off-flavors. However, through the consumption of oxygen by the microbiota, rancidity may be delayed.

Regarding yeasts, the proteolytic and lipolytic activities as well as the production of aroma compounds primarily by *D. hansenii* and secondarily by *S. cerevisiae* have been studied to some extent and a potential contribution to the improvement of the sensorial properties has been exhibited (Bolumar et al. 2003a,b, 2005, 2008, Chaves-Lopez et al. 2011, Corral et al. 2014, Cano-Garcia et al. 2014a,b). Since these properties are strain-dependent, the need for selection of the appropriate strain as well as the adjustment of technological parameters to allow for improved survival and activity have been adequately highlighted (Flores et al. 2015, Corral et al. 2015).

4. Safety Aspects

Regarding safety of fermented meat products, research in recent years was mainly focused on biogenic amine accumulation and bacteriocin production and utilization. Both may be produced by the technological biota employed and therefore should be taken into consideration for the design of proper starter cultures.

4.1 Biogenic Amine Accumulation

Biogenic amines are produced through decarboxylation of amino acids catalyzed by microbial enzymes. The increased interest in these compounds results from the toxic effects that are exerted upon their ingestion; this intoxication may even be fatal.

Several studies have addressed occurrence of biogenic amines in fermented meat products. Detection of tyramine, histamine, phenylethylamine, tryptamine, putrescine, cadaverine, spermine and spermidine has been reported in a variety of products analyzed (Brink et al. 1990, Hernandez-Jover et al. 1997a, Eerola et al. 1998, Bover-Cid et al. 1999a, Montel et al. 1999, Parente et al. 2001a, Ekici et al. 2004, Riebroy et al. 2004, Ruiz-Capillas and Jimenez-Comener 2004, Miguelez-Arrizado et al. 2006, Latorre-Moratalla et al. 2008, Papavergou 2011, Ikonic et al. 2013, Gardini et al. 2013, Dos Santos et al. 2015, Gianotti et al. 2015). The most abundant biogenic amine in fermented meat products is tyramine, the accumulation of which is mostly assigned to the LAB, i.e., the technological biota. On the other hand, enterobacteria and pseudomonads, which flourish during storage of the raw materials under improper hygienic and environmental conditions, are considered responsible for histamine, putrescine and cadaverine accumulation. Taking into consideration the constant release of precursor amino acids through the activity of endogenous and microbial proteases, fermentation is rightfully considered as a procedure that promotes biogenic amine accumulation. Reduction of biogenic amine content may occur by avoiding accumulation and adopting removal strategies. The former may take place by using raw materials of high quality and starter cultures that do not exhibit decarboxylase activity. Indeed, a significant reduction in the biogenic amine levels by using proper starter cultures has been exhibited (Bover-Cid et al. 2000b, Gonzalez-Fernandez et al. 2003, Latorre-Moratalla et al. 2006). Removal of biogenic amines from fermented meat products has also been studied to some extent. The utilization of microorganisms that exhibit amine oxidase activity and therefore degrade biogenic amines seems to be a promising approach. Several strains belonging to technologically relevant species have been studied (Martuscelli et al. 2000, Fadda et al. 2001, Gardini et al. 2002). However, the results obtained *in situ* revealed limited effectiveness, most probably due to the low population obtained by the amine-oxidizing strains as well as the limited supply of oxygen (Leuschner and Hammes 1998).

4.2 Production of Antimicrobial Compounds

Production of antimicrobial compounds by bacterial isolates from fermented meat products has been extensively studied. A wide range of bacteriocinogenic isolates from a variety of species including *Carnobacterium divergens*, *Cb. piscicola*, *Enterococcus faecium*, *Lactococcus lactis*, *Lb. curvatus*, *Lb. plantarum*, *Lb. sakei*, *Leuconostoc carnosum*, *Ln. gelidum*, *Ln. mesenteroides*, *Pediococcus acidilactici*, *Pd. parvulus*, *Pd. pentosaceus*, *St. warneri* and *St. xylosus* has been studied (Ahn and Stiles 1990, Harding and Shaw 1990, Mortvedt et al. 1991, Holck et al. 1996, Villani et al. 1997, Mataragas et al. 2003b, Noonpakdee et al. 2003, Drosinos et al. 2005b, Prema et al. 2006, Schneider et al. 2006, Albano et al. 2007a,b, Osmanagaoglu 2007, Todorov et al. 2013, Paramithiotis et al. 2014, Engelhardt et al. 2015, Amadoro et al. 2015, Fontana et al. 2015, Casaburi et al. 2016). Isolation of a bacteriocinogenic strain is usually followed by studies on the parameters that may affect bacteriocin production. Thus, the effect of the physiological state of the bacteriocinogenic strain (Drosinos et al. 2005b, 2006, Xiraphi et al. 2006), the growth medium in terms of carbon source (Mataragas et al. 2004, Drosinos et al. 2005c), nitrogen source (Mataragas et al. 2004) and micro-elements (Parente and Ricciardi 1999), pH value (Mataragas et al. 2003a), temperature (Krier et al. 1998, Lejeune et al. 1998) and specific food ingredients such as sodium chloride, nitrate, nitrite and spices have been assessed in detail (Aymerich et al. 2000a, Hugas et al. 2002, Verluyten et al. 2003, 2004a,b). Similarly, the *in situ* application of the bacteriocins either as such or through the incorporation of the producer strain in the starter culture, has been subjected to intensive study as well (Lahti et al. 2001, Dicks et al. 2004, Nieto-Lozano et al. 2006, Albano et al. 2007b). Although the results reported are promising, the rather narrow antimicrobial spectrum, the uneven distribution within the product and the reduced solubility and stability, limit their *in situ* application. However, bacteriocins may be useful within the context of the hurdle concept. Indeed, several studies have revealed enhancement or decrease of the bacteriocin activity by the various ingredients used. More accurately, the negative effect of sodium chloride (Jydegaard et al. 2000) and the synergistic effect of nitrite, organic acids and chelating agents, essential oils, thermal treatment, as well as packaging have already been established (Nilsson et al. 2000, Abriouel et al. 2002, Gill and Holley 2003, Garcia et al. 2003, 2004a,b, Ananou et al. 2004, 2005a,b, Bakes et al. 2004, Grande et al. 2007, 2006) highlighting the necessity of adjustments in the formulation and technology applied once this hurdle is to be incorporated.

5. Nutritional Improvement

Modification of the formulation towards healthier products has been extensively studied over the last decade. The modification of the lipid fraction, the reduction of sodium and nitrite as well as the delivery of probiotics have been at the epicenter of intensive research with promising results.

5.1 Lipid Fraction Modification

The role of the lipid fraction is fundamental regarding both technological functions and organoleptic quality. More accurately, during drying it assists continuous moisture release by keeping the structure loose whereas in the final product it contributes to flavor, juiciness and mouthfeel (Wirth 1988). The pork backfat that is usually employed is rich in saturated fatty acids and cholesterol, which in turn have been correlated with increased risk of coronary heart disease (Bhupathiraju and Tucker 2011). Therefore, a trend towards substitution of this type of fat with vegetable oils, which have a higher ratio of unsaturated to saturated fatty acids, has been established. However, it should be kept in mind that this substitution requires the use of suitable antioxidants since vegetable oils are susceptible to oxidative and rancidity processes.

The replacement of pork back fat with vegetable oils or dietary fibers, which may be used as fat mimetics has been extensively studied. Olive oil seemed an attractive alternative due to the important nutritional properties. Direct addition of olive oil resulted in very soft products with unacceptable appearance (Bloukas et al. 1997) therefore pre-emulsification seemed a prerequisite. Substitution of up to 25% of the pork back fat with pre-emulsified olive oil has been reported and in most of the cases the substitution resulted in improved sensorial scores without any rancidity problems or other physico-chemical defects (Bloukas et al. 1997, Muguerza et al. 2001, 2002, Kayaardi and Gok 2003, Severini et al. 2003, Ansorena and Astiasaran 2004a, Koutsopoulos et al. 2008, Del Nobile et al. 2009). Moreover, application of a variety of pre-emulsified oils such as hazelnut (Yildiz-Turp and Serdaroglu 2008), soy (Muguerza et al. 2003), linseed (Ansorena and Astiasaran 2004b, Valencia et al. 2006a), algae (oil from the microalgae *Schizochytrium* sp.) (Valencia et al. 2007), deodorized fish oil (Valencia et al. 2006b), flaxseed and canola (Pelser et al. 2007) has been reported. In many cases, stability to the oxidation was guaranteed with the use of antioxidants and/or specific storage conditions. Several attempts for the incorporation of dietary fiber as partial replacement of pork back fat have been reported. The addition of various amounts of dietary fiber from cereals such as wheat and oat (Garcia et al. 2002), fruits such as peach, apple and orange (Garcia et al. 2002, Fernandez-Lopez et al. 2008, Yalinkilic et al. 2012), roots such as carrot (Eim et al. 2008, 2012) as well as fructooligosaccharides (Salazar et al. 2009, Alves dos Santos et al. 2012) and powdered inulin (Mendoza et al. 2001) has been studied with promising results. In all cases the nutritional properties were significantly enhanced; however an effect on textural parameters was recorded in some cases.

5.2 Sodium Reduction

Addition of sodium chloride serves many purposes. Through the reduction of water activity it contributes to the microbial stability of the final product. In the case of spontaneous fermentations, it contributes to the selection of the dominant microbial consortium. It affects the organoleptic quality in many ways: it imparts a salty taste and improves juiciness and mouthfeel by increasing the water retention ability. Moreover, it increases the solubility of the myofibrillar protein fraction by increasing the ionic strength of the solution. Finally, its concentration affects enzyme activity and lipid oxidation. However, sodium intake has been correlated with high blood pressure, occurrence of cardiovascular diseases in developed countries and is considered a risk factor for a number of other diseases such as obesity, osteoporosis, etc. (Cappuccio et al. 2000, He and MacGregor 2010). Although many attempts to reduce or substitute sodium have been made, it seems that it remains more or less trivial due to the significant defects caused by the replacement of salts. More accurately, 50% substitution with potassium chloride led to higher hetero-fermentative activity, nitrosation intensity and bitterness (Ibanez et al. 1995, Gou et al. 1996). Regarding the latter, it was avoided when substitution of sodium chloride with potassium chloride took place to less than 30% (Gou et al. 1996). Substitution of more than 40% of sodium chloride with potassium lactate affected negatively color stability and product consistency (Gou et al. 1996). Moreover, utilization of potassium lactate resulted in a delay in the drop of the pH value which could be compensated by the antimicrobial action of the salt itself and thus did not pose spoilage risk. Finally, more than 40% with glycine resulted in unacceptable sweet taste and inconsistency (Gou et al. 1996). Gimeno et al. (1999) studied the partial replacement with a mixture of potassium and calcium chlorides and reported a reduction in organoleptic scores, hardness, cohesiveness, gumminess and chewiness. Reduction in hardness and gumminess was also reported by Gimeno et al. (2001) as a result of the partial replacement of sodium chloride with calcium ascorbate. Similarly, partial replacement with a mixture of potassium, magnesium and calcium chlorides resulted in modification of the color, most probably due to changes in nitrosation process triggered by a decrease in Micrococcaceae population (Gimeno et al. 1998).

5.3 Nitrite Reduction

The role of nitrite is manifold; it affects organoleptic quality through the formation of nitrosomyoglobin and the typical curing flavor and it also controls growth of pathogenic and spoilage microbiota (Hammes 2012). However, several concerns have been raised due to the methaemoglobinemia induction resulting from excessive intake as well as the formation of N-nitrosamines upon reaction with secondary or tertiary amines (Scanlan 1983), compounds whose occurrence in fermented meat products has been discussed in a previous paragraph. Therefore, addition of nitrites and nitrates is strictly regulated (EU 2006). Presence of N-nitrosamines in fermented meat products has been studied to some extent. Yurchenko and Molder (2007) reported the presence of N-nitrosodimethylamine, N-nitrosodiethylamine, N-nitrosopyrrolidine, N-nitrosopiperidine and N-nitrosodibutylamine in products from the Estonian market. However, their quantities were low since their mean concentration was less than 0.93 ug kg^{-1}. The occurrence of volatile and non-volatile N-nitrosamines in products from the Danish and the Belgian markets was studied by Herrmann et al. (2015). Generally, the mean levels of the volatile N-nitrosamines were low (less than 0.8 ug kg^{-1}) whereas the respective ones of the non-volatile was significantly higher (less than 118 ug kg^{-1}). Regarding the former, N-nitrosodimethylamine, N-nitrosomorpholine, N-nitrosopyrrolidine and N-nitrosopiperidine were detected in all products, N-nitrosodiethylamine only in Danish ones and N-nitrosomethylethylamine in none. As far as the non-volatile N-nitrosamines were concerned, presence of N-nitrosoproline, N-nitrosomethylaniline, N-nitrosothiazolidine-4-carboxylic acid and N-nitroso-2-methylthiazolidine-4-carboxylic acid was verified in all samples and N-nitrososarcosine only in Danish ones. Presence of N-nitrosopiperidine and N-nitrosomorpholine in products from the Belgian market was also reported by De Mey et al. (2014). However, their presence could not be linked to biogenic amine content or any specific product type. On the contrary, occurrence of N-nitrosopiperidine was potentially attributed to the presence of piperidine due to the use of pepper.

Several alternatives to nitrite have been studied and proposed (Shahidi and Pegg 1995, Sebranek and Bacus 2007). Moreover, the striking growth of the natural and organic foods market has created a demand for nitrite-free meat products. With regard to color formation, De Maere et al. (2016) provided very promising results for the development of the typical fermented sausages color without the addition of nitrate/nitrite. The effect of reduced levels of nitrate and nitrite on the survival of *Listeria innocua, Salmonella* Typhimurium and toxin production by *Clostridium botulinum* Group I and Group II was studied by Hospital et al. (2012, 2014, 2016). The importance of their addition was highlighted in the first two studies. It was stated that effective control of *L. innocua* and *S.* Typhimurium could only be achieved by nitrate/nitrite addition, even when pH and aw values were below the ones considered as minimum, at least regarding *S.* Typhimurium. On the contrary, no botulinum neurotoxin was detected in the absence of nitrate/nitrite, indicating either the efficacy of pH, a$_w$, temperature and competitive microbiota to control toxin production or that the time during which germination and growth were favored was not enough for the production of detectable toxin amounts. Given the above, substitution of nitrite/nitrate in fermented meat products seems to remain elusive and thus the source of nitrite rather than the addition *per se* is the subject of debate.

5.4 Functional Meat Products

The improvement of the functional value of meat and meat products has also been at the epicenter of some research. The delivery of probiotics has been the main vehicle for many years towards this aim. Therefore, significant amount of research has taken place regarding the *in vitro* evaluation of the probiotic potential of autochthonous to fermented meat microecosystem strains (Erkkila and Petaja 2000, Klingberg et al. 2005, Villani et al. 2005, Pennacchia et al. 2004, 2006, Klingberg and Budde 2006, Rebucci et al. 2007). The criteria applied are continuously updated and involve several technological and functional properties (Giraffa 2012, Kolozyn-Krajewska and Dolatowski 2012); their delivery may take place either in their native form or microencapsulated form (Cavalheiro et al. 2015, De Prisco and Mauriello 2016). The results obtained are very promising but further research is still necessary, especially regarding intake of the potentially probiotic products and concomitant modulation of host immunity.

Recently, the possibility of meat functional properties improvement through the delivery of bioactive peptides has been recognized (Arihara 2006). Although only a limited amount of studies have been performed (Stadnik and Keska 2015), the potential of meat-borne lactobacilli such as *Lb. sakei* and *Lb. curvatus* to liberate bioactive peptides with angiotensin I converting enzyme (ACE) inhibitory activity through their proteolytic activities has been exhibited (Castellano et al. 2013).

6. Conclusions and Future Perspectives

Fermented meat products have been extensively studied, mostly due to the increased commercial importance. Development of the microecosystem, as well as a range of properties that may affect the product from a technological perspective have been adequately assessed and well understood. However, the intensive criticism that some technologically important ingredients have been subjected to, due to the correlation with adverse health effects, resulted in a trend towards their substitution. Moreover, further enhancement of the nutritional value through the delivery of bioactive peptides is also very attractive. Since these are issues of increased interest, it is very likely that in the next few years research will focus on them.

Keywords: Fermented sausages, microbiota, flavor, safety, nutritional improvement

References

Abriouel, H., Maqueda, M., Galvez, A., Martinez-Bueno, M. and Valdivia, E. (2002). Inhibition of bacterial growth, enterotoxin production, and spore outgrowth in strains of *Bacillus cereus* by bacteriocin AS-48. Applied and Environmental Microbiology 68: 1473–1477.

Ahn, C. and Stiles, M.E. (1990). Plasmid-associated bacteriocin production by a strain of *Carnobacteriumpiscicola* from meat. Applied and Environmental Microbiology 56: 2503–2510.

Albano, H., Oliveira, M., Aroso, R., Cubero, N., Hogg, T. and Teixeira, P. (2007b). Antilisterial activity of lactic acid bacteria isolated from 'Alheiras' (traditional Portuguese fermented sausages): *In situ* assays. Meat Science 76: 796–800.

Albano, H., Todorov, S.D., van Reenen, C.A., Hogg, T., Dicks, L.M.T. and Teixeira, P. (2007a). Characterization of two bacteriocins produced by *Pediococcus acidilactici* isolated from 'Alheira', a fermented sausage traditionally produced in Portugal. International Journal of Food Microbiology 116: 239–247.

Alves dos Santos, B., Campagno, P.C.B., Pacheco, M.T.B. and Pollonio, M.A.R. (2012). Fructo-oligosaccharides as a fat replacer in fermented cooked sausages. International Journal of Food Science and Technology 47: 1183–1192.

Amadoro, C., Rossi, F., Piccirilli, M. and Colavita, G. (2015). Features of *Lactobacillus sakei* isolated from Italian sausages: Focus on strains from Ventricina del Vastese. Italian Journal of Food Safety 4: 5449.

Ananou, S., Maqueda, M., Martinez-Bueno, M., Galvez, A. and Valdivia, E. (2005b). Control of *Staphylococcus aureus* in sausages by enterocin AS-48. Meat Science 71: 549–556.

Ananou, S., Galvez, A., Martinez-Bueno, M., Maqueda, M. and Valdivia, E. (2005a). Synergistic effect of enterocin AS-48 in combination with outer membrane permeabilizing treatments against *Escherichia coli* O157:H7. Journal of Applied Microbiology 99: 1364–1372.

Ananou, S., Valdivia, E., Martinez Bueno, M., Galvez, A. and Maqueda, M. (2004). Effect of combined physico-chemical preservatives on enterocin AS-48 activity against the enterotoxigenic *Staphylococcus aureus* CECT 976 strain. Journal of Applied Microbiology 97: 48–56.

Anba-Mondoloni, J., Chaillou, S., Zagorec, M. and Champomier-Verges, M.C. (2013). Catabolism of N-acetylneuraminic acid, a fitness function of the food-borne lactic acid bacterium *Lactobacillus sakei*, involves two newly characterized proteins. Applied and Environmental Microbiology 79: 2012–2018.

Ansorena, D. and Astiasaran, I. (2004b). The use of linseed oil improves nutritional quality of the lipid fraction of dry-fermented sausages. Food Chemistry 87: 69–74.

Ansorena, D. and Astiasaran, I. (2004a). Effect of storage and packaging on fatty acid composition and oxidation in dry fermented sausages made with added olive oil and antioxidants. Meat Science 67: 237–244.

Arihara, K. (2006). Strategies for designing novel functional meat products. Meat Science 74: 219–229.

Aymerich, T., Artigas, M.G., Garriga, M., Monfort, J.M. and Hugas, M. (2000a). Effect of sausage ingredients and additives on the production of enterocins A and B by *Enterococcus faecium* CTC492. Optimization of *in vitro* production and anti-listerial effect in dry fermented sausages. Journal of Applied Microbiology 88: 686–694.

Bakes, S.H., Kitis, F.Y.E., Quattlebaum, R.G. and Barefoot, S.F. (2004). Sensitization of Gram negative and Gram-positive bacteria to jenseniin G by sublethal injury. Journal of Food Protection 67: 1009–1013.

Baruzzi, F., Matarante, A., Caputoa, L. and Morea, M. (2006). Molecular and physiological characterization of natural microbial communities isolated from a traditional Southern Italian processed sausage. Meat Science 72: 261–269.

Bhupathiraju, S.N. and Tucker, K.L. (2011). Coronary heart disease prevention: nutrients, foods, and dietary patterns. Clinica Chimica Acta 412: 1493–1514.

Bloukas, J.G., Paneras, E.D. and Fournitzis, G.C. (1997). Effect of replacing pork backfat with olive oil on processing and quality characteristics of fermented sausages. Meat Science 45: 133–144.

Bolumar, T., Sanz, Y., Aristoy, M.C. and Toldra, F. (2008). Purification and characterisation of proteases A and D from *Debaryomyces hansenii*. International Journal of Food Microbiology 124: 135–141.

Bolumar, T., Sanz, Y., Aristoy, M.C. and Toldra, F. (2005). Protease B from *Debaryomyces hansenii*: purification and biochemical properties. International Journal of Food Microbiology 98: 167–177.

Bolumar, T., Sanz, Y., Aristoy, M.C. and Toldra, F. (2003b). Purification and properties of an arginyl aminopeptidase from *Debaryomyces hansenii*. International Journal of Food Microbiology 86: 141–151.

Bolumar, T., Sanz, Y., Aristoy, M.C. and Toldra, F. (2003a). Purification and characterization of a prolyl aminopeptidase from *Debaryomyces hansenii*. Applied and Environmental Microbiology 69: 227–232.

Bouvard, V., Loomis, D., Guyton, K.Z., Grosse, Y., El Ghissassi, F., Benbrahim-Tallaa, L., Guha, N., Mattock, H. and Straif, K. (2015). Carcinogenicity of consumption of red and processed meat. The Lancet Oncology 16: 1599–1600.

Bover-Cid, S., Hugas, M., Izquierdo-Pulido, M. and Vidal-Carou, M.C. (2000b). Reduction of biogenic amine formation using a negative amino acid-decarboxylase starter culture for fermentation of fuet sausages. Journal of Food Protection 63: 237–243.

Bover-Cid, S., Schoppen, S., Izquierdo-Pulido, M. and Vidal-Carou, M.C. (1999a). Relationship between biogenic amine contents and the size of dry fermented sausages. Meat Science 51: 305–311.

Brink, B., Damink, C., Joosten, H.M. and Huis in't Veld, J.H. (1990). Occurrence and formation of biologically active amines in foods. International Journal of Food Microbiology 11: 73–84.

Cano-Garcia, L., Rivera-Jimenez, S., Belloch, C. and Flores, M. (2014b). Generation of aroma compounds in a fermented sausage meat model system by *Debaryomyces hansenii* strains. Food Chemistry 151: 364–373.

Cano-Garcia, L., Belloch, C. and Flores, M. (2014a). Impact of *Debaryomyces hansenii* strains inoculation on the quality of slow dry-cured fermented sausages. Meat Science 96: 1469–1477.

Cappuccio, F.P., Kalaitzidis, R., Duneclift, A. and Eastwood, J.B. (2000).Unravelling the links between calcium excretion, salt intake, hypertension, kidney stones and bone metabolism. Journal of Nephrology 13: 169–177.

Casaburi, A., Di Martino, V., Ferranti, P., Picariello, L. and Villani, F. (2016). Technological properties and bacteriocins production by *Lactobacillus curvatus* 54M16 and its use as starter culture for fermented sausage manufacture. Food Control 59: 31–45.

Casaburi, A., Aristoy, M.C., Cavella, S., Di Monaco, R., Ercolini, D. and Toldra, F. (2007). Biochemical and sensory characteristics of traditional fermented sausages of Vallo di Diano (Southern Italy) as affected by the use of starter cultures. Meat Science 76: 295–307.

Castellano, P., Aristoy, M.-C., Sentandreu, M.A., Vignolo, G. and Toldra, F. (2013). Peptides with angiotensin I converting enzyme (ACE) inhibitory activity generated from porcine skeletal muscle proteins by the action of meat-borne *Lactobacillus*. Journal of Proteomics 89: 183–190.

Castellano, P., Aristoy, M.C., Sentandreu, M.A., Vignolo, G. and Toldra, F. (2012). *Lactobacillus sakei* CRL1862 improves safety and protein hydrolysis in meat systems. Journal of Applied Microbiology 113: 1407–1416.

Cavalheiro, C.P., Ruiz-Capillas, C., Herrero, A.M., Jimenez-Colmenero, F., de Menezes, C.R. and Fries, L.L.M. (2015). Application of probiotic delivery systems in meat products. Trends in Food Science and Technology 46: 120–131.

Chaillou, S., Champomier-Verges, M.C., Cornet, M., Crutz-Le Coq, A.M., Dudez, A.M., Martin, V., Beaufils, S., Darbon-Rongère, E., Bossy, R., Loux, V. and Zagorec, M. (2005). The complete genome sequence of the meat-borne lactic acid bacterium *Lactobacillus sakei* 23 K. Nature Biotechnology 23: 1527–1533.

Champomier-Verges, M.-C., Zuniga, M., Morel-Deville, F., Perez-Martinez, G., Zagorec, M. and Ehrlich, S.D. (1999). Relationships between arginine degradation, pH and survival in *Lactobacillus sakei*. FEMS Microbiology Letters 180: 297–304.

Chaves-Lopez, C., Paparella, A., Tofalo, R. and Suzzi, G. (2011). Proteolytic activity of *Saccharomyces cerevisiae* strains associated with Italian dry-fermented sausages in a model system. International Journal of Food Microbiology 150: 50–58.

Comi, G., Urso, R., Iacumin, L., Rantsiou, K., Cattaneo, P., Cantoni, C. and Cocolin, L. (2005). Characterisation of naturally fermented sausages produced in the North East of Italy. Meat Science 69: 381–392.

Coppola, R., Giagnacovo, B., Iorizzo, M. and Grazia, L. (1998). Characterization of lactobacilli involved in the ripening of soppressata molisana, a typical southern Italy fermented sausage. Food Microbiology15: 347–353.

Coppola, R., Iorizzo, M., Saotta, R., Sorrentino, E. and Grazia, L. (1997). Characterization of micrococci and staphylococci isolated from soppressata molisana, a Southern Italy fermented sausage. Food Microbiology 14: 47–53.

Corral, S., Salvador, A., Belloch, C. and Flores, M. (2015). Improvement the aroma of reduced fat and salt fermented sausages by *Debaryomyceshansenii* inoculation. Food Control 47: 526–535.

Corral, S., Salvador, A., Belloch, C. and Flores, M. (2014). Effect of fat and salt reduction on the sensory quality of slow fermented sausages inoculated with *Debaryomyces hansenii* yeast. Food Control 45: 1–7.

De Maere, H., Fraeye, I., De Mey, E., Dewulf, L., Michiels, C., Paelinck, H. and Chollet, S. (2016). Formation of naturally occurring pigments during the production of nitrite-free dry fermented sausages. Meat Science 114: 1–7.

De Mey, E., De Klerck, K., De Maere, H., Dewulf, L., Derdelinckx, G., Peeters, M.-C., Fraeye, I., van der Heyden, Y. and Paelinck, H. (2014). The occurrence of N-nitrosamines, residual nitrite and biogenic amines in commercial dry fermented sausages and evaluation of their occasional relation. Meat Science 96: 821–828.

De Prisco, A. and Mauriello, G. (2016). Probiotication of foods: A focus on microencapsulation tool. Trends in Food Science and Technology 48: 27–39.

Del Nobile, M.A., Conte, A., Incoronato, A.L., Panza, O., Sevi, A. and Marino, R. (2009). New strategies for reducing the pork back-fat content in typical Italian salami. Meat Science 81: 263–269.

Dicks, L.M.T., Mellett, F.D. and Hoffman, L.C. (2004). Use of bacteriocin-producing starter cultures of *Lactobacillus plantarum* and *Lactobacillus curvatus* in production of ostrich meat salami. Meat Science 66: 703–708.

Dos Santos, L.F.L., Marsico, E.T., Lazaro, C.A., Teixeira, R., Doro, L., Conte Junior, C.A. (2015). Evaluation of biogenic amines levels, and biochemical and microbiological characterization of Italian-type salami sold in Rio de Janeiro. Brazil Italian Journal of Food Safety 4: 4048.

Drosinos, E.H., Paramithiotis, S., Kolovos, G., Tsikouras, I. and Metaxopoulos, I. (2007). Phenotypic and technological diversity of lactic acid bacteria and staphylococci isolated from traditionally fermented sausages in Greece. Food Microbiology 24: 260–270.

Drosinos, E.H., Mataragas, M. and Metaxopoulos, J. (2006). Modeling of growth and bacteriocin production by *Leuconostoc mesenteroides* E131. Meat Science 74: 690–696.

Drosinos, E.H., Mataragas, M. and Metaxopoulos, J. (2005c). Biopreservation: A new direction towards food safety. pp. 31–64. In: A.P. Riley (ed.). New Developments in Food Policy, Control and Research Nova Science Publishers, NY, USA.

Drosinos, E.H., Mataragas, M., Nasis, P., Galiotou, M. and Metaxopoulos, J. (2005b). Growth and bacteriocin production kinetics of *Leuconostoc mesenteroides* E131. Journal of Applied Microbiology 99: 1314–1323.

Drosinos, E.H., Mataragas, M., Xiraphi, N., Moschonas, G., Gaitis, F. and Metaxopoulos, J. (2005a). Characterization of the microbial flora from a traditional Greek fermented sausage. Meat Science 69: 307–317.

Eerola, H.S., Roig-Sagues, A.X. and Hirvi, T.K. (1998). Biogenic amines in Finnish dry sausages. Journal of Food Safety 18: 127–138.

Eim, V.S., Garcia-Perez, J.V., Rossello, C., Femenia, A. and Simal, S. (2012). Influence of the addition of dietary fiber on the drying curves and microstructure of a dry fermented sausage (sobrassada). Drying Technology 30: 146–153.

Eim, V.S., Simal, S., Rossello, C. and Femenia, A. (2008). Effects of addition of carrot dietary fibre on the ripening process of a dry fermented sausage (sobrassada). Meat Science 80: 173–182.

Ekici, K., Sekeroglu, R., Sancak, Y.C. and Noyan, T. (2004). Note on histamine levels in Turkish style fermented sausages. Meat Science 68: 123–125.

El Soda, M., Korayem, M. and Ezzat, N. (1986). The esterolytic and lipolytic activities of lactobacilli. III: Detection and characterisation of the lipase system. Milchwissenschaft 41: 353–355.

Engelhardt, T., Albano, H., Kisko, G., Mohacsi-Farkas, C. and Teixeira, P. (2015). Antilisterial activity of bacteriocinogenic *Pediococcus acidilactici* HA6111-2 and *Lactobacillus plantarum* ESB 202 grown under pH and osmotic stress conditions. Food Microbiology 48: 109–115.

Erkkila, S. and Petaja, E. (2000). Screening of commercial meat starter cultures at low pH and in the presence of bile salts for potential probiotic use. Meat Science 55: 297–300.

EU (2006). Directive 2006/52/EC food additives other than colours and sweeteners. Official Journal of the European Union 10–22.

Fadda, S., Vignolo, G. and Oliver, G. (2001). Tyramine degradation and tyramine/histamine production by lactic acid bacteria and *Kocuria* strains. Biotechnology Letters 23: 2015–2019.

Fadda, S., Sanz, Y., Vignolo, G., Aristoy, M.-C., Oliver, G. and Toldra, F. (1999). Characterization of muscle sarcoplasmic and myofibrillar protein hydrolysis caused by *Lactobacillus plantarum*. Applied and Environmental Microbiology 65: 3540–3546.

Fernandez-Lopez, J., Sendra, E., Sayas-Barbera, E., Navarro, C. and Perez-Alvarez, J.A. (2008). Physico-chemical and microbiological profiles of salchichon (Spanish dry-fermented sausage) enriched with orange fiber. Meat Science 80: 410–417.

Flores, M., Corral, S., Cano-García, L., Salvador, A. and Belloch, C. (2015). Yeast strains as potential aroma enhancers in dry fermented sausages. International Journal of Food Microbiology 212: 16–24.

Flores, M. and Toldra, F. (2011). Microbial enzymatic activities for improved fermented meats. Trends in Food Science and Technology 22: 81–90.

Fontana, C., Cocconcelli, P.S., Vignolo, G. and Saavedra, L. (2015). Occurrence of antilisterial structural bacteriocins genes in meat borne lactic acid bacteria. Food Control 47: 53–59.

Garcia-Garcia, M.I., Gil-Ortiz, F., Garcia-Carmona, F. and Sanchez-Ferrer, A. (2014). First functional and mutational analysis of group 3 N-acetylneuraminate lyases from *Lactobacillus antri* and *Lactobacillus sakei* 23 K. PLoS ONE 9(5): e96676.

Garcia, M.T., Martinez Canamero, M., Lucas, R., Ben Omar, N., Perez Pulido, R. and Galvez, A. (2004b). Inhibition of *Listeria monocytogenes* by enterocin EJ97 produced by *Enterococcus faecalis* EJ97. International Journal of Food Microbiology 90: 161–170.

Garcia, M.T., Lucas, R., Abriouel, H., Ben Omar, N., Perez, R., Grande, M.J., Martinez-Canamero, M. and Galvez, A. (2004a). Antimicrobial activity of enterocin EJ97 against 'Bacillus macroides/Bacillusmaroccanus' isolated from zucchini puree. Journal of Applied Microbiology 97: 731–737.

Garcia, M.T., Ben Omar, N., Lucas, R., Perez-Pulido, R., Castro, A., Grande, M.J., Martinez-Canamero, M. and Galvez, A. (2003). Antimicrobial activity of enterocin EJ97 on *Bacilluscoagulans* CECT 12. Food Microbiology 20: 533–536.

Garcia, M.L., Dominguez, R., Galvez, M.D., Casas, C. and Selgas, M.D. (2002). Utilization of cereal and fruit fibres in low fat dry fermented sausages. Meat Science 60: 227–236.

Garcia-Varona, M., Santos, E.M., Jaime, I. and Rovira, J. (2000). Characterisation of *Micrococcaceae* isolated from different varieties of chorizo. International Journal of Food Microbiology 54: 189–195.

Gardini, F., Tabanelli, G., Lanciotti, R., Montanari, C., Luppi, M., Coloretti, F., Chiavari, C. and Grazia, L. (2013). Biogenic amine content and aromatic profile of Salama da sugo, a typical cooked fermented sausage produced in Emilia Romagna Region (Italy). Food Control 32: 638–643.

Gardini, F., Martuscelli, M., Crudele, M.A., Paparella, A. and Suzzi, G. (2002). Use of *Staphylococcus xylosus* as a starter culture in dried sausages: effect on the biogenic amine content. Meat Science 61: 275–283.

Gardini, F., Suzzi, G., Lombardi, A., Galgano, F., Crudele, M.A., Andrighetto, C., Schirone, M. and Tofalo, R.A. (2001a). Survey of yeasts in traditional sausages of southern Italy. FEMS Yeast Research 1: 161–167.

Gianotti, V., Panseri, S., Robotti, E., Benzi, M. , Mazzucco, E., Gosetti, F., Frascarolo, P., Oddone, M., Baldizzone, M., Marengo, E. and Chiesa, L.M. (2015). Chemical and microbiological characterization for PDO labelling of typical east piedmont (Italy) Salami. Journal of Chemistry 597471.

Gill, A.O. and Holley, R.A. (2003). Interactive inhibition of meat spoilage and pathogenic bacteria by lysozyme, nisin and EDTA in the presence of nitrite and sodium chloride at 24°C. International Journal of Food Microbiology 80: 251–259.

Gimeno, O., Astiasaran, I. and Bello, J. (2001). Calcium ascorbate as a potential partial substitute for NaCl in dry fermented sausages: effect on color, texture and hygienic quality at different concentrations. Meat Science 57: 23–29.

Gimeno, O., Astiasaran, I. and Bello, J. (1999). Influence of partial replacement of NaCl with KCl and CaCl$_2$ on texture and color of dry fermented sausages. Journal of Agricultural and Food Chemistry 47: 873–877.

Gimeno, O., Astiasaran, I. and Bello, J. (1998). A mixture of potassium, magnesium and calcium chlorides as a partial replacement of sodium chloride in dry fermented sausages. Journal of Agricultural and Food Chemistry 46: 4372–4375.

Giraffa, G. (2012). Selection and design of lactic acid bacteria probiotic cultures Engineering in Life Sciences 12: 391–398.

Gonzalez-Fernandez, C., Santos, E., Jaime, I. and Rovira, J. (2003). Influence of starter cultures and sugar concentrations of biogenic amine contents in chorizo dry sausage. Food Microbiology 20: 275–284.

Gou, P., Guerrero, L., Gelabert, J. and Arnau, J. (1996). Potassium chloride, potassium lactate and glycine as sodium chloride substitutes in fermented sausages and in dry-cured pork loin. Meat Science 42: 37–48.

Grande, Ma.J., Lucas, R., Abriouel, H., Valdivia, E., Ben Omar, N., Maqueda, M., Martinez-Canamero, M. and Galvez, A. (2007). Treatment of vegetable sauces with enterocin AS-48 aloneor in combination with phenolic compounds to inhibit proliferation of *Staphylococcus aureus*. Journal of Food Protection 70: 405–411.

Grande, Ma.J., Lucas, R., Abriouel, H., Valdivia, E., Ben Omar, N., Maqueda, M., Martinez-Bueno, M., Martinez-Canamero, M. and Galvez, A. (2006). Inhibition of toxicogenic *Bacillus cereus* in rice-based foods by enterocinAS-48. International Journal of Food Microbiology 106: 185–194.

Greco, M., Mazzette, R., De Santis, E.P.L., Corona, A. and Cosseddu, A.M. (2005). Evolution and identification of lactic acid bacteria isolated during the ripening of Sardinian sausages. Meat Science 69: 733–739.

Hammes, W.P. (2012). Metabolism of nitrate in fermented meats: The characteristic feature of a specific group of fermented foods. Food Microbiology 29: 151–156.

Harding, C.D. and Saw, B.G. (1990). Antimicrobial activity of *Leuconostocgelidum* against closely related species and *Listeriamonocytogenes*. Journal of Applied Bacteriology 69: 648–654.

He, F.J. and Macgregor, G.A. (2010). Reducing population salt intake worldwide: from evidence to implementation. Progress in Cardiovascular Diseases 52: 363–382.

Hebert, E.M., Saavedra, L., Taranto, M.P., Mozzi, F., Magni, C., Nader, M.E., de Valdez, G.F., Sesma, F., Vignolo, G. and Raya, R.R. (2012). Genome sequence of the bacteriocin-producing *Lactobacillus curvatus* strain CRL705. Journal of Bacteriology 194: 538–539.

Hernandez-Jover, T., Izquierdo-Pulido, M., Veciana-Nogues, M.T., Marine-Font, A. and Vidal Carou, M.C. (1997a). Biogenic amine and polyamine contents in meat and meat products. Journal of Agricultural and Food Chemistry 45: 2098–2102.

Herrmann, S.S., Duedahl-Olesen, L. and Granby, K. (2015). Occurrence of volatile and non-volatile N-nitrosamines in processed meat products and the role of heat treatment. Food Control 48: 163-169.

Holck, A., Axelsson, L. and Schillinger, U. (1996). Divergicin 750, a novel bacteriocin produced by *Carnobacteriumdivergens* 750. FEMS Microbiology Letters 136: 163–168.

Hospital, X., Hierro, E., Stringer, S. and Fernandez, M. (2016). A study on the toxigenesis by *Clostridiumbotulinum* in nitrate and nitrite-reduced dry fermented sausages. International Journal of Food Microbiology 218: 66–70.

Hospital, X.F., Hierro, E. and Fernandez, M. (2012). Survival of *Listeriainnocua* in dry fermented sausages and changes in the typical microbiota and volatile profile as affected by the concentration of nitrate and nitrite. International Journal of Food Microbiology 153: 395–401.

Hospital, X.F., Hierro, E. and Fernandez, M. (2014). Effect of reducing nitrate and nitrite added to dry fermented sausages on the survival of *Salmonella* Typhimurium. Food Research International 62: 410–415.

Hugas, M., Garriga, M., Pascual, M., Aymerich, M.T. and Monfort, J.M. (2002). Enhancement of sakacin K activity against *Listeria monocytogenes* in fermented sausages with pepper or manganese as ingredients. Food Microbiology 19: 519–528.

Ibanez, C., Quintanilla, L., Irigoyen, A., Garcia-Jalon, I., Cid, C., Astiasaran, I. and Bello, J. (1995). Partial replacement of sodium chloride with potassium chloride in dry fermented sausages: influence on carbohydrate fermentation and the nitrosation process. Meat Science 40: 45–53.

Ikonic, P., Tasic, T., Petrovic, L., Skaljac, S., Jokanovic, M., Mandic, A. and Ikonic, B. (2013). Proteolysis and biogenic amines formation during the ripening of Petrovská klobása, traditional dry-fermented sausage from Northern Serbia. Food Control 30: 69–75.

Jydegaard, A.-M., Gravesen, A. and Knochel, S. (2000). Growth condition-related response of *Listeria monocytogenes* 412 to bacteriocin inactivation. Letters in Applied Microbiology 31: 68–72.

Kayaardi, S. and Gok, V. (2003). Effect of replacing beef fat with olive oil on quality characteristics of Turkish soudjouk (sucuk). Meat Science 66: 249–257.

Kenneally, P.M., Leuschner, R.G. and Arendt, E.K. (1998). Evaluation of the lipolytic activity of starter cultures for meat fermentation purposes. Journal of Applied Microbiology 84: 839–846.

Klingberg, T.D. and Budde, B.B. (2006). The survival and persistence in the human gastrointestinal tract of five potential probiotic lactobacilli consumed as freeze-dried cultures or as probiotic sausage. International Journal of Food Microbiology 109: 157–159.

Klingberg, T.D., Axelsson, L., Naterstad, K., Elsser, D. and Budde, B.B. (2005). Identification of potential probiotic starter cultures for Scandinavian-type fermented sausages. International Journal of Food Microbiology 105: 419–431.

Kolozyn-Krajewska, D. and Dolatowski, Z.J. (2012). Probiotic meat products and human nutrition. Process Biochemistry 47: 1761–1772.

Koutsopoulos, D.A., Koutsimanis, G.E. and Bloukas, J.G. (2008). Effect of carrageenan level and packaging during ripening on processing and quality characteristics of low-fat fermented sausages produced with olive oil. Meat Science 79: 188–197.

Krier, F., Revol-Junelles, A.M. and Germain, P. (1998). Influence of temperature and pH production of two bacteriocins by *Leuconostoc mesenteroides* subsp. *mesenteroides* FR52 during batch fermentation. Applied Microbiology and Biotechnology 50: 359–363.

Lahti, E., Johansson, T., Honkanen-Buzalski, T., Hill, P. and Nurmi, E. (2001). Survival and detection of *Escherichia coli* O157:H7 and *Listeria monocytogenes* during the manufacture of dry sausage using two different starter cultures. Food Microbiology 18: 75–85.

Latorre-Moratalla, M.L., Veciana-Nogues, T., Bover-Cid, S., Garriga, M., Aymerich, T., Zanardi, E. and Vidal-Carou, M.C. (2008). Biogenic amines in traditional fermented sausages produced in selected European countries. Food Chemistry 107: 912–921.

Latorre-Moratalla, M.L., Bover-Cid, S., Aymerich, T., Marcos, B., Vidal-Carou, M.C. and Garriga, M. (2006). Aminogenesis control in fermented sausages manufactured with pressurized meat batter and starter culture. Meat Science 75: 460–469.

Lejeune, R., Callewaert, R., Crabbe, K. and De Vuyst, L. (1998). Modelling the growth and bacteriocin production by *Lactobacillus amylovorus* DCE 471 in batch cultivation. Journal of Applied Microbiology 84: 159–168.

Leroy, F., Verluyten, J. and De Vuyst, L. (2006). Functional meat starter cultures for improved sausage fermentation. International Journal of Food Microbiology 106: 270–285.

Leuschner, R.G.K. and Hammes, W.P. (1998). Tyramine degradation by micrococci during ripening of fermented sausage. Meat Science 49: 189–196.

Lizaso, G., Chasco, M. and Beriain, J. (1999). Microbiological and biochemical changes during ripening of salchichon, a Spanish dry cured sausage. Food Microbiology 16: 219–228.

Luecke, F.K. (2000). Fermented meats. pp. 420–444. *In*: B.M. Lund, T.C. Baird-Parker and G.W. Gould (eds.). The Microbiological Safety and Quality of Food, vol II. Aspen Publishers, Gaithersburg, MD.

Martuscelli, M., Crudele, M.A., Gardini, F. and Suzzi, G. (2000). Biogenic amine formation and oxidation by *Staphylococcus xylosus* strains from artisanal fermented sausages. Letters in Applied Microbiology 31: 228–232.

Mataragas, M., Drosinos, E.H., Tsakalidou, E. and Metaxopoulos, J. (2004). Influence of nutrients on growth and bacteriocin production by *Leuconostoc mesenteroides* L124 and *Lactobacillus curvatus* L442. Antonie van Leeuwenhoek 85: 191–198.

Mataragas, M., Metaxopoulos, J., Galiotou, M. and Drosinos, E.H. (2003b). Influence of pH and temperature on growth and bacteriocin production by *Leuconostoc mesenteroides* L124 and *Lactobacillus curvatus* L442. Meat Science 64: 265–271.

Mataragas, M., Drosinos, E.H. and Metaxopoulos, J. (2003a). Antagonistic activity of lactic acid bacteria against *Listeria monocytogenes* in sliced cooked cured pork shoulder stored under vacuum or modified atmosphere at 4°C. Food Microbiology 20: 259–265.

Mauriello, G., Casaburi, A., Blaiotta, G. and Villani, F. (2004). Isolation and technological properties of coagulase negative staphylococci from fermented sausages of Southern Italy. Meat Science 67: 149–158.

Mauriello, G., Casaburi, A. and Villani, F. (2002). Proteolytic activity of *Staphylococcus xylosus* strains on pork myofibrillar and sarcoplasmic proteins and use of selected strains in the production of Naples type salami. Journal of Applied Microbiology 92: 482–490.

Mendoza, E., Garcia, M.L., Casas, C. and Selgas, M.D. (2001). Inulin as fat substitute in low fat, dry fermented sausages. Meat Science 57: 387–393.

Miguelez-Arrizado, M.J., Bover-Cid, S. and Vidal-Carou, M.C. (2006). Biogenic amine contents in Spanish fermented sausages of different acidification degree as a result of artisanal or industrial manufacture. Journal of the Science of Food and Agriculture 86: 549–557.

Miralles, M.C., Flores, J. and Perez-Martinez, G. (1996). Biochemical tests for the selection of *Staphylococcus* strains as potential meat starter cultures. Food Microbiology 13: 27–236.

Molly, K., Demeyer, D.I., Johansson, G., Raemaekers, M., Ghistelinck, M. and Geenen, I. (1997). The importance of meat enzymes in ripening and flavor generation in dry fermented sausages. First results of a European project. Food Chemistry 54: 539–545.

Molly, K., Demeyer, D.I., Civera, T. and Verplaetse, A. (1996). Lipolysis in Belgian sausages: Relative importance of endogenous and bacterial enzymes. Meat Science 43: 235–244.

Montel, M., Masson, F. and Talon, R. (1999). Comparison of biogenic amine content in traditional and industrial French dry sausages. Sciences des Aliments 19: 247–254.

Montel, M.C., Masson, F. and Talon, R. (1998). Bacterial role in flavour development. Meat Science 49: S111–S123.

Mortvedt, C.I., Nissen-Meyer, J., Sletten, K. and Nes, I.F. (1991). Purification and amino acid sequence of lactocin S, a bacteriocin produced by *Lactobacillus sake* L45. Applied and Environmental Microbiology 57: 1829–1834.

Muguerza, E., Ansorena, D. and Astiasaran, I. (2003). Improvement of nutritional properties of Chorizo de Pamplona by replacement of pork backfat with soy oil. Meat Science 65: 1361–1367.

Muguerza, E., Fista, G., Ansorena, D., Astiasaran, I. and Bloukas, J.G. (2002). Effect of fat level and partial replacement of pork backfat with olive oil on processing and quality characteristics of fermented sausages. Meat Science 61: 397–404.

Muguerza, E., Gimeno, O., Ansorena, D., Bloukas, J.G. and Astiasaran, I. (2001). Effect of replacing pork backfat with pre-emulsified olive oil on lipid fraction and sensory quality of Chorizo de Pamplona—a traditional Spanish fermented sausage. Meat Science 59: 251–258.

Nieto-Lozano, J.C., Reguera-Useros, J.I., del C. Pelaez-Martinez, M. and de la Torr, A.H. (2006). Effect of a bacteriocin produced by *Pediococcus acidilactici* against *Listeria monocytogenes* and *Clostridium perfringens* on Spanish raw meat. Meat Science 72: 57–61.

Nilsson, L., Chen, Y., Chikindas, M.L., Huss, H.H., Gram, L. and Montville, T.J. (2000). Carbon dioxide and nisin act synergistically on *Listeria monocytogenes*. Applied and Environmental Microbiology 66: 769–774.

Noonpakdee, W., Santivarngkna, C., Jumriangrit, P., Sonomoto, K. and Panyim, S. (2003). Isolation of nisin-producing *Lactococcus lactis* WNC 20 strain from nham, a traditional Thai fermented sausage. International Journal of Food Microbiology 81: 137–145.

Osmanagaoglu, O. (2007). Detection and characterization of Leucocin OZ, a new anti-listerial bacteriocin produced by *Leuconostoc carnosum* with a broad spectrum of activity. Food Control 18: 118–123.

Papamanoli, E., Tzanetakis, N., Litopoulou-Tzanetaki, E. and Kotzekidou, P. (2003). Characterization of lactic acid bacteria isolated from a Greek dry-fermented sausage in respect of their technological and probiotic properties. Meat Science 65: 859–867.

Papavergou, E.J. (2011). Biogenic amine levels in dry fermented sausages produced and sold in Greece. Procedia Food Science 1: 1126–1131.

Paramithiotis, S., Vlontartzik, E. and Drosinos, E.H. (2014). Enterocin production by *Enterococcus faecium* strains isolated from Greek spontaneously fermented sausages. Italian Journal of Food Science 26: 12–17.

Parente, E., Grieco, S. and Crudele, M.A. (2001b). Phenotypic diversity of lactic acid bacteria isolated from fermented sausages produced in Basilicata (Southern Italy). Journal of Applied Microbiology 90: 943–952.

Parente, E., Martuscelli, M., Gardini, F., Grieco, S., Crudele, M. and Suzzi, G. (2001a). Evolution of microbial populations and biogenic amine production in dry sausages produced in Southern Italy. Journal of Applied Microbiology 90: 882–891.

Parente, E. and Ricciardi, A. (1999). Production, recovery and purification of bacteriocins from lactic acid bacteria. Applied Microbiology and Biotechnology 52: 628–638.

Pelser, W.M., Linssen, J.P.H., Legger, A. and Houben, J.H. (2007). Lipid oxidation in n-3 fatty acid enriched Dutch style fermented sausages. Meat Science 75: 1–11.

Pennacchia, C., Vaughan, E.E. and Villani, F. (2006). Potential probiotic *Lactobacillus* strains from fermented sausages: Further investigations on their probiotic properties. Meat Science 73: 90–101.

Pennacchia, C., Ercolini, D., Blaiotta, G., Pepe, O., Mauriello, G. and Villani, F. (2004). Selection of *Lactobacillus* strains from fermented sausages for their potential use as probiotics. Meat Science 67: 309–317.

Plumbridge, J. and Vimr, E.R. (1999). Convergent pathways for utilization of the amino sugars N-acetylglucosamine, *N*-acetylmannosamine, and Nacetylneuraminic acid by *Escherichia coli*. Journal of Bacteriology 181: 47–54.

Prema, P., Bharathy, S., Palavesam, A., Sivasubramanian, M. and Immanuel, G. (2006). Detection, purification and efficacy of warnerin produced by *Staphylococcus warneri*. World Journal of Microbiology and Biotechnology 22: 865–872.

Rebucci, R., Sangalli, L., Fava, M., Bersani, C., Cantoni, C. and Baldi, A. (2007). Evaluation of functional aspects in *Lactobacillus* strains isolated from dry fermented sausages. Journal of Food Quality 30: 187–201.

Riebroy, S., Benjakul, S., Visessanguan, W., Kijrongrojana, K. and Tanaka, M. (2004). Some characteristics of commercial Som-fug produced in Thailand. Food Chemistry 88: 527–535.

Rimaux, T., Riviere, A., Hebert, E.M., Mozzi, F., Weckx, S., De Vuyst, L. and Leroy, F. (2013). A putative transport protein is involved in citrulline excretion and re-uptake during arginine deiminase pathway activity by *Lactobacillus sakei*. Research in Microbiology 164: 216–225.

Rimaux, T., Riviere, A., Illeghems, K., Weckx, S., De Vuyst, L. and Leroy, F. (2012). Expression of the arginine deiminase pathway genes in *Lactobacillus sakei* is strain dependent and is affected by the environmental pH. Applied and Environmental Microbiology 78: 4874–4883.

Rimaux, T., Vrancken, G., Pothakos, V., Maes, D., De Vuyst, L. and Leroy, F. (2011b). The kinetics of the arginine deiminase pathway in the meat starter culture *Lactobacillus sakei* CTC 494 are pH-dependent. Food Microbiology 28: 597–604.

Rimaux, T., Vrancken, G., Vuylsteke, B., De Vuyst, L. and Leroy, F. (2011a). The pentose moiety of adenosine and inosine is an important energy source for the fermented-meat starter culture *Lactobacillus sakei* CTC 494. Applied and Environmental Microbiology 77: 6539–6550.

Rosenstein, R., Nerz, C., Biswas, L., Resch, A., Raddatz, G., Schuster, S.C. and Gotz, F. (2009). Genome analysis of the meat starter culture bacterium *Staphylococcus carnosus* TM300. Applied and Environmental Microbiology 75: 811–822.

Ruiz-Capillas, C. and Jimenez-Colmenero, F. (2004). Biogenic amine content in Spanish retail market meat products treated with protective atmosphere and high pressure. European Food Research and Technology 218: 237–241.

Salazar, P., Garcia, M.L. and Selgas, M.D. (2009). Short-chain fructooligosaccharides as potential functional ingredient in dry fermented sausages with different fat levels. International Journal of Food Science and Technology 44: 1100–1107.

Sanz, B., Selgas, D., Parejo, I. and Ordonez, J.A. (1998). Characteristics of lactobacilli isolated from dry fermented sausages. International Journal of Food Microbiology 6: 199–205.

Scanlan, R.A. (1983). Formation and occurrence of nitrosamines in food. Cancer Research 43: 2435s–2440s.

Schneider, R., Fernandez, F.J., Aquilar, M.B., Guerrero-Legarreta, I., Alpuche-Solis, A. and Ponce-Alquicira, E. (2006). Partial characterization of a class IIa pediocin produced by *Pediococcus parvulus* 133 strain isolated from meat (Mexical 'chorizo'). Food Control 17: 909–915.

Sebranek, J.G. and Bacus, J.N. (2007). Cured meat products without direct addition of nitrate or nitrite: what are the issues? Meat Science 77: 136–147.

Severini, C., De Pilli, T. and Baiano, A. (2003). Partial substitution of pork backfat with extra-virgin olive oil in 'salami' products: effects on chemical, physical and sensorial quality. Meat Science 64: 323–331.

Shahidi, F. and Pegg, R.B. (1995). Nitrite alternatives for processed meats. pp. 1223–241. *In*: G. Charalambous (ed.). Food Flavors: Generation, Analysis and Process Influence. Elsevier Science B.V., Amsterdam, Netherlands.

Siezen, R.J., Francke, C., Renckens, B., Boekhorst, J., Wels, M., Kleerebezem, M. and van Hijum, S.A. (2012). Complete resequencing and reannotation of the *Lactobacillus plantarum* WCFS1 genome. Journal of Bacteriology 194: 195–196.

Stadnik, J. and Keska, P. (2015). Meat and fermented meat products as a source of bioactive peptides. Acta Scientiarum Polonorum Technologia Alimentaria 14: 181–190.

Tao, F., Zhang, Y., Ma, C. and Xu, P. (2010). Biotechnological production and applications of N-acetyl-D-neuraminic acid: current state and perspectives. Applied Microbiology and Biotechnology 87: 1281–1289.

Todorov, S.D., Vaz-Velho, M., de Melo Franco, B.D.G. and Holzapfel, W.H. (2013). Partial characterization of bacteriocins produced by three strains of *Lactobacillus sakei*, isolated from salpicao, a fermented meat product from North-West of Portugal. Food Control 30: 111–121.

Toldra, F. (1998). Proteolysis and lipolysis in flavour development of dry-cured meat products. Meat Science 49: S101–S110.

Valencia, I., Ansorena, D. and Astiasaran, I. (2007). Development of dry fermented sausages rich in docosahexaenoic acid with oil from the microalgae *Schizochytrium* sp.: Influence on nutritional properties, sensorial quality and oxidation stability. Food Chemistry 104: 1087–1096.

Valencia, I., Ansorena, D. and Astiasaran, I. (2006b). Nutritional and sensory properties of dry fermented sausages enriched with n-3 PUFAs. Meat Science 72: 727–733.

Valencia, I., Ansorena, D. and Astiasaran, I. (2006a). Stability of linseed oil and antioxidants containing dry fermented sausages: A study of the lipid fraction during different storage conditions. Meat Science 73: 269–277.

Verluyten, J., Leroy, F. and de Vuyst, L. (2004b). Effects of different spices used in production of fermented sausages on growth of and curvacin A production by *Lactobacillus curvatus* LTH1174. Applied and Environmental Microbiology 70: 4807–4813.

Verluyten, J., Messens, W. and De Vuyst, L. (2004a). Sodium chloride reduces production of curvacin A, a bacteriocin produced by *Lactobacillus curvatus* strain LTH 1174, originating from fermented sausage. Applied and Environmental Microbiology 70: 2271–2278.

Verluyten, J., Messens, W. and De Vuyst, L. (2003). The curing agent sodium nitrite, used in the production of fermented sausages, is less inhibiting to the bacteriocin-producing meat starter culture *Lactobacillus curvatus* LTH 1174 under anaerobic conditions. Applied and Environmental Microbiology 69: 3833–3839.

Villani, F., Casaburi, A., Pennacchia, C., Filosa, L., Russo, F. and Ercolini, D. (2007). Microbial ecology of the Soppressata of Vallo di Diano, a traditional dry fermented sausage from Southern Italy, and *in vitro* and *in situ* selection of autochthonous starter cultures. Applied and Environmental Microbiology 73: 5453–5463.

Villani, F., Mauriello, G., Pepe, O., Blaiotta, G., Ercolini, D., Casaburi, A., Pennachia, C. and Russo, F. (2005). The SIQUALTECA project: Technological and probiotic characteristics of *Lactobacillus* and coagulase negative *Staphylococcus* strains as starter for fermented sausage manufacture. Italian Journal of Animal Science 4: 498.

Villani, F., Sannino, L., Moschetti, G., Mauriello, G., Pepe, O., Amodio-Cocchieri, R. and Coppola, S. (1997). Partial characterization of an antagonistic substance produced by *Staphylococcus xylosus* 1E and determination of the effectiveness of the producer strain to inhibit *Listeria monocytogenes* in Italian sausages. Food Microbiology 14: 555–566.

Wirth, F. (1988). Technologies for making fat-reduce meat products. What possibilities are there? Fleischwirtschaft 68(9): 1153–1156.

Xiraphi, N., Georgalaki, M., Van Driessche, G., Devreese, B., Van Beeumen, J., Tsakalidou, E., Metaxopoulos, J. and Drosinos, E.H. (2006). Purification and characterization of curvaticin L442, a bacteriocin produced by *Lactobacillus curvatus* L442. Antonie van Leeuwenhoek 89: 19–26.

Yalinkilic, B., Kaban, G. and Kaya, M. (2012). The effects of different levels of orange fiber and fat on microbiological, physical, chemical and sensorial properties of sucuk. Food Microbiology 29: 255–259.

Yildiz-Turp, G. and Serdaroglu, M. (2008). Effect of replacing beef fat with hazelnut oil on quality characteristics of sucuk—A Turkish fermented sausage. Meat Science 78: 447–454.

Yurchenko, S. and Molder, U. (2007). The occurrence of volatile N-nitrosamines in Estonian meat products. Food Chemistry 100: 1713–1721.

Zuniga, M., Perez, G. and Gonzalez-Candelas, F. (2002). Evolution of arginine deiminase (ADI) pathway genes. Molecular Phylogenetics and Evolution 25: 429–444.

22

Solid-State Fermentation Applications in the Food Industry

Maria Papagianni

1. Introduction

The main characteristics of solid-state fermentation (SSF) are the involvement of a solid matrix which could be either the nutrient substrate itself or a support impregnated with the nutrient medium, and the very low moisture content which makes this type of process suitable for a limited number of microorganisms, mainly fungi, yeasts and a small number of bacteria. SSF resembles the natural habitat of microorganisms and reproduces natural microbiological processes like composting and ensiling. In industrial applications, SSF can be employed in a controlled way to produce desired metabolites like bulk chemicals and enzymes. Established SSF processes span from the production of industrial enzymes (Alkorta et al. 1998, Papagianni et al. 1999, 2001, Ellaiah et al. 2002, Rosés and Guerra 2009) to value-added fine products, such as ethanol, single-cell protein, organic acids, amino acids, secondary metabolites and others (Barrios-Gonzalez et al. 1993, Gutierrez-Rozas et al. 1995, Mass et al. 2006, Couto and Sanromán 2006).

Department of Hygiene and Technology of Food of Animal Origin, School of Veterinary Medicine, Aristotle University of Thessaloniki, Thessaloniki 54124, Greece.
E-mail: mp2000@vet.auth.gr

SSF is known since ancient times and it has been used extensively in the production of fermented foods. The koji process (fermentation of rice by *Aspergillus oryzae*) is a typical example, while the production of soy sauce, vinegar, various cheeses and Chinese wine, are only few examples of foods that continue to be produced since ancient times through long-established SSF processes. Today, a large number of microbial fermentation products are increasingly utilized as food additives or supplements. The most common among these include substances that serve as preservatives, antioxidants, colorants, flavourings, and sweeteners. Production in SSF systems may be inferior as a process compared to submerged fermentation (SmF) but higher yields, lower costs or even better product characteristics have been reported (Singhania et al. 2009).

The type of the solid substrate is the parameter with the greatest influence on process productivities in SSF systems. The solid substrate provides nutrients and serves as immobilization matrix for cells. Several factors influence microbial growth in a particular substrate but the most critical among them are the provided area and particle size and the water activity. Smaller substrate particles provide a larger area for microbial colonization but if they are too small they tend to agglomerate and growth is poor. Larger particles provide limited surface area but the provided aeration is better. Substrate availability and costs underline SSF' process economics. Research in this area is mainly focused on the selection of suitable substrates among various agro-industrial residues (Singhania et al. 2009). The utilization of agro-industrial wastes in SSF processes provides a useful and environment-friendly alternative way for their disposal which, along with their availability and low cost, makes them particularly desirable as substrates for SSF applications. Filamentous fungi can utilize even the hardest of solid substrates offering the advantage of efficient utilization of low-cost substrates and value addition for wastes (Akpan et al. 1999, Moddles et al. 2001, Rosés and Guerra 2009).

Comparisons between SSF and SmF systems can be found in the literature but they are general and effective mostly on a laboratory scale. SSF systems require simpler technology and have less energy requirements compared to the highly sophisticated SmF systems. Also, fermentation productivities and product titers can be higher, catabolic repression phenomena may be limited, product stability can be increased and the demand for sterility can be also lower due to the low water activity environment of SSF. On the other hand, apparent difficulties in mixing, scale-up, and control of process parameters like pH, temperature and moisture, represent serious limitations on SSF applications. Separation of biomass and product recovery is also a big challenge in SSF. Product recovery and purification processes are difficult and expensive when natural supports are used as substrates. Their utilization however, supposes a significant cost reduction and usually much higher final product titers are obtained. Therefore, an economical evaluation of an overall process should be done in order to decide upon its feasibility for a specific application. Extensive research in the area of biochemical engineering aims to reduce the impact of these problems (Mitchell et al. 2000). Process modelling is important in throwing light on major mass and heat transfer problems. Novel bioreactor designs have been proposed which, along with process optimization studies, contribute to the establishment of more effective design and control criteria of SSF processes (Singhania et al. 2009, Mitchell et al. 2003, Durand 2003).

The aim of the present work is to review the application of SSF processes—established or potential, for the production of important metabolites in the food industry. Organic acids (citric and lactic acid), enzymes (α-amylase, invertase, lipases, pectinases, and proteases), flavour compounds and gums' production through SSF will be discussed. Production of fungal spores in SSF systems will be discussed as well.

2. Production of Organic Acids

Fermentation processes are the most commonly used method in the production of organic acid and is globally dominated by submerged fermentation. SSF however, offers a simpler alternative with numerous advantages, the most important among others is the use of agro-industrial residues as substrates.

2.1 Citric Acid (CA)

CA is the most important organic acid produced at industrial level and the world's second largest fermentation product produced after industrial ethanol. It has a broad range of applications in the chemical industry, agriculture, food industry, pharmaceutical industry, biotechnology, nanotechnology and others. CA is also the most commonly used organic acid in the food industry. It is used as acidifier, flavour enhancer, antioxidant and antifoam agent.

The world's existing demand for CA is almost entirely covered (over 99%) by fermentation processes. Although CA can be produced by chemical synthesis, the method is expensive and fermentation has long gained preference as the method of industrial production. Currently, CA is mainly produced by the filamentous fungus *Aspergillus niger* in SmF (Papagianni 2003, Papagianni 2007). Beet or cane molasses, sucrose or glucose syrups serve as carbon sources in media formulations for the CA fermentation. Although the particular process is highly efficient, the CA industry is currently facing economic challenges due to increasing substrate and energy costs. At the same time, the demand for CA is increasing due to an increasing number of new applications. The need for the development of economical alternatives has led to growing attention on the alternative commercial method of production, the SSF, and the use of low-cost agro-industrial wastes as substrates. *A. niger* is again the most employed microorganism in solid-state CA fermentation because of its ability to grow efficiently on a variety of solid substrates while producing significant amounts of the acid. Among the solid substrates used are wheat and rice bran, fruit wastes, sugarcane bagasse, coffee husk, corncob, brewery wastes, and others (Mussatto et al. 2012).

As with the submerged process, production of CA in SSF depends on the employed strain and process parameters. Aeration is critical and the increased oxygen demand of the process is met through forced aeration and agitation (Pintado et al. 1998, Prado et al. 2004). Agitation, apart from the distribution of oxygen in the medium, facilitates the removal of carbon dioxide and excess heat, and protects the medium against local desiccation and excessive moistening. Agitation is generally beneficial; however, it is questionable whether it should be applied continuously or intermittently. Dhillon and co-workers (Dhillon et al. 2011) showed that intermittent agitation during SSF with *A. niger* resulted in higher yields of CA production compared to continuous agitation. It is well known that the applied mechanical forces influence fungal morphology and CA production, while in excess they cause fragmentation of hyphae and reduce growth and product formation. Maintaining the optimum moisture content during the SSF, CA process is also critical for growth and product formation. Increased moisture reduces the porosity of the substrate and enhances aerial growth. Care therefore should be taken to avoid the formation of clumps of solid particles and this can be managed by using the appropriate medium formulations. Dhillon et al. (2011) for example supplemented the apple pomace substrate of SSF with rice husk to increase the porosity and prevent the formation of clumps.

Various solid substrates have been tested—among these are a large variety of agro-industrial wastes, for their potential use as substrates for the solid-state CA fermentation (Vandenberghe 2000). By using inexpensive solid substrates, such as orange and pineapple wastes, apple and grape pomace, kiwi fruit peel, rice and wheat bran, carrot waste, carob pod, cassava bagasse, coffee husk, and mussel processing wastes (Hang and Woodams 1984, 1985, 1987, Aravantinos-Zafiris et al. 1994, Pandey et al. 2000a, Vandenberghe 2000, Prado et al. 2005), the costs of the process can be reduced. Raw materials are often pre-treated and supplemented with nitrogen and phosphorus sources to assist their utilization by the fungus (Prado et al. 2005, Mussatto et al. 2012). Addition of trace metals, e.g., copper, magnesium and zinc, has been shown to enhance CA production. The same was reported for the addition of lower alcohols, e.g., methanol,

ethanol and isopropanol (Krishna 2005, Dhillon et al. 2013). Tran et al. (1998) observed a drastic increase in CA production (from 261.2 g/kg to 380.9 g/kg) with 4% methanol addition in SSF of citrus pulp. Dhillon et al. (2011) reported a significant increase in yield following supplementation of dry apple pomace with 3% (v/w) ethanol or 4% (v/w) methanol.

In most of this type of works, optimization studies were performed using central composite designs or other statistical methods. The reported yields in many cases exceeded 70%, based on the consumed sugars. Pandey et al. (2000a), using cassava bagasse and *A. niger* NRRL 2001, obtained 27 g CA per 100 g dry substrate under optimum fermentation conditions, corresponding to a 70% yield. Prado et al. (2005) using another strain of *A. niger* and thermally pre-treated cassava bagasse obtained a similar yield. Immediate comparisons of reported results however are difficult as various types of reactors and different fermentation techniques have been used for CA production in SSF. Flasks, plastic trays, columns, packed bed and rotating drum-type solid-state bioreactors have all been used in processes with different solid substrates and fermentation conditions. Dhillon and co-workers (2013) compared CA production using various solid substrates and different fermentation techniques. As reported, increased yields in rotating drum bioreactors can be achieved only when the drum is filled up to 30% of its capacity to ensure efficient agitation (Couto and Sanromán 2006, Dhillon et al. 2013). Using a rotating drum-type bioreactor and apple pomace supplemented with rice husk as substrate, Dhillon et al. (2013) reported the yield of 29.4 g per 100 g of dry substrate. The applied optimum conditions were 75% (v/w) moisture, 3% (w/v) methanol, intermittent agitation of 1 hr every 12 hr at 2 rpm and 1 vvm air flow rate. Tran et al. (1998) also compared CA production in flasks, trays and a rotating drum reactor, using pineapple peels and various *A. niger* strains. Higher production was achieved in flasks after 3 days of fermentation: 140 g per kg of dry substrate with 74% yield per kg of sugar consumed. The optimum conditions for that were 65% (w/w) initial moisture content, 3% (v/w) methanol, 30°C, initial pH 3.4 (unadjusted) and 2 mm particle size. Vandenberghe et al. (2004) compared various types of reactors, e.g., flasks, trays, columns and a horizontal drum reactor, and reported the highest production of 30.9 g CA per 100 g dry substrate in a column fermenter using cassava bagasse as solid substrate.

2.2 Lactic Acid

Lactic acid (LA) is an industrially important product with wide use in the food, pharmaceuticals, textiles, and polymers industries. In the food industry, LA is used mainly as acidulant and preservative of many foods, e.g., cheese, meat, beer, and jellies. It is also used as flavouring and pH buffering agent in a wide variety of processed foods. Currently, LA is in great demand due to its use as starting material in the production of biodegradable polymers used in industrial, medical and consumer products.

LA can be produced either by chemical synthesis or by microbial fermentation. It has two enantiomer forms, L(+) and D(−), of which only the L(+) can be metabolized by the human organism because of the presence of the enzyme of L-lactate dehydrogenase and therefore, it is preferred where use for food is concerned. Today, almost all LA produced worldwide is the product of microbial fermentation processes (Wee et al. 2006). Fermentation offers the advantage of an optically highly pure product by employing appropriate strains, while chemical synthesis always results in racemic mixtures of the two isomers (Hofvendahl and Hahn-Hägerdal 2000). In addition, fermentation offers the advantage of the utilization of inexpensive substrates and lower temperatures, resulting in more economical production processes. Batch, repeated batch, fed-batch, and continuous submerged fermentations are the most frequently employed methods in LA production. Although the economics of a fermentation process for LA production depends on many factors, the cost of raw materials remain the major one and currently, research is focused on the potential of utilization of various renewable resources and especially lignocellulosic materials, as substrates (Hofvendahl and Hahn-Hägerdal 2000, Ray et al. 2009, Wee et al. 2006, John et al. 2009, Abdel-Rachman et al. 2011). Fermentative LA production from such resources comprises a number of steps including: hydrolysis of substrates to sugars, fermentation of sugars to LA, extraction of LA from the broth and purification. The subject of fermentative LA production from renewable resources has been treated extensively in the reviews of Hofvendahl and Hahn-Hägerdal (2000), John et al. (2009) and Abdel-Rahman et al. (2011).

Production of LA through SSF was studied in several cases using lactic acid bacteria (LAB) or fungi and various substrates. LAB are used traditionally in SSF-based food fermentations and ensiling of various agricultural residues which are able to grow in SSF systems. Their potential as effective producers of LA in SSF systems was first investigated in 1994 by Xavier and Lonsane who used *Lactobacillus casei* and sugar cane pressmud as solid substrate. Production reached a maximum of 5.27 g per 100 g dry sugar cane pressmud, which was a high yield. In the same year, Richter and Träger (1994) reported the L(+)-lactic acid production in SSF by *Lb. paracasei* using sweet sorghum as the solid substrate. *Lb. casei* was also used by Rojan et al. (2005) for the production of L(+)-lactic acid in SSF with sugarcane bagasse as the inner support soaked with cassava starch hydrolysate. Production of LA approximated 49 g per 100 g starch, corresponding to a 97% conversion of sugar to LA. Naveena et al. (2005a,b) used *Lb. amylophilus* and wheat bran as both support and substrate. Most recent works report the production of LA in SSF with *Lb. plantarum* grown on tea wastes (Gowdhaman et al. 2012), and the use of various *Lactobacillus strains* (e.g., *Lb. delbrueckii*, *Lb. pentosus*) grown on pine needles (Ghosh and Gosh 2012a) or wheat bran beds, in pure cultures or co-cultures (Ghosh and Gosh 2012b). The described systems (Gowdhaman et al. 2012, Ghosh and Gosh 2012a,b) confirmed the ability of LAB to grow efficiently on the solid substrates employed and the reported conversions of sugar to LA approximated 97%.

Fungal *Rhizopus* species have long been recognized as suitable producers of LA that, unlike the LAB, generate only the L(+) form of the acid (Zhang et al. 2007). LA production from *Rhizopus* spp. through simultaneous saccharification and fermentation of renewable materials has been studied extensively and reviewed by Zhang et al. (2007). Reports however, on LA production from *Rhizopus* in SSF systems are rather rare. Soccol et al. (1994) studied the production of LA by *R. oryzae* in SSF using sugarcane bagasse as a support and reported a slightly higher productivity than in submerged cultivation. In a recent study, *R. oryzae* was cultivated on cassava pulp and was found to be able to produce 206.20 mg LA per g initial dry pulp (Parichat et al. 2012).

3. Production of Enzymes

Many different classes of industrial enzymes can be produced effectively by SSF processes Panda et al. (2016).

3.1 Proteases

Proteases (proteolytic enzymes) are enzymes that catalyze the hydrolysis of peptide bonds in proteins and polypeptides. They occur widely in all living organisms and thousands of them have been isolated and characterized. Proteases, mostly of microbial origin are used extensively by the industry. In the food technology area proteases are used in the dairy and baking industries, the fish and seafood processing industry, in animal and plant protein processing, as well as in yeast hydrolysis. Food-grade proteases are produced efficiently by several fungal species of *Aspergillus*, *Penicillium*, *Rhizopus* and *Mucor*, *Bacillus*, *Streptomyces*, *Streptococcus* and *Pseudomonas* genera also include important bacterial protease producer species (Sumantha et al. 2006). Microbial proteases are produced both in SmF and SSF systems.

Neutral proteases of fungal origin are the most important component of commercial protease preparations with applications in baking, food processing and protein modification. *A. oryzae* is the main producer of this type of proteases (Nakadai et al. 1972, 1973). A comparative evaluation of neutral protease production by *A. oryzae* in SmF and SSF was reported by Sandhia et al. (2005). The superiority of SSF over the alternative SmF was demonstrated for this particular fermentation as enzyme production was 3.5 fold higher in the SSF system. Similar observations were also made for protease production by *Aspergillus* and *Mucor* spp. in SSF. Higher yields were obtained in SSF using wheat bran as solid substrate than in SmF (Pandey et al. 2000a,b,c).

Literature information on protease production by bacteria in SSF is very limited. Soares et al. (2005) reported increased protease production by the bacterium *B. subtilis* in SSF with soy cake as the solid substrate. Compared to SmF, SSF enzyme productivity was 45% higher.

3.2 α-Amylases

Today, a large number of fungal and bacterial amylases are available commercially, with α-amylases ranking first in terms of commercial applications (Regulapati et al. 2007). α-Amylases (EC 3.2.1.1) are extracellular enzymes that cleave the 1,4-α linkages between adjacent glucose units in the linear amylase chain and generate glucose, maltose and maltotriose units. Since the 1950s, amylases of fungal origin have been used in the production of sugar syrups containing specific mixtures of sugars that could not be produced by the conventional acid hydrolysis of starch. Apart from their major application in the manufacture of glucose and sucrose syrups, α-amylases are used extensively by the food-processing industry, as well as the beverage, textiles and detergent industries. Industrial production of α-amylases is carried out mainly by bacteria of the genus *Bacillus*. *B. subtilis*, *B. stearothermophilus*, *B. licheniformis* and *B. amyloliquefaciens* and are all efficient producers of α-amylases in SmF. Several fungal species of the genus *Aspergillus* are also important producers of α-amylases in SmF (Vihinen and Mantasala 1989, Pandey et al. 2000d). Industrial production of α-amylases has generally been carried out using SmF; however, the potential of SSF systems, as a means of production of either bacterial or fungal α-amylases has been shown to be promising in many cases.

Among species of the genus *Bacillus* used, *B. subtilis*, *B. licheniformis*, *B. cereus*, *B. mesentericus*, *B. vulgatus*, *B. polymyxa*, *B. mycoides*, *B. coagulans*, *B. atterimus*, *B. megaterium*, have all been shown to be effective producers of thermostable α-amylases in SSF (Babu and Satyanarayana 1995, Sodhi et al. 2005, Anto et al. 2006, Couto and Sanromán 2006, Shukla and Kar 2006). Various substrates were employed, e.g., wheat, corn and rice bran, fruit stalks, potato peel, cassava bagasse and others. The reported yields varied and a direct comparison is difficult as they were expressed in different units. Babu and Satyanarayana (1995) used *B. coagulans* and wheat bran and reported a maximum α-amylase production of 26,350 units per g dry substrate in an aerated reactor within 48 hours of cultivation. Sodhi et al. (2005) studied the production of α-amylases by several *B. subtilis* species in SSF and the influence of a number of process parameters such as the nature of the solid substrate (wheat, rice, corn bran and combinations of two of them) and the nature of the moistening agent, the moisture content and the incubation temperature, the presence or absence of surfactants and the supplementation of carbon, nitrogen, minerals, amino acids and vitamins. Maximum enzyme production was obtained with wheat bran supplemented with glycerol (1%, w/w), soybean meal (1%, w/w), L-proline (0.1%, w/w) and vitamin B complex (0.01%, w/w), while the solid substrate was moistened with tap water containing Tween-40 (1%, v/v). Krishna and Chandrasekaran (1996) studied the production of α-amylase by *B. subtilis* in SSF with banana fruit stalks as substrate. The influence of a large number of process parameters was evaluated in that work in order to optimise the process. The effects of thermal treatment time and temperature, incubation temperature, initial moisture content, pH, particle size, inoculum size and the duration of fermentation, were all shown to influence the gross yield of the produced enzymes. Anto and co-workers (2006) reported results on α-amylase production in SSF with *B. cereus* and wheat bran and rice flake manufacturing waste as substrates. Process parameters were optimised as inoculum size 10% (v/w) and substrate:moisture ratio 1:1. Among various carbon sources supplemented, glucose at 0.04 g/g showed enhanced enzyme production (~122 units/g) while supplementation of different nitrogen sources (0.02 g/g) led to decreased enzyme production. Optimum enzyme activity was observed at 55°C and pH 5.0.

Fungal α-amylases can be produced effectively in SSF systems by various fungi. Sugathi et al. (2011) cultivated *A. niger* BAN3E, a strain isolated from bread, on a variety of solid substrates like rice, wheat and black gram bran, coconut oil cake, gingerly oil cake and groundnut oil cake, and reported specific activities of 311 units per mg dry substrate after 6 days of incubation at 37°C with groundnut oil cake. Supplementation of the solid substrate with ammonium nitrate led to significant increases in yield. In general, nitrogen supplementation enhances α-amylase production by *Aspergillus* cultivated on solid substrates, as was shown in the works of Pandey et al. (1994), Anupama and Ravindra (2001) and Rosés and Guerra (2009) with *A. niger*, Got et al. (1998) with *A. fumigatus*, and Sivaramakrishnan et al. (2007) and Ramachandran et al. (2004) with *A. oryzae*.

The filamentous fungus *Penicillium chrysogenum* was employed by Balkan and Ertan (2007) for the production of α-amylase in SSF carried out using agricultural by-products like corncob leaf, rye and wheat straw and wheat bran. The effects of moisture level, particle size and inoculum concentration were investigated while the highest enzyme production (160 units/ml) was obtained with wheat bran under the conditions identified as optimum at 65% moisture, 1 mm particle size and 20% (v/w), inoculum concentration.

The thermophilic fungus *Thermomyces lanuginosus* is another fungus tested for α-amylase production in SSF by Kunamneni et al. (2005). A large variety of solid substrates were examined, e.g., wheat, rice, molasses, barley bran, maize meal, crushed maize, millet, wheat flakes, crushed wheat and corncobs. The highest α-amylase activity of 534 U/g was obtained using wheat bran supplemented with soluble starch (1%, w/w) and peptone (1% w/w) under the optimum conditions of 90% initial moisture content, 50°C, pH 6.0, inoculum level of 10% (v/w), a ratio of substrate weight to flask volume of 1:100 and incubation for 120 hours.

3.3 Lipases

Lipases (triaglycerol acylhydrolases, EC 3.1.1.3) catalyse both the hydrolysis and the synthesis of esters formed from glycerol and long-chain fatty acids. Because of their numerous applications either as hydrolases or as synthetases, they are used widely by the food, pharmaceutical, cosmetics, and chemical industries (Jaeger and Reetz 1998). Microbial lipases are produced by bacteria, yeasts and fungi and most studies on their production are carried out in SmF systems. In recent years however, high yields of microbial lipases have been reported using SSF processes and commercial substrates (Dominguez et al. 2003, Adinarayana et al. 2004, Vaseghi et al. 2013). Rivera-Munoz et al. (1991), Ohnisi et al. (1994), Christen et al. (1995) and Benjamin and Pandey (1997, 1998) compared SmF and SSF systems for lipases production to find that yields, as well as enzyme stability, were higher with SSF.

Extracellular lipase production is affected by several factors, e.g., medium composition, aeration, pH, and temperature. The presence of triglycerides or fatty acids in the substrate has been reported to induce lipolytic enzyme secretion in a number of microorganisms (Marek and Bednarski 1996, Couto and Sanromán 2006). Therefore, the type of substrate in SSF can be utilized to enhance production of lipases and substrates such as food and agro-industrial wastes rich in fatty acids, triglycerides and sugars are particularly suitable (Rathi et al. 2002, Dominguez et al. 2003). López and co-workers (2010) investigated the feasibility of several food-processing wastes as support substrates for lipolytic enzymes production by the fungus *R. oryzae* under solid-state conditions. Optimization experiments were carried out to select the variables that allow for high levels of lipolytic enzyme activity, e.g., the use of inert and non-inert solid materials and lipidic and surfactant compounds. It was observed that addition of Triton X-100 together with barley bran resulted in lipolytic production values tenfold higher than the cultures exclusively grown on an inert support. In addition, from preliminary thermo-inactivation kinetics studies, it was concluded that the strategy proposed in that investigation provided another benefit in terms of resistance of the produced enzymes against thermo-inactivation. Several cheap agro-industrial wastes are used as solid substrates and among these, coconut, groundnut and olive oil cakes, fruit wastes, sugarcane bagasse and various types of bran supplemented with oils, most commonly, olive oil.

Filamentous fungi and yeasts are important producers of lipases in SSF systems (Treichel et al. 2010). Strains belonging to the genera of *Aspergillus*, *Rhizopus*, *Penicillium*, *Geotrichum* and *Rhizomucor* have been reported to produce significant amounts of lipases in SSF systems (Kamini et al. 1998, Ul-Haq et al. 2002, Colen et al. 2006). Among yeasts, *Candida rugosa* has been shown in several cases to produce high amounts of lipase in SSF systems (Benjamin and Pandey 1997, 1998, Santis-Navarro et al. 2011), with the lipolytic activity reaching up to 120,000 AU per g of dry substrate (Santis-Navarro et al. 2011). *C. utilis* has also been shown to produce high amounts of lipases in SSF systems using solid wastes from the oil processing industry (Moftah et al. 2012). *Yarrowia lipolytica* is another lipase producer yeast that has been tested by Farias et al. (2014) in SSF using different agro-industrial residues such as cottonseed cake, soybean bran and its sludge. Production of lipase reached 139 ± 3 U per g of dry substrate in 14 hours of fermentation with cottonseed cake, corresponding to 9.9 U/g/h of productivity. The cottonseed cake was also used as substrate in SSF and gave a very good level of enzyme activity, 102 ± 6 units per g of dry substrate in 28 h of fermentation.

Bacteria of the genus *Bacillus*, such as *B. coagulans*, *B. subtilis*, and *B. stearothermophilus* (Alkan et al. 2007, Singh et al. 2010, Sabat et al. 2012) are lipase producers and amounts as high as 148,932 U per g of dry substrate have been reported by using SSF systems (Alkan et al. 2007). Sahoo et al. (2014) investigated lipase production by *Pseudomonas* sp. S1 in nonsterile SSF with olive oil cake as solid substrate. Production reached 58 IU per g dry substrate while in SmF *Pseudomonas* sp. S1 produced the maximum of 49 IU/ml of lipase using olive oil as substrate. The produced bacterial lipase is a thermophilic organic solvent tolerant enzyme suitable for the production of biodiesel as well as treatment of oily waste water.

3.4 Pectinases

Pectinases are a heterogenous group of enzymes involved in the breakdown of pectin from a variety of plants. They are classified according to their substrate specificity (pectin, pectic acid and oligo-D-galacturonate), the degradation mechanism (hydrolysis and transelimination) and the type of cleavage (random and terminal) (Kashyap et al. 2001). Pectinases are used widely by the food industry to clarify fruit juices, wine and alcoholic beverages, to improve oil extraction, to increase the firmness of fruits and in many other applications (Kashyap et al. 2001). Commercial grade preparations of pectinases are produced from fungi and mainly *Aspergillus* spp. (Jayani et al. 2005). Yeasts (*Saccharomyces, Kluyveromyces*) and bacteria (*Bacillus, Streptomyces*) also produce pectinases. SmF and SSF processes have been used widely for production of pectinases by different types of microorganisms and several studies have shown that higher productivities are obtained in SSF systems (Nakadai and Nasuno 1988, Trejo-Hernandez et al. 1991, Solis-Pereira et al. 1993, Acuña-Argüelles et al. 1995, Favela-Torres et al. 2006). Another interesting finding is that pectinases produced by SSF have more stable properties such as a higher stability to pH and temperature and are less affected by catabolic repression than pectinases produced by SmF (Acuña-Argüelles et al. 1995).

Solis-Pereira et al. (1993) studied polygalactorunase (PGase) production by *A. niger* CH4 in SmF and SSF and reported an 11-fold higher production in the latter. Similarly, Kapoor and Kuhad (2002) in a comparative study of production of alkaline PGases from a *Bacillus* sp., demonstrated a more than 60-times higher enzyme titer per gram of dry matter in SSF than per ml of culture medium in SmF. A 14.2-fold higher enzyme production was attained in SSF with *B. pumilus* dcsr1 as compared to that in submerged fermentation in the work of Sharma and Satyanarayana (2012). In that work, maximum enzyme titer (348.0 ± 11.8 U/g) in SSF was attained, when a mixture of agro-industrial residues (sesame oilseed cake, wheat bran, and citrus pectin, 1:1:0.01) was moistened with mineral salt solution (a_w 0.92, pH 9.0) at a substrate-to-moistening agent ratio of 1:2.5 and inoculated with 25% of 24 hr-old inoculum, in 144 hr at 40°C. Parametric optimization in SSF resulted in 1.7-fold enhancement in the enzyme production as compared to that recorded under unoptimized conditions.

The production of exo-polygalacturonase (exo-PG) and endo-PG by *A. awamori* grown on wheat in SSF was studied by Blandino and co-workers (2002). Endo- and exo-PG activities were detected after 24 hr of inoculation. Glucose released from starch hydrolysis acted as a catabolite repressor for the exo-PG enzyme. In contrast, endo-PG production was not affected by glucose repression. When milled grains were used, the particle-size distribution and the chemical composition of the medium influenced the rate of microorganism growth and therefore the trend followed by endo- and exo-PG production. Moisture contents of 60% were found to provide a higher yield of pectinases by *A. awamori*.

Several agricultural and agro-industrial residues have been used as solid substrates for pectinase production, e.g., wheat, rice and soy bran, sugarcane and orange bagasse, citrus fruit peels and pomace, cocoa and coffee pulp and husk, and others (Couto and Sanromán 2006, Favela-Torres et al. 2006, Liu et al. 2012, Tariq and Reyaz 2012). The highest reported pectinase activities were obtained by *A. carbonarius* cultivated on wheat bran, approximating 500 IU per g of dry substrate (Singh et al. 1999, Kavitha and Umesh-Kumar 2000).

Attempts for genetic improvement of producer microorganisms for increased pectinase production using mutagenesis techniques or protoplast fusion have been successful (Solis et al. 1997, Hadj-Taieb et al. 2002). Kavitha and Umesh-Kumar (2000) obtained, through interspecific protoplast fusion, hybrids of *A. niger* and *A. carbonarius* that produced 6-fold higher pectinase activities and had improved growth compared to the parent strain of *A. carbonarius* on SSF, using wheat bran as the solid substrate and the sole source of nutrients.

4. Production of Flavour and Aroma Compounds

Flavouring compounds represent over a quarter of the global market for food additives and most of them are products of chemical synthesis or plant extraction. Fermentation can provide an alternative source of flavours since several microorganisms are able to synthesize aroma compounds. Microbial cultures are used to produce flavour compounds either *in situ* as part of food fermentations or in fermentation for the production of food additives. Longo and Sanromán (2006) reviewed the microbial-based methodologies for the production of food aroma compounds and concluded that the production using microbial cultures offers several advantages over traditional methodologies, while the use of SSF can give higher yields or better product characteristics than SmF with lower costs.

SSF applications in food aroma production have been described for fungi, yeasts and bacteria. Several methylketones, e.g., 2-undecanone, 2-nonanone and 2-heptanone, are produced commercially by SSF from *A. niger* using coconut fat as substrate with a yield of 40% (Krings and Berger 1998, Vandamme and Soetaert 2002). *A. oryzae* (Ito et al. 1990) and *R. ozyzae* (Bramorski et al. 1998a, Christen et al. 2000) produce volatile compounds such as acetaldehyde, methylbutanol and others, when cultivated in rice koji and agro-industrial substrates, respectively. *Ceratocystis fimbriata* strains produce fruity aromas. Bramorski et al. (1998b) reported the production of strong fruity aroma by *C. fimbriata* in solid-state cultures using cassava bagasse, apple pomace and soybean. Soares et al. (2000) cultivated *C. fimbriata* on coffee husk and reported the production of strong pineapple aroma. Major flavour compounds identified were acetaldehyde, ethanol, ethyl acetate and ethyl isobutyrate. A strong banana odor can be produced by adding leucine in the solid substrate that further enhanced the production of ethyl acetate and isoamyl acetate. Medeiros et al. (2003) evaluated the ability of two different strains of *C. fimbriata* for fruity aroma production by SSF carried out using coffee pulp and coffee husk complemented with glucose as substrates. *Neurospora* spp. was also used to produce fruity aromas in SSF of pre-gelatinized rice (Pastore et al. 1994). The basidiomycete *Pycnoporus cinnabarinus* has been used for the direct conversion of ferulic acid, enzymatically released from agricultural wastes such as beet pulp and cereal bran (maize, wheat), to vanillin (Mathew and Abraham 2005). A two-step process has also been described using first *A. niger* to transform ferulic into vanillic acid, then *P. cinnabarinus* to obtain vanillin from vanillic acid (Bonnin et al. 2001, Zamzuri and Abd-Aziz 2013).

Production of coconut aroma by *Trichoderma* species in SSF has been investigated in several cases (Longo and Sanromán 2006, Ramos et al. 2008, Penha et al. 2012). The responsible volatile compound δ-lactone 6-pentyl-α-pyrone (6-PP) is produced in cultures of several *Trichoderma* species. Recently, Fadel and co-workers (2015) evaluated the production of coconut aroma by *T. viride* EMCC-107 in SSF by using sugarcane bagasse as solid substrate. The unsaturated lactone 6-pentyl-α-pyrone (6-PP) was the major identified volatile compound (3.62 mg/g dry substrate). Saturated lactones, δ-octalactone, γ-nonalactone, γ-undecalactone, γ-dodecalactone and δ-dodecalactone, were also identified in the coconut aroma produced during the induction period (12 days). The effect of varying the concentration of sugarcane bagasse on 6-PP production and biomass growth was also evaluated in that work. The results revealed high 6-PP production at 4.5 g sugarcane bagasse whereas the biomass showed significant increase by increasing the concentration of sugarcane bagasse.

Sugawara et al. (1994) studied the formation of HEMF (4-hydroxy-2 (or 5)-ethyl-5(or 2)-methyl-3 (2H)-furanone) aroma components by the yeast *Zygosaccharomyces rouxii* in miso. Production of aroma compounds by SSF has also been reported by Medeiros et al. (2001) with *Kluyveromyces marxianus* cultivated on various solid substrates, such as cassava bagasse, giant palm bran, apple pomace, sugarcane bagasse and sunflower seeds. *K. marxianus* was able to produce several different alcohols, esters and aldehydes, while monoterpene alcohols and isoamyl acetate were responsible for fruity aromas.

Several bacterial species have been found to be suitable producers of flavour compounds in SSF. Pyrazines, the food additives with the nutty and roasty flavour, were produced by *B. subtilis* cultivated on soybeans (Besson et al. 1997, Larroche et al. 1999). The lactic acid bacteria *Pediococcus pentosaceus* and *L. acidophilus* produced significant amounts of butter-flavoured compounds in semi-solid, maize-based, cultures (Escamilla-Hurtado et al. 2005).

5. Production of Xanthan Gum

Xanthan gum is an extracellular polysaccharide (EPS) produced by the bacterium *Xanthomonas campestris* in submerged fermentation commonly using sucrose or glucose as carbon sources (Papagianni 2004). Xanthan is the most important microbial EPS with several industrial applications in the food, pharmaceutical, cosmetics and textile industries. Xanthan production in SSF was first reported by Stredanski and Conti (1999) and Stredanski et al. (1999). A variety of agro-industrial solid substrates, e.g., apple and grape pomace, spent malt grains and citrus peel, were used and the obtained yields were comparable with those obtained from SmF processes. Recently, Vidhyalakshmi et al. (2012) reported increased yields of xanthan (2.9 g/50 g solid substrate) in SSF by *X. citri* using potato peels as the solid substrate. These studies show that xanthan gum can be produced effectively using inexpensive raw materials although impurities may raise the need for extra purification steps in order to obtain a food-grade product.

6. Production of Fungal Spores

Fungal spores production for applications in the food industry, e.g., in the production of blue cheese and fermented sausages, have been carried out predominantly in SSF due to better yields of pure and homogeneous spores (Larroche and Gros 1989). Literature information however on SSF and spore production is scarce. *Penicillium nalgiovense* is the typical species used as starter culture on the surface of fermented sausages (Papagianni and Papamichael 2007) and its use requires large quantities of viable spores. Ludemann et al. (2010) examined spore production by *P. nalgiovense* in SSF with different substrates, e.g., soy beans, wheat bran and maize kernels. Among them, wheat bran was the best substrate for spore production and viability. Hölker et al. (2004) discussed the industrial spore production of *P. nalgiovense* spores in 100 g SSF batches with bread as solid substrate resulting in spore counts of $1-2 \times 10^9$ per g of solid substrate after 18 days of cultivation. Hölker (2003), using a SSF bioreactor of 5 kg working volume reported spore counts of 1.6×10^9 per g of solid substrate after 14 days of cultivation.

7. Conclusions and Perspectives

Analysis of the literature shows that production of organic acids, industrial enzymes and other microbial metabolites of importance to the food industry by SSF offers several advantages. Metabolite titres produced in SSF systems are very often many-fold more than in SmF systems while their stability is higher. SSF technology results in certain processing advantages of significant potential economic and environmental importance as compared with SmF. Lower energy and sterility requirements, as well as the increasing utilization of agro-industrial wastes as solid substrates are the key characteristics in SSF for lower processing costs compared to sophisticated and expensive SmF processes. Production scale however, remains small for the majority of SSF processes and the reason lies in inherent difficulties regarding scale up.

The build-up of gradients in moisture, temperature, pH, oxygen, substrate and inoculum are difficult to control. There has been intensive research and significant progress regarding biochemical engineering aspects and bioreactor design with the goal of process scaling up. However, an area that remains a "black box" in SSF biotechnology is that of microbial physiology. Work needs to be carried out to elucidate the physiological and molecular background behind the phenomena and the different growth behavior of microorganisms when cultivated on solid or liquid substrate.

Keywords: Solid-state fermentation, food production, organic acids, enzymes, aroma compounds, fungal spores

References

Abdel-Rachman, M.A., Tashiro, Y. and Sonomoto, K. (2011). Lactic acid production from lignocellulose-derived sugars using lactic acid bacteria: Overview and limits. Journal of Biotechnology 156: 286–301.

Acuna-Arguelles, M.E., Cutierrez-Rojas, M., Viniegra-Gonzales, G. and Favela-Torres, E. (1995). Production and properties of three pectinolytic activities produced by *Aspergillus niger* in submerged and solid-state fermentation. Applied Microbiology and Biotechnology 43: 808–814.

Adinarayana, K., Raju, K., Zargar, M.I, Devi, R.B., Lakshmi, P.J. and Ellaiah, P. (2004). Optimization of process parameters for production of lipase in solid-state fermentation by newly isolated *Aspergillus* species. Indian Journal of Biotechnology 3: 65–69.

Akpan, I., Bankjole, M.O., Adesermowo, A.M. and Landtunde-Data, A. (1999). Production of α-amylase by *Aspergillus niger* in a cheap solid medium using rice bran and agricultural material. Tropical Science 39: 77–79.

Alkan, H., Baysal, Z., Uyar, F. and Dogru, M. (2007). Production of lipase by a newly isolated *Bacillus coagulans* under solid-state fermentation using melon wastes. Applied Biochemistry and Biotechnology 136: 183–192.

Alkorta, I., Garbisu, G., Llama, M.J. and Serra, J.L. (1998). Industrial applications of pectic enzymes: A review. Process Biochemistry 33: 21–28.

Anto, H., Trivedi, U. and Patel, K. (2006). Alpha amylase production by *Bacillus cereus* MTCC 1305 using solid-state fermentation. Food Technology and Biotechnology 44: 241–245.

Anupama, A. and Ravindra, P. (2001). Studies on production of single cell protein by *Aspergillus niger* in solid-state fermentation of rice bran. Brazilian Archives of Biology and Technology 11: 79–88.

Aravantinos-Zafiris, G., Tzia, C., Oreopoulou, V. and Thomopoulos, C.D. (1994). Fermentation of orange processing wastes for citric acid production. Journal of the Science of Food and Agriculture 65: 117–120.

Babu, K.R. and Satyanarayana, T. (1995). α-amylase production by thermophilic *Bacillus coagulans* in solid-state fermentation. Process Biochemistry 30: 305–309.

Balkan, B. and Ertan, F. (2007). Production of α-amylase from *Penicillium chrysogenum* under solid-state fermentation by using some agricultural by-products. Food Technology and Biotechnology 45: 439–442.

Barrios-Gonzalez, J., Gonzales, H. and Mejia, A. (1993). Effect of particle size, packing density and agitation on penicillin production in solid-state fermentation. Biotechnology Advances 11: 539–547.

Benjamin, S. and Pandey, A. (1998). Mixed solid substrate fermentation: A novel process for enhanced lipase production by *Candida rugosa*. Acta Biotechnologica 18: 315–324.

Benjamin, S. and Pandey, A. (1997). Coconut cake: A potent substrate for production of lipase by *Candida rugosa* in solid-state fermentation. Acta Biotechnologica 17: 241–251.

Besson, I., Creuly, C., Gros, J.B. and Larroche, C. (1997). Pyrazine production by *Bacillus subtilis* in solid-state fermentation on soybeans. Applied Microbiology and Biotechnology 47: 489–495.

Blandino, A., Iqbalsyah, T., Pandiella, S.S., Cantero, D. and Webb, C. (2002). Polygalacturonase production by *Aspergillus awamori* on wheat in solid-state fermentation. Applied Microbiology and Biotechnology 58: 164–169.

Bonnin, E., Brunel, M., Gouy, Y., Lesage-Meessen, L., Asther, M. and Thibault, J.F. (2001). *Aspergillus niger* I–1472 and *Pycnoporus cinnabarinus* MUCL39533, selected for the biotransformation of ferulic acid to vanillin, are also able to produce cell wall polysaccharide-degrading enzymes and feruloyl esterases. Enzyme Microbial Technology 28: 70–80.

Bramorski, A., Christen, P., Ramirez, M., Soccol, C.R. and Revah, S. (1998). Production of volatile compounds by the edible fungus *Rhizopus oryzae* during solid-state cultivation on tropical agro-industrial substrates. Biotechnology Letters 20: 359–362.

Bramorski, A., Soccol, C.R., Christen, P. and Revah, S. (1998). Fruit aroma production by *Ceratocystis fimbriata* in static cultures from solid agro-industrial wastes. Revista de Microbiologia 28: 208–212.

Christen, P., Bramorski, A., Revah, S. and Soccol, C.R. (2000). Characterization of volatile compounds produced by *Rhizopus* strains grown on agro-industrial solid wastes. Bioresource Technology 71: 211–215.

Christen, P., Angeles, N., Corzo, G., Farres, A. and Revah, S. (1995). Microbial lipase production on a polymeric resin. Biotechnology Techniques 9: 597–600.

Colen, G., Junqueira, R.G. and Moraes-Santos, T. (2006). Isolation and screening of alkaline lipase-producing fungi from Brazilian savanna soil. Journal of Microbiology and Biotechnology 22: 881–885.

Couto, S.R. and Sanromán, M.A. (2006). Application of solid-state fermentation to food industry-A review. Journal of Food Engineering 76: 291–302.

Dhillon, G.S., Brar, S.K., Kaur, S. and Verma, M. (2013). Bioproduction and extraction optimization of citric acid from *Aspergillus niger* by rotating drum type solid-state bioreactor. Industrial Crops and Products 41: 78–84.

Dhillon, G.S., Brar, S.K., Verma, M. and Tyagi, R.D. (2011). Enhanced solid-state citric acid bioproduction using apple pomace waste through response surface methodology. Journal of Applied Microbiology 110: 1045–1055.

Dominguez, A., Costas, M., Longo, M.A. and Saronman, A. (2003). A novel application of solid-state culture: Production of lipases by *Yarrowia Lipolytica*. Biotechnology Letters 25: 1225–1229.

Durand, A. (2003). Bioreactor designs for solid-state fermentation. Biochemical Engineering Journal 13: 113–125.

Ellaiah, P., Adinayana, K., Bhavani, Y., Padmaja, P. and Srinivasulu, B. (2002). Optimization of process parameters for glucoamylase production under solid-state fermentation by a newly isolated *Aspergillus* species. Process Biochemistry 38: 615–620.

Escamilla-Hurtado, M.L., Valdes-Martinez, S.E., Soriano-Santos, J., Gomez-Pliego, R., Verde-Calvo, J.R., Reyes-Dorantes, A. and Tomasini-Campocosio, A. (2005). Effect of culture conditions on production of butter flavor compounds by *Pediococcus pentosaceus* and *Lactobacillus acidophilus* in semisolid maize-based cultures. International Journal of Food Microbiology 105: 305–316.

Fadel, H.H.M., Mahmoud, M.G., Asker, M.M.S. and Lofty, S.N. (2015). Characterization and evaluation of coconut aroma produced by *Trichoderma viride* EMCC-107 in solid-state fermentation on sugarcane bagasse. Electronic Journal of Biotechnology 18: 5–9.

Farias, M.A., Valoni, E.A., Castro, A.M. and Coelho, M.A.Z. (2014). Lipase production by *Yarrowia lipolytica* in solid-state fermentation using different agro industrial residues. Chemical Engineering Transactions 38: 301–306.

Favela-Torres, E., Volke-Sepulveda, T. and Viniegra-Gonzalez, G. (2006). Production of hydrolytic depolymerising pectinases. Food Technology and Biotechnology 44: 221–227.

Ghosh, M.K. and Ghosh, U.K. (2012b). Utilization of wheat bran as bed material in solid-state bacterial production of lactic acid with various nitrogen sources. International Journal of Medical and Biological Sciences 6: 159–162.

Ghosh, M.K. and Ghosh, U.K. (2012a). Utilization of pine needles as a bed material in solid-state fermentation for production of lactic acid by *Lactobacillus* strains. BioResources 6: 1556–1575.

Got, C.E., Barbosa, E.P., Kistner, L.C.I., Gandra, R.F., Arrias, V.L. and Peralta, R.M. (1998). Production of amylase by *Aspergillus fumigatus*. Revista de Microbiologia 28: 99–103.

Gowdhaman, D., Sugumaran, K.R. and Ponnusami, V. (2012). Optimization of lactic acid production from tea waste by *Lactobacillus plantarum* MTCC 6161 in solid-state fermentation by central composite design. International Journal of ChemTech Research 4: 143–148.

Gutierrez-Rozas, M., Cordova, J., Auria, R., Revah, S. and Favela-Torres, E. (1995). Citric acid and polyols production by *Aspergillus niger* at high glucose concentration in solid-state fermentation on inert support. Biotechnology Letters 17: 219–224.

Hadj-Taieb, N., Ayadi, M., Trigui, S., Bouabdallah, F. and Gargouri, A. (2002). Hyperproduction of pectinase activities by a fully constitutive mutant (CT1) of *Penicillium occitanis*. Enzyme Microbial Technology 30: 662–666.

Hang, Y.D. and Woodams, E.E. (1984). Apple pomace: A potential substrate for citric acid production by *Aspergillus niger*. Biotechnology Letters 6: 763–764.

Hang, Y.D. and Woodams, E.E. (1985). Grape pomace: A novel substrate for microbial production of citric acid", Biotechnology Letters 7: 253–254.

Hang, Y.D. and Woodams, E.E. (1987). Microbial production of citric acid by solid-state fermentation of kiwifruit peel. Journal of Food Science 52: 226–227.

Hofvendahl, K. and Hahn-Hägerdal, B. (2000). Factors affecting the fermentative lactic acid production from renewable recources", Enzyme Microbial Technology 26: 87–107.

Hölker, U. (2003). Fermentation auf festen Substraten. Bio Tec 3–4: 32–33.

Hölker, U., Höfer, M. and Lenz, J. (2004). Biotechnological advances of laboratory scale solid-state fermentation with fungi. Applied Microbiology and Biotechnology 64: 175–186.

Ito, K., Yoshida, K., Ishikawa, T. and Kobayashi, S. (1990). Volatile compounds produced by fungus *Aspergillus oryzae* in rice koji and their changes during cultivation. Journal of Fermentation and Bioengineering 70: 169–172.

Jaeger, K.E. and Reetz, M.T. (1998). Microbial lipases form versatile tools for biotechnology. Trends in Biotechnology 16: 396–402.

Jayani, R.S., Saxena, S. and Gupta, R. (2005). Microbial pectinolytic enzymes: A review. Process Biochemistry 40: 2931–2944.

John, R.P., Anisha, G.S., Nanpoothiri, K.M. and Pandey, A. (2009). Direct lactic acid fermentation: Focus on simultaneous saccharification and lactic acid production. Biotechnology Advances 27: 145–152.

Kamini, N.R., Mala, J.G.S. and Puvanakrishnan, R. (1998). Lipase production from *Aspergillus niger* by solid-state fermentation using gingelly oil cake. Process Biochemistry 33: 505–511.

Kapoor, M. and Kuhad, R.C. (2002). Improved polygalacturonase production from *Bacillus* sp. MG-cp-2 under submerged (SmF) and solid-state (SSF) fermentation. Letters in Applied Microbiology 34: 317–322.

Kashyap, D.R., Vohra, P.K., Chopra, S. and Tewari, R. (2001). Applications of pectinases in the commercial sector: A review. Bioresource Technology 77: 215–227.

Kavitha, R. and Umesh-Kumar, S. (2000). Genetic improvement of *Aspergillus carbonarius* for pectinase overproduction during solid-state growth. Biotechnology and Bioengineering 67: 121–125.

Krings, U. and Berger, R.G. (1998). Biotechnological production of flavours and fragrances. Applied Microbiology and Biotechnology 49: 1–8.

Krishna, C. (2005). Solid-state fermentation systems—An overview. Critical Reviews in Biotechnology 25: 1–30.

Krishna, C. and Chandrasekaran, M. (1996). Banana waste as substrate for α-amylase production by *Bacillus subtilis* (CBTK 106) under solid-state fermentation. Applied Microbiology and Biotechnology 46: 106–111.

Kunamneni, A., Perumal, K. and Singh, S. (2005). Amylase production and solid-state fermentation by the thermophilic fungus *Thermomyces langinosus*. Journal of Bioscience and Bioengineering 2: 168–171.

Larroche, C., Besson, I. and Gros, J.B. (1999). High pyrazine production by *Bacillus subtilis* in solid substrate fermentation on ground soy-beans. Process Biochemistry 34: 67–74.

Larroche, C. and Gros, J.B. (1989). Strategies for spore production by *Penicillium roquefortii* using solid-state fermentation techniques. Process Biochemistry 24: 97–103.

Liu, M.Q., Guan, R.F., Dai, X.J., Bai, L.F. and Pan, L. (2012). Optimization of solid-state fermentation for acidophilic pectinase production by *Aspergillus niger* Jl-15 using response surface methodology and oligogalacturonate preparation. American Journal of Food Technology 71: 656–667.

Longo, M.A. and Sanromán, M.A. (2006). Production of food aroma compounds: Microbial and enzymatic methodologies. Food Technology and Biotechnology 44: 335–353.

Ludemann, V., Greco, M., Paz Rodríguez, M., Carlos Basílico, J.C. and Pardo, G.A. (2010). Conidial production by *Penicillium nalgiovense* for use as starter cultures in dry fermented sausages by solid-state fermentation. LWT-Food Science and Technology 43: 315–318.

López, E., Deive, F.J., Longo, M.A. and Sanromán, M.A. (2010). Strategies for utilisation of food-processing wastes to produce lipases in solid-state cultures of *Rhizopus oryzae*. Bioprocess and Biosystems Engineering 33: 929–935.

Marek, A. and Bednarski, W. (1996). Some factors affecting lipase production by yeasts and filamentous fungi. Biotechnology Letters 18: 1155–1160.

Mass, R.H.W., Baker, R.R., Eggink, G. and Weusthuis, R.A. (2006). Lactic acid production from xylose by the fungus *Rhizopus oryzae*. Applied Microbiology and Biotechnology 72: 861–868.

Mathew, S. and Abraham, T.E. (2005). Studies on the production of feruloyl esterase from cereal brans and sugar cane bagasse by microbial fermentation. Enzyme Microbial Technology 36: 565–570.

Medeiros, A.B.P., Christen, P., Roussos, S., Gern, J.C. and Soccol, C.R. (2003). Coffee residues as substrates for aroma production by *Ceratocystis fimbriata* in solid-state fermentation. Brazilian Journal of Microbiology 34: 245–248.

Medeiros, A.B.P., Pandey, A., Christen, P., Fontoura, P.S.G., Freitas, R.J.S. and Soccol, C.R. (2001). Aroma compounds produced by *Kluyveromyces marxianus* in solid-state fermentation on packed bed column bioreactor. World Journal of Microbiology and Biotechnology 17: 767–771.

Mitchell, D.A., von Meien, O.F. and Krieger, N. (2003). Recent developments in modeling of solid-state fermentation: heat and mass transfer in bioreactors. Biochemical Engineering Journal 13: 137–147.

Mitchell, D.A., Berovic, M. and Krieger, N. (2000). Biochemical engineering aspects of solid-state bioprocessing. Advances in Biochemical Engineering 68: 61–138.

Moddles, A.N., Alonso, J.L. and Parajo, J.C. (2001). Strategies to improve the bioconversion of processed wood into lactic acid by simultaneous saccharification and fermentation. Journal of Chemical Technology and Biotechnology 76: 279–284.

Moftah, O.A., Grbavčić, S., Zuža, M., Luković, N., Bezbradica, D. and Knežević-Jugović, Z. (2012). Adding value to the oil cake as a waste from oil processing industry: production of lipase and protease by *Candida utilis* in solid-state fermentation. Applied Biochemistry and Biotechnology 166: 348–364.

Mussatto, S.I., Ballesteros, L.F., Martins, S. and Texeira, J.A. (2012). Use of agro-industrial wastes in solid-state fermentation processes. pp. 121–140. *In*: K.Y. Show (ed.). Industrial Waste. In Tech-Open Access Publisher, Rijeka, Croatia.

Nakadai, T. and Nasuno, S. (1988). Culture conditions of *Aspergillus oryzae* for production of enzyme preparation. Journal of Fermentation Technology 5: 525–533.

Nakadai, T., Nasuno, S. and Iguchi, N. (1973). Purification and properties of alkaline proteinase from *A. oryzae*. Agricultural and Biological Chemistry 37: 2685–2694.

Nakadai, T., Nasuno, S. and Iguchi, N. (1972). The action of peptidases from *A. oryzae* in digestion of soybean proteins. Agricultural and Biological Chemistry 36: 261–268.

Naveena, B.J., Altaf, M., Bhadrayya, K., Madhavendra, S.S. and Reddy, G. (2005b). Selection of medium components by Plackett-Burman design for production of to L(+) lactic acid by *Lactobacillus amylophilus* GV6 in SSF using wheat bran. Bioresource Technology 96: 485–490.

Naveena, B.J., Altaf, M., Bhadrayya, K., Madhavendra, S.S. and Reddy, G. (2005a). Direct fermentation of starch to L(+) lactic acid by *Lactobacillus amylophilus* GV6 in SSF using wheat bran as support and substrate: Medium optimizations using RSM. Process Biochemistry 40: 681–690.

Ohnisi, K., Yoshida, Y. and Sekiguchi, J. (1994). Lipase production of *Aspergillus oryzae*. Journal of Fermentation and Bioengineering 77: 490–495.

Panda, S.K., Mishra, S.S., Kayitesi, E and Ray, R.C. (2016). Microbial-processing of fruit and vegetable wastes for production of vital enzymes and organic acids: Biotechnology and scopes. Environmental Research 146: 161–172.

Pandey, A., Soccol, C.R., Nigam, P. and Soccol, V.T. (2000d). Biotechnological potential of agro-industrial residues: I. Sugarcane bagasse. Bioresource Technology 74: 69–80.

Pandey, A., Soccol, C.R., Nigam, P., Brand, D., Mohan, R. and Roussos, S. (2000c). Biotechnological potential of coffee pulp and coffee husk for bioprocesses. Biochemical Engineering Journal 6: 153–162.

Pandey, A., Nigam, P., Soccol, C.R., Soccol, V.T., Singh, D. and Mohan, R. (2000b). Advances in microbial amylases. Biotechnology and Applied Biochemistry 31: 135–152.

Pandey, A., Soccol, C.R., Nigam, P., Soccol, V.T. and Mohan, R. (2000a). Biotechnological potential of agro-industrial residues. II. Cassava bagasse. Bioresource Technology 74: 81–87.

Pandey, A., Selvakumar, P. and Ashakumari, L. (1994). Glucose production by *Aspergillus niger* on rice bran is improved by addition of nitrogen sources. World Journal of Microbiology and Biotechnology 10: 348–384.

Papagianni, M. (2007). Advances in citric acid fermentation by *Aspergillus niger*: Biochemical aspects, membrane transport and modeling. Biotechnology Advances 25: 244–263.

Papagianni, M. and Papamichael, E.M. (2007). Modeling growth, substrate consumption and product formation of *Penicillium nalgiovense* grown on meat simulation medium in submerged batch culture. Journal of Industrial Microbiology and Biotechnology 34: 225–231.

Papagianni, M. (2004). Xanthan gum production. pp. 315–323. *In*: A. Pandey (ed.). Concise Encyclopedia of Bioresource Technology. The Haworth Press Inc., Binghamton, NY, USA.

Papagianni, M. (2003). Fungal morphology and metabolite production in submerged mycelial processes. Biotechnology Advances 22: 189–259.

Papagianni, M., Nokes, S.E. and Filer, K. (2001). Submerged and solid-state phytase fermentation by *Aspergillus niger*: Effects of agitation and medium viscosity on phytase production, fungal morphology and inoculum performance. Food Technology and Biotechnology 39: 319–326.

Papagianni, M., Nokes, S.E. and Filer, K. (1999). Production of phytase by *Aspergillus niger* in submerged and solid-state fermentation. Process Biochemistry 35: 397–402.

Parichat, P., Songsri, K., Sarintip, S. and Nuttha, T. (2012). Direct fermentation of L(+)-lactic acid from cassava pulp by solid-state culture of *Rhizopus oryzae*. Bioprocess and Biosystems Engineering 35: 1429–1434.

Pastore, G.M., Park, Y.K. and Min, D.B. (1994). Production of a fruity aroma by *Neurospora* from beiju. Mycological Research 98: 25–35.

Penha, M.P., Rocha-Leao, M.H.M. and Leite, S.G.F. (2012). Sugarcane bagasse as support for the production of coconut aroma by solid-state fermentation (SSF). Bioresources 7: 2366–2375.

Pintado, J., Lonsane, B.K., Gaime-Perraud, I. and Roussos, S. (1998). On-line monitoring of citric acid production in solid-state culture by respirometry. Process Biochemistry 33: 513–518.

Prado, F.C., Vandenberghe, L.P.S., Woiciechowski, A.L., Rodrígues-León, J.A. and Soccol, C.R. (2005). Citric acid production by solid-state fermentation. Brazilian Journal of Chemical Engineering 22: 547–555.

Prado, F.C., Vandenberghe, L.P.S., Lisboa, C., Paca, J., Pandey, A. and Soccol, C.R. (2004). Relation between citric acid production and respiration rate of *Aspergillus niger* in solid-state fermentation. Engineering in Life Sciences 4: 179–186.

Ramachandran, S., Patel, K.A., Nampoothiri, K.M., Francis, F., Nagy, V., Szakacs, G. and Pandey, A. (2004) Coconut oil cake—A potential raw material for the production of α-amylase. Bioresource Technology 93: 169–174.

Ramos, A.S., Fiaux, S.B. and Leite, S.G.F. (2008). Production of 6-pentyl-α-pyrone by *Trichoderma harzianum* in solid-state fermentation. Brazilian Journal of Microbiology 39: 712–717.

Rathi, P., Goswami, V.K., Sahai, V. and Gupta, R. (2002). Statistical medium optimization and production of a thermostable lipase from *Burkholderia cepapcia* in a bioreactor. Journal of Applied Microbiology 93: 930–945.

Ray, R.C., Sharma, P. and Panda, S.H. (2009). Lactic acid production from cassava fibrous residue using *Lactobacillus plantarum* MTCC 1407. Journal of Environmental Biology 30: 847–852.

Regulapati, R., Malav, P. and Gummadi, S. (2007). Production of thermostable α-amylases by solid-state fermentation. A review. American Journal of Food Technology 2: 1–11.

Richter, K. and Träger, A. (1994). L(+) lactic acid from sweet sorghum by submerged and solid-state fermentations. Acta Biotechnologica 14: 367–378.

Rivera-Munoz, G., Tinoko-Valencia, J.R., Sanchez, S. and Farres, A. (1991). Production of microbial lipases in a solid-state fermentation system. Biotechnology Letters 13: 277–280.

Rojan, P.J., Nampoothiri, K.M., Nair, A.S. and Pandey, A. (2005). L(+)-lactic acid production using *Lactobacillus casei* in solid-state fermentation. Biotechnology Letters 27: 1685–1688.

Rosés, R.P. and Guerra, N.P. (2009). Optimization of amylase production by *Aspergillus niger* in solid-state fermentation using sugarcane bagasse as solid support material. World Journal of Microbiology and Biotechnology 25: 1929–1939.

Sabat, S., Krishna Murthy, V., Pavithra, M., Mayur, P. and Chandavar, A. (2012). Production and characterisation of extracellular lipase from *Bacillus stearothermophilus* MTCC 37 under different fermentation conditions. International Journal of Engineering Research and Applications 2: 1775–1781.

Sahoo, R.K., Subudhi, E. and Kumar, M. (2014). Quantitative approach to track lipase producing *Pseudomonas* sp. S1 in nonsterilized solid-state fermentation. Letters in Applied Microbiology 58: 610–616.

Sandhia, C., Sumantha, A., Szakacs, G. and Pandey, A. (2005). Comparative evaluation of neutral protease production by *Aspergillus oryzae* in submerged and solid-state fermentation. Process Biochemistry 40: 2689–2694.

Santis-Navarro, A., Gea, T., Barrena, R. and Sanchez, A. (2011). Production of lipases by solid-state fermentation using vegetable oil-refining wastes. Bioresource Technology 102: 10080–10084.

Sharma, D.C. and Satyanarayana, T. (2012). Biotechnological potential of agro residues for economical production of thermoalkali-stable pectinase by *Bacillus pumilus* dcsr1 by solid-state fermentation and its efficacy in the treatment of ramie fibres. Enzyme Research 2012: 281384.

Shukla, J. and Kar, R. (2006). Potato peel as a solid-state substrate for thermostable α-amylase production by thermophilic *Bacillus* isolates. World Journal of Microbiology and Biotechnology 22: 417–422.

Singh, M., Saurav, K., Srivastava, N. and Kannabiran, K. (2010). Lipase production by *Bacillus subtilis* OCR-4 in solid-state fermentation using ground nut oil cakes as substrate. Current Research Journal of Biological Sciences 2: 241–245.

Singh, S.A., Ramakrishna, M. and Rao, A.G.A. (1999). Optimisation of downstream processing parameters for the recovery of pectinase from the fermented bran of *Aspergillus carbonarius*. Process Biochemistry 35: 411–417.

Singhania, R.R., Patel, A.K., Soccol, C.R. and Pandey, A. (2009). Recent advances in solid-state fermentation. Biochemical Engineering Journal 44: 13–18.

Sivaramakrishnan, S., Gangadharan, D., Nampoothiri, K.M., Soccol, C.R. and Pandey, A. (2007). Alpha amylase production by *Aspergillus oryzae* employing solid-state fermentation. Journal of Scientific and Industrial Research 66: 621–626.

Soares, V.F., Castilho, L.R., Bon, E.P.S. and Freire, D.M.G. (2005). High-yield *Bacillus subtilis* protease production by solid-state fermentation. Applied Biochemistry and Biotechnology 121-124: 311–319.

Soares, M., Christen, P., Pandey, A. and Soccol, C.R. (2000). Fruity flavour production by *Ceratocystis fimbriata* grown on coffee husk in solid-state fermentation. Process Biochemistry 35: 857–861.

Soccol, C.R., Marin, B., Raimbault, M. and Lebeault, J.M. (1994). Potential of solid-state fermentation for production of L(+) lactic acid by *Rhizopus oryzae*. Applied Microbiology and Biotechnology 41: 286–290.

Sodhi, H.K., Sharma, K., Gupta, J.K. and Soni, S.K. (2005). Production of a thermostable α-amylase form *Bacillus* sp. PS-7 by solid-state fermentation and its synergisitic use in the hydrolysis of malt starch for alcohol production. Process Biochemistry 40: 525–534.

Solis, S., Flores, M.E. and Huitron, C. (1997). Improvement of pectinase production by interspecific hybrids of *Aspergillus* strains. Letters in Applied Microbiology 24: 77–81.

Solis-Pereira, S., Favela-Torres, E., Viniegra-Gonzalez, G., Gutierrez-Rojas, M. (1993). Effect of different carbon sources on the synthesis of pectinase by *Aspergillus niger* in submerged and solid-state fermentations. Applied Microbiology and Biotechnology 39: 36–41.

Sugathi, R., Benazir, J.F., Santhi, R., Ramesh Kumar, V., Hari, A., Meenaksi, N., Nidhiya, K.A., Kavitha, G. and Lakshmi, R. (2011). Amylase production by *Aspergillus niger* under solid-state fermentation using agroindustrial wastes. International Journal of Engineering Science and Technology 3: 1756–1763.

Sugawara, E., Hashimoto, S., Sakurai, Y. and Kobayashi, A. (1994). Formation by yeast of the HEMF (4-hydroxy-2 (or 5)-ethyl-5 (or 2)-methyl-3 (2H)-furanone) aroma components in Miso with aging. Bioscience, Biotechnology and Biochemistry 58: 1134–1135.

Sumantha, A., Larroche, C. and Pandey, A. (2006). Microbiology and industrial biotechnology of food-grade proteases: A perspective. Food Technology and Biotechnology 44: 211–220.

Stredanski, M. and Conti, E. (1999). Xanthan production by solid-state fermentation. Process Biochemistry 34: 581–587.

Stredanski, M., Conti, E., Navarini, L. and Bertocci, C. (1999). Production of bacterial polysaccharides by solid-state fermentation. Process Biochemistry 34: 11–16.

Tariq, A.L. and Reyaz, A.L. (2012). The influence of carbon and nitrogen sources on pectinase productivity of *Penicillium chrysogenum* in solid-state fermentation. International Research Journal of Microbiology 3: 202–207.

Tran, C.T., Sly, L.I. and Mitchell, D.A. (1998). Selection of a strain of *Aspergillus* for the production of citric acid from pineapple waste in solid-state fermentation. World Journal of Microbiology and Biotechnology 14: 399–404.

Treichel, H., de Oliveira, D., Mazouti, M.A., Di Luccio, M. and Oliveira, J.V. (2010). A review on microbial lipases production. Food and Bioprocess Technology 3: 182–196.

Trejo-Hernandez, M.R., Oriol, E., Lopez-Canales, A., Roussos, S., Viniegra, G. and Raimbault, M. (1991). Production of pectinases by *Aspergillus niger* by solid-state fermentation on support. Micologia Neotropical Aplicada 4: 49–62.

Ul-Haq, I., Idrees, S. and Rajoka, M.I. (2002). Production of lipases by *Rhizopus oligosporus* by solid-state fermentation. Process Biochemistry 37: 637–641.

Vandamme, E.J. and Soetaert, W. (2002). Bioflavours and fragrances *via* fermentation and biocatalysis. Journal of Chemical Technology and Biotechnology 77: 1323–1332.

Vandenberghe, L.P.S. (2000). Développement d'un Procédé pour la Production d'Acide Citrique par Fermentation en Milieu Solide à partir de Résidus de l'Agro-industrie du Manioc. Ph.D. Thesis, Université de Technologie de Compiègne. Compiègne, France.

Vandenberghe, L.P., Soccol, C.R., Prado, F.C. and Pandey, A. (2004). Comparison of citric acid production by solid-state fermentation in flask, column, tray, and drum bioreactors. Applied Biochemistry and Biotechnology 118: 293–303.

Vaseghi, Z., Najafpoor, G.D., Mohseni, S. and Mahjoub, S. (2013). Production of active lipase by *Rhizopus oryzae* from sugarcane bagasse: Solid-state fermentation in a tray bioreactor. International Journal of Food Science and Technology 48: 283–289.

Vidhyalakshmi, R., Vallinachiyar, C. and Radhika, R. (2012). Production of Xanthan from agro-industrial waste. Journal of Advanced Scientific Research 3: 56–59.

Vihinen, M. and Mantasala, P. (1989). Microbial amylolytic enzymes. Critical Reviews in Biochemistry and Molecular Biology 24: 329–418.

Wee, Y.J., Kim, J.N. and Ryu, H.W. (2006). Biotechnological production of lactic acid and its recent applications. Food Technology and Biotechnology 44: 163–172.

Xavier, S. and Lonsane, B.K. (1994). Sugar-cane pressmud as a novel and inexpensive substrate for production of lactic acid in a solid-state fermentation system. Applied Microbiology and Biotechnology 41: 291–295.

Zhang, Z.Y., Jin, B. and Kelly, J.M. (2007). Production of lactic acid from renewable materials by *Rhizopus* fungi. Biochemical Engineering Journal 35: 251–263.

Zamzuri, N.A. and Abd-Aziz, S. (2013). Biovanillin from agro wastes as an alternative food flavour. Journal of the Science of Food and Agriculture 93: 429–438.

Index

Printed and bound by CPI Group (UK) Ltd, Croydon, CR0 4YY

01/11/2024

01782624-0019